Mechanical Engineering Series

Frederick F. Ling
Series Editor

Springer
New York
Berlin
Heidelberg
Barcelona
Budapest
Hong Kong
London
Milan
Paris
Santa Clara
Singapore
Tokyo

Mechanical Engineering Series

Introductory Attitude Dynamics
F.P. Rimrott

Balancing of High-Speed Machinery
M.S. Darlow

Theory of Wire Rope
G.A. Costello

Theory of Vibration: An Introduction, 2nd ed.
A.A. Shabana

Theory of Vibration: Discrete and Continuous Systems, 2nd ed.
A.A. Shabana

Laser Machining: Theory and Practice
G. Chryssolouris

Underconstrained Structural Systems
E.N. Kuznetsov

Principles of Heat Transfer in Porous Media, 2nd ed.
M. Kaviany

Mechatronics: Electromechanics and Contromechanics
D.K. Miu

Structural Analysis of Printed Circuit Board Systems
P.A. Engel

Kinematic and Dynamic Simulation of Multibody Systems: The Real-Time Challenge
J. García de Jalón and E. Bayo

High Sensitivity Moiré: Experimental Analysis for Mechanics and Materials
D. Post, B. Han, and P. Ifju

Principles of Convective Heat Transfer
M. Kaviany

(continued after index)

A.A. Shabana

Vibration of Discrete and Continuous Systems

Second Edition

With 147 Figures

Springer

A.A. Shabana
Department of Mechanical Engineering
University of Illinois at Chicago
P.O. Box 4348
Chicago, IL 60680
USA

Series Editors
Frederick F. Ling
Ernest F. Gloyna Regents Chair in Engineering
Department of Mechanical Engineering
The University of Texas at Austin
Austin, TX 78712-1063 USA
and
William Howard Hart Professor Emeritus
Department of Mechanical Engineering,
Aeronautical Engineering and Mechanics
Rensselaer Polytechnic Institute
Troy, NY 12180-3590 USA

Library of Congress Cataloging-in-Publication Data
Shabana, Ahmed A.
Vibration of discrete and continuous systems, second edition)/A.A. Shabana
p. cm.—(Mechanical engineering series)
Includes bibliographical references and index.
Contents: v. 1. An introduction—v. 2. Discrete and continuous systems
ISBN 0-387-94744-2 (hardcover: alk. paper)
1. Vibration. I. Title. II. Series: Mechanical engineering series (Berlin, Germany)
QA865.S488 1996
531′.32—dc20 96-12476

Printed on acid-free paper.

Production coordinated by Publishing Network and managed by Francine McNeill; manufacturing supervised by Jeffrey Taub.
Typeset by Asco Trade Typesetting Ltd., Hong Kong.
Printed and bound by Maple-Vail Book Manufacturing Group, York, PA.
Printed in the United States of America.

9 8 7 6 5 4 3 2 1

ISBN 0-387-94744-2 Springer-Verlag New York Berlin Heidelberg SPIN 10534700

Dedicated to the Memory of Professor M.M. Nigm

Mechanical Engineering Series

Frederick F. Ling
Series Editor

Series Preface

Mechanical engineering, an engineering discipline borne of the needs of the industrial revolution, is once again asked to do its substantial share in the call for industrial renewal. The general call is urgent as we face profound issues of productivity and competitiveness that require engineering solutions, among others. The Mechanical Engineering Series features graduate texts and research monographs intended to address the need for information in contemporary areas of mechanical engineering.

The series is conceived as a comprehensive one that covers a broad range of concentrations important to mechanical engineering graduate education and research. We are fortunate to have a distinguished roster of consulting editors on the advisory board, each an expert in one of the areas of concentration. The names of the consulting editors are listed on the next page of this volume. The areas of concentration are: applied mechanics; biomechanics; computational mechanics; dynamic systems and control; energetics; mechanics of materials; processing; thermal science; and tribology.

Professor Marshek, the consulting editor for dynamic systems and control, and I are pleased to present the second edition of *Vibration of Discrete and Continuous Systems* by Professor Shabana. We note that this is the second of two volumes. The first deals with the *theory of vibration.*

Austin, Texas — Frederick F. Ling

Preface

The theory of vibration of single and two degree of freedom systems is covered in the first volume of this book. In the treatment presented in the first volume, the author assumed only a basic knowledge of mathematics and dynamics on the part of the student. Therefore, the first volume can serve as a textbook for a first one-semester undergraduate course on the theory of vibration. The second volume contains material for a one-semester graduate course that covers the theory of multi-degree of freedom and continuous systems. An introduction to the finite-element method is also presented in this volume. In the first and the second volumes, the author attempts to cover only the basic elements of the theory of vibration that students should learn before taking more advanced courses on this subject. Each volume, however, represents a separate entity and can be used without reference to the other. This gives the instructor the flexibility of using one of these volumes with other books in a sequence of two courses on the theory of vibration. For this volume to serve as an independent text, several sections from the first volume are used in Chapters 1 and 5 of this book.

SECOND EDITION

Several important additions and corrections have been made in the second edition of the book. Several new examples also have been provided in several sections. The most important additions in this new edition can be summarized as follows: Three new sections are included in Chapter 1 in order to review some of the basic concepts used in dynamics and in order to demonstrate the assumptions used to obtain the single degree of freedom linear model from the more general multi-degree of freedom nonlinear model. Section 3.7, which discusses the case of proportional damping in multi-degree of freedom systems has been significantly modified in order to provide a detailed discussion on *experimental modal analysis techniques* which are widely used in the vibration analysis of complex structural and mechanical systems. A new section, Section 5.10, has been introduced in order to demonstrate the use of the finite-element method in the large rotation and deformation analysis of mechanical and structural systems. The *absolute nodal coordinate formulation*, described in this section, can be used to efficiently solve many vibration problems such as the vibrations of cables and flexible space antennas. A new chapter, Chapter 6, was added to provide a discussion on the subject of *similarity transformation* which is important in understanding the numerical methods used in the large scale computations of the eigenvalue problem. In this new chapter, the Jacobi method and the **QR** decomposition method, which are used to determine the natural frequencies and mode shapes, are also discussed.

CONTENTS OF THE BOOK

The book contains six chapters and an appendix. In the appendix, some of the basic operations in vector and matrix algebra, which are repeatedly used in this book, are reviewed. The contents of the chapters can be summarized as follows:

Chapter 1 of this volume covers some of the basic concepts and definitions used in dynamics, in general, and in the analysis of single degree of freedom systems, in particular. These concepts and definitions are also of fundamental importance in the vibration analysis of multi-degree of freedom and continuous systems. Chapter 1 is of an introductory nature and can serve to review the materials covered in the first volume of this book.

In Chapter 2, a brief introduction to Lagrangian dynamics is presented. The concepts of generalized coordinates, virtual work, and generalized forces are first introduced. Using these concepts, Lagrange's equation of motion is then derived for multi-degree of freedom systems in terms of scalar energy and work quantities. The kinetic and strain energy expressions for vibratory systems are also presented in a matrix form. Hamilton's principle is discussed in Section 6 of this chapter, while general energy conservation theorems are presented in Section 7. Chapter 2 is concluded with a discussion on the use of the principle of virtual work in dynamics.

Matrix methods for the vibration analysis of multi-degree of freedom systems are presented in Chapter 3. The use of both Newton's second law and Lagrange's equation of motion for deriving the equations of motion of multi-degree of freedom systems is demonstrated. Applications related to angular oscillations and torsional vibrations are provided. Undamped free vibration is first presented, and the orthogonality of the mode shapes is discussed. Forced vibration of the undamped multi-degree of freedom systems is discussed in Section 6. The vibration of viscously damped multi-degree of freedom systems using proportional damping is examined in Section 7, and the case of general viscous damping is presented in Section 8. Coordinate reduction methods using the modal transformation are discussed in Section 9. Numerical methods for determining the mode shapes and natural frequencies are discussed in Sections 10 and 11.

Chapter 4 deals with the vibration of continuous systems. Free and forced vibrations of continuous systems are discussed. The analysis of longitudinal, torsional, and transverse vibrations of continuous systems is presented. The orthogonality relationships of the mode shapes are developed and are used to define the modal mass and stiffness coefficients. The use of both elementary dynamic equilibrium conditions and Lagrange's equations in deriving the equations of motion of continuous systems is demonstrated. The use of approximation methods as a means of reducing the number of coordinates of continuous systems to a finite set is also examined in this chapter.

In Chapter 5 an introduction to the finite-element method is presented. The assumed displacement field, connectivity between elements, and the formulation of the mass and stiffness matrices using the finite-element method are discussed. The procedure for assembling the element matrices in order to obtain the structure equations of motion is outlined. The convergence of the finite-element solution is examined, and the use of higher order and spatial elements in the vibration analysis of structural systems is demonstrated. This chapter is concluded with a discussion on the use of the *absolute nodal coordinate formulation* in the large finite-element rotation and deformation analysis.

Chapter 6 is devoted to the eigenvalue analysis and to a more detailed discussion on the similarity transformation. The results presented in this chapter can be used to determine whether or not a matrix has a complete set of independent eigenvectors associated with repeated eigenvalues. The definition of Jordan matrices and the concept of the generalized eigenvectors are introduced, and several computer methods for solving the eigenvalue problem are presented.

ACKNOWLEDGMENT

I would like to thank many of the teachers, colleagues, and students who contributed, directly or indirectly, to this book. In particular, I would like to thank my students Drs. D.C. Chen and W.H. Gau, who have made major contributions to the development of this book. My special thanks go to Ms. Denise Burt for the excellent job in typing the manuscript of this book. The editorial and production staffs of Springer-Verlag deserve special thanks for their cooperation and their thorough professional work. Finally, I thank my family for the patience and encouragement during the time of preparation of this book.

Chicago, Illinois Ahmed A. Shabana

Contents

1 Introduction

The theory of vibration of single degree of freedom systems serves as one of the fundamental building blocks in the theory of vibration of discrete and continuous systems. As will be shown in later chapters, the concepts introduced and the techniques developed for the analysis of single degree of freedom systems can be generalized to study discrete systems with multi-degrees of freedom as well as continuous systems. For this volume to serve as an independent text, several of the important concepts and techniques used in the analysis of single degree of freedom systems are briefly discussed in this chapter. First, the methods of formulating the kinematic and dynamic equations are reviewed in the first three sections. It is also shown in these sections that the dynamic equation of a single degree of freedom system can be obtained as a special case of the equations of the multi-degree of freedom systems. The free vibrations of the single degree of freedom systems are reviewed in Sections 4 and 5, and the significant effect of viscous, structural, and Coulomb damping is discussed and demonstrated. Section 6 is devoted to the analysis of the forced vibrations of single degree of freedom systems subject to harmonic excitations, while the impulse response and the response to an arbitrary forcing function are discussed, respectively, in Sections 7 and 8.

1.1 KINEMATICS OF RIGID BODIES

The dynamic equations of motion of multi-degree of freedom mechanical and structural systems can be formulated using the Newtonian or the Lagrangian method. Using either method, the formulation of the kinematic position and velocity equations is a necessary step which is required in developing the inertia, elastic, and applied forces of mechanical and structural systems. When the approach of the Newtonian mechanics is used, one must also formulate the kinematic acceleration equations in addition to the position and velocity equations. In the Lagrangian method described in the following chapter, the equations of motion are formulated using the scalar energy quantities and, as such, the formulation of the acceleration equations is not necessary. In the case of a general rigid body displacement, the kinematic position, velocity, and

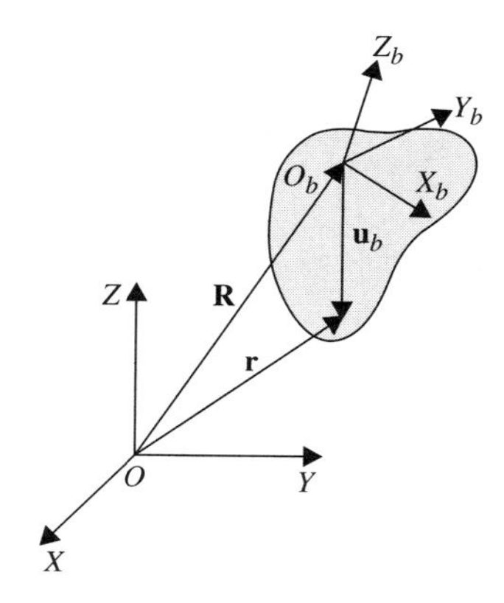

FIG. 1.1. Coordinates of the rigid body.

acceleration equations can be expressed in terms of six independent parameters: three parameters describe the absolute translation of a reference point on the rigid body, and three rotation coordinates define the body orientation in a selected global frame of reference. Using these six independent parameters, the position, velocity, and acceleration vectors of an arbitrary point on the rigid body can be systematically developed. Figure 1 depicts a rigid body that has a body coordinate system denoted as $X_b Y_b Z_b$. The global position vector of an arbitrary point on this rigid body can be described using the translation of the reference point O_b as

$$\mathbf{r} = \mathbf{R} + \mathbf{A}\mathbf{u}_b \tag{1.1}$$

where $\mathbf{R}$ is the global position vector of the reference point as shown in the figure, $\mathbf{A}$ is the transformation matrix that defines the orientation of the body in the global coordinate system, and $\mathbf{u}_b$ is a constant vector that defines the location of the arbitrary point with respect to the reference point. The vector $\mathbf{u}_b$ is defined in terms of its constant components as

$$\mathbf{u}_b = [x \quad y \quad z]^{\mathrm{T}} \tag{1.2}$$

Planar Kinematics In the case of planar motion, the rotation of the rigid body can be described by one parameter only, while the translation can be described by two parameters. In this case, the vector $\mathbf{R}$ has two time-varying components, while the transformation matrix is a function of one angle that defines the rotation of the rigid body about the axis of rotation. Without any loss of generality, we select the axis of rotation to be the Z axis. If the angle of rotation is denoted as θ, the transformation matrix $\mathbf{A}$ can be written in this special case as

$$\mathbf{A} = \begin{bmatrix} \cos\theta & -\sin\theta & 0 \\ \sin\theta & \cos\theta & 0 \\ 0 & 0 & 1 \end{bmatrix}$$

while the vector $\mathbf{R}$ is defined as

$$\mathbf{R} = [R_x \quad R_y \quad 0]^{\mathrm{T}}$$

Spatial Kinematics In the case of a general three-dimensional displacement, the vector $\mathbf{R}$ is a function of three time-dependent coordinates, while the transformation matrix is a function of three independent parameters that define the rotations about the three perpendicular axes of the body coordinate system. A simple rotation ϕ about the X_b-axis is defined by the matrix

$$\mathbf{A}_x = \begin{bmatrix} 1 & 0 & 0 \\ 0 & \cos\phi & -\sin\phi \\ 0 & \sin\phi & \cos\phi \end{bmatrix}$$

A simple rotation θ about the body Y_b-axis can be described using the rotation matrix

$$\mathbf{A}_y = \begin{bmatrix} \cos\theta & 0 & \sin\theta \\ 0 & 1 & 0 \\ -\sin\theta & 0 & \cos\theta \end{bmatrix}$$

Similarly, a simple rotation ψ about the body Z_b-axis is described by the rotation matrix

$$\mathbf{A}_z = \begin{bmatrix} \cos\psi & -\sin\psi & 0 \\ \sin\psi & \cos\psi & 0 \\ 0 & 0 & 1 \end{bmatrix}$$

The three independent rotations ϕ, θ, and ψ, shown in Fig. 2, can be used to define an arbitrary orientation of a rigid body in space. Since these angles are defined about the moving-body axes, the sequence of the three transformations previously defined can be used to define the transformation matrix that

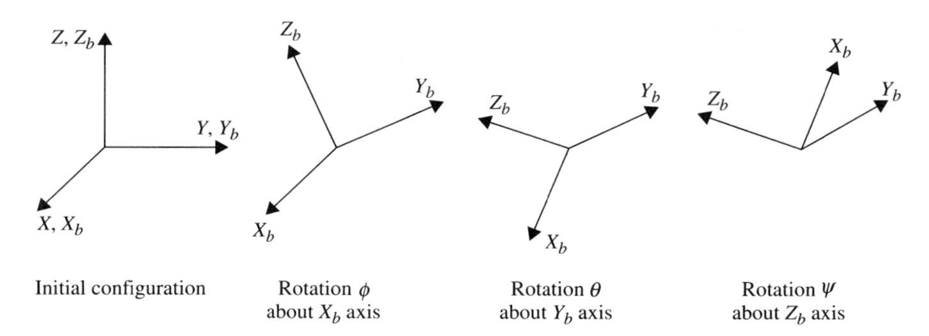

FIG. 1.2. Three-dimensional rotations.

defines the final orientation of the body as

$$\mathbf{A} = \mathbf{A}_x\mathbf{A}_y\mathbf{A}_z$$

$$= \begin{bmatrix} \cos\theta\cos\psi & -\cos\theta\sin\psi & \sin\theta \\ \sin\phi\sin\theta\cos\psi + \cos\phi\sin\psi & -\sin\phi\sin\theta\sin\psi + \cos\phi\cos\psi & -\sin\phi\cos\theta \\ -\cos\phi\sin\theta\cos\psi + \sin\phi\sin\psi & \cos\phi\sin\theta\sin\psi + \sin\phi\cos\psi & \cos\phi\cos\theta \end{bmatrix} \tag{1.3}$$

The columns of this transformation matrix define three unit vectors along the axes of the body coordinate system. It can be demonstrated that this matrix satisfies the following *orthogonality condition*:

$$\mathbf{A}^{\mathrm{T}}\mathbf{A} = \mathbf{A}\mathbf{A}^{\mathrm{T}} = \mathbf{I}$$

where $\mathbf{I}$ is the 3×3 identity matrix. The orthogonality of $\mathbf{A}$ implies that the inverse of $\mathbf{A}$ is equal to its transpose. Furthermore, since the transformation matrix in planar kinematics is a special case of the spatial transformation matrix, the planar transformation matrix also must satisfy the preceding orthogonality condition.

In the spatial kinematics, different sequences and angles of rotations can be used to define the body orientation. Therefore, different transformation matrices that depend on the selection of the sequence and angles of rotations can be obtained, as demonstrated by the following example.

Example 1.1

In this section, the rotations ϕ, θ, and ψ are defined about the axes of the body coordinate system. If these rotations are performed about the axes of the global coordinate system, a different sequence of multiplication must be used to define the final transformation matrix. In this case, the final orientation of the body is determined by using a sequence of three rotations about the fixed axes. The transformation matrix $\mathbf{A}$ is defined in this case as follows:

$$\mathbf{A} = \mathbf{A}_z\mathbf{A}_y\mathbf{A}_x$$

with which, upon using the previously obtained expressions for the simple rotation matrices $\mathbf{A}_x$, $\mathbf{A}_y$, and $\mathbf{A}_z$, one obtains

$$\mathbf{A} = \begin{bmatrix} \cos\theta\cos\psi & -\cos\phi\sin\psi + \sin\phi\sin\theta\cos\psi & \sin\phi\sin\psi + \cos\phi\sin\theta\cos\psi \\ \cos\theta\sin\psi & \cos\phi\cos\psi + \sin\phi\sin\theta\sin\psi & -\sin\phi\cos\psi + \cos\phi\sin\theta\sin\psi \\ -\sin\theta & \sin\phi\cos\theta & \cos\phi\cos\theta \end{bmatrix} \tag{1.4}$$

It can be shown that this transformation matrix is also orthogonal.

Velocity Vector The velocity vector can be obtained by differentiating the position vector with respect to time. This leads to

$$\mathbf{v} = \dot{\mathbf{r}} = \dot{\mathbf{R}} + \dot{\mathbf{A}}\mathbf{u}_b \tag{1.5}$$

The velocity vector also can be written in the familiar form

$$\mathbf{v} = \dot{\mathbf{R}} + \boldsymbol{\omega} \times \mathbf{u} \tag{1.6}$$

where

$$\boldsymbol{\omega} = [\omega_x \quad \omega_y \quad \omega_z]^{\mathrm{T}}$$

is the angular velocity vector and

$$\mathbf{u} = \mathbf{A}\mathbf{u}_b$$

One can prove the following cross-product identity:

$$\boldsymbol{\omega} \times \mathbf{u} = \tilde{\boldsymbol{\omega}}\mathbf{u}$$

where

$$\tilde{\boldsymbol{\omega}} = \begin{bmatrix} 0 & -\omega_z & \omega_y \\ \omega_z & 0 & -\omega_x \\ -\omega_y & \omega_x & 0 \end{bmatrix}$$

The velocity vector then can be written as

$$\mathbf{v} = \dot{\mathbf{R}} + \tilde{\boldsymbol{\omega}}\mathbf{A}\mathbf{u}_b \tag{1.7}$$

It follows from Eqs. 5 and 7 that

$$\tilde{\boldsymbol{\omega}}\mathbf{A} = \dot{\mathbf{A}}$$

or

$$\tilde{\boldsymbol{\omega}} = \dot{\mathbf{A}}\mathbf{A}^{\mathrm{T}}$$

Using this identity and the transformation matrix of Eq. 3, it can be shown that the angular velocity vector $\boldsymbol{\omega}$ can be expressed in terms of the angles φ, θ, and ψ and their time derivatives as (Shabana, 1989 and 1994)

$$\boldsymbol{\omega} = \mathbf{G}\dot{\boldsymbol{\beta}} \tag{1.8}$$

where

$$\mathbf{G} = \begin{bmatrix} 1 & 0 & \sin\theta \\ 0 & \cos\phi & -\sin\phi\cos\theta \\ 0 & \sin\phi & \cos\phi\cos\theta \end{bmatrix}$$

$$\boldsymbol{\beta} = [\phi \quad \theta \quad \psi]^{\mathrm{T}}$$

Note that the components of the angular velocity vector are not the time derivatives of the orientation coordinates.

The angular velocity vector also can be defined in the body coordinate system as

$$\boldsymbol{\omega}_b = \mathbf{A}^{\mathrm{T}}\boldsymbol{\omega}$$

where $\boldsymbol{\omega}_b$ is the angular velocity vector defined in the body coordinate system. This vector can be written in terms of the orientation coordinates and their time derivatives as

$$\boldsymbol{\omega}_b = \mathbf{G}_b\dot{\boldsymbol{\beta}} \tag{1.9}$$

Using the angles that define the transformation matrix of Eq. 3, one can show that

$$\mathbf{G}_b = \begin{bmatrix} \cos\theta\cos\psi & \sin\psi & 0 \\ -\cos\theta\sin\psi & \cos\psi & 0 \\ \sin\theta & 0 & 1 \end{bmatrix}$$

Example 1.2

Using the different definition of the sequence of rotations and the resulting transformation matrix defined in Example 1, and using the identity

$$\tilde{\boldsymbol{\omega}} = \dot{\mathbf{A}}\mathbf{A}^{\mathrm{T}}$$

the angular velocity vector can be defined in terms of the orientation coordinates and their derivatives as

$$\boldsymbol{\omega} = \mathbf{G}\dot{\boldsymbol{\beta}}$$

where

$$\mathbf{G} = \begin{bmatrix} \cos\theta\cos\psi & -\sin\psi & 0 \\ \cos\theta\sin\psi & \cos\psi & 0 \\ -\sin\theta & 0 & 1 \end{bmatrix}$$

$$\boldsymbol{\beta} = [\phi \quad \theta \quad \psi]^{\mathrm{T}}$$

It also can be shown, in this case, that the angular velocity vector defined in the body coordinate system is given by

$$\boldsymbol{\omega}_b = \mathbf{G}_b\dot{\boldsymbol{\beta}}$$

where

$$\mathbf{G}_b = \begin{bmatrix} 1 & 0 & \sin\theta \\ 0 & \cos\phi & \sin\phi\cos\theta \\ 0 & -\sin\phi & \cos\phi\cos\theta \end{bmatrix}$$

Angular Acceleration Vector The angular acceleration vector $\boldsymbol{\alpha}$ is defined as the time derivative of the absolute angular velocity vector $\boldsymbol{\omega}$, that is,

$$\boldsymbol{\alpha} = \dot{\boldsymbol{\omega}}$$

which also can be written in terms of the derivatives of the orientation coordinates as

$$\boldsymbol{\alpha} = \mathbf{G}\ddot{\boldsymbol{\beta}} + \dot{\mathbf{G}}\dot{\boldsymbol{\beta}}$$

The angular acceleration vector also can be defined in the body coordinate system as

$$\boldsymbol{\alpha}_b = \mathbf{A}^{\mathrm{T}}\boldsymbol{\alpha}$$

This equation also can be written as

$$\boldsymbol{\alpha}_b = \mathbf{A}^{\mathrm{T}}\dot{\boldsymbol{\omega}} = \mathbf{A}^{\mathrm{T}}(\dot{\mathbf{A}}\boldsymbol{\omega}_b + \mathbf{A}\dot{\boldsymbol{\omega}}_b)$$

The following identities can be verified (Shabana, 1994):

$$\mathbf{A}^{\mathrm{T}}\dot{\mathbf{A}} = \tilde{\boldsymbol{\omega}}_b, \qquad \tilde{\boldsymbol{\omega}}_b\boldsymbol{\omega}_b = \boldsymbol{\omega}_b \times \boldsymbol{\omega}_b = \mathbf{0}$$

It follows, upon the use of these identities and the orthogonality property of the transformation matrix, that

$$\boldsymbol{\alpha}_b = \dot{\boldsymbol{\omega}}_b$$

1.2 DYNAMIC EQUATIONS

The three-dimensional motion of a rigid body can be described using six equations which can be written in terms of the velocities and accelerations. In the *Newton–Euler formulation*, a centroidal body coordinate system that has an origin rigidly attached to the body center of mass is used. The Newton–Euler equations can be written as (Shabana, 1994)

$$\begin{bmatrix} m\mathbf{I} & \mathbf{0} \\ \mathbf{0} & \mathbf{I}_b \end{bmatrix} \begin{bmatrix} \ddot{\mathbf{R}} \\ \boldsymbol{\alpha}_b \end{bmatrix} = \begin{bmatrix} \mathbf{F}_R \\ \mathbf{F}_\alpha \end{bmatrix} + \begin{bmatrix} \mathbf{0} \\ -\boldsymbol{\omega}_b \times (\mathbf{I}_b\boldsymbol{\omega}_b) \end{bmatrix} \tag{1.10}$$

where m is the total mass of the rigid body, $\mathbf{I}_b$ is the inertia tensor, and $\mathbf{F}_R$ and $\mathbf{F}_\alpha$ are the vectors of forces and moments acting at the center of mass of the rigid body. The inertia tensor is defined as

$$\mathbf{I}_b = \begin{bmatrix} I_{xx} & I_{xy} & I_{xz} \\ I_{xy} & I_{yy} & I_{yz} \\ I_{xz} & I_{yz} & I_{zz} \end{bmatrix} \tag{1.11}$$

where

$$I_{xx} = \int_V \rho(y^2 + z^2)\, dV, \quad I_{yy} = \int_V \rho(x^2 + z^2)\, dV, \quad I_{zz} = \int_V \rho(x^2 + y^2)\, dV,$$

$$I_{xy} = -\int_V \rho xy\, dV, \qquad I_{xz} = -\int_V \rho xz\, dV, \qquad I_{yz} = -\int_V \rho yz\, dV$$

in which V and ρ are, respectively, the volume and mass density of the rigid body. In the Newton–Euler equations, the vector $\mathbf{R}$ is the global position vector of the center of mass of the body. As a result of using the center of mass as the reference point, there is no inertia or dynamic coupling between the translation and the rotation of the body in the Newton–Euler formulation.

Constrained Motion A mechanical system that consists of n_b rigid bodies can have at most 6 n_b independent coordinates. The number of independent coordinates or the *degrees of freedom* of the system is

$$n_d = 6n_b - n_c$$

where n_d is the number of the system degrees of freedom, and n_c is the number of constraint functions that describe the relationships between the 6 n_b coordinates. The dynamics of a mechanical system that has n_d degrees of freedom can be described using n_d independent differential equations only. In order to demonstrate this fact, we consider the case in which the rigid body rotates about a fixed axis. Without any loss of generality, we assume that the axis of rotation is the Z_b-axis of the body coordinate system. In this special case, the angular velocity and acceleration vectors are

$$\boldsymbol{\omega}_b = [0 \quad 0 \quad \dot{\psi}]^{\mathrm{T}}, \qquad \boldsymbol{\alpha}_b = [0 \quad 0 \quad \ddot{\psi}]^{\mathrm{T}}$$

In this case, the vector of centrifugal forces can be written as

$$\boldsymbol{\omega}_b \times (\mathbf{I}_b \boldsymbol{\omega}_b) = \dot{\psi}^2[-I_{yz} \quad I_{xz} \quad 0]^{\mathrm{T}}$$

and the Newton–Euler equations of the rigid body can be written as

$$\begin{bmatrix} m & 0 & 0 & 0 & 0 & 0 \\ 0 & m & 0 & 0 & 0 & 0 \\ 0 & 0 & m & 0 & 0 & 0 \\ 0 & 0 & 0 & I_{xx} & I_{xy} & I_{xz} \\ 0 & 0 & 0 & I_{xy} & I_{yy} & I_{yz} \\ 0 & 0 & 0 & I_{xz} & I_{yz} & I_{zz} \end{bmatrix} \begin{bmatrix} \ddot{R}_x \\ \ddot{R}_y \\ \ddot{R}_z \\ \alpha_{bx} \\ \alpha_{by} \\ \alpha_{bz} \end{bmatrix} = \begin{bmatrix} F_{Rx} \\ F_{Ry} \\ F_{Rz} \\ F_{\alpha x} \\ F_{\alpha y} \\ F_{\alpha z} \end{bmatrix} - \dot{\psi}^2 \begin{bmatrix} 0 \\ 0 \\ 0 \\ -I_{yz} \\ I_{xz} \\ 0 \end{bmatrix}$$

Since the rotation is about a fixed axis, there is only one independent orientation coordinate, and the rigid body has four independent coordinates. The accelerations of the rigid body can be expressed in terms of the independent accelerations as

$$\begin{bmatrix} \ddot{R}_x \\ \ddot{R}_y \\ \ddot{R}_z \\ \alpha_{bx} \\ \alpha_{by} \\ \alpha_{bz} \end{bmatrix} = \begin{bmatrix} 1 & 0 & 0 & 0 \\ 0 & 1 & 0 & 0 \\ 0 & 0 & 1 & 0 \\ 0 & 0 & 0 & 0 \\ 0 & 0 & 0 & 0 \\ 0 & 0 & 0 & 1 \end{bmatrix} \begin{bmatrix} \ddot{R}_x \\ \ddot{R}_y \\ \ddot{R}_z \\ \ddot{\psi} \end{bmatrix}$$

This matrix can be written compactly as

$$\ddot{\mathbf{q}} = \mathbf{B}_r \ddot{\mathbf{q}}_i \tag{1.12}$$

where

$$\ddot{\mathbf{q}} = [\ddot{R}_x \quad \ddot{R}_y \quad \ddot{R}_z \quad \alpha_{bx} \quad \alpha_{by} \quad \alpha_{bz}]^{\mathrm{T}}$$

$$\ddot{\mathbf{q}}_i = [\ddot{R}_x \quad \ddot{R}_y \quad \ddot{R}_z \quad \ddot{\psi}]^{\mathrm{T}}$$

$$\mathbf{B}_r = \begin{bmatrix} 1 & 0 & 0 & 0 \\ 0 & 1 & 0 & 0 \\ 0 & 0 & 1 & 0 \\ 0 & 0 & 0 & 0 \\ 0 & 0 & 0 & 0 \\ 0 & 0 & 0 & 1 \end{bmatrix}$$

Substituting the constraints of Eq. 12 into the Newton–Euler equations, and premultiplying by the transpose of the matrix $\mathbf{B}_r$, one obtains

$$\begin{bmatrix} m & 0 & 0 & 0 \\ 0 & m & 0 & 0 \\ 0 & 0 & m & 0 \\ 0 & 0 & 0 & I_{zz} \end{bmatrix} \begin{bmatrix} \ddot{R}_x \\ \ddot{R}_y \\ \ddot{R}_z \\ \ddot{\psi} \end{bmatrix} = \begin{bmatrix} F_{Rx} \\ F_{Ry} \\ F_{Rz} \\ F_{\alpha z} \end{bmatrix}$$

If we further assume that the body does not translate along the Z-axis, the number of degrees of freedom is reduced by one, and we obtain the Newton–Euler equations which govern the *planar motion* of the rigid body as

$$\begin{bmatrix} m & 0 & 0 \\ 0 & m & 0 \\ 0 & 0 & I_{zz} \end{bmatrix} \begin{bmatrix} \ddot{R}_x \\ \ddot{R}_y \\ \ddot{\psi} \end{bmatrix} = \begin{bmatrix} F_{Rx} \\ F_{Ry} \\ F_{\alpha z} \end{bmatrix} \tag{1.13}$$

Example 1.3

We consider the case of the spherical pendulum shown in Fig. 3. The rod is assumed to be uniform with mass m, and length l. The centroidal coordinate system of the rod is assumed to have principal axes such that the products of inertia of the rod are all equal to zero. In this case, the inertia tensor of the rod is defined as

$$\mathbf{I}_b = \begin{bmatrix} I_{xx} & 0 & 0 \\ 0 & I_{yy} & 0 \\ 0 & 0 & I_{zz} \end{bmatrix}$$

Since point O is assumed to be a fixed point, the absolute acceleration of this point is equal to zero, and therefore,

$$\mathbf{a}_O = \ddot{\mathbf{R}} + \mathbf{A}(\boldsymbol{\alpha}_b \times \mathbf{u}_{bO}) + \mathbf{A}(\boldsymbol{\omega}_b \times (\boldsymbol{\omega}_b \times \mathbf{u}_{bO})) = \mathbf{0}$$

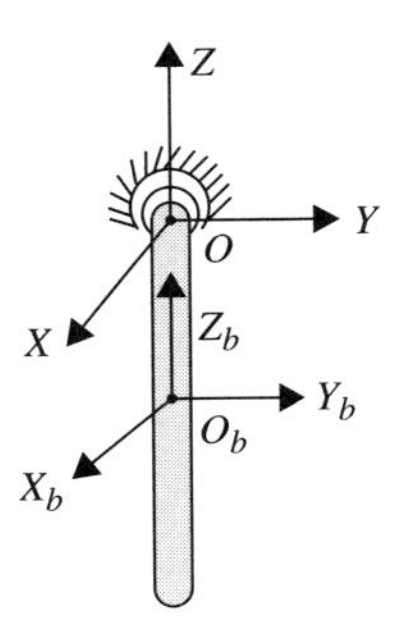

FIG. 1.3. Spherical pendulum.

where

$$\mathbf{u}_{bO} = \begin{bmatrix} 0 & 0 & \frac{l}{2} \end{bmatrix}^{\mathrm{T}}$$

is the position of point O with respect to the center of mass of the rod. It follows that

$$\ddot{\mathbf{R}} = \mathbf{A}\tilde{\mathbf{u}}_{bO}\boldsymbol{\alpha}_b + \boldsymbol{\gamma}$$

where

$$\boldsymbol{\gamma} = -\mathbf{A}\{\boldsymbol{\omega}_b \times (\boldsymbol{\omega}_b \times \mathbf{u}_{bO})\} = -\mathbf{A}\tilde{\boldsymbol{\omega}}_b^2\mathbf{u}_{bO}$$

Using the acceleration constraint equations, the accelerations of the rod can be expressed in terms of the angular accelerations as

$$\begin{bmatrix} \ddot{\mathbf{R}} \\ \boldsymbol{\alpha}_b \end{bmatrix} = \begin{bmatrix} \mathbf{A}\tilde{\mathbf{u}}_{bO} \\ \mathbf{I} \end{bmatrix}\boldsymbol{\alpha}_b + \begin{bmatrix} \boldsymbol{\gamma} \\ \mathbf{0} \end{bmatrix} = \mathbf{B}_r\boldsymbol{\alpha}_b + \boldsymbol{\gamma}_t$$

where

$$\mathbf{B}_r = \begin{bmatrix} \mathbf{A}\tilde{\mathbf{u}}_{bO} \\ \mathbf{I} \end{bmatrix}, \qquad \boldsymbol{\gamma}_t = \begin{bmatrix} \boldsymbol{\gamma} \\ \mathbf{0} \end{bmatrix}$$

Substituting into Eq. 10 and premultiplying by the transpose of the matrix $\mathbf{B}_r$, one obtains

$$(m\tilde{\mathbf{u}}_{bO}^{\mathrm{T}}\tilde{\mathbf{u}}_{bO} + \mathbf{I}_b)\boldsymbol{\alpha}_b + m\tilde{\mathbf{u}}_{bO}^{\mathrm{T}}\mathbf{A}^{\mathrm{T}}\boldsymbol{\gamma} = \tilde{\mathbf{u}}_{bO}^{\mathrm{T}}\mathbf{A}^{\mathrm{T}}\mathbf{F}_R + \mathbf{F}_\alpha - \boldsymbol{\omega}_b \times (\mathbf{I}_b\boldsymbol{\omega}_b) \tag{1.14}$$

It can be shown that

$$(m\tilde{\mathbf{u}}_{bO}^{\mathrm{T}}\tilde{\mathbf{u}}_{bO} + \mathbf{I}_b) = \begin{bmatrix} I_{xx} + \frac{ml^2}{4} & 0 & 0 \\ 0 & I_{yy} + \frac{ml^2}{4} & 0 \\ 0 & 0 & I_{zz} \end{bmatrix}$$

$$m\tilde{\mathbf{u}}_{bO}^{\mathrm{T}}\mathbf{A}^{\mathrm{T}}\boldsymbol{\gamma} = \frac{ml^2}{4}\begin{bmatrix} \omega_{by}\omega_{bz} \\ -\omega_{bx}\omega_{bz} \\ 0 \end{bmatrix}$$

It also can be shown that the vector of gyroscopic forces is given in terms of the components of the angular velocity vector as

$$-\boldsymbol{\omega}_b \times (\mathbf{I}_b \boldsymbol{\omega}_b) = \begin{bmatrix} \omega_{by}\omega_{bz}(I_{yy} - I_{zz}) \\ \omega_{bx}\omega_{bz}(I_{zz} - I_{xx}) \\ \omega_{bx}\omega_{by}(I_{xx} - I_{yy}) \end{bmatrix}$$

If the gravity is the only force acting on the pendulum, one has

$$\mathbf{F}_R = [0 \quad 0 \quad -mg]^{\mathrm{T}}, \qquad \mathbf{F}_\alpha = [0 \quad 0 \quad 0]^{\mathrm{T}}$$

It follows that

$$\tilde{\mathbf{u}}_{bO}^{\mathrm{T}} \mathbf{A}^{\mathrm{T}} \mathbf{F}_R = -mg\frac{l}{2}[a_{32} \quad -a_{31} \quad 0]^{\mathrm{T}}$$

where a_{ij} are the elements of the transformation matrix $\mathbf{A}$. Therefore, the equations of motion of the spherical pendulum, defined by Eq. 14, can be written explicitly in a matrix form as

$$\begin{bmatrix} I_{xx} + \dfrac{ml^2}{4} & 0 & 0 \\ 0 & I_{yy} + \dfrac{ml^2}{4} & 0 \\ 0 & 0 & I_{zz} \end{bmatrix} \begin{bmatrix} \alpha_{bx} \\ \alpha_{by} \\ \alpha_{bz} \end{bmatrix}$$

$$= -\frac{ml^2}{4} \begin{bmatrix} \omega_{by}\omega_{bz} \\ -\omega_{bx}\omega_{bz} \\ 0 \end{bmatrix}$$

$$- mg\frac{l}{2} \begin{bmatrix} a_{32} \\ -a_{31} \\ 0 \end{bmatrix} + \begin{bmatrix} \omega_{by}\omega_{bz}(I_{yy} - I_{zz}) \\ \omega_{bx}\omega_{bz}(I_{zz} - I_{xx}) \\ \omega_{bx}\omega_{by}(I_{xx} - I_{yy}) \end{bmatrix}$$

1.3 SINGLE DEGREE OF FREEDOM SYSTEMS

Single degree of freedom systems are special cases of multi-degree of freedom systems or continuous systems. The dynamics of these systems is governed by one differential equation which can be obtained by imposing more motion constraints. Clearly, a more detailed and accurate model requires, in many applications, the use of more degrees of freedom. Nonetheless, the theory of vibration of single degree of freedom systems remains one of the fundamental building blocks in the analysis of multi-degree of freedom and continuous systems. In order to demonstrate how the mathematical model of a single degree of freedom system can be obtained as a special case of a multi-degree of freedom model, we use Eq. 13, which defines Newton–Euler equations for the planar motion of rigid bodies. If we assume that one point on the rigid body is fixed such that its velocity remains equal to zero, as in the case of a simple pendulum, one has

$$\mathbf{v} = \dot{\mathbf{R}} + \dot{\mathbf{A}}\mathbf{u}_b = \mathbf{0}$$

where in this special case

$$\mathbf{R} = [R_x \quad R_y]^{\mathrm{T}}, \qquad \mathbf{u}_b = [x \quad y]^{\mathrm{T}}$$

$$\dot{\mathbf{A}} = \dot{\psi}\begin{bmatrix} -\sin\psi & -\cos\psi \\ \cos\psi & -\sin\psi \end{bmatrix}$$

Note that $\mathbf{R}$ is the global position vector of the center of mass of the body, and $\mathbf{u}_b$ is the position vector of the fixed point with respect to the center of mass of the body. The two velocity constraints of the fixed point can be differentiated with respect to time. The result of the differentiation can be used to write the acceleration vector of the center of mass in terms of the angular acceleration as

$$\ddot{\mathbf{R}} = \begin{bmatrix} \ddot{R}_x \\ \ddot{R}_y \end{bmatrix} = \ddot{\psi}\begin{bmatrix} x\sin\psi + y\cos\psi \\ -x\cos\psi + y\sin\psi \end{bmatrix} + \dot{\psi}^2\begin{bmatrix} x\cos\psi - y\sin\psi \\ x\sin\psi + y\cos\psi \end{bmatrix}$$

These constraint equations can be used to write the accelerations of the body in terms of the independent angular acceleration and velocity as

$$\ddot{\mathbf{q}} = \begin{bmatrix} \ddot{R}_x \\ \ddot{R}_y \\ \ddot{\psi} \end{bmatrix} = \ddot{\psi}\begin{bmatrix} x\sin\psi + y\cos\psi \\ -x\cos\psi + y\sin\psi \\ 1 \end{bmatrix} + \dot{\psi}^2\begin{bmatrix} x\cos\psi - y\sin\psi \\ x\sin\psi + y\cos\psi \\ 0 \end{bmatrix}$$

The coefficient vectors of the angular velocity and accelerations on the right-hand side of this equation are orthogonal as the result of the orthogonality of the tangential and normal components of the acceleration of the center of mass. Substituting the preceding equation into Eq. 13, which defines the Newton–Euler matrix equation in the case of planar motion, and premultiplying by the transpose of the coefficient vector of the angular acceleration defined in the preceding equation, one obtains the independent differential equation of motion of the single degree of freedom system as

$$(I_{zz} + ml^2)\ddot{\psi} = F_{\alpha z} + F_{Rx}(x\sin\psi + y\cos\psi) + F_{Ry}(-x\cos\psi + y\sin\psi)$$

where

$$l = \sqrt{x^2 + y^2}$$

Example 1.4

It was shown in Example 3 that the equations of motion of the spherical pendulum are

$$\begin{bmatrix} I_{xx} + \frac{ml^2}{4} & 0 & 0 \\ 0 & I_{yy} + \frac{ml^2}{4} & 0 \\ 0 & 0 & I_{zz} \end{bmatrix} \begin{bmatrix} \alpha_{bx} \\ \alpha_{by} \\ \alpha_{bz} \end{bmatrix}$$

$$= -\frac{ml^2}{4} \begin{bmatrix} \omega_{by}\omega_{bz} \\ -\omega_{bx}\omega_{bz} \\ 0 \end{bmatrix}$$

$$- mg\frac{l}{2} \begin{bmatrix} a_{32} \\ -a_{31} \\ 0 \end{bmatrix} + \begin{bmatrix} \omega_{by}\omega_{bz}(I_{yy} - I_{zz}) \\ \omega_{bx}\omega_{bz}(I_{zz} - I_{xx}) \\ \omega_{bx}\omega_{by}(I_{xx} - I_{yy}) \end{bmatrix}$$

The case of a simple pendulum can be obtained as a special case of the spherical pendulum by using the following assumptions:

$$\alpha_{by} = \alpha_{bz} = 0, \qquad \omega_{by} = \omega_{bz} = 0$$

Using these assumptions, the matrix equation of motion of the spherical pendulum reduces to

$$\left(I_{xx} + \frac{ml^2}{4}\right)\alpha_{bx} = -mga_{32}\frac{l}{2}$$

Since the rotation is assumed to be about the X-axis only, one has

$$a_{32} = \sin\phi, \qquad \alpha_{bx} = \ddot{\phi}$$

The equation of free vibration of the single degree of freedom simple pendulum reduces to

$$I_O\ddot{\phi} + mg\frac{l}{2}\sin\phi = 0$$

where I_O is the mass moment of inertia about point O and is defined as

$$I_O = I_{xx} + \frac{ml^2}{4}$$

Linear and Nonlinear Oscillations The results of the preceding example demonstrate that the dynamics of simple single degree of freedom systems

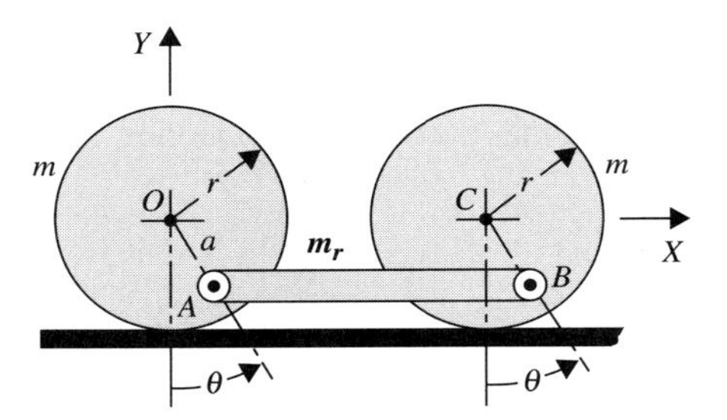

FIG. 1.4. Multi-body system.

can be governed by second-order nonlinear differential equations. Nonlinearities can be due to nonlinear restoring, damping, or inertia forces. Because of the kinematic constraints, the inertia forces associated with the system degrees of freedom can take a complex nonlinear form. In order to demonstrate this, we consider the multibody system shown in Fig. 4. The system consists of two homogeneous circular cylinders, each of mass m and centroidal mass moment of inertia I, and a connecting rod AB of mass m_r and length l. The cylinders, which have radius r, are assumed to roll without slipping. Because of the rolling conditions, the velocities of the centers of mass of the cylinders are equal and given by

$$\mathbf{v}_O = \mathbf{v}_C = [-r\dot{\theta} \quad 0]^{\mathrm{T}}$$

The absolute velocities of points A and B are equal since the angular velocity of the connecting rod is equal to zero. These velocities, which are also equal to the absolute velocity of the center of mass of the rod, are defined as

$$\mathbf{v}_B = \mathbf{v}_A = \mathbf{v}_O + \mathbf{v}_{AO} = \mathbf{v}_O + \boldsymbol{\omega} \times \mathbf{r}_{AO} = \begin{bmatrix} -(r - a\cos\theta)\dot{\theta} \\ a\dot{\theta}\sin\theta \end{bmatrix}$$

where $\mathbf{v}_{AO}$ is the velocity of point A with respect to point O, $\boldsymbol{\omega}$ is the angular velocity vector of the cylinders, and $\mathbf{r}_{AO}$ is the position vector of point A with respect to point O and is defined as

$$\mathbf{r}_{AO} = [a\sin\theta \quad -a\cos\theta]^{\mathrm{T}}$$

The absolute acceleration of the center of mass of the cylinders and the acceleration of the center of mass of the rod $\mathbf{a}_r$, which is equal to the accelerations of points A and B, are

$$\mathbf{a}_O = \mathbf{a}_C = [-r\ddot{\theta} \quad 0]^{\mathrm{T}}$$

$$\mathbf{a}_r = \begin{bmatrix} a_{rx} \\ a_{ry} \end{bmatrix} = \mathbf{a}_A = \mathbf{a}_B = \begin{bmatrix} -(r - a\cos\theta)\ddot{\theta} - a\dot{\theta}^2\sin\theta \\ a\ddot{\theta}\sin\theta + a\dot{\theta}^2\cos\theta \end{bmatrix}$$

The multi-body system shown in Fig. 4 has one degree of freedom, and

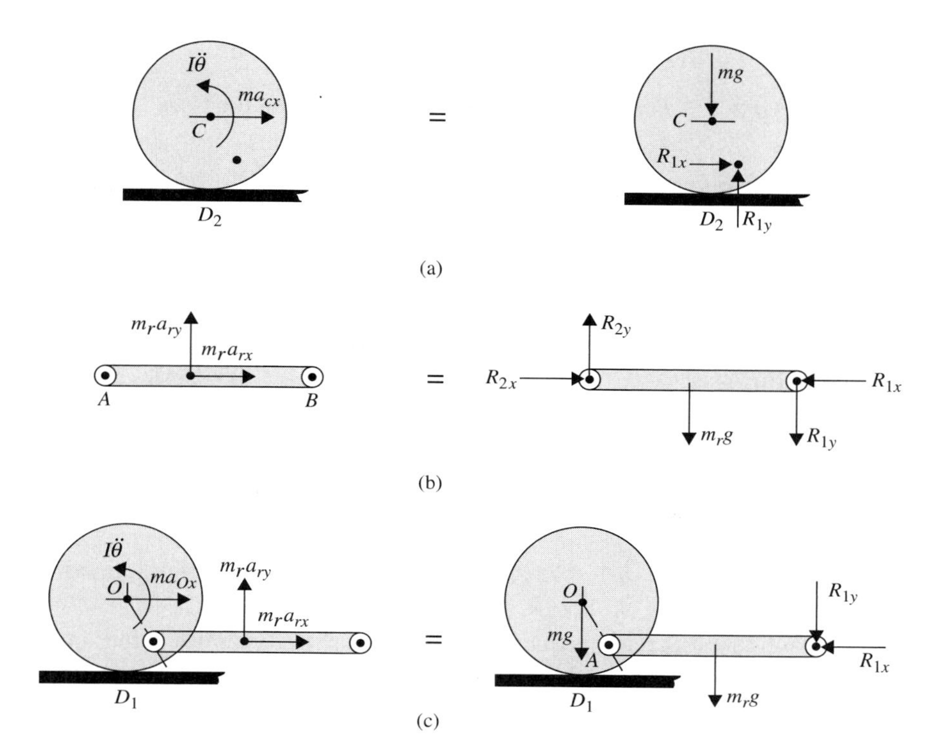

FIG. 1.5. Dynamics of the subsystems.

therefore one must be able to obtain one differential equation that does not include the constraint forces. This equation can be obtained by using D'Alembert's principle (Shabana, 1994). To this end, we consider the free body diagram of the right cylinder shown in Fig. 5a, where R_{ix} and R_{iy} will be used to denote the components of the reaction forces. By taking the moments of the inertia forces of the right cylinder about point D_2 and equating the results with the moments of the applied and joint forces, one obtains

$$I\ddot{\theta} + mr^2\ddot{\theta} = -R_{1x}(r - a\cos\theta) + R_{1y}a\sin\theta$$

We can also consider the forces acting on the connecting rod AB. Using the free body diagram shown in Fig. 5b and taking the moments of the inertia and applied forces about point A, we obtain

$$m_r a_{ry}\frac{l}{2} = -R_{1y}l - m_r g\frac{l}{2}$$

Next we consider the equilibrium of the subsystem shown in Fig. 5c. We apply again D'Alembert's principle by taking the moments of the inertia and applied forces about point D_1. The result is

$$I\ddot{\theta} + mr^2\ddot{\theta} - m_r a_{rx}(r - a\cos\theta) + m_r a_{ry}\left(\frac{l}{2} + a\sin\theta\right)$$

$$= R_{1x}(r - a\cos\theta) - R_{1y}(l + a\sin\theta) - m_r g\left(\frac{l}{2} + a\sin\theta\right)$$

Adding the first and third of the last preceding three equations, subtracting the second equation from the result, and using the expression for the acceleration of the center of mass of the connecting rod, one can eliminate the joint reaction forces and obtain the equation of motion of the multi-body system shown in Fig. 4 as

$$2\left\{I + mr^2 + \frac{1}{2}m_r(r^2 + a^2 - 2ra\cos\theta)\right\}\ddot{\theta} + m_r ra\dot{\theta}^2\sin\theta + m_r ga\sin\theta = 0$$

Since for a cylinder $I = mr^2/2$, the preceding equation reduces to

$$2\left\{\frac{3}{2}mr^2 + \frac{1}{2}m_r(r^2 + a^2 - 2ra\cos\theta)\right\}\ddot{\theta} + m_r ra\dot{\theta}^2\sin\theta + m_r ga\sin\theta = 0$$

This a nonlinear differential equation that governs the dynamics of the single degree of freedom multi-body system shown in Fig. 4. If we use the assumption of small oscillations, the preceding equation becomes a linear second-order ordinary differential equation given by

$$2\left\{\frac{3}{2}mr^2 + \frac{1}{2}m_r(r - a)^2\right\}\ddot{\theta} + m_r ga\theta = 0$$

In this book, we will focus on the analysis of linear systems. The vibration analysis of complex nonlinear multi-body systems is discussed in the multi-body literature (Shabana, 1989). In the remainder of this chapter, we briefly review the theory of vibration of single degree of freedom systems, which is important in the vibration analysis of multi-degree of freedom systems as well as continuous systems. The cases of free and forced vibrations as well as the effect of the damping on the dynamics of single degree of freedom systems will be examined.

1.4 OSCILLATORY AND NONOSCILLATORY MOTION

In this section, we study the effect of viscous damping on the free vibration of single degree of freedom systems. The differential equation of such systems will be developed, solved, and examined, and it will be seen from the theoretical development and the examples presented in this section that the damping force has a pronounced effect on the stability of the systems.

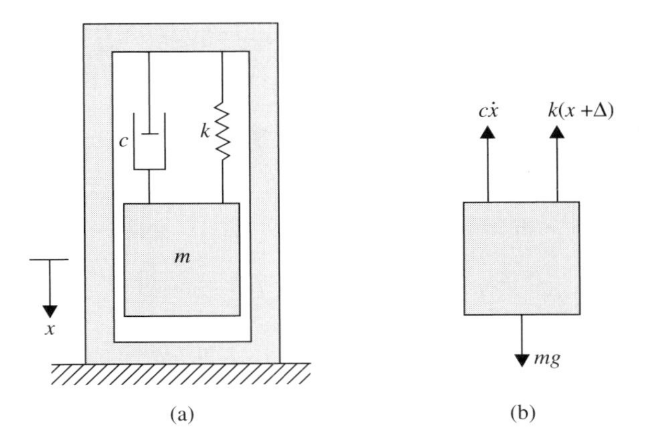

FIG. 1.6. Damped single degree of freedom system.

Figure 6(a) depicts a single degree of freedom system which consists of a mass m supported by a spring and a damper. The stiffness coefficient of the spring is k and the viscous damping coefficient of the damper is c. If the system is set in motion because of an initial displacement and/or an initial velocity, the mass will vibrate freely. At an arbitrary position x of the mass from the equilibrium position, the restoring spring force is equal to kx and the viscous damping force is proportional to the velocity and is equal to $c\dot{x}$, where the displacement x is taken as positive downward from the equilibrium position. Using the free body diagram shown in Fig. 6(b), the differential equation of motion can be written as

$$m\ddot{x} = mg - c\dot{x} - k(x + \Delta) \tag{1.15}$$

where Δ is the static deflection at the equilibrium position. Since the damper does not exert force at the static equilibrium position, the condition for the static equilibrium can be written as

$$mg = k\Delta$$

Substituting this equation into Eq. 15 yields

$$m\ddot{x} = -c\dot{x} - kx$$

or

$$m\ddot{x} + c\dot{x} + kx = 0 \tag{1.16}$$

This is the standard form of the second-order differential equation of motion that governs the linear vibration of damped single degree of freedom systems. A solution of this equation is in the form

$$x = Ae^{pt}$$

Substituting this solution into the differential equation yields

$$(mp^2 + cp + k)Ae^{pt} = 0$$

From which the characteristic equation is defined as

$$mp^2 + cp + k = 0 \tag{1.17}$$

The roots of this equation are given by

$$p_1 = -\frac{c}{2m} + \frac{1}{2m}\sqrt{c^2 - 4mk} \tag{1.18a}$$

$$p_2 = -\frac{c}{2m} - \frac{1}{2m}\sqrt{c^2 - 4mk} \tag{1.18b}$$

Define the following dimensionless quantity

$$\xi = \frac{c}{C_c} \tag{1.19}$$

where ξ is called the *damping factor* and C_c is called the *critical damping coefficient* defined as

$$C_c = 2m\omega = 2\sqrt{km} \tag{1.20}$$

where ω is the system *circular* or *natural frequency* defined as

$$\omega = \sqrt{k/m} \tag{1.21}$$

The roots p_1 and p_2 of the characteristic equation can be expressed in terms of the damping factor ξ as

$$p_1 = -\xi\omega + \omega\sqrt{\xi^2 - 1} \tag{1.22a}$$

$$p_2 = -\xi\omega - \omega\sqrt{\xi^2 - 1} \tag{1.22b}$$

If ξ is a greater than one, the roots p_1 and p_2 are real and distinct. If ξ is equal to one, the root p_1 is equal to p_2 and both roots are real. If ξ is less than one, the roots p_1 and p_2 are complex conjugates. The damping factor ξ is greater than one if the damping coefficient c is greater than the critical damping coefficient C_c. This is the case of an *overdamped* system. The damping factor ξ is equal to one when the damping coefficient c is equal to the critical damping coefficient C_c. In this case, the system is said to be *critically damped*. The damping factor ξ is less than one if the damping coefficient c is less than the critical damping coefficient C_c, and in this case, the system is said to be *underdamped*. In the following, the three cases of overdamped, critically damped, and underdamped systems are discussed in more detail.

Overdamped System In the overdamped case, the roots p_1 and p_2 of Eq. 22 are real. The response of the single degree of freedom system can be written as

$$x(t) = A_1 e^{p_1 t} + A_2 e^{p_2 t} \tag{1.23}$$

where A_1 and A_2 are arbitrary constants. Thus the solution, in this case, is the sum of two exponential functions and the motion of the system is non-oscillatory. The velocity can be obtained by differentiating Eq. 23 with respect to time as

$$\dot{x}(t) = p_1 A_1 e^{p_1 t} + p_2 A_2 e^{p_2 t} \tag{1.24}$$

The constants A_1 and A_2 can be determined from the initial conditions. For instance, if x_0 and $\dot{x}_0$ are, respectively, the initial displacement and velocity, one has from Eqs. 23 and 24

$$x_0 = A_1 + A_2$$

$$\dot{x}_0 = p_1 A_1 + p_2 A_2$$

from which A_1 and A_2 are

$$A_1 = \frac{x_0 p_2 - \dot{x}_0}{p_2 - p_1} \tag{1.25}$$

$$A_2 = \frac{\dot{x}_0 - p_1 x_0}{p_2 - p_1} \tag{1.26}$$

provided that $(p_1 - p_2)$ is not equal to zero. The displacement $x(t)$ can then be written in terms of the initial conditions as

$$x(t) = \frac{1}{p_2 - p_1}[(x_0 p_2 - \dot{x}_0)e^{p_1 t} + (\dot{x}_0 - p_1 x_0)e^{p_2 t}] \tag{1.27}$$

Example 1.5

The damped mass–spring system shown in Fig. 6 has mass $m = 10$ kg, stiffness coefficient $k = 1000$ N/m, and damping coefficient $c = 300$ N · s/m. Determine the displacement of the mass as a function of time.

Solution. The natural frequency ω of the system is

$$\omega = \sqrt{\frac{k}{m}} = \sqrt{\frac{1000}{10}} = 10 \text{ rad/s}$$

The critical damping coefficient C_c is

$$C_c = 2m\omega = 2(10)(10) = 200 \text{ N} \cdot \text{s/m}$$

The damping factor ξ is given by

$$\xi = \frac{c}{C_c} = \frac{300}{200} = 1.5$$

Since $\xi > 1$, the system is overdamped and the solution is given by

$$x(t) = A_1 e^{p_1 t} + A_2 e^{p_2 t}$$

where p_1 and p_2 can be determined using Eq. 22 as

$$p_1 = -\xi\omega + \omega\sqrt{\xi^2 - 1} = -(1.5)(10) + (10)\sqrt{(1.5)^2 - 1} = -3.8197$$

$$p_2 = -\xi\omega - \omega\sqrt{\xi^2 - 1} = -(1.5)(10) - (10)\sqrt{(1.5)^2 - 1} = -26.1803$$

The solution $x(t)$ is then given by

$$x(t) = A_1 e^{p_1 t} + A_2 e^{p_2 t} = A_1 e^{-3.8197t} + A_2 e^{-26.1803t}$$

The constants A_1 and A_2 can be determined from the initial conditions.

Critically Damped Systems For critically damped systems, the damping coefficient c is equal to the critical damping coefficient C_c. In this case, the damping factor ξ is equal to one, and the roots p_1 and p_2 of the characteristic equation are equal and are given by

$$p_1 = p_2 = p = -\omega$$

The solution, in this case is given by

$$x(t) = (A_1 + A_2 t)e^{-\omega t} \tag{1.28}$$

where A_1 and A_2 are arbitrary constants. It is clear from the above equation that the solution $x(t)$ is nonoscillatory and it is the product of a linear function of time and an exponential decay. The form of the solution depends on the constants A_1 and A_2 or, equivalently, on the initial conditions. The velocity $\dot{x}$ can be obtained by differentiating Eq. 28 with respect to time as

$$\dot{x}(t) = [A_2 - \omega(A_1 + A_2 t)]e^{-\omega t} \tag{1.29}$$

The constants A_1 and A_2 can be determined from the initial conditions. For instance, given the initial displacement x_0 and the initial velocity $\dot{x}_0$, Eqs. 28 and 29 yield

$$x_0 = A_1$$

$$\dot{x}_0 = A_2 - \omega A_1$$

from which

$$A_1 = x_0 \tag{1.30}$$

$$A_2 = \dot{x}_0 + \omega x_0 \tag{1.31}$$

The displacement can then be written in terms of the initial conditions as

$$x(t) = [x_0 + (\dot{x}_0 + \omega x_0)t]e^{-\omega t} \tag{1.32}$$

Example 1.6

The damped mass–spring system shown in Fig. 6 has mass $m = 10$ kg, stiffness coefficient $k = 1000$ N/m, and damping coefficient $c = 200$ N·s/m. Determine the displacement of the mass as a function of time.

Solution. The natural frequency ω of the system is

$$\omega = \sqrt{\frac{k}{m}} = \sqrt{\frac{1000}{10}} = 10 \text{ rad/s}$$

The critical damping coefficient C_c is given by

$$C_c = 2m\omega = 2(10)(10) = 200 \text{ N}\cdot\text{s/m}$$

The damping factor ξ is given by

$$\xi = \frac{c}{C_c} = \frac{200}{200} = 1$$

Since $\xi = 1$, the system is critically damped and the solution is given by Eq. 28 as

$$x(t) = (A_1 + A_2 t)e^{-10t}$$

where the constants A_1 and A_2 can be determined from the initial conditions by using Eqs. 30 and 31.

Underdamped Systems In the case of underdamped systems, the damping coefficient c is less than the critical damping coefficient C_c. In this case, the damping factor ξ is less than one and the roots of the characteristic equation p_1 and p_2, defined by Eq. 22, are complex conjugates. Let us define the *damped natural frequency* ω_d as

$$\omega_d = \omega\sqrt{1 - \xi^2} \tag{1.33}$$

Using this equation, the roots p_1 and p_2 of the characteristic equation given by Eq. 22 are defined as

$$p_1 = -\xi\omega + i\omega_d \tag{1.34a}$$

$$p_2 = -\xi\omega - i\omega_d \tag{1.34b}$$

In this case, one can show that the solution $x(t)$ of the underdamped system can be written as

$$x(t) = Xe^{-\xi\omega t}\sin(\omega_d t + \phi) \tag{1.35}$$

where the amplitude X and the phase angle ϕ are constant and can be determined from the initial conditions. The solution $x(t)$ is the product of an exponential decay and a harmonic function. Unlike the preceding two cases of overdamped and critically damped systems, the motion of the underdamped system is oscillatory.

The velocity $\dot{x}(t)$ is obtained by differentiating Eq. 35 with respect to time. This leads to

$$\dot{x}(t) = Xe^{-\xi\omega t}[-\xi\omega \sin(\omega_d t + \phi) + \omega_d \cos(\omega_d t + \phi)] \tag{1.36}$$

The peaks of the displacement curve can be obtained by setting $\dot{x}(t)$ equal to zero, that is,

$$Xe^{-\xi\omega t_i}[-\xi\omega \sin(\omega_d t_i + \phi) + \omega_d \cos(\omega_d t_i + \phi)] = 0$$

where t_i is the time at which peak i occurs. The above equation yields

$$\tan(\omega_d t_i + \phi) = \frac{\omega_d}{\xi\omega} = \frac{\sqrt{1 - \xi^2}}{\xi} \tag{1.37}$$

Using the trigonometric identity

$$\sin\theta = \frac{\tan\theta}{\sqrt{1 + \tan^2\theta}},$$

Eq. 37 yields

$$\sin(\omega_d t_i + \phi) = \sqrt{1 - \xi^2} \tag{1.38}$$

Equations 35 and 38 can be used to define the displacement of the peak i as

$$x_i = \sqrt{1 - \xi^2}\, Xe^{-\xi\omega t_i} \tag{1.39}$$

Undamped Vibration Note that in the case of undamped systems, $c = 0$, and Eqs. 35 and 36 reduce in this special case to

$$x(t) = X \sin(\omega t + \phi) \tag{1.40}$$

$$\dot{x}(t) = \omega X \cos(\omega t + \phi) \tag{1.41}$$

Example 1.7

The damped mass–spring system shown in Fig. 6 has mass $m = 10$ kg, stiffness coefficient $k = 1000$ N/m, and damping coefficient $c = 10$ N·s/m. Determine the displacement of the mass as a function of time.

Solution. The circular frequency ω of the system is

$$\omega = \sqrt{\frac{k}{m}} = \sqrt{\frac{1000}{10}} = 10 \text{ rad/s}$$

and the critical damping factor ξ is given by

$$C_c = 2m\omega = 2(10)(10) = 200 \text{ N}\cdot\text{s/m}$$

Therefore, the damping factor ξ is given by

$$\xi = \frac{c}{C_c} = \frac{10}{200} = 0.05$$

The damped natural frequency ω_d is given by

$$\omega_d = \omega\sqrt{1-\xi^2} = 10\sqrt{1-(0.05)^2} = 9.9875 \text{ rad/s}$$

Substituting ω, ξ, and ω_d into Eq. 35, the solution for the undamped single degree of freedom system can be expressed as

$$x = Xe^{-0.5t}\sin(9.9875t + \phi)$$

where X and ϕ are constants which can be determined from the initial conditions.

Equivalent Coefficients The linear differential equation of free vibration of the damped single degree of freedom system can, in general, be written in the following form

$$m_e\ddot{x} + c_e\dot{x} + k_e x = 0 \tag{1.42}$$

where m_e, c_e, and k_e are equivalent inertia, damping, and stiffness coefficients, and the dependent variable x can be a linear or angular displacement. In this general case, m_e, c_e, and k_e must have consistent units. The natural frequency ω, the critical damping coefficient C_c, and the damping factor ξ are defined, in this case, as

$$\omega = \sqrt{\frac{k_e}{m_e}} \tag{1.43}$$

$$C_c = 2m_e\omega = 2\sqrt{m_e k_e} \tag{1.44}$$

$$\xi = \frac{c_e}{C_c} \tag{1.45}$$

Note that Eqs. 42, 43, 44, and 45 reduce, respectively, to Eqs. 16, 21, 20, and 19 in the simple case of damped mass–spring systems. The use of Eqs. 43–45 is demonstrated by the following example.

Example 1.8

Assuming small oscillations, obtain the differential equation of the free vibration of the pendulum shown in Fig. 7. Determine the circular frequency, the critical damping coefficient, and the damping factor of this system, assuming that the rod is massless.

Solution. As shown in the figure, let R_x and R_y be the components of the reaction force at the pin joint. The moments of the externally applied forces about O are

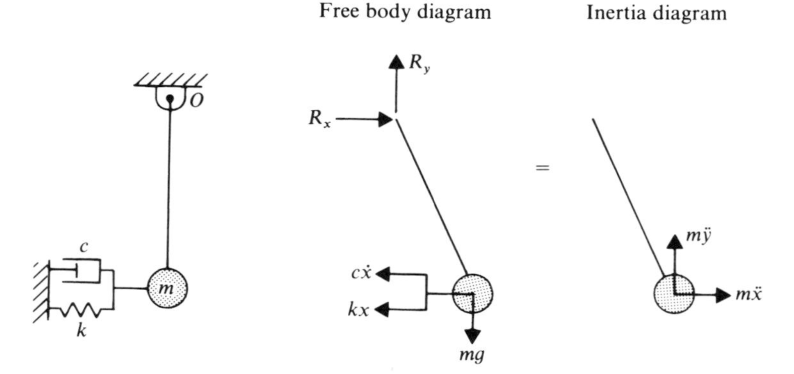

FIG. 1.7. Angular oscillations.

$$M_a = -(kl \sin \theta)l \cos \theta - (cl\dot{\theta} \cos \theta)l \cos \theta - mgl \sin \theta$$

For small oscillations, $\sin \theta \approx \theta$ and $\cos \theta \approx 1$. In this case, M_a reduces to

$$M_a = -kl^2\theta - cl^2\dot{\theta} - mgl\theta$$

One can show that the moment of the inertia (effective) forces about O is given by

$$M_{\text{eff}} = ml^2\ddot{\theta}$$

Therefore, the second-order differential equation of motion of the free vibration is given by

$$-kl^2\theta - cl^2\dot{\theta} - mgl\theta = ml^2\ddot{\theta}$$

or

$$ml^2\ddot{\theta} + cl^2\dot{\theta} + (kl + mg)l\theta = 0 \tag{1.46}$$

which can be written in the general form of Eq. 42 as

$$m_e\ddot{\theta} + c_e\dot{\theta} + k_e\theta = 0$$

where

$$m_e = ml^2$$

$$c_e = cl^2$$

$$k_e = (kl + mg)l$$

where the units of m_e are $\text{kg} \cdot \text{m}^2$ or, equivalently, $\text{N} \cdot \text{m} \cdot \text{s}^2$, the units of the equivalent damping coefficient c_e are $\text{N} \cdot \text{m} \cdot \text{s}$, and the units of the equivalent stiffness coefficient k_e are $\text{N} \cdot \text{m}$. The natural frequency ω is

$$\omega = \sqrt{\frac{k_e}{m_e}} = \sqrt{\frac{(kl + mg)l}{ml^2}} = \sqrt{\frac{kl + mg}{ml}} \quad \text{rad/s}$$

The critical damping coefficient C_c is

$$C_c = 2m_e\omega = 2ml^2\sqrt{\frac{kl+mg}{ml}}$$

$$= 2\sqrt{ml^3(kl+mg)}$$

The damping factor ξ of this system is

$$\xi = \frac{c_e}{C_c} = \frac{cl^2}{2\sqrt{ml^3(kl+mg)}}$$

1.5 OTHER TYPES OF DAMPING

Thus far, we have considered the case of a viscous damping force which is proportional to the velocity. In many cases, such simple expressions for the damping forces are not directly available. It is, however, possible to obtain an equivalent viscous damping coefficient by equating energy expressions that represent the dissipated energy during the motion. In this section, we consider the case of *structural damping* which is sometimes referred to as *hysteretic damping* and the case of *Coulomb* or *dry friction* damping.

Structural Damping The influence of the structural damping can be seen in the vibration of solid materials which, in general, are not perfectly elastic. When solids vibrate, there is an energy dissipation due to internal friction, which results from the relative motion between the particles during deformation. It was observed that there is a phase lag between the applied force F and the displacement x, as shown by the hysteresis loop curve in Fig. 8, which

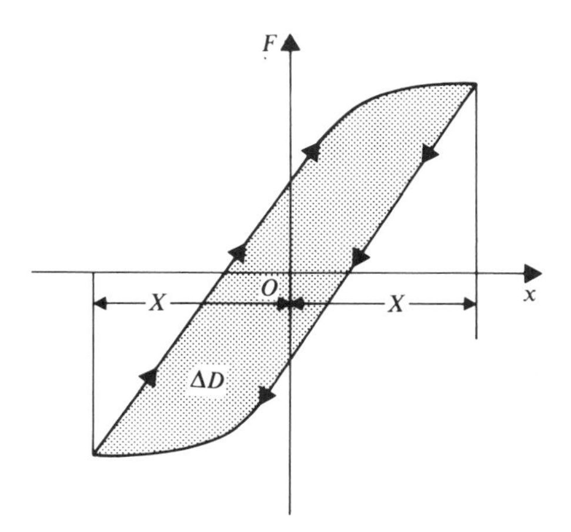

FIG. 1.8. Hysteresis loop.

clearly demonstrates that the effect of the force does not suddenly disappear when the force is removed. The energy loss during one cycle can be obtained as the enclosed area in the hysteresis loop, and can be expressed mathematically using the following integral

$$\Delta D = \int F \, dx \tag{1.47}$$

It was also observed experimentally that the energy loss during one cycle is proportional to the stiffness of the material k and the square of the amplitude of the displacement X, and can be expressed in the following simple form

$$\Delta D = \pi c_s k X^2 \tag{1.48}$$

where c_s is a dimensionless *structural damping coefficient* and the factor π is included for convenience. Equation 48 can be used to obtain an equivalent viscous damping coefficient, if we assume simple harmonic oscillations in the form

$$x = X \sin(\omega t + \phi)$$

The force exerted by a viscous damper can then be written as

$$F_d = c_e \dot{x} = c_e X \omega \cos(\omega t + \phi) \tag{1.49}$$

and the energy loss per cycle can be written as

$$\Delta D = \int F_d \, dx = \int c_e \dot{x} \, dx \tag{1.50}$$

Since $\dot{x} = dx/dt$, we have

$$dx = \dot{x} \, dt \tag{1.51}$$

Substituting Eqs. 49 and 51 into Eq. 50 yields

$$\begin{aligned} \Delta D &= \int_0^\tau c_e \dot{x}^2 \, dt = \int_0^\tau c_e \omega^2 X^2 \cos^2(\omega t + \phi) \, dt \\ &= \pi c_e \omega X^2 \end{aligned} \tag{1.52}$$

where τ is the periodic time defined as

$$\tau = \frac{2\pi}{\omega}$$

Equating Eqs. 48 and 52 yields the equivalent viscous damping coefficient as

$$c_e = \frac{k c_s}{\omega} \tag{1.53}$$

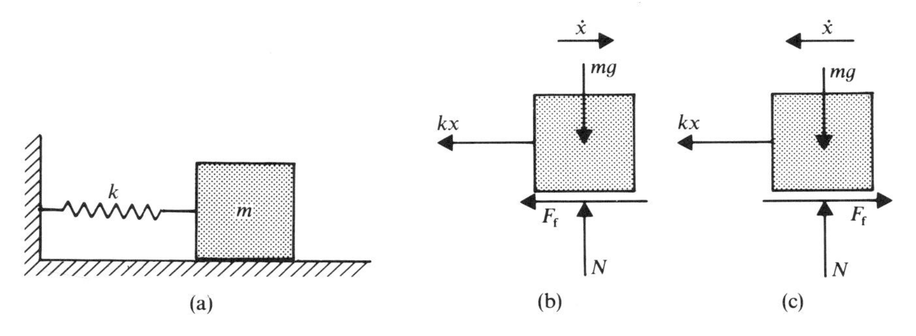

FIG. 1.9. Coulomb damping.

Coulomb Damping Consider the mass–spring system shown in Fig. 9. In the case of Coulomb damping, the friction force always acts in a direction opposite to the direction of the motion of the mass, and this friction force can be written as

$$F_f = \mu N \tag{1.54}$$

where μ is the *coefficient of sliding friction* and N is the normal reaction force. If the motion of the mass is to the right, $\dot{x} > 0$, and the friction force F_f is negative, as shown in Fig. 9(b). If the motion of the mass is to the left, $\dot{x} < 0$, and the friction force F_f is positive, as shown in Fig. 9(c). Therefore, the vibration of the system is governed by two differential equations which depend on the direction of motion. From the free body diagram shown in Fig. 9(b), it is clear that if the mass moves to the right, the differential equation of motion is

$$m\ddot{x} = -kx - F_f, \qquad \dot{x} > 0 \tag{1.55}$$

Similarly, the free body diagram of Fig. 9(c) shows that the differential equation of motion, when the mass moves to the left, is

$$m\ddot{x} = -kx + F_f, \qquad \dot{x} < 0 \tag{1.56}$$

Equation 55 and 56 can be combined in one equation as

$$m\ddot{x} + kx = \mp F_f \tag{1.57}$$

where the negative sign is used when the mass moves to the right and the positive sign is used when the mass moves to the left. Equation 57 is a nonhomogeneous differential equation, and its solution consists of two parts; the homogeneous solution or the complementary function and the forced or the particular solution. Since the force F_f is constant, the particular solution x_p is assumed as

$$x_p = C$$

where C is a constant. Substituting this solution into Eq. 57 yields

$$x_p = \mp \frac{F_f}{k}$$

Therefore, the solution of Eq. 57 can be written as

$$x(t) = A_1 \sin \omega t + A_2 \cos \omega t - \frac{F_f}{k}, \qquad \dot{x} \geq 0 \tag{1.58}$$

$$x(t) = B_1 \sin \omega t + B_2 \cos \omega t + \frac{F_f}{k}, \qquad \dot{x} < 0 \tag{1.59}$$

where ω is the natural frequency defined as

$$\omega = \sqrt{\frac{k}{m}}$$

and A_1 and A_2 are constants that depend on the initial conditions of motion to the right, and B_1 and B_2 are constants that depend on the initial conditions of motion to the left.

Let us now consider the case in which the mass was given an initial displacement x_0 to the right and zero initial veloctiy. Equation 59 can then be used to yield the following algebraic equations

$$x_0 = B_2 + \frac{F_f}{k}$$

$$0 = \omega B_1$$

which imply that

$$B_1 = 0, \qquad B_2 = x_0 - \frac{F_f}{k}$$

that is,

$$x(t) = \left(x_0 - \frac{F_f}{k}\right) \cos \omega t + \frac{F_f}{k} \tag{1.60}$$

and

$$\dot{x}(t) = -\omega \left(x_0 - \frac{F_f}{k}\right) \sin \omega t \tag{1.61}$$

The direction of motion will change when $\dot{x} = 0$. The above equation then yields the time t_1 at which the velocity starts to be positive. The time t_1 can be obtained from this equation as

$$0 = -\omega \left(x_0 - \frac{F_f}{k}\right) \sin \omega t_1$$

or

$$t_1 = \frac{\pi}{\omega}$$

At this time, the displacement is determined from Eq. 60 as

$$x(t_1) = x\left(\frac{\pi}{\omega}\right) = -x_0 + \frac{2F_f}{k} \tag{1.62}$$

which shows that the amplitude in the first half-cycle is reduced by the amount $2F_f/k$, as the result of dry friction.

In the second half-cycle, the mass moves to the right and the motion is governed by Eq. 58 with the initial conditions

$$x\left(\frac{\pi}{\omega}\right) = -x_0 + \frac{2F_f}{k}$$

$$\dot{x}\left(\frac{\pi}{\omega}\right) = 0$$

Substituting these initial conditions into Eq. 58 yields

$$A_1 = 0$$

and

$$A_2 = x_0 - 3\frac{F_f}{k}$$

The displacement $x(t)$ in the second half-cycle can then be written as

$$x(t) = \left(x_0 - 3\frac{F_f}{k}\right)\cos \omega t - \frac{F_f}{k} \tag{1.63}$$

and the velocity is

$$\dot{x}(t) = -\left(x_0 - 3\frac{F_f}{k}\right)\omega \sin \omega t \tag{1.64}$$

Observe that the velocity is zero at time t_2 when $t_2 = 2\pi/\omega = \tau$, where τ is the periodic time of the natural oscillations. At time t_2, the end of the first cycle, the displacement is

$$x(t_2) = x\left(\frac{2\pi}{\omega}\right) = x_0 - \left(\frac{4F_f}{k}\right) \tag{1.65}$$

which shows that the amplitude decreases in the second half-cycle by the amount $2F_f/k$, as shown in Fig. 10. By continuing in this manner, one can verify that there is a constant decrease in the amplitude of $2F_f/k$ every half-cycle of motion. Furthermore, unlike the case of viscous damping, the frequency of oscillation is not affected by the Coulomb damping. It is also important to point out, in the case of Coulomb friction, that it is not necessary that the system comes to rest at the undeformed spring position. The final position will be at an amplitude X_f, at which the spring force $F_s = kX_f$ is less than or equal to the friction force.

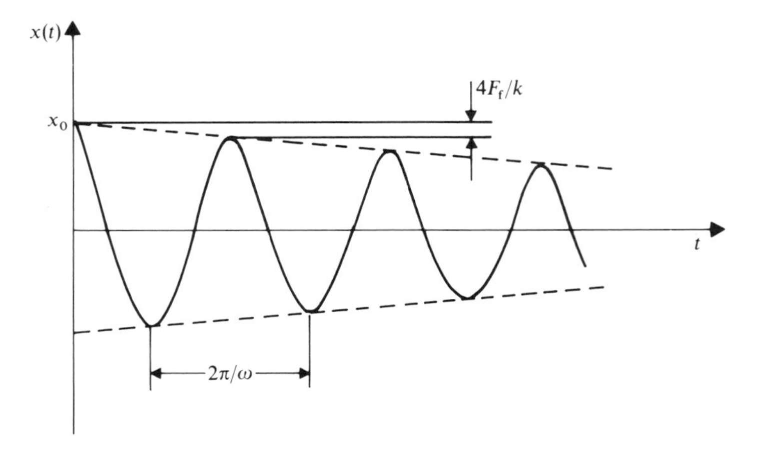

FIG. 1.10. Effect of the Coulomb damping.

1.6 FORCED VIBRATION

Figure 11 depicts a viscously damped single degree of freedom mass–spring system subjected to a forcing function $F(t)$. By applying Newton's second law, the differential equation of motion can be written as

$$m\ddot{x} + c\dot{x} + kx = F(t) \tag{1.66}$$

where m is the mass, c is the damping coefficient, k is the stiffness coefficient, and x is the displacement of the mass. Equation 66, which is a nonhomogeneous second-order ordinary differential equation with constant coefficients, has solution that consists of two parts; the complementary function x_h

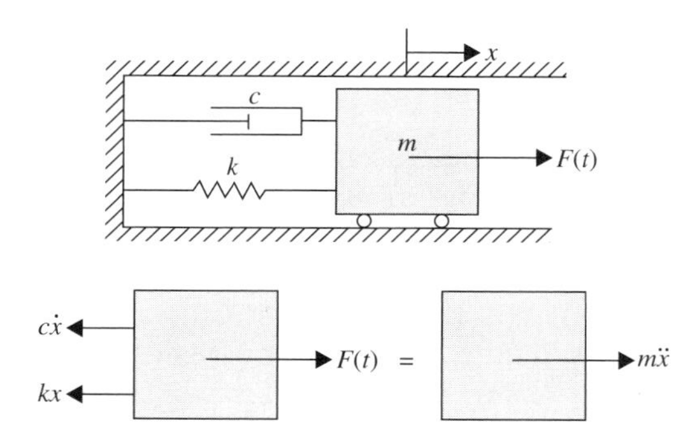

FIG. 1.11. Forced vibration of single degree of freedom systems.

and the particular solution x_p, that is,

$$x = x_h + x_p \tag{1.67}$$

where the complementary function x_h is the solution of the homogeneous equation

$$m\ddot{x}_h + c\dot{x}_h + kx_h = 0 \tag{1.68}$$

The complementary function x_h is sometimes called the *transient solution* since in the presence of damping this solution dies out. Methods for obtaining the transient response were discussed in the preceding sections. The particular solution x_p represents the response of the system to the forcing function, and is sometimes called the *steady state solution* because it exists long after the transient vibration disappears. The transient solution contains two arbitrary constants, while the steady state solution does not contain any arbitrary constants. Therefore, the solution of Eq. 67 contains two arbitrary constants which can be determined by using the initial conditions.

In the analysis presented in this section, we consider the case of harmonic excitation in which the forcing function $F(t)$ can be expressed in the form

$$F(t) = F_0 \sin \omega_f t \tag{1.69}$$

Substituting this equation into Eq. 66 yields

$$m\ddot{x} + c\dot{x} + kx = F_0 \sin \omega_f t \tag{1.70}$$

The steady state solution x_p can be assumed in the form

$$x_p = A_1 \sin \omega_f t + A_2 \cos \omega_f t \tag{1.71}$$

which yields the following expressions for the velocity and acceleration

$$\dot{x}_p = \omega_f A_1 \cos \omega_f t - \omega_f A_2 \sin \omega_f t \tag{1.72}$$

$$\ddot{x}_p = -\omega_f^2 A_1 \sin \omega_f t - \omega_f^2 A_2 \cos \omega_f t = -\omega_f^2 x_p \tag{1.73}$$

Substituting Eqs. 71–73 into Eq. 70 and rearranging terms yields

$$(k - \omega_f^2 m)(A_1 \sin \omega_f t + A_2 \cos \omega_f t) + c\omega_f(A_1 \cos \omega_f t - A_2 \sin \omega_f t)$$
$$= F_0 \sin \omega_f t$$

or

$$[(k - \omega_f^2 m) A_1 - c\omega_f A_2] \sin \omega_f t + [c\omega_f A_1 + (k - \omega_f^2 m) A_2] \cos \omega_f t$$
$$= F_0 \sin \omega_f t$$

This equation yields the following two algebraic equations in A_1 and A_2

$$(k - \omega_f^2 m) A_1 - c\omega_f A_2 = F_0 \tag{1.74}$$

$$c\omega_f A_1 + (k - \omega_f^2 m) A_2 = 0 \tag{1.75}$$

Dividing these two equations by the stiffness coefficient k yields

$$(1 - r^2) A_1 - 2r\xi A_2 = X_0 \tag{1.76}$$

$$2r\xi A_1 + (1 - r^2) A_2 = 0 \tag{1.77}$$

where

$$r = \frac{\omega_f}{\omega} \tag{1.78}$$

$$\xi = \frac{c}{C_c} = \frac{c}{2m\omega} \tag{1.79}$$

$$X_0 = \frac{F_0}{k} \tag{1.80}$$

in which $C_c = 2m\omega$ is the *critical damping coefficient*. The two algebraic equations of Eqs. 76 and 77 can be solved using *Cramer's rule* in order to obtain the constants A_1 and A_2 as

$$A_1 = \frac{\begin{vmatrix} X_0 & -2r\xi \\ 0 & 1 - r^2 \end{vmatrix}}{(1 - r^2)^2 + (2r\xi)^2} = \frac{(1 - r^2) X_0}{(1 - r^2)^2 + (2r\xi)^2} \tag{1.81}$$

$$A_2 = \frac{\begin{vmatrix} 1 - r^2 & X_0 \\ 2r\xi & 0 \end{vmatrix}}{(1 - r^2)^2 + (2r\xi)^2} = \frac{-(2r\xi) X_0}{(1 - r^2)^2 + (2r\xi)^2} \tag{1.82}$$

The steady state solution x_p of Eq. 71 can then be written as

$$x_p = \frac{X_0}{(1 - r^2)^2 + (2r\xi)^2} [(1 - r^2) \sin \omega_f t - (2r\xi) \cos \omega_f t] \tag{1.83}$$

which can be written as

$$x_p = \frac{X_0}{\sqrt{(1 - r^2)^2 + (2r\xi)^2}} \sin(\omega_f t - \psi) \tag{1.84}$$

where ψ is the phase angle defined by

$$\psi = \tan^{-1}\left(\frac{2r\xi}{1 - r^2}\right) \tag{1.85}$$

Equation 84 can be written in a more compact form as

$$x_p = X_0 \beta \sin(\omega_f t - \psi) \tag{1.86}$$

where β is the *magnification factor* defined in the case of damped systems as

$$\beta = \frac{1}{\sqrt{(1 - r^2)^2 + (2r\xi)^2}} \tag{1.87}$$

If the damping factor ξ is equal to zero, the magnification factor β reduces to

$$\beta = \frac{1}{1 - r^2}$$

When $r = 1$, that is $\omega_f = \omega$, the magnification factor in the case of undamped systems approaches infinity. This case is known as *resonance.*

The magnification factor β and the phase angle ψ are shown, respectively, in Figs. 12 and 13 as functions of the frequency ratio r for different damping factors ξ. It is clear from these figures, that for damped systems, the system does not attain infinite displacement at resonance, since for $\omega_f = \omega$, which corresponds to the case in which the frequency ratio $r = 1$, the magnification factor reduces to

$$\beta = \frac{1}{2\xi} \tag{1.88}$$

Furthermore, at resonance the magnification factor β does not have the maximum value, which can be obtained by differentiating β of Eq. 87 with

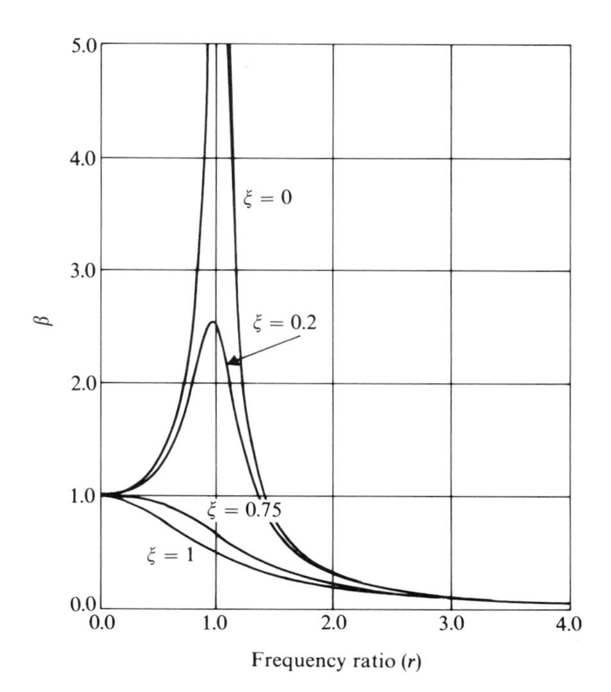

FIG. 1.12. Magnification factor.

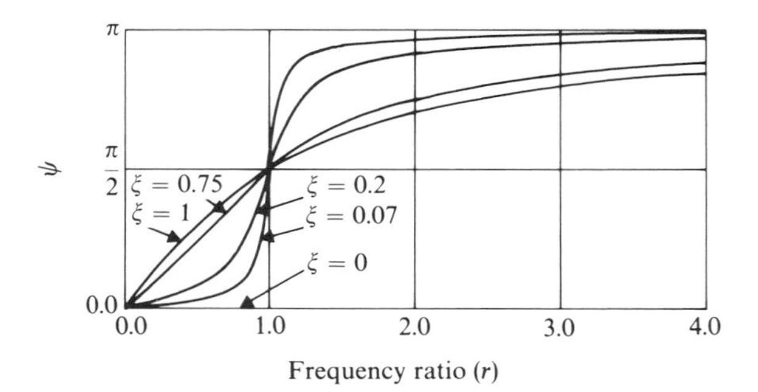

FIG. 1.13. Phase angle.

respect to r and setting the result equal to zero. This leads to an algebraic equation which can be solved for the frequency ratio r at which the magnification factor β is maximum. By so doing, one can show that the magnification factor β is maximum when

$$r = \sqrt{1 - 2\xi^2} \tag{1.89}$$

At this value of the frequency ratio, the maximum magnification factor is given by

$$\beta_{\max} = \frac{1}{2\xi\sqrt{1 - \xi^2}} \tag{1.90}$$

Force Transmission From Eq. 84 and Fig. 12 it is clear that by increasing the spring stiffness k and the damping coefficient c the amplitude of vibration decreases. The increase in the stiffness and damping coefficients, however, may have an adverse effect on the force transmitted to the support. In order to reduce this force, the stiffness and damping coefficients must be properly selected. Figure 14 shows a free body diagram for the mass and the support system. The force transmitted to the support in the steady state can be written as

$$F_t = kx_p + c\dot{x}_p \tag{1.91}$$

From Eq. 86, $\dot{x}_p$ is

$$\dot{x}_p = \omega_f X_0 \beta \cos(\omega_f t - \psi)$$

Equation 91 can then be written as

$$\begin{aligned} F_t &= kX_0\beta \sin(\omega_f t - \psi) + c\omega_f X_0 \beta \cos(\omega_f t - \psi) \\ &= X_0 \beta \sqrt{k^2 + (c\omega_f)^2} \sin(\omega_f t - \bar{\psi}) \end{aligned} \tag{1.92}$$

where

$$\bar{\psi} = \psi - \psi_t \tag{1.93}$$

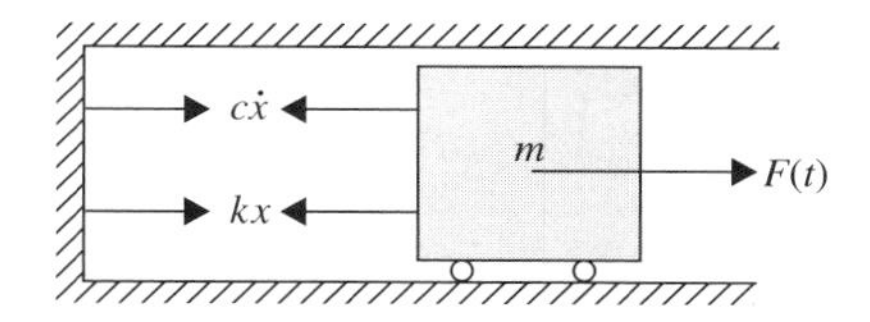

FIG. 1.14. Transmitted force.

and ψ_t is a phase angle defined as

$$\psi_t = \tan^{-1}\left(\frac{c\omega_f}{k}\right) = \tan^{-1}(2r\xi) \tag{1.94}$$

Equation 92 can also be written as

$$F_t = X_0 k\beta\sqrt{1 + (2r\xi)^2}\sin(\omega_f t - \bar{\psi}) \tag{1.95}$$

Since $X_0 = F_0/k$, the above equation can be written as

$$\begin{aligned} F_t &= F_0\beta\sqrt{1 + (2r\xi)^2}\sin(\omega_f t - \bar{\psi}) \\ &= F_0\beta_t \sin(\omega_f t - \bar{\psi}) \end{aligned} \tag{1.96}$$

where

$$\begin{aligned} \beta_t &= \beta\sqrt{1 + (2r\xi)^2} \\ &= \frac{\sqrt{1 + (2r\xi)^2}}{\sqrt{(1 - r^2)^2 + (2r\xi)^2}} \end{aligned} \tag{1.97}$$

Note that β_t represents the ratio between the amplitude of the transmitted force and the amplitude of the applied force. β_t is called the *transmissibility* and is plotted in Fig. 15 versus the frequency ratio r for different values of the damping factor ξ. It is clear from Fig. 15 that $\beta_t > 1$ for $r < \sqrt{2}$, that is, in this region the amplitude of the transmitted force is greater than the amplitude of the applied force. Furthermore, for $r < \sqrt{2}$, the transmitted force to the support can be reduced by increasing the damping factor ξ. For $r > \sqrt{2}$, $\beta_t < 1$, and in this region the amplitude of the transmitted force is less than the amplitude of the applied force. In this region, the amplitude of the transmitted force increases by increasing the damping factor ξ.

Work Per Cycle Equation 86, which defines the steady state response to a harmonic excitation in the presence of damping, implies that, for a given frequency ratio r and a given damping factor ξ, the amplitude of vibration remains constant. This can be achieved only if the energy input to the system, as the result of the work done by the external harmonic force, is equal to the energy dissipated as the result of the presence of damping. In order to see this, we first evaluate the work of the harmonic force as

$$dW_e = F(t)\,dx = F(t)\dot{x}\,dt$$

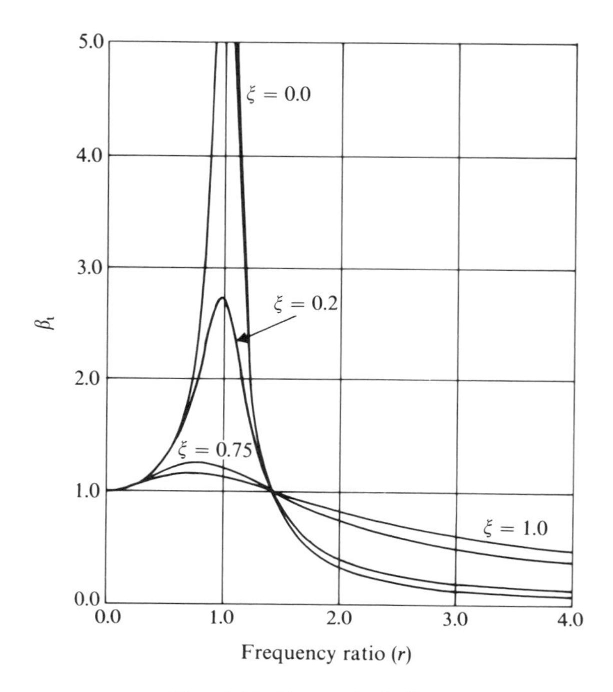

FIG. 1.15. Transmissibility.

where W_e is the work done by the external force per cycle, and is given by

$$
\begin{aligned}
W_e &= \int_0^{2\pi/\omega_f} F(t)\dot{x}\,dt \\
&= \int_0^{2\pi/\omega_f} F_0 \sin \omega_f t\, X_0 \beta \omega_f \cos(\omega_f t - \psi)\,dt \\
&= F_0 X_0 \beta \int_0^{2\pi} \sin \omega_f t \cos(\omega_f t - \psi)\,d(\omega_f t)
\end{aligned}
$$

which upon integration yields

$$
W_e = \pi F_0 X_0 \beta \sin \psi
$$

Similarly, one can evaluate the energy dissipated per cycle, as the result of the damping force, as

$$
\begin{aligned}
W_d &= \int_0^{2\pi/\omega_f} c\dot{x}\dot{x}\,dt \\
&= cX_0^2 \beta^2 \omega_f \int_0^{2\pi} \cos^2(\omega_f t - \psi)\,d(\omega_f t)
\end{aligned}
$$

which upon integration yields

$$W_d = \pi c X_0^2 \beta^2 \omega_f$$

Note that the input energy to the system is a linear function of the amplitude of the steady state vibration $X_0\beta$, while the energy dissipated as the result of the damping force is a quadratic function of the amplitude. Since at the steady state they must be equal, one has

$$W_e = W_d$$

or

$$\pi F_0 X_0 \beta \sin\psi = \pi c X_0^2 \beta^2 \omega_f$$

which defines the magnification factor β as

$$\beta = \frac{F_0/X_0}{c\omega_f} \sin\psi$$

Using the definition of X_0 and the phase angle ψ given, respectively, by Eqs. 80 and 85, the magnification factor β can be written as

$$\beta = \frac{F_0/X_0}{c\omega_f} \sin\psi = \frac{k}{c\omega_f} \frac{2r\xi}{\sqrt{(1-r^2)^2 + (2r\xi)^2}}$$

and since $c\omega_f/k = 2r\xi$, the above equation reduces to

$$\beta = \frac{1}{\sqrt{(1-r^2)^2 + (2r\xi)^2}}$$

which is the same definition of the magnification factor obtained by solving the differential equation. It is obtained here from equating the input energy, resulting from the work done by the harmonic force, to the energy dissipated as the result of the damping force. In fact, this must be the case, since the change in the strain energy in a complete cycle must be equal to zero owing to the fact that the spring takes the same elongation after a complete cycle. This can also be demonstrated mathematically by using the definition of the work done by the spring force as

$$\begin{aligned} W_s &= \int_0^{2\pi/\omega_f} kx\dot{x}\, dt \\ &= kX_0^2\beta^2 \int_0^{2\pi} \sin(\omega_f t - \psi)\cos(\omega_f t - \psi)\, d(\omega_f t) \end{aligned}$$

which upon integration yields

$$W_s = 0$$

Example 1.9

A damped single degree of freedom mass–spring system has mass $m = 10$ kg, spring coefficient $k = 4000$ N/m, and damping coefficient $c = 40$ N·s/m. The amplitude of the forcing function $F_0 = 60$ N, and the frequency $\omega_f = 40$ rad/s. Determine the displacement of the mass as a function of time. Determine also the transmissibility and the amplitude of the force transmitted to the support.

Solution. The circular frequency of the system is

$$\omega = \sqrt{\frac{k}{m}} = \sqrt{\frac{4000}{10}} = 20 \text{ rad/s}$$

The frequency ratio r is given by

$$r = \frac{\omega_f}{\omega} = \frac{40}{20} = 2$$

The critical damping coefficient C_c is defined as

$$C_c = 2m\omega = 2(10)(20) = 400 \text{ N}\cdot\text{s/m}$$

The damping factor ξ is then given by

$$\xi = \frac{c}{C_c} = \frac{40}{400} = 0.1$$

which is the case of an underdamped system. One can then write the complete solution in the following form

$$\begin{aligned} x(t) &= x_h + x_p \\ &= Xe^{-\xi\omega t}\sin(\omega_d t + \phi) + X_0\beta\sin(\omega_f t - \psi) \end{aligned}$$

where ω_d is the damped circular frequency

$$\omega_d = \omega\sqrt{1 - \xi^2} = 20\sqrt{1 - (0.1)^2} = 19.8997 \text{ rad/s}$$

The constants X_0, β, and ψ are

$$X_0 = \frac{F_0}{k} = \frac{60}{4000} = 0.015 \text{ m}$$

$$\begin{aligned} \beta &= \frac{1}{\sqrt{(1 - r^2)^2 + (2r\zeta)^2}} = \frac{1}{\sqrt{(1 - (2)^2)^2 + (2 \times 2 \times 0.1)^2}} = \frac{1}{\sqrt{9 + 0.16}} \\ &= 0.3304 \end{aligned}$$

$$\psi = \tan^{-1}\left(\frac{2r\xi}{1 - r^2}\right) = \tan^{-1}\left(\frac{2(2)(0.1)}{1 - (2)^2}\right) = \tan^{-1}(0.13333) = 0.13255 \text{ rad}$$

The displacement can then be written as a function of time as

$$x(t) = Xe^{-2t}\sin(19.8997t + \phi) + 0.004956\sin(40t - 0.13255)$$

The constants X and ϕ can be determined using the initial conditions. The transmissibility β_t is defined as

$$\beta_t = \frac{\sqrt{1 + (2r\xi)^2}}{\sqrt{(1 - r^2)^2 + (2r\xi)^2}} = \frac{\sqrt{1 + (2 \times 2 \times 0.1)^2}}{\sqrt{[1 - (2)^2]^2 + (2 \times 2 \times 0.1)^2}} = 0.35585$$

The amplitude of the force transmitted is given by

$$|F_t| = F_0\beta = (60)(0.35585) = 21.351 \text{ N}$$

1.7 IMPULSE RESPONSE

An impulsive force is defined as a force which has a large magnitude, and acts during a very short time duration such that the time integral of this force is finite. If the impulsive force $F(t)$ shown in Fig. 16 acts on the single degree of freedom system shown in Fig. 17, the differential equation of motion of this system can be written as

$$m\ddot{x} + c\dot{x} + kx = F(t)$$

Integrating this equation over the very short interval (t_1, t_2), one obtains

$$\int_{t_1}^{t_2} m\ddot{x}\, dt + \int_{t_1}^{t_2} c\dot{x}\, dt + \int_{t_1}^{t_2} kx\, dt = \int_{t_1}^{t_2} F(t)\, dt \tag{1.98}$$

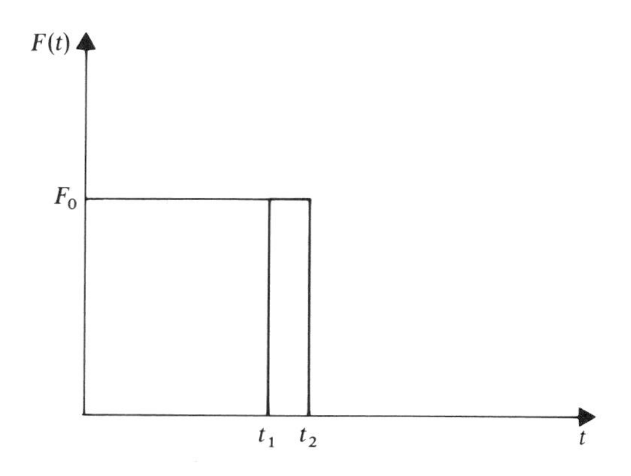

FIG. 1.16. Impulsive force.

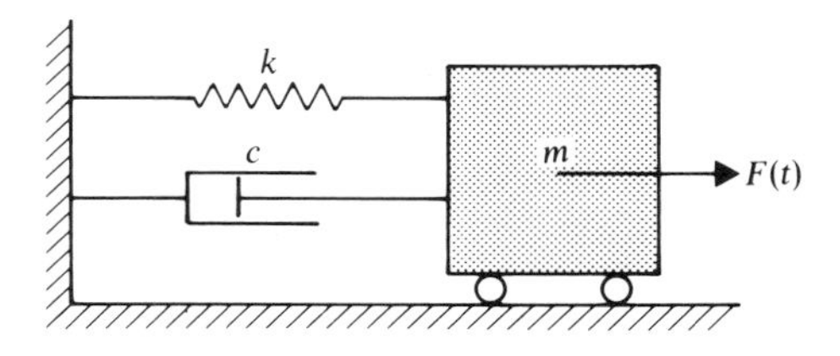

FIG. 1.17. Single degree of freedom system under the effect of impulsive force $F(t)$.

Since the time interval (t_1, t_2) is assumed to be very small, we assume that x does not change appreciably, and we also assume that the change in the velocity $\dot{x}$ is finite. One, therefore, has

$$\lim_{t_1 \to t_2} \int_{t_1}^{t_2} c\dot{x}\, dt = 0$$

$$\lim_{t_1 \to t_2} \int_{t_1}^{t_2} kx\, dt = 0$$

Therefore, if t_1 approaches t_2, Eq. 98 yields

$$\int_{t_1}^{t_2} m\ddot{x}\, dt = \int_{t_1}^{t_2} F(t)\, dt \tag{1.99}$$

Since $\ddot{x} = d\dot{x}/dt$, Eq. 99 can be written as

$$\int_{\dot{x}_1}^{\dot{x}_2} m\, d\dot{x} = \int_{t_1}^{t_2} F(t)\, dt \tag{1.100}$$

where $\dot{x}_1$ and $\dot{x}_2$ are, respectively, the velocities at t_1 and t_2. Equation 100 yields

$$m(\dot{x}_2 - \dot{x}_1) = \int_{t_1}^{t_2} F(t)\, dt$$

The preceding equation defines

$$\Delta\dot{x} = \dot{x}_2 - \dot{x}_1 = \frac{1}{m}\int_{t_1}^{t_2} F(t)\, dt \tag{1.101}$$

where $\Delta\dot{x}$ is the jump discontinuity in the velocity of the mass due to the impulsive force. The time integral in Eq. 101 is called the *linear impulse* I and is defined by

$$I = \int_{t_1}^{t_2} F(t)\, dt \tag{1.102}$$

In the particular case in which the linear impulse is equal to one, I is called the *unit impulse*.

Equation 101 can be written as

$$\Delta\dot{x} = \dot{x}_2 - \dot{x}_1 = \frac{I}{m} \tag{1.103}$$

This result indicates that the effect of the impulsive force, which acts over a very short time duration on a system which is initially at rest, can be accounted for by considering the motion of the system with initial velocity I/m and zero initial displacement. That is, in the case of impulsive motion, we consider the system vibrating freely as the result of the initial velocity defined by Eq. 103.

The free vibration of the underdamped single degree of freedom system shown in Fig. 17 is governed by the equations

$$x(t) = Xe^{-\xi\omega t}\sin(\omega_d t + \phi) \tag{1.104}$$

$$\dot{x}(t) = -\xi\omega X e^{-\xi\omega t}\sin(\omega_d t + \phi) + \omega_d X e^{-\xi\omega t}\cos(\omega_d t + \phi) \tag{1.105}$$

where X and ϕ are constants to be determined from the initial conditions, ω is the natural frequency, ξ is the damping factor, and ω_d is the damped natural frequency

$$\omega_d = \omega\sqrt{1 - \xi^2}$$

As the result of applying an impulsive force with a linear impulse I at $t = 0$, the initial conditions are

$$x(t = 0) = 0, \qquad \dot{x}(t = 0) = \frac{I}{m}$$

Since the initial displacement is zero, Eq. 104 yields

$$\phi = 0$$

Using Eq. 105 and the initial velocity, it is an easy matter to verify that

$$x(t) = \frac{I}{m\omega_d} e^{-\xi\omega t}\sin\omega_d t \tag{1.106}$$

which can be written as

$$x(t) = IH(t) \tag{1.107}$$

where $H(t)$ is called the *impulse response function* and is defined as

$$H(t) = \frac{1}{m\omega_d} e^{-\xi\omega t}\sin\omega_d t \tag{1.108}$$

Example 1.10

Find the response of the single degree of freedom system shown in Fig. 17 to the rectangular impulsive force shown in Fig. 16, where $m = 10$ kg, $k = 9{,}000$ N/m, $c = 18$ N·s/m, and $F_0 = 10{,}000$ N. The force is assumed to act at time $t = 0$ and the impact interval is assumed to be 0.005 s.

Solution. The linear impulse I is given by

$$I = \int_{t_1}^{t_2} F(t)\,dt = \int_0^{0.005} 10{,}000\,dt = 10{,}000(0.005) = 50 \text{ N·s}$$

The natural frequency of the system ω is given by

$$\omega = \sqrt{\frac{k}{m}} = \sqrt{\frac{9000}{10}} = 30 \text{ rad/s}$$

The critical damping coefficient C_c is

$$C_c = 2m\omega = 2(10)(30) = 600$$

The damping factor ξ is

$$\xi = \frac{c}{C_c} = \frac{18}{600} = 0.03$$

The damped natural frequency ω_d is

$$\omega_d = \omega\sqrt{1 - \xi^2} = 30\sqrt{1 - (0.03)^2} = 29.986 \text{ rad/s}$$

The system response to the impulsive force is then given by

$$\begin{aligned} x(t) &= \frac{I}{m\omega_d} e^{-\xi\omega t} \sin \omega_d t \\ &= \frac{50}{(10)(29.986)} e^{-(0.03)(30)t} \sin 29.986t \\ &= 0.1667e^{-0.9t} \sin 29.986t \end{aligned}$$

1.8 RESPONSE TO AN ARBITRARY FORCING FUNCTION

In this section, we consider the response of the single degree of freedom system to an arbitrary forcing function $F(t)$, shown in Fig. 18. The procedure described in the preceding section for obtaining the impulse response can be

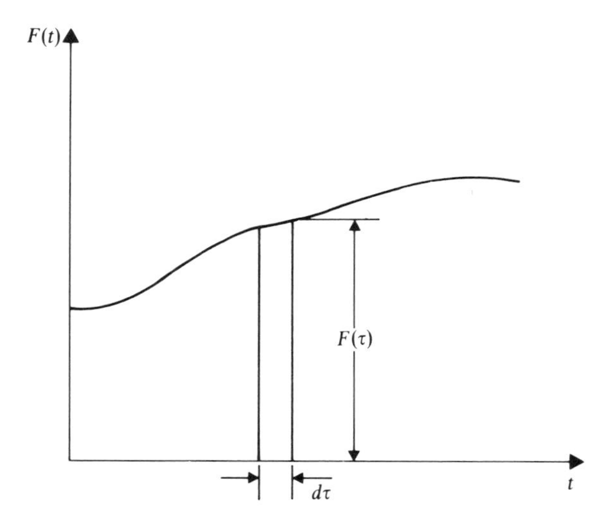

FIG. 1.18. Arbitrary forcing function $F(t)$.

used as a basis for developing a general expression for the response of the system to an arbitrary forcing function. The arbitrary forcing function $F(t)$ can be regarded as a series of impulsive forces $F(\tau)$ acting over a very short-lived interval $d\tau$. The force $F(\tau)$ then produces the short duration impulse $F(\tau)\,d\tau$, and the response of the system to this impulse for all $t > \tau$ is given by

$$dx = F(\tau)\,d\tau\,H(t-\tau) \tag{1.109}$$

where $H(t)$ is the impulse response function defined by Eq. 108. It follows that

$$dx = F(\tau)\,d\tau \cdot \frac{1}{m\omega_{\mathrm{d}}} e^{-\xi\omega(t-\tau)} \sin \omega_{\mathrm{d}}(t-\tau) \tag{1.110}$$

In this equation, dx represents the incremental response of the damped single degree of freedom system to the incremental impulse $F(\tau)\,d\tau$ for $t > \tau$. In order to determine the total response, we integrate Eq. 110 over the entire interval

$$x(t) = \int_0^t F(\tau)H(t-\tau)\,d\tau \tag{1.111}$$

or

$$x(t) = \frac{1}{m\omega_{\mathrm{d}}} \int_0^t F(\tau)e^{-\xi\omega(t-\tau)} \sin \omega_{\mathrm{d}}(t-\tau)\,d\tau \tag{1.112}$$

Equation 111 or Eq. 112 is called the *Duhamel integral* or the *convolution integral.* It is important to emphasize, however, that in obtaining the convolution integral we made use of the principle of superposition which is valid only for linear systems. Furthermore, in deriving the convolution integral, no mention was given to the initial conditions and, accordingly, the integral of Eq. 111, or Eq. 112, provides only the forced response. If the initial conditions are not equal to zero, that is,

$$x_0 = x(t=0) \neq 0 \qquad \text{and/or} \qquad \dot{x}_0 = \dot{x}(t=0) \neq 0$$

then Eq. 112 must be modified to include the effect of the initial conditions. To this end, we first define the homogeneous solution and determine the arbitrary constants in the case of free vibration as the result of these initial conditions, and then use the principle of superposition to add the homogeneous function to the forced response.

Special case A special case of the preceding development is the case of an *undamped single degree of freedom system.* In this case, $\omega_{\mathrm{d}} = \omega$ and $\xi = 0$ and the impulse response function of Eq. 108 reduces to

$$H(t) = \frac{1}{m\omega} \sin \omega t \tag{1.113}$$

The forced response, in this special case, is given by

$$x(t) = \frac{1}{m\omega} \int_0^t F(\tau) \sin \omega(t - \tau) \, d\tau \tag{1.114}$$

If the effect of the initial conditions is considered, the general solution is given by

$$x(t) = \frac{\dot{x}_0}{\omega} \sin \omega t + x_0 \cos \omega t + \frac{1}{m\omega} \int_0^t F(\tau) \sin \omega(t - \tau) \, d\tau \tag{1.115}$$

Example 1.11

Find the forced response of the damped single degree of freedom system to the step function shown in Fig. 19.

Solution. The forced response of the damped single degree of freedom system to an arbitrary forcing function is

$$x(t) = \frac{1}{m\omega_d} \int_0^t F(\tau) e^{-\xi\omega(t-\tau)} \sin \omega_d(t - \tau) \, d\tau$$

In the case of a step function, the forcing function $F(t)$ is defined as

$$F(t) = F_0; \qquad t > 0$$

that is,

$$\begin{aligned} x(t) &= \frac{1}{m\omega_d} \int_0^t F_0 e^{-\xi\omega(t-\tau)} \sin \omega_d(t - \tau) \, d\tau \\ &= \frac{F_0}{m\omega_d} \int_0^t e^{-\xi\omega(t-\tau)} \sin \omega_d(t - \tau) \, d\tau \\ &= \frac{F_0}{k} \left[1 - \frac{e^{-\xi\omega t}}{\sqrt{1 - \xi^2}} \cos(\omega_d t - \psi) \right] \end{aligned}$$

FIG. 1.19. Step function.

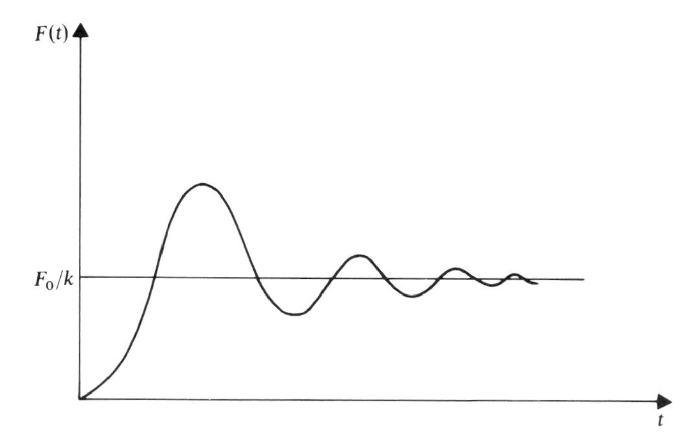

FIG. 1.20. Response of damped single degree of freedom system to a step forcing function.

where the angle ψ is defined as

$$\psi = \tan^{-1}\left(\frac{\xi}{\sqrt{1-\xi^2}}\right)$$

The response of this system is shown in Fig. 20.

Problems

1.1. A single degree of freedom mass–spring system consists of a 10 kg mass suspended by a linear spring which has a stiffness coefficient of 6×10^3 N/m. The mass is given an initial displacement of 0.04 m and it is released from rest. Determine the differential equation of motion, and the natural frequency of the system. Determine also the maximum velocity.

1.2. The oscillatory motion of an undamped single degree of freedom system is such that the mass has maximum acceleration of 50 m/s^2 and has natural frequency of 30 Hz. Determine the amplitude of vibration and the maximum velocity.

1.3. A single degree of freedom undamped mass–spring system is subjected to an impact loading which results in an initial velocity of 5 m/s. If the mass is equal to 10 kg and the spring stiffness is equal to 6×10^3 N/m, determine the system response as a function of time.

1.4. The undamped single degree of freedom system of Problem 1 is subjected to the initial conditions $x_0 = 0.02$ m and $\dot{x}_0 = 3$ m/s. Determine the system response as a function of time. Also determine the maximum velocity and the total energy of the system.

1.5. A single degree of freedom system consists of a mass m which is suspended by a linear spring of stiffness k. The static equilibrium deflection of the spring was found to be 0.02 m. Determine the system natural frequency and the response of the system as a function of time if the initial displacement is 0.03 m and the initial velocity is zero. What is the total energy of the system?

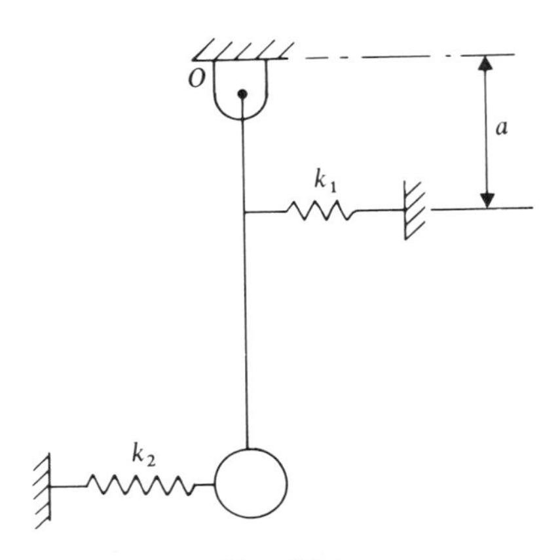

FIG. P1.1

1.6. The system shown in Fig. P1 consists of a mass m and a massless rod of length l. The system is supported by two springs which have stiffness coefficients k_1 and k_2, as shown in the figure. Derive the system differential equation of motion assuming small oscillations. Determine the natural frequency of the system.

1.7. If the two springs k_1 and k_2 in Problem 6 are to be replaced by an equivalent spring which is connected at the middle of the rod, determine the stiffness coefficient k_e of the new spring.

1.8. If the system in Problem 6 is given an initial angular displacement θ_0 counterclockwise, determine the system response as a function of time assuming that the initial angular velocity is zero. Determine also the maximum angular velocity.

1.9. In the system shown in Fig. P2, $m = 5$ kg, $k_1 = k_5 = k_6 = 1000$ N/m, $k_3 = k_4 = 1500$ N/m, and $k_2 = 3000$ N/m. The motion of the mass is assumed to be in the vertical direction. If the mass is subjected to an impact such that the motion starts with an initial upward velocity of 5 m/s, determine the displacement, velocity, and acceleration of the mass after 2 s.

1.10. The system shown in Fig. P3 consists of a mass m and a massless rod of length l. The system is supported by a spring which has a stiffness coefficient k. Obtain the differential equation of motion and discuss the stability of the system. Determine the stiffness coefficient k at which the system becomes unstable.

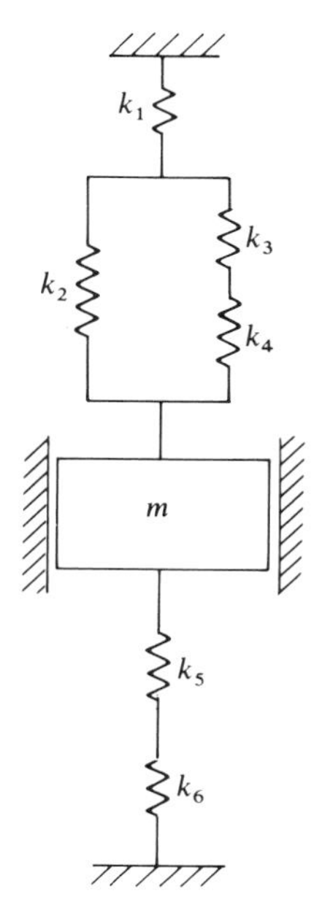

FIG. P1.2

1.11. The system shown in Fig. P4 consists of a uniform rod which has length l, mass m, and mass moment of inertia about its mass center I. The rod is supported by two springs which have stiffness coefficients k, as shown in the figure. Determine the system differential equation of motion for small oscillations. Determine also the system natural frequency.

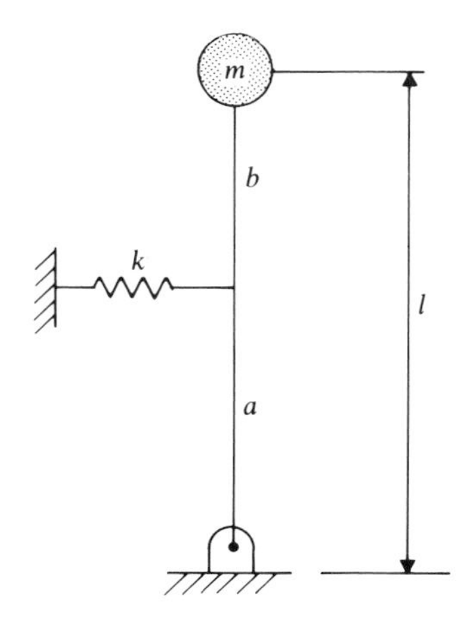

FIG. P1.3

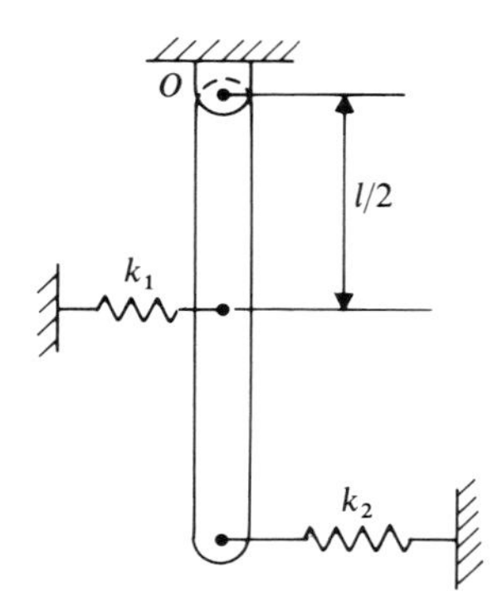

FIG. P1.4

1.12. A damped single degree of freedom mass–spring system has $m = 4$ kg, $k =$ 3000 N/m, and $c = 300$ N · s/m. Determine the equation of motion of the system.

1.13. A damped single degree of freedom mass–spring system has $m = 0.5$ kg, $k =$ 1000 N/m, and $c = 10$ N · s/m. The mass is set in motion with initial conditions $x_0 = 0.05$ m and $\dot{x} = 0.5$ m/s. Determine the displacement, velocity, and acceleration of the mass after 0.3 s.

1.14. A viscously damped single degree of freedom mass–spring system has a mass m of 2 kg, a spring coefficient k of 2000 N/m, and a damping constant c of 5 N · s/m. Determine (a) the damping factor ξ, (b) the natural frequency ω, (c) the damped natural frequency ω_d, and (d) the spring coefficient needed to obtain a critically damped system.

1.15. An overdamped single degree of freedom mass–spring system has a damping factor $\xi = 1.5$ and a natural frequency $\omega = 20$ rad/s. Determine the equation of motion and plot the displacement and velocity versus time for the following initial conditions: (a) $x_0 = 0$, $\dot{x}_0 = 1$ m/s; (b) $x_0 = 0.05$ m, $\dot{x}_0 = 0$; and (c) $x_0 =$ 0.05 m, $\dot{x}_0 = 1$ m/s.

1.16. Repeat Problem 15 if the system is critically damped.

1.17. A mass equal to 0.5 kg attached to a linear spring elongates it 0.008 m. Determine the system natural frequency.

1.18. For the system shown in Fig. P5, let $m = 0.5$ kg, $a = 0.2$ m, $l = 0.4$ m, $k =$ 1000 N/m. Determine the damping coefficient c if the system is to be critically damped. If the system has an initial angular velocity of 5 rad/s counterclockwise, determine the angular displacement and angular velocity after 0.3 s. Assume small oscillations.

1.19. For the system shown in Fig. P6, let $m = 0.5$ kg, $l = 0.5$ m, $a = 0.2$ m, and $k = 3000$ N/m. Determine the damping coefficient c if: (a) the system is underdamped with $\xi = 0.09$, (b) the system is critically damped; and (c) the system is overdamped with $\xi = 1.2$.

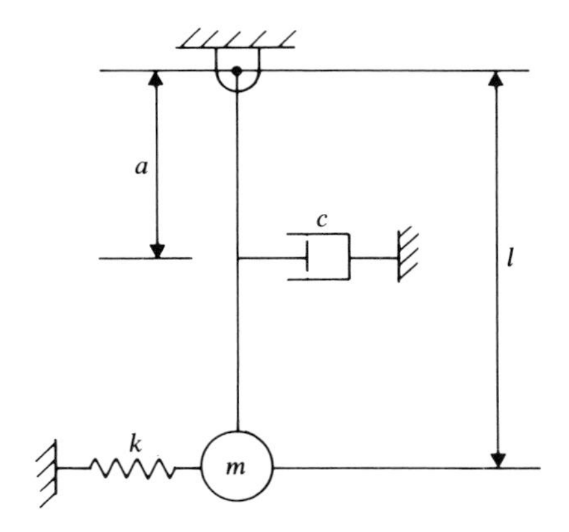

FIG. P1.5

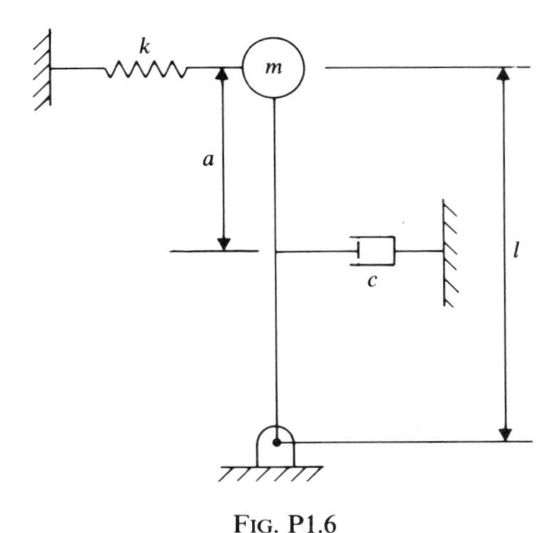

FIG. P1.6

1.20. The system shown in Fig. P7 consists of a uniform bar of length l, mass m, and mass moment of inertia I. The bar is supported by a spring and damper which have stiffness and damping coefficients k and c, respectively. Derive the differential equation of motion and determine the system natural frequency and the critical damping coefficient.

1.21. A single degree of freedom mass–spring system has the following parameters, $m = 0.5$ kg, $k = 2 \times 10^3$ N/m, coefficient of dry friction $\mu = 0.15$, initial displacement $x_0 = 0.1$ m, and initial velocity $\dot{x}_0 = 0$. Determine:
(1) the decrease in amplitude per cycle;
(2) the number of half-cycles completed before the mass comes to rest;
(3) the displacement of the mass at time $t = 0.1$ s;
(4) the location of the mass when oscillation stops.

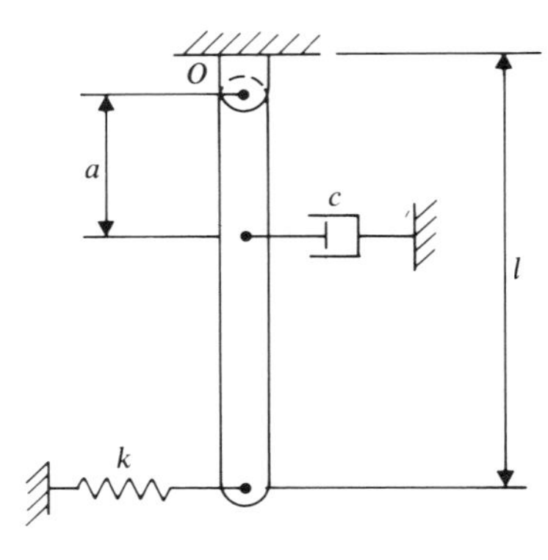

FIG. P1.7

1.22. A spring–mass system is subjected to a harmonic force which has an amplitude 30 N and frequency 20 rad/s. The system has mass $m = 5$ kg, and stiffness coefficient $k = 2 \times 10^3$ N. The initial conditions are such that $x_0 = 0$, $\dot{x}_0 = 2$ m/s. Determine the displacement, velocity, and acceleration of the mass after 0.5, 1, 1.5 s.

1.23. A single degree of freedom mass–spring system has a mass $m = 3$ kg, a spring stiffness $k = 2700$ N/m, and a damping coefficient $c = 18$ N·s/m. The mass is subjected to a harmonic force which has an amplitude $F_0 = 20$ N and a frequency $\omega_f = 15$ rad/s. The initial conditions are $x_0 = 1$ cm and $\dot{x}_0 = 0$. Determine the displacement, velocity, and acceleration of the mass after time $t = 0.5$ s.

1.24. For the system shown in Fig. P8, let $m = 3$ kg, $k_1 = k_2 = 1350$ N/m, $c = 40$ N·s/m, and $y = 0.04 \sin 15t$. The initial conditions are such that $x_0 = 5$ mm and $\dot{x}_0 = 0$. Determine the displacement, velocity, and acceleration of the mass after time $t = 1$ s.

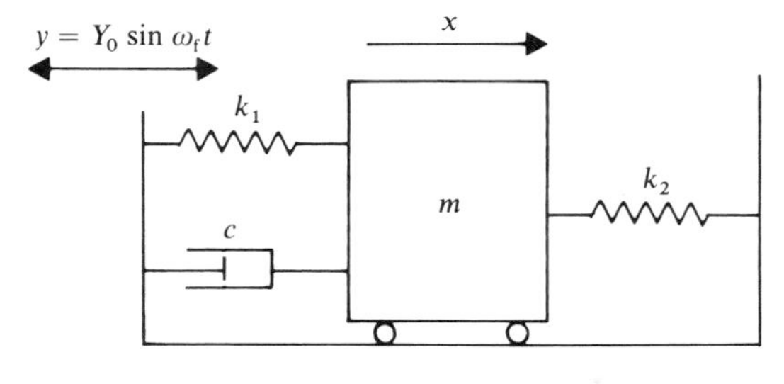

FIG. P1.8

1.25. For the system shown in Fig. P9, let $m = 3$ kg, $k_1 = k_2 = 1350$ N/m, $c = 40$ N·s/m, and $y = 0.02 \sin 15t$. The initial conditions are $x_0 = 5$ mm and $\dot{x}_0 = 0$. Determine the displacement, velocity, and acceleration of the mass after time $t = 1$ s.

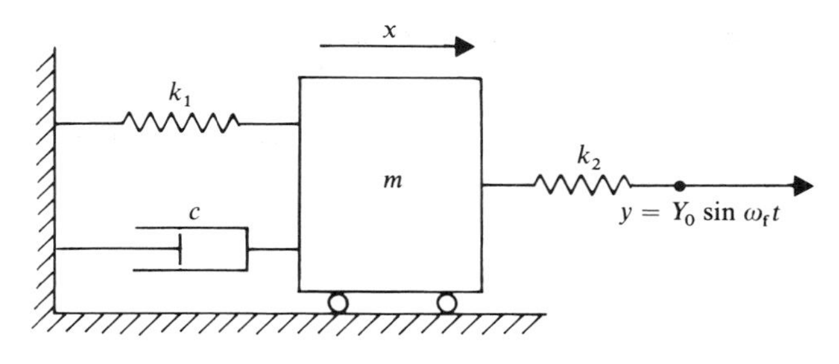

FIG. P1.9

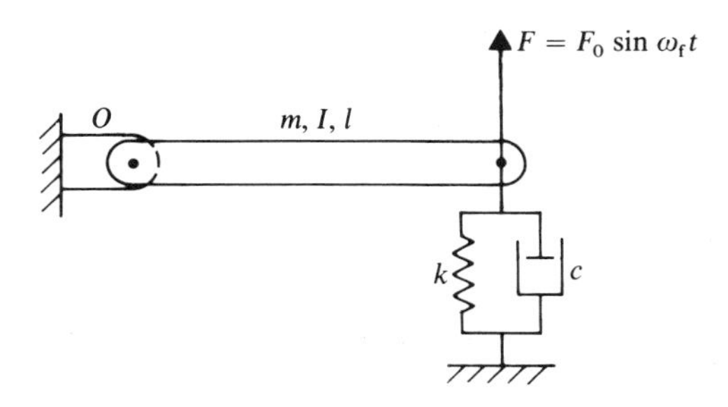

FIG. P1.10

1.26. Derive the differential equation of motion of the system shown in Fig. P10, assuming small angular oscillations. Determine also the steady state response of this system.

1.27. In Problem 26, let the rod be uniform and slender with mass $m = 0.5$ kg and $l = 0.5$ m. Let $k = 2000$ N/m, $c = 20$ N·s/m, and $F = 10 \sin 10t$ N. The initial conditions are $\theta_0 = 0$ and $\dot{\theta}_0 = 3$ rad/s. Determine the displacement equation of the beam as a function of time.

1.28. Determine the forced response of the undamped single degree of freedom spring–mass system to the forcing function shown in Fig. P11.

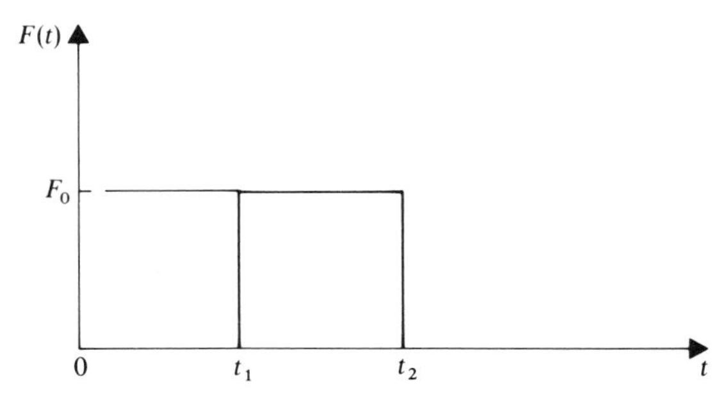

FIG. P1.11

1.29. Determine the forced response of the damped single degree of freedom mass–spring system to the forcing function shown in Fig. P11.

1.30. Determine the response of the damped single degree of freedom mass–spring system to the forcing function shown in Fig. P12.

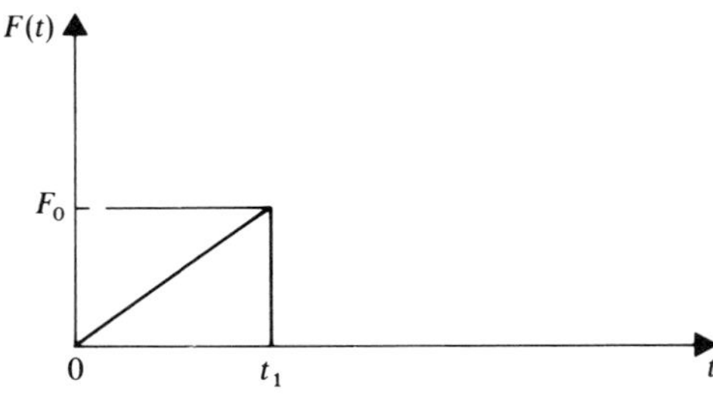

FIG. P1.12

1.31. Obtain the response of the damped mass–spring system to the forcing function shown in Fig. P13.

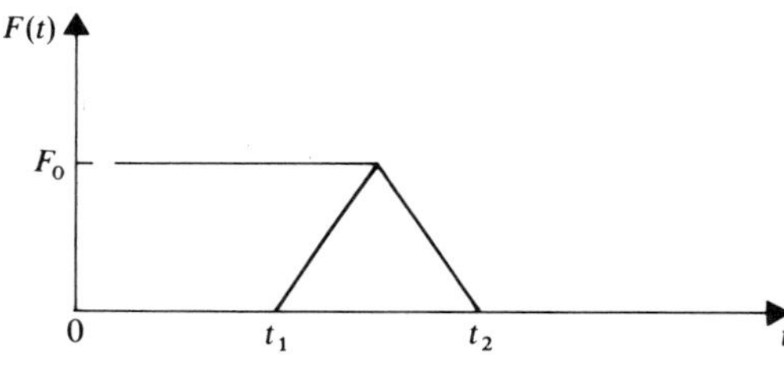

FIG. P1.13

2
Lagrangian Dynamics

The differential equations of motion of single and multi-degree of freedom systems can be developed using the vector approach of Newtonian mechanics. Another alternative for developing the system differential equations of motion from scalar quantities is the *Lagrangian approach* where scalars such as the kinetic energy, strain energy, and virtual work are used. In this chapter, the use of *Lagrange's equation* to formulate the dynamic differential equations of motion is discussed. The use of Lagrange's equation is convenient in developing the dynamic relationships of multi-degree of freedom systems. Important concepts and definitions, however, have to be first introduced. In the first section of this chapter, we introduce the concept of the system *generalized coordinates*, and in Section 2 the *virtual work* is used to develop the generalized forces associated with the system generalized coordinates. The concepts and definitions presented in the first two sections are then used in Sections 3–5 to develop Lagrange's equation of motion for multi-degree of freedom systems in terms of scalar quantities such as the kinetic energy, strain energy, and virtual work. An alternate approach for deriving the dynamic equations of motion using scalar quantities is *Hamilton's principle* which is discussed in Section 6. Hamilton's principle can be used to derive Lagrange's equation and, consequently, both techniques lead to the same results when the same set of coordinates is used. The use of conservation of energy to obtain the differential equations of motion of *conservative systems* is discussed in Section 7, where general *conservation theorems* are developed and their use is demonstrated using simple examples.

2.1 GENERALIZED COORDINATES

In the Lagrangian dynamics, the configuration of a mechanical system is identified by a set of variables called coordinates or *generalized coordinates* that completely define the location and orientation of each component in the system. The configuration of a particle in space is defined using three coordinates that describe the translation of this particle with respect to the three perpendicular axes of the inertial frame. No rotational coordinates are needed to describe the motion of the particle and, therefore, the three translational coordinates completely define the particle position. This simplified description of the particle kinematics is the result of the assumption that the particle has such small dimensions that a point in the three-dimensional space is sufficient to describe the position of the particle. This assumption is not valid, however, when rigid bodies are considered. The configuration of an unconstrained rigid body, as demonstrated in the preceding chapter, can be completely described using six independent coordinates, three coordinates describe the location of the origin of the body axes and three rotational coordinates describe the orientation of the body with respect to the fixed frame. Once this set of coordinates is identified, the global position of an arbitrary point on the rigid body can be defined.

In this chapter and subsequent chapters, the vector $\mathbf{q}$ will be used to denote the system generalized coordinates. If the system has n generalized coordinates, the vector $\mathbf{q}$ is given by

$$\mathbf{q} = [q_1 \quad q_2 \quad \cdots \quad q_n]^{\mathrm{T}} \tag{2.1}$$

In this book we will assume that the generalized coordinates $q_1, q_2, \ldots, q_n$ are independent, and as such, the number of system generalized coordinates is equal to the number of system *degrees of freedom*.

Figure 1 shows some examples of vibratory systems, which have different

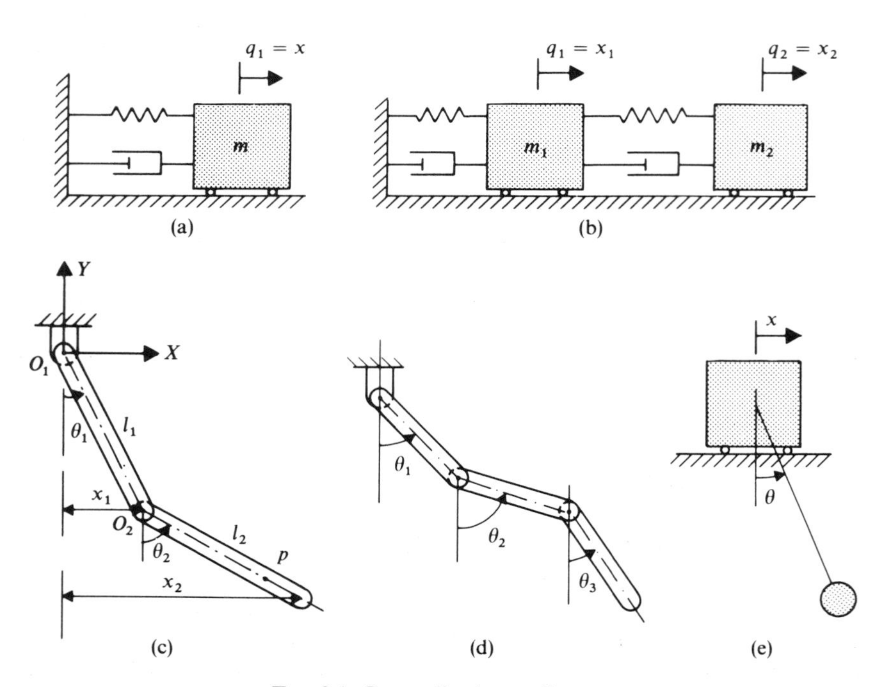

FIG. 2.1. Generalized coordinates.

types or numbers of generalized coordinates. The system shown in Fig. 1(a) is a single degree of freedom system, and accordingly the vector of generalized coordinates $\mathbf{q}$ reduces to a scalar since $\mathbf{q} = q_1 = x$. The configuration of this system can be completely defined in terms of the generalized coordinate q_1. The configuration of the two degree of freedom system shown in Fig. 1(b), on the other hand, is completely defined by the coordinates $q_1 = x_1$ and $q_2 = x_2$, that is,

$$\mathbf{q} = [q_1 \quad q_2]^{\mathrm{T}} = [x_1 \quad x_2]^{\mathrm{T}}$$

Similarly, the vector of the generalized coordinates of the system shown in Fig. 1(c) is

$$\mathbf{q} = [q_1 \quad q_2]^{\mathrm{T}} = [\theta_1 \quad \theta_2]^{\mathrm{T}}$$

The position coordinates of an arbitrary point p in this system can be written in terms of these generalized coordinates as

$$x_p = l_1 \sin \theta_1 + d \sin \theta_2$$

$$y_p = -l_1 \cos \theta_1 - d \cos \theta_2$$

where d is the position of point p with respect to point O_2 measured along the rod axis.

The system shown in Fig. 1(d) has more than two degrees of freedom. One can verify that the vector of generalized coordinates of this system is given by

$$\mathbf{q} = [q_1 \quad q_2 \quad q_3]^{\mathrm{T}} = [\theta_1 \quad \theta_2 \quad \theta_3]^{\mathrm{T}}$$

It is important to emphasize at this point that the set of generalized coordinates $q_1, q_2, \ldots, q_n$ is not unique. For instance, the configuration of the system shown in Fig. 1(c) can also be described by the generalized coordinates $q_1 = x_1$ and $q_2 = x_2$ as shown in the figure. In many cases, simple relationships can be found between different sets of generalized coordinates of the system. For example, in Fig. 1(c) one can verify that

$$x_1 = l_1 \sin \theta_1$$

$$x_2 = l_1 \sin \theta_1 + l_2 \sin \theta_2$$

The inverse relationship can also be obtained as

$$\sin \theta_1 = \frac{x_1}{l_1}$$

$$\sin \theta_2 = \frac{x_2 - x_1}{l_2}$$

or

$$\theta_1 = \sin^{-1}\left(\frac{x_1}{l_1}\right)$$

$$\theta_2 = \sin^{-1}\left(\frac{x_2 - x_1}{l_2}\right)$$

The coordinates θ_1 and θ_2 or x_1 and x_2 in the example of Fig. 1(c) are independent since they can be arbitrarily changed. Any coordinate which can be expressed in terms of the other coordinates is not an independent one and cannot be used as a degree of freedom. Similarly, a coordinate which is specified as a function of time is not a degree of freedom since it has a specified value and cannot be changed arbitrarily. For instance, consider the system shown in Fig. 1(e) which has the two degrees of freedom $q_1 = x$ and $q_2 = \theta$. If q_1 is specified as a function of time, say $q_1 = x = X_1 \sin \omega_f t$ where X_1 and ω_f are, respectively, given amplitude and frequency, then the system has only one degree of freedom which is $q_2 = \theta$. The relationship $q_1 = x = X_1 \sin \omega_f t$ can be considered as a kinematic constraint imposed on the motion of the system, and every such kinematic constraint relationship, as demonstrated in the preceding chapter, eliminates one of the system degrees of freedom. In other words, the number of system degrees of freedom is equal to the number of coordinates minus the number of kinematic constraint equations, provided that these kinematic constraint equations are linearly independent.

Example 2.1

Discuss different alternatives for selecting the generalized coordinates of the two degree of freedom system shown in Fig. 2.

Solution. Since the system has only two degrees of freedom, mainly the vertical displacement and the rotation, only two generalized coordinates are required to identify the configuration of the system. As shown in Fig. 2, one set of coordinates is given by

$$\mathbf{q} = [q_1 \quad q_2]^{\mathrm{T}} = [y_c \quad \theta]^{\mathrm{T}}$$

where y_c is the vertical displacement of the center of mass of the beam and θ is the angular oscillation of the beam. The coordinates y_c and θ can be used as the system degrees of freedom since they are independent and are sufficient to define the configuration of the system. For instance, given an arbitrary point p at a distance l_p from the center of mass of the beam, the vertical displacement of the arbitrary point p can be expressed in terms of the coordinates y_c and θ as

$$y_p = y_c + l_p \sin \theta$$

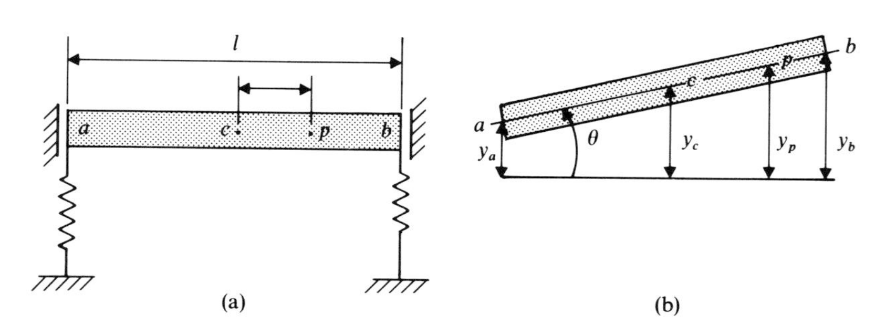

FIG. 2.2. Two degree of freedom system.

As a second choice for the generalized coordinates, one may select the following set:

$$\mathbf{q} = [q_1 \quad q_2]^{\mathrm{T}} = [y_a \quad \theta]^{\mathrm{T}}$$

where y_a is the vertical displacement of the end point a. The coordinates y_a and θ are also independent and can be used to define the displacement of an arbitrary point on the beam. The vertical displacement of the point p can be expressed in terms of the generalized coordinates y_a and θ as

$$y_p = y_a + \left(\frac{l}{2} + l_p\right) \sin \theta$$

where l is the length of the beam. Observe that the vertical displacement of the center of mass y_c which was previously used as a generalized coordinate can be expressed in terms of the new set of generalized coordinates y_a and θ as

$$y_c = y_a + \frac{l}{2} \sin \theta$$

An alternate set of generalized coordinates is also the following:

$$\mathbf{q} = [q_1 \quad q_2]^{\mathrm{T}} = [y_a \quad y_b]^{\mathrm{T}}$$

where y_b is the vertical displacement of the end point b. In terms of these coordinates, the vertical displacement of the center of mass and the arbitrary point p are given, respectively, by

$$y_c = \frac{y_b - y_a}{2}$$

$$y_p = y_a + (y_b - y_a)\frac{l/2 + l_p}{l}$$

Also, the angular orientation θ can be expressed in terms of the new set y_a and y_b as

$$\sin \theta = \frac{y_b - y_a}{l}$$

$$\theta = \sin^{-1}\left(\frac{y_b - y_a}{l}\right)$$

Clearly, for a simple system such as the one used in this example, there are an infinite number of different sets of generalized coordinates. In many applications, proper selection of the generalized coordinates can significantly simplify the dynamic formulation.

2.2 VIRTUAL WORK AND GENERALIZED FORCES

In this section, the important concept of the *virtual work* is introduced and used to provide definitions for the *generalized forces* associated with the system generalized coordinates. As will be seen in the following section, the definition of the generalized forces is an important step in the Lagrangian formulation, wherein the dynamic equations of motion are formulated in terms of the generalized coordinates.

Since the generalized coordinates are sufficient to completely identify the configuration of a system of particles or rigid bodies, the position vector of an arbitrary point in the system can be expressed in terms of the generalized coordinates as

$$\mathbf{r} = \mathbf{r}(q_1, q_2, \ldots, q_n) = \mathbf{r}(\mathbf{q}) \tag{2.2}$$

where $\mathbf{r}$ is the position vector of an arbitrary point in the system, and $\mathbf{q} = [q_1 \; q_2 \; \cdots \; q_n]^{\mathrm{T}}$ is the vector of generalized coordinates which are assumed to be independent.

At this point, we introduce the concept of the *virtual displacement* which refers to a change in the configuration of the system as the result of an arbitrary infinitesimal change in the vector of system generalized coordinates $\mathbf{q}$, consistent with the forces and constraints imposed on the system at the given instant of time t. The displacement is called virtual to distinguish it from an actual displacement of the system occurring in a time interval Δt, during which the time and constraints may be changing. For a virtual change in the system coordinates, Eq. 2 yields

$$\delta\mathbf{r} = \frac{\partial \mathbf{r}}{\partial q_1}\delta q_1 + \frac{\partial \mathbf{r}}{\partial q_2}\delta q_2 + \cdots + \frac{\partial \mathbf{r}}{\partial q_n}\delta q_n \tag{2.3}$$

where $\delta\mathbf{r}$ is the virtual change in the position vector of Eq. 2, and δq_i is the virtual change in the generalized coordinate q_i. Equation 3 can be written in a compact form as

$$\delta\mathbf{r} = \sum_{j=1}^{n} \frac{\partial \mathbf{r}}{\partial q_j}\delta q_j \tag{2.4}$$

Let $\mathbf{F}$ be the vector of forces that act at the point whose position vector is defined by the vector $\mathbf{r}$ of Eq. 2, then the virtual work due to this force vector is defined by the dot product

$$\delta W = \mathbf{F} \cdot \delta\mathbf{r} = \mathbf{F}^{\mathrm{T}}\delta\mathbf{r} \tag{2.5}$$

which upon using Eqs. 3 and 4 yields

$$\begin{aligned}\delta W &= \mathbf{F}^{\mathrm{T}}\frac{\partial \mathbf{r}}{\partial q_1}\delta q_1 + \mathbf{F}^{\mathrm{T}}\frac{\partial \mathbf{r}}{\partial q_2}\delta q_2 + \cdots + \mathbf{F}^{\mathrm{T}}\frac{\partial \mathbf{r}}{\partial q_n}\delta q_n \\ &= \sum_{j=1}^{n} \mathbf{F}^{\mathrm{T}}\frac{\partial \mathbf{r}}{\partial q_j}\delta q_j \end{aligned} \tag{2.6}$$

Now we define the generalized force Q_j associated with the jth system generalized coordinate q_j as

$$Q_j = \mathbf{F}^{\mathrm{T}}\frac{\partial \mathbf{r}}{\partial q_j}, \qquad j = 1, 2, \ldots, n \tag{2.7}$$

Using this definition, the virtual work of Eq. 6 can be expressed in terms of the generalized forces Q_j as

$$\delta W = Q_1\delta q_1 + Q_2\delta q_2 + \cdots + Q_n\delta q_n$$
$$= \sum_{j=1}^{n} Q_j\delta q_j \tag{2.8}$$

which can be written in a vector form as

$$\delta W = \mathbf{Q}^{\mathrm{T}}\delta\mathbf{q} \tag{2.9}$$

where $\mathbf{Q}$ is the vector of generalized forces defined as

$$\mathbf{Q} = [Q_1 \quad Q_2 \quad \cdots \quad Q_n]^{\mathrm{T}} \tag{2.10}$$

Example 2.2

Derive the generalized forces due to the spring, damper, and the external force $F(t)$ of the damped single degree of freedom system shown in Fig. 3.

Solution. For this single degree of freedom system, the vector of generalized coordinates $\mathbf{q}$ reduces to a scalar given by

$$q_1 = x$$

It follows that

$$\delta x = \delta q_1$$

The forces that act on this system are shown in the figure and given by

$$f = F(t) - kx - c\dot{x}$$

The virtual work is then defined as

$$\delta W = f\delta x = (F(t) - kx - c\dot{x})\delta x$$

Since in this example $x = q_1$ and $\dot{x} = \dot{q}_1$, in terms of the generalized coordinate q_1, the virtual work δW can be written as

$$\delta W = (F(t) - kq_1 - c\dot{q}_1)\delta q_1$$
$$= Q_1\delta q_1$$

where Q_1 is the generalized force associated with the generalized coordinate q_1 and is defined as

$$Q_1 = F(t) - kq_1 - c\dot{q}_1$$

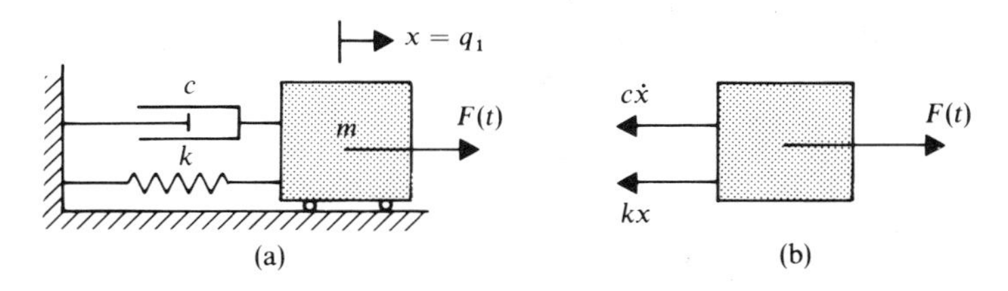

FIG. 2.3. Single degree of freedom system.

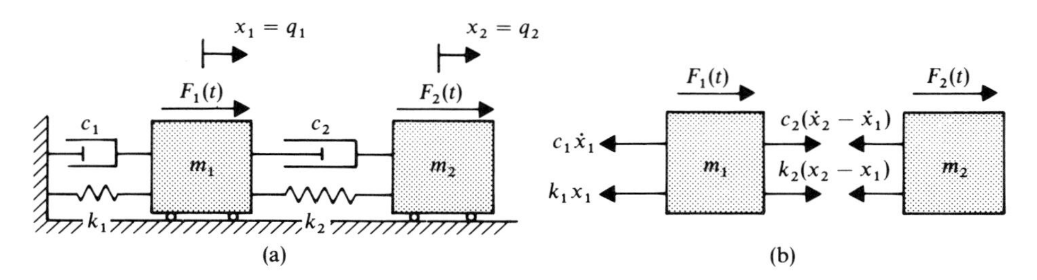

FIG. 2.4. Two degree of freedom system.

Example 2.3

For the two degree of freedom system shown in Fig. 4, obtain the generalized forces associated with the generalized coordinates $q_1 = x_1$ and $q_2 = x_2$.

Solution. As shown in the figure, the following force

$$f_1 = -k_1x_1 - c_1\dot{x}_1 + k_2(x_2 - x_1) + c_2(\dot{x}_2 - \dot{x}_1) + F_1(t)$$

acts at a point whose displacement is $x_1 = q_1$. The force f_2 given by

$$f_2 = -k_2(x_2 - x_1) - c_2(\dot{x}_2 - \dot{x}_1) + F_2(t)$$

acts at a point whose displacement is $x_2 = q_2$. Therefore, the virtual work δW is given by

$$\delta W = f_1\delta x_1 + f_2\delta x_2$$

Using the fact that $x_1 = q_1$, $\dot{x}_1 = \dot{q}_1$, $x_2 = q_2$, and $\dot{x}_2 = \dot{q}_2$, the virtual work δW can be expressed in terms of the generalized coordinates as

$$\begin{aligned}\delta W = \{&-k_1q_1 - c_1\dot{q}_1 + k_2(q_2 - q_1) + c_2(\dot{q}_2 - \dot{q}_1) + F_1(t)\}\delta q_1 \\ &+ \{-k_2(q_2 - q_1) - c_2(\dot{q}_2 - \dot{q}_1) + F_2(t)\}\delta q_2\end{aligned}$$

or

$$\begin{aligned}\delta W = \{&F_1(t) - (k_1 + k_2)q_1 - (c_1 + c_2)\dot{q}_1 + k_2q_2 + c_2\dot{q}_2\}\delta q_1 \\ &+ \{F_2(t) - k_2q_2 - c_2\dot{q}_2 + k_2q_1 + c_2\dot{q}_1\}\delta q_2\end{aligned}$$

Example 2.4

For the three degree of freedom pendulum system shown in Fig. 5, obtain the generalized forces associated with the generalized coordinates $q_1 = \theta_1, q_2 = \theta_2$, and $q_3 = \theta_3$, where **F** and **M** are defined as

$$\mathbf{F} = [F_1 \quad F_2]^{\mathrm{T}}, \qquad \mathbf{M} = [M_1 \quad M_2 \quad M_3]^{\mathrm{T}}$$

Solution. The position vector of the point of application of the force **F** is given by the vector

$$\mathbf{r} = \begin{bmatrix} r_1 \\ r_2 \end{bmatrix} = \begin{bmatrix} l_1 \sin\theta_1 + l_2 \sin\theta_2 + l_3 \sin\theta_3 \\ -(l_1 \cos\theta_1 + l_2 \cos\theta_2 + l_3 \cos\theta_3) \end{bmatrix}$$

where l_1, l_2, and l_3 are the lengths of the rods. The virtual change in the position

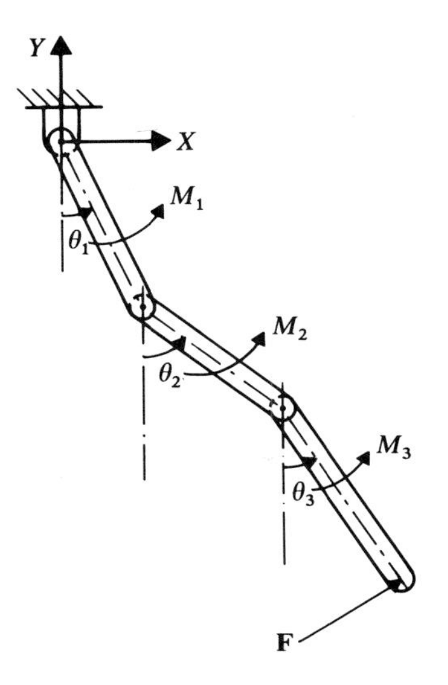

FIG. 2.5. Three degree of freedom system.

vector $\mathbf{r}$ can be written as

$$\delta \mathbf{r} = \begin{bmatrix} l_1 \cos \theta_1 \delta\theta_1 + l_2 \cos \theta_2 \delta\theta_2 + l_3 \cos \theta_3 \delta\theta_3 \\ l_1 \sin \theta_1 \delta\theta_1 + l_2 \sin \theta_2 \delta\theta_2 + l_3 \sin \theta_3 \delta\theta_3 \end{bmatrix}$$

which can be written as

$$\delta \mathbf{r} = \begin{bmatrix} \delta r_1 \\ \delta r_2 \end{bmatrix} = \begin{bmatrix} l_1 \cos \theta_1 & l_2 \cos \theta_2 & l_3 \cos \theta_3 \\ l_1 \sin \theta_1 & l_2 \sin \theta_2 & l_3 \sin \theta_3 \end{bmatrix} \begin{bmatrix} \delta\theta_1 \\ \delta\theta_2 \\ \delta\theta_3 \end{bmatrix}$$

The virtual work of the component M_i of the moment is $M_i \delta\theta_i$. Therefore, the virtual work δW of the force vector $\mathbf{F}$ and the moment $\mathbf{M}$ is given by

$$\delta W = \mathbf{F}^{\mathrm{T}} \delta \mathbf{r} + \mathbf{M}^{\mathrm{T}} \delta \boldsymbol{\theta}$$

where

$$\delta \boldsymbol{\theta} = [\delta\theta_1 \quad \delta\theta_2 \quad \delta\theta_3]^{\mathrm{T}}$$

The virtual work δW is then given in a more explicit form by

$$\begin{aligned} \delta W = {} & [F_1 \quad F_2] \begin{bmatrix} l_1 \cos \theta_1 & l_2 \cos \theta_2 & l_3 \cos \theta_3 \\ l_1 \sin \theta_1 & l_2 \sin \theta_2 & l_3 \sin \theta_3 \end{bmatrix} \begin{bmatrix} \delta\theta_1 \\ \delta\theta_2 \\ \delta\theta_3 \end{bmatrix} \\ & + [M_1 \quad M_2 \quad M_3] \begin{bmatrix} \delta\theta_1 \\ \delta\theta_2 \\ \delta\theta_3 \end{bmatrix} \\ = {} & [M_1 + l_1(F_1 \cos \theta_1 + F_2 \sin \theta_1) \qquad M_2 + l_2(F_1 \cos \theta_2 + F_2 \sin \theta_2) \\ & M_3 + l_3(F_1 \cos \theta_3 + F_2 \sin \theta_3)] \begin{bmatrix} \delta\theta_1 \\ \delta\theta_2 \\ \delta\theta_3 \end{bmatrix} \end{aligned}$$

which can also be written as

$$\delta W = [Q_1 \quad Q_2 \quad Q_3] \begin{bmatrix} \delta\theta_1 \\ \delta\theta_2 \\ \delta\theta_3 \end{bmatrix} = \mathbf{Q}^{\mathrm{T}} \delta\boldsymbol{\theta}$$

where the generalized forces Q_1, Q_2, and Q_3 associated, respectively, with the generalized coordinates θ_1, θ_2, and θ_3 are defined as

$$Q_1 = M_1 + l_1(F_1 \cos\theta_1 + F_2 \sin\theta_1)$$
$$Q_2 = M_2 + l_2(F_1 \cos\theta_2 + F_2 \sin\theta_2)$$
$$Q_3 = M_3 + l_3(F_1 \cos\theta_3 + F_2 \sin\theta_3)$$

2.3 LAGRANGE'S EQUATION

In this section, we utilize the concepts introduced in the preceding section to develop Lagrange's equation of motion for a system of particles. The obtained Lagrange's equation, however, can be applied to the dynamics of rigid bodies as well, since a rigid body can be considered as a collection of a large number of particles.

In the following discussion, we assume that the system consists of n_p particles. The displacement $\mathbf{r}^i$ of the ith particle is assumed to depend on a set of generalized coordinates q_j, where $j = 1, 2, \ldots, n$. Hence

$$\mathbf{r}^i = \mathbf{r}^i(q_1, q_2, \ldots, q_n, t) \tag{2.11}$$

Differentiating Eq. 11 with respect to time using the chain rule of differentiation yields

$$\begin{aligned} \dot{\mathbf{r}}^i &= \frac{\partial \mathbf{r}^i}{\partial q_1}\dot{q}_1 + \frac{\partial \mathbf{r}^i}{\partial q_2}\dot{q}_2 + \cdots + \frac{\partial \mathbf{r}^i}{\partial q_n}\dot{q}_n + \frac{\partial \mathbf{r}^i}{\partial t} \\ &= \sum_{j=1}^{n} \frac{\partial \mathbf{r}^i}{\partial q_j}\dot{q}_j + \frac{\partial \mathbf{r}^i}{\partial t} \end{aligned} \tag{2.12}$$

The virtual displacement $\delta\mathbf{r}^i$ can be expressed in terms of the virtual change of the generalized coordinates as

$$\delta\mathbf{r}^i = \sum_{j=1}^{n} \frac{\partial \mathbf{r}^i}{\partial q_j}\delta q_j \tag{2.13}$$

The dynamic equilibrium of the particle can be defined using Newton's second law as

$$\dot{\mathbf{p}}^i = \mathbf{F}^i \tag{2.14}$$

where $\mathbf{p}^i$ is the vector of linear momentum and $\mathbf{F}^i$ is the vector of the total forces acting on the particle i. The linear momentum vector $\mathbf{p}^i$ is defined as

$$\mathbf{p}^i = m^i\dot{\mathbf{r}}^i$$

where m^i is the mass of the particle i which is assumed to be constant. Consequently,

$$\dot{\mathbf{p}}^i = m^i \ddot{\mathbf{r}}^i \tag{2.15}$$

Substituting Eq. 15 into Eq. 14 yields

$$m^i \ddot{\mathbf{r}}^i - \mathbf{F}^i = \mathbf{0}$$

which yields

$$(m^i \ddot{\mathbf{r}}^i - \mathbf{F}^i)^{\mathrm{T}} \delta \mathbf{r}^i = 0 \tag{2.16}$$

By summing up these expressions for the particles and using the definition of $\delta \mathbf{r}^i$ given by Eq. 13, one obtains

$$\sum_{i=1}^{n_p} (m^i \ddot{\mathbf{r}}^i - \mathbf{F}^i)^{\mathrm{T}} \left(\sum_{j=1}^{n} \frac{\partial \mathbf{r}^i}{\partial q_j} \delta q_j \right) = 0$$

which can also be written as

$$\sum_{j=1}^{n} \sum_{i=1}^{n_p} (m^i \ddot{\mathbf{r}}^i - \mathbf{F}^i)^{\mathrm{T}} \left(\frac{\partial \mathbf{r}^i}{\partial q_j} \delta q_j \right) = 0 \tag{2.17}$$

Define the generalized force Q_j associated with the generalized coordinate q_j as

$$Q_j = \sum_{i=1}^{n_p} \mathbf{F}^{i\mathrm{T}} \frac{\partial \mathbf{r}^i}{\partial q_j} \tag{2.18}$$

One can also show that

$$\sum_{i=1}^{n_p} \left(m^i \ddot{\mathbf{r}}^{i\mathrm{T}} \frac{\partial \mathbf{r}^i}{\partial q_j} \right) = \sum_{i=1}^{n_p} \left[\frac{d}{dt} \left(m^i \dot{\mathbf{r}}^{i\mathrm{T}} \frac{\partial \mathbf{r}^i}{\partial q_j} \right) - m^i \dot{\mathbf{r}}^{i\mathrm{T}} \frac{d}{dt} \left(\frac{\partial \mathbf{r}^i}{\partial q_j} \right) \right] \tag{2.19}$$

It is, however, clear from Eq. 12 that

$$\frac{\partial \dot{\mathbf{r}}^i}{\partial \dot{q}_j} = \frac{\partial \mathbf{r}^i}{\partial q_j} \tag{2.20}$$

Furthermore,

$$\frac{d}{dt} \left(\frac{\partial \mathbf{r}^i}{\partial q_j} \right) = \sum_{k=1}^{n} \frac{\partial^2 \mathbf{r}^i}{\partial q_j \partial q_k} \dot{q}_k + \frac{\partial^2 \mathbf{r}^i}{\partial q_j \partial t} = \frac{\partial \dot{\mathbf{r}}^i}{\partial q_j} \tag{2.21}$$

Substituting the results of Eqs. 20 and 21 into Eq. 19 yields

$$\begin{aligned} \sum_{i=1}^{n_p} m^i \ddot{\mathbf{r}}^{i\mathrm{T}} \frac{\partial \mathbf{r}^i}{\partial q_j} &= \sum_{i=1}^{n_p} \frac{d}{dt} \left[\frac{\partial}{\partial \dot{q}_j} \left(\tfrac{1}{2} m^i \dot{\mathbf{r}}^{i\mathrm{T}} \dot{\mathbf{r}}^i\right) \right] - \frac{\partial}{\partial q_j} \left(\tfrac{1}{2} m^i \dot{\mathbf{r}}^{i\mathrm{T}} \dot{\mathbf{r}}^i\right) \\ &= \sum_{i=1}^{n_p} \frac{d}{dt} \left(\frac{\partial T^i}{\partial \dot{q}_j} \right) - \frac{\partial T^i}{\partial q_j} \end{aligned} \tag{2.22}$$

where T^i is the kinetic energy of the particle i defined as

$$T^i = \tfrac{1}{2} m^i \dot{\mathbf{r}}^{i\mathrm{T}} \dot{\mathbf{r}}^i \tag{2.23}$$

The kinetic energy of the system of particles is given by

$$T = \sum_{i=1}^{n_p} T^i \tag{2.24}$$

Therefore, Eq. 22 can be written in terms of the kinetic energy of the system of particles as

$$\sum_{i=1}^{n_p} m^i \ddot{\mathbf{r}}^{iT} \frac{\partial \mathbf{r}^i}{\partial q_j} = \frac{d}{dt}\left(\frac{\partial T}{\partial \dot{q}_j}\right) - \frac{\partial T}{\partial q_j} \tag{2.25}$$

Substituting Eqs. 18 and 25 into Eq. 17 yields

$$\sum_{j=1}^{n} \left[\frac{d}{dt}\left(\frac{\partial T}{\partial \dot{q}_j}\right) - \frac{\partial T}{\partial q_j} - Q_j\right] \delta q_j = 0 \tag{2.26}$$

Since the generalized coordinates $q_1, q_2, \ldots, q_n$ are assumed to be linearly independent, Eq. 26 yields a set of n equations defined as

$$\frac{d}{dt}\left(\frac{\partial T}{\partial \dot{q}_j}\right) - \frac{\partial T}{\partial q_j} - Q_j = 0, \qquad j = 1, 2, \ldots, n$$

or

$$\frac{d}{dt}\left(\frac{\partial T}{\partial \dot{q}_j}\right) - \frac{\partial T}{\partial q_j} = Q_j, \qquad j = 1, 2, \ldots, n \tag{2.27a}$$

This equation is called *Lagrange's equation of motion.* Clearly, there are as many equations as the number of generalized coordinates. These equations can be derived using scalar quantities, mainly the kinetic energy of the system and virtual work of the applied and elastic forces. The use of Lagrange's equation for developing the differential equations of motion for single and two degree of freedom systems is demonstrated by the following examples.

Example 2.5

Using Lagrange's equation, develop the differential equation of motion of the single degree of freedom system given in Example 2 and shown in Fig. 3.

Solution. The system of Example 2 has one degree of freedom and only one differential equation results from the application of Lagrange's equation which can be stated in this case as

$$\frac{d}{dt}\left(\frac{\partial T}{\partial \dot{x}}\right) - \frac{\partial T}{\partial x} = Q$$

It was shown in Example 2 that the generalized force Q is given by

$$Q = F(t) - kx - c\dot{x}$$

The kinetic energy of this system is given by

$$T = \tfrac{1}{2} m \dot{x}^2$$

which yields

$$\frac{\partial T}{\partial \dot{x}} = m\dot{x}, \qquad \frac{d}{dt}\left(\frac{\partial T}{\partial \dot{x}}\right) = m\ddot{x}, \quad \text{and} \quad \frac{\partial T}{\partial x} = 0$$

Using Lagrange's equation, one obtains

$$m\ddot{x} = F(t) - kx - c\dot{x}$$

or

$$m\ddot{x} + c\dot{x} + kx = F(t)$$

which is the same differential equation obtained for this system by applying Newton's second law.

Example 2.6

Using Lagrange's equation, derive the differential equations of motion of the two degree of freedom system given in Example 3 and shown in Fig. 4.

Solution. Since the system has the two degrees of freedom $q_1 = x_1$ and $q_2 = x_2$ as generalized coordinates, the application of Lagrange's equation yields the two equations

$$\frac{d}{dt}\left(\frac{\partial T}{\partial \dot{q}_1}\right) - \frac{\partial T}{\partial q_1} = Q_1$$

$$\frac{d}{dt}\left(\frac{\partial T}{\partial \dot{q}_2}\right) - \frac{\partial T}{\partial q_2} = Q_2$$

where T is the total kinetic energy of the system and Q_1 and Q_2 are the generalized forces associated, respectively, with the system generalized coordinates. The kinetic energy T is defined as

$$T = \tfrac{1}{2}m_1\dot{q}_1^2 + \tfrac{1}{2}m_2\dot{q}_2^2 = \tfrac{1}{2}[\dot{q}_1 \quad \dot{q}_2]\begin{bmatrix} m_1 & 0 \\ 0 & m_2 \end{bmatrix}\begin{bmatrix} \dot{q}_1 \\ \dot{q}_2 \end{bmatrix}$$

It was shown in Example 3 that the generalized forces Q_1 and Q_2 are given by

$$Q_1 = F_1(t) - (k_1 + k_2)q_1 - (c_1 + c_2)\dot{q}_1 + k_2q_2 + c_2\dot{q}_2$$

$$Q_2 = F_2(t) - k_2q_2 - c_2\dot{q}_2 + k_2q_1 + c_2\dot{q}_1$$

By using the definition of the kinetic energy, one obtains

$$\frac{\partial T}{\partial \dot{q}_1} = m_1\dot{q}_1, \qquad \frac{d}{dt}\left(\frac{\partial T}{\partial \dot{q}_1}\right) = m_1\ddot{q}_1, \qquad \frac{\partial T}{\partial q_1} = 0$$

$$\frac{\partial T}{\partial \dot{q}_2} = m_2\dot{q}_2, \qquad \frac{d}{dt}\left(\frac{\partial T}{\partial \dot{q}_2}\right) = m_2\ddot{q}_2, \qquad \frac{\partial T}{\partial q_2} = 0$$

Substituting into Lagrange's equation for q_1 and q_2, yields, respectively, the following two differential equations:

$$m_1\ddot{q}_1 = F_1(t) - (k_1 + k_2)q_1 - (c_1 + c_2)\dot{q}_1 + k_2q_2 + c_2\dot{q}_2$$

$$m_2\ddot{q}_2 = F_2(t) - k_2q_2 - c_2\dot{q}_2 + k_2q_1 + c_2\dot{q}_1$$

or

$$m_1\ddot{q}_1 + (c_1 + c_2)\dot{q}_1 - c_2\dot{q}_2 + (k_1 + k_2)q_1 - k_2 q_2 = F_1(t)$$

$$m_2\ddot{q}_2 + c_2\dot{q}_2 - c_2\dot{q}_1 + k_2 q_2 - k_2 q_1 = F_2(t)$$

which can be written in a matrix form as

$$\begin{bmatrix} m_1 & 0 \\ 0 & m_2 \end{bmatrix}\begin{bmatrix} \ddot{q}_1 \\ \ddot{q}_2 \end{bmatrix} + \begin{bmatrix} c_1 + c_2 & -c_2 \\ -c_2 & c_2 \end{bmatrix}\begin{bmatrix} \dot{q}_1 \\ \dot{q}_2 \end{bmatrix} + \begin{bmatrix} k_1 + k_2 & -k_2 \\ -k_2 & k_2 \end{bmatrix}\begin{bmatrix} q_1 \\ q_2 \end{bmatrix} = \begin{bmatrix} F_1(t) \\ F_2(t) \end{bmatrix}$$

which is the same matrix equation that can be obtained for this two degree of freedom system by applying Newton's second law.

Example 2.7

Using Lagrange's equation, derive the differential equation of motion of the two degree of freedom system shown in Fig. 6. Use the assumption of small oscillations.

Solution. We select x and θ as the system generalized coordinates, that is,

$$\mathbf{q} = [x \quad \theta]^{\mathrm{T}}$$

The kinetic energy of the system is

$$T = \tfrac{1}{2}m\dot{x}^2 + \tfrac{1}{2}m_r\dot{x}_c^2 + \tfrac{1}{2}I\dot{\theta}^2$$

where m_r is the total mass of the rod, $\dot{x}_c$ is the absolute velocity of the center of mass of the rod, and I is the mass moment of inertia of the rod about its mass center, that is,

$$I = \frac{m_r l^2}{12}$$

where l is the length of the rod. The absolute velocity of the center of mass of the

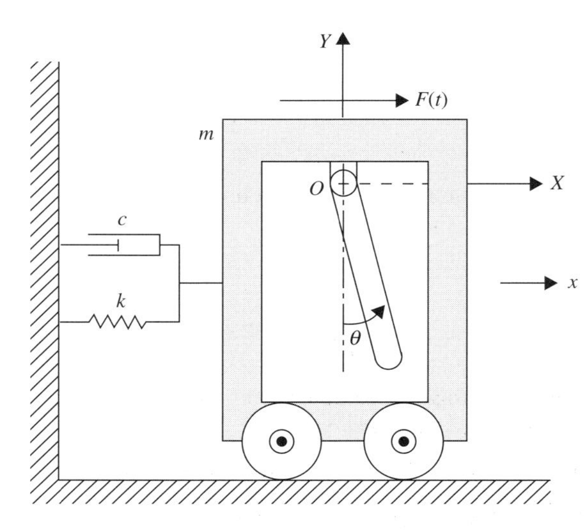

FIG. 2.6. Pendulum with moving base.

rod can be expressed in terms of the generalized velocities as

$$\dot{x}_c \approx \dot{x} + \dot{\theta}\frac{l}{2}$$

where the assumption of small oscillation is utilized.

The kinetic energy T can then be written in terms of the generalized velocities as

$$\begin{aligned} T &= \tfrac{1}{2}m\dot{x}^2 + \tfrac{1}{2}m_r\left(\dot{x} + \dot{\theta}\frac{l}{2}\right)^2 + \tfrac{1}{2}I\dot{\theta}^2 \\ &= \tfrac{1}{2}(m + m_r)\dot{x}^2 + m_r\frac{l}{2}\dot{x}\dot{\theta} + \tfrac{1}{2}\left(I + m_r\frac{l^2}{4}\right)\dot{\theta}^2 \\ &= \tfrac{1}{2}(m + m_r)\dot{x}^2 + m_r\frac{l}{2}\dot{x}\dot{\theta} + \tfrac{1}{2}I_O\dot{\theta}^2 \\ &= \tfrac{1}{2}[\dot{x} \quad \dot{\theta}]\begin{bmatrix} m + m_r & m_r(l/2) \\ m_r(l/2) & I_O \end{bmatrix}\begin{bmatrix} \dot{x} \\ \dot{\theta} \end{bmatrix} \end{aligned}$$

where I_O is the mass moment of inertia of the rod about point O and is given by

$$I_O = I + m_r\frac{l^2}{4}$$

The virtual work of all the forces acting on the system is given by

$$\delta W = [F(t) - kx - c\dot{x}]\delta x - m_r g\delta y_c$$

where g is the gravitational constant, and δy_c is the virtual displacement of the center of mass in the vertical direction which is given by

$$\delta y_c = \frac{l}{2}\sin\theta\delta\theta$$

Therefore,

$$\begin{aligned} \delta W &= [F(t) - kx - c\dot{x}]\delta x - m_r g\frac{l}{2}\sin\theta\delta\theta \\ &= Q_x\delta x + Q_\theta\delta\theta \end{aligned}$$

where Q_x and Q_θ are the generalized forces associated, respectively, with the generalized coordinates x and θ and defined as

$$\begin{aligned} Q_x &= F(t) - kx - c\dot{x} \\ Q_\theta &= -m_r g\frac{l}{2}\sin\theta \approx -m_r g\frac{l}{2}\theta \end{aligned}$$

The application of Lagrange's equation yields the following two equations:

$$\begin{aligned} \frac{d}{dt}\left(\frac{\partial T}{\partial \dot{x}}\right) - \frac{\partial T}{\partial x} &= Q_x \\ \frac{d}{dt}\left(\frac{\partial T}{\partial \dot{\theta}}\right) - \frac{\partial T}{\partial \theta} &= Q_\theta \end{aligned}$$

Differentiation of the kinetic energy with respect to time and with respect to the generalized coordinates yields

$$\frac{\partial T}{\partial \dot{x}} = (m + m_r)\dot{x} + m_r \frac{l}{2}\dot{\theta}$$

$$\frac{d}{dt}\left(\frac{\partial T}{\partial \dot{x}}\right) = (m + m_r)\ddot{x} + m_r \frac{l}{2}\ddot{\theta}, \qquad \frac{\partial T}{\partial x} = 0$$

$$\frac{\partial T}{\partial \dot{\theta}} = m_r \frac{l}{2}\dot{x} + I_O\dot{\theta}, \qquad \frac{d}{dt}\left(\frac{\partial T}{\partial \dot{\theta}}\right) = m_r \frac{l}{2}\ddot{x} + I_O\ddot{\theta}$$

$$\frac{\partial T}{\partial \theta} = 0$$

Substituting these equations in Lagrange's equation and using the expressions for the generalized forces yields

$$(m + m_r)\ddot{x} + m_r \frac{l}{2}\ddot{\theta} = F(t) - kx - c\dot{x}$$

$$m_r \frac{l}{2}\ddot{x} + I_O\ddot{\theta} = -m_r g \frac{l}{2}\theta$$

or

$$(m + m_r)\ddot{x} + m_r \frac{l}{2}\ddot{\theta} + c\dot{x} + kx = F(t)$$

$$m_r \frac{l}{2}\ddot{x} + I_O\ddot{\theta} + m_r g \frac{l}{2}\theta = 0$$

which can be written in a matrix form as

$$\begin{bmatrix} m + m_r & m_r(l/2) \\ m_r(l/2) & I_O \end{bmatrix}\begin{bmatrix} \ddot{x} \\ \ddot{\theta} \end{bmatrix} + \begin{bmatrix} c & 0 \\ 0 & 0 \end{bmatrix}\begin{bmatrix} \dot{x} \\ \dot{\theta} \end{bmatrix} + \begin{bmatrix} k & 0 \\ 0 & m_r g(l/2) \end{bmatrix}\begin{bmatrix} x \\ \theta \end{bmatrix} = \begin{bmatrix} F(t) \\ 0 \end{bmatrix}$$

Remarks Lagrange's equation as defined by Eq. 27a states that the generalized external force Q_j associated with the jth generalized coordinate is equal to the generalized inertia force associated with the same coordinate. This generalized inertia force is defined in terms of the system kinetic energy as

$$(Q_i)_j = \frac{d}{dt}\left(\frac{\partial T}{\partial \dot{q}_j}\right) - \frac{\partial T}{\partial q_j}, \qquad j = 1, 2, \ldots, n \tag{2.27b}$$

Therefore, Lagrange's equation as defined by Eq. 27a can be written as

$$(Q_i)_j = Q_j, \qquad j = 1, 2, \ldots, n \tag{2.27c}$$

Note that the generalized inertia force $(Q_i)_j$ was originally defined by Eq. 25 as

$$(Q_i)_j = \sum_{i=1}^{n_p} m^i \ddot{\mathbf{r}}^{iT} \frac{\partial \mathbf{r}^i}{\partial q_j} \tag{2.27d}$$

which is the same as (see Eq. 20)

$$(Q_i)_j = \sum_{i=1}^{n_p} m^i \ddot{\mathbf{r}}^{i\mathrm{T}} \frac{\partial \dot{\mathbf{r}}^i}{\partial \dot{q}_j} \tag{2.27e}$$

Therefore, in the Lagrangian formulation, one can use Eq. 27b, Eq. 27d, or Eq. 27e to obtain the generalized inertia forces associated with the system generalized coordinates. These equations can be applied to systems of particles as well as rigid-body systems. In rigid-body dynamics, there are inertia forces as well as inertia moments. Therefore, in the case of a system that consists of particles and rigid bodies, Eqs. 27d and 27e must be modified to

$$(Q_i)_j = \sum_{i=1}^{n_p} m^i \ddot{\mathbf{r}}^{i\mathrm{T}} \frac{\partial \mathbf{r}^i}{\partial q_j} + \sum_{i=1}^{n_b} \left[m^i \ddot{\mathbf{r}}_{\mathrm{c}}^{i\mathrm{T}} \frac{\partial \mathbf{r}_{\mathrm{c}}^i}{\partial q_j} + I^i \ddot{\theta}^i \frac{\partial \theta^i}{\partial q_j} \right] \tag{2.27f}$$

and

$$(Q_i)_j = \sum_{i=1}^{n_p} m^i \ddot{\mathbf{r}}^{i\mathrm{T}} \frac{\partial \dot{\mathbf{r}}^i}{\partial \dot{q}_j} + \sum_{i=1}^{n_b} \left[m^i \ddot{\mathbf{r}}_{\mathrm{c}}^{i\mathrm{T}} \frac{\partial \dot{\mathbf{r}}_{\mathrm{c}}^i}{\partial \dot{q}_j} + I^i \ddot{\theta}^i \frac{\partial \dot{\theta}^i}{\partial \dot{q}_j} \right] \tag{2.27g}$$

where $\mathbf{r}_{\mathrm{c}}^i$ is the vector of coordinates of the center of mass of the rigid body i, θ^i is its angular orientation, m^i and I^i are, respectively, its mass and mass moment of inertia about the center of mass, and n_{b} is the total number of rigid bodies in the system.

In order to demonstrate the use of the equations presented in this section in the case of rigid-body systems, we consider the system shown in Fig. 6. It was shown in the preceding example that the use of Eq. 27b leads to the following generalized inertia forces associated with the coördinates x and θ, respectively

$$(Q_i)_x = (m + m_r)\ddot{x} + m_r \frac{l}{2} \ddot{\theta}$$

$$(Q_i)_\theta = m_r \frac{l}{2} \ddot{x} + I_O \ddot{\theta}$$

where m is the mass of the block, m_r and l are, respectively, the mass and length of the rod, and I_O is the mass moment of inertia of the rod about point O. Without linearizing the kinematic relationship, the coordinates of the center of mass of the rod can be written as

$$\mathbf{r}_{\mathrm{c}} = \begin{bmatrix} x_c \\ y_c \end{bmatrix} = \begin{bmatrix} x + \frac{l}{2} \sin\theta \\ -\frac{l}{2} \cos\theta \end{bmatrix}$$

Differentiating these kinematic equations once and twice with repect to time leads to the velocity

$$\dot{\mathbf{r}}_c = \begin{bmatrix} \dot{x}_c \\ \dot{y}_c \end{bmatrix} = \begin{bmatrix} \dot{x} + \dot{\theta}\frac{l}{2}\cos\theta \\ \dot{\theta}\frac{l}{2}\sin\theta \end{bmatrix},$$

and the acceleration

$$\ddot{\mathbf{r}}_c = \begin{bmatrix} \ddot{x}_c \\ \ddot{y}_c \end{bmatrix} = \begin{bmatrix} \ddot{x} + \ddot{\theta}\frac{l}{2}\cos\theta - \dot{\theta}^2\frac{l}{2}\sin\theta \\ \ddot{\theta}\frac{l}{2}\sin\theta + \dot{\theta}^2\frac{l}{2}\cos\theta \end{bmatrix}$$

Note that

$$\frac{\partial \mathbf{r}_c}{\partial x} = \frac{\partial \dot{\mathbf{r}}_c}{\partial \dot{x}} = \frac{\partial \ddot{\mathbf{r}}_c}{\partial \ddot{x}} = \begin{bmatrix} 1 \\ 0 \end{bmatrix}$$

$$\frac{\partial \mathbf{r}_c}{\partial \theta} = \frac{\partial \dot{\mathbf{r}}_c}{\partial \dot{\theta}} = \frac{\partial \ddot{\mathbf{r}}_c}{\partial \ddot{\theta}} = \begin{bmatrix} \frac{l}{2}\cos\theta \\ \frac{l}{2}\sin\theta \end{bmatrix}$$

Therefore, the use of Eq. 27f leads to

$$(Q_i)_x = m\ddot{x}\frac{\partial x}{\partial x} + m_r\ddot{\mathbf{r}}_c^{\mathrm{T}}\frac{\partial \mathbf{r}_c}{\partial x} + I\ddot{\theta}\frac{\partial \theta}{\partial x}$$

Since θ is an independent coordinate, $\partial\theta/\partial x = 0$ and the preceding equation leads to

$$\begin{aligned} (Q_i)_x &= m\ddot{x} + m_r\ddot{\mathbf{r}}_c^{\mathrm{T}}\frac{\partial \mathbf{r}_c}{\partial x} \\ &= m\ddot{x} + m_r\begin{bmatrix} \ddot{x} + \ddot{\theta}\frac{l}{2}\cos\theta - \dot{\theta}^2\frac{l}{2}\sin\theta \\ \ddot{\theta}\frac{l}{2}\sin\theta + \dot{\theta}^2\frac{l}{2}\cos\theta \end{bmatrix}^{\mathrm{T}}\begin{bmatrix} 1 \\ 0 \end{bmatrix} \\ &= (m + m_r)\ddot{x} + m_r\ddot{\theta}\frac{l}{2}\cos\theta - m_r\dot{\theta}^2\frac{l}{2}\sin\theta \end{aligned}$$

Assuming small oscillations, the generalized inertia force associated with the coordinate x becomes

$$(Q_i)_x = (m + m_r)\ddot{x} + m_r\frac{l}{2}\ddot{\theta}$$

which is the same generalized inertia force previously obtained. Similarly, the generalized inertia force $(Q_i)_\theta$ associated with the coordinate θ, can be obtained using Eq. 27f as

$$(Q_i)_\theta = m\ddot{x}\frac{\partial x}{\partial \theta} + m_r\ddot{\mathbf{r}}_c^{\mathrm{T}}\frac{\partial \mathbf{r}_c}{\partial \theta} + I\ddot{\theta}\frac{\partial \theta}{\partial \theta}$$

$$= m_r\ddot{\mathbf{r}}_c^{\mathrm{T}}\frac{\partial \mathbf{r}_c}{\partial \theta} + I\ddot{\theta}$$

which upon using the definition of $\ddot{\mathbf{r}}_c$ and $\partial \mathbf{r}_c/\partial\theta$ yields

$$(Q_i)_\theta = m_r\begin{bmatrix} \ddot{x} + \ddot{\theta}\frac{l}{2}\cos\theta - \dot{\theta}^2\frac{l}{2}\sin\theta \\ \ddot{\theta}\frac{l}{2}\sin\theta + \dot{\theta}^2\frac{l}{2}\cos\theta \end{bmatrix}^{\mathrm{T}}\begin{bmatrix} \frac{l}{2}\cos\theta \\ \frac{l}{2}\sin\theta \end{bmatrix} + I\ddot{\theta}$$

$$= m_r\ddot{x}\frac{l}{2}\cos\theta + m_r\frac{l^2}{4}\ddot{\theta} + I\ddot{\theta}$$

$$= m_r\ddot{x}\frac{l}{2}\cos\theta + I_O\ddot{\theta}$$

where $I_O = I + m_r(l^2/4)$.

As in the case of the preceding example, we assume small oscillations. In this case $\cos\theta \approx 1$, and consequently $(Q_i)_\theta$ reduces to

$$(Q_i)_\theta = m_r\frac{l}{2}\ddot{x} + I\ddot{\theta}$$

which is the same as the generalized inertia force obtained by using the kinetic energy. Note also as the result of Eq. 20, the generalized external force Q_j can also be written as

$$Q_j = \sum_{i=1}^{n_b} \mathbf{F}^{i\mathrm{T}}\frac{\partial \dot{\mathbf{r}}^i}{\partial \dot{q}_j}$$

2.4 KINETIC ENERGY

As was shown in the preceding section, one alternative for formulating the dynamic equations is to obtain an expression for the kinetic energy of the system and use it with Lagrange's equation, as defined by Eq. 27a. In this section, the formulation of the kinetic energy using vector and matrix notation is discussed.

System of Particles In many applications, such as the examples presented in the preceding sections, the position vector of the particle i in the system can be expressed in terms of the generalized coordinates as

$$\mathbf{r}^i = \mathbf{r}^i(q_1, q_2, \ldots, q_n) \tag{2.28}$$

and the velocity vector as

$$\dot{\mathbf{r}}^i = \frac{\partial \mathbf{r}^i}{\partial q_1}\dot{q}_1 + \frac{\partial \mathbf{r}^i}{\partial q_2}\dot{q}_2 + \cdots + \frac{\partial \mathbf{r}^i}{\partial q_n}\dot{q}_n = \sum_{j=1}^{n} \frac{\partial \mathbf{r}^i}{\partial q_j}\dot{q}_j \tag{2.29}$$

which can be written using matrix notation as

$$\dot{\mathbf{r}}^i = \begin{bmatrix} \frac{\partial \mathbf{r}^i}{\partial q_1} & \frac{\partial \mathbf{r}^i}{\partial q_2} & \cdots & \frac{\partial \mathbf{r}^i}{\partial q_n} \end{bmatrix} \begin{bmatrix} \dot{q}_1 \\ \dot{q}_2 \\ \vdots \\ \dot{q}_n \end{bmatrix} \tag{2.30}$$

This equation can be written in a compact form as

$$\dot{\mathbf{r}}^i = \mathbf{B}^i\dot{\mathbf{q}} \tag{2.31}$$

where

$$\mathbf{B}^i = \begin{bmatrix} \frac{\partial \mathbf{r}^i}{\partial q_1} & \frac{\partial \mathbf{r}^i}{\partial q_2} & \cdots & \frac{\partial \mathbf{r}^i}{\partial q_n} \end{bmatrix} \tag{2.32}$$

$$\mathbf{q} = [q_1 \quad q_2 \quad \cdots \quad q_n]^{\mathrm{T}} \tag{2.33}$$

The kinetic energy of the system of particles can also be obtained as

$$T^i = \tfrac{1}{2}m^i\dot{\mathbf{r}}^{i\mathrm{T}}\dot{\mathbf{r}}^i \tag{2.34}$$

where m^i is the mass of the particle i. Substituting Eq. 31 into Eq. 34 yields

$$T^i = \tfrac{1}{2}m^i(\mathbf{B}^i\dot{\mathbf{q}})^{\mathrm{T}}(\mathbf{B}^i\dot{\mathbf{q}}) = \tfrac{1}{2}m^i\dot{\mathbf{q}}^{\mathrm{T}}\mathbf{B}^{i\mathrm{T}}\mathbf{B}^i\dot{\mathbf{q}} = \tfrac{1}{2}\dot{\mathbf{q}}^{\mathrm{T}}(m^i\mathbf{B}^{i\mathrm{T}}\mathbf{B}^i)\dot{\mathbf{q}} \tag{2.35}$$

The kinetic energy of the system of particles can also be obtained as

$$T = \sum_{i=1}^{n_{\mathrm{p}}} T^i = \sum_{i=1}^{n_{\mathrm{p}}} \tfrac{1}{2}\dot{\mathbf{q}}^{\mathrm{T}}(m^i\mathbf{B}^{i\mathrm{T}}\mathbf{B}^i)\dot{\mathbf{q}} = \tfrac{1}{2}\dot{\mathbf{q}}^{\mathrm{T}}\sum_{i=1}^{n_{\mathrm{p}}}(m^i\mathbf{B}^{i\mathrm{T}}\mathbf{B}^i)\dot{\mathbf{q}}$$

which can be written as

$$T = \tfrac{1}{2}\dot{\mathbf{q}}^{\mathrm{T}}\mathbf{M}\dot{\mathbf{q}} \tag{2.36}$$

where $\mathbf{M}$ is the mass matrix of the system of particles, defined as

$$\mathbf{M} = \sum_{i=1}^{n_{\mathrm{p}}}(m^i\mathbf{B}^{i\mathrm{T}}\mathbf{B}^i) \tag{2.37}$$

The kinetic energy of Eq. 36 is a quadratic function in the generalized velocity vector $\dot{\mathbf{q}}$, and the mass matrix $\mathbf{M}$ is symmetric and may be constant or may depend on the system generalized coordinates.

Rigid Bodies The kinetic energy of rigid bodies can also be expressed in a form similar to Eq. 36. For instance, for a rigid body i in the system, let x_c^i and y_c^i be the x- and y-coordinates of the center of mass of the body and let θ^i be the angular orientation of the body. The kinetic energy of the body can be written as

$$T^i = \tfrac{1}{2}m^i\dot{x}_c^{i2} + \tfrac{1}{2}m^i\dot{y}_c^{i2} + \tfrac{1}{2}I^i\dot{\theta}^{i2} \tag{2.38a}$$

where m^i and I^i are, respectively, the mass and the mass moment of inertia of the body, and the mass moment of inertia is defined with respect to the center of mass of the body. Equation 38a can also be written as

$$\begin{aligned} T^i &= \tfrac{1}{2}[\dot{x}_c^i \quad \dot{y}_c^i \quad \dot{\theta}^i]\begin{bmatrix} m^i & 0 & 0 \\ 0 & m^i & 0 \\ 0 & 0 & I^i \end{bmatrix}\begin{bmatrix} \dot{x}_c^i \\ \dot{y}_c^i \\ \dot{\theta}^i \end{bmatrix} \\ &= \tfrac{1}{2}\dot{\mathbf{q}}_r^{iT}\mathbf{M}_r^i\dot{\mathbf{q}}_r^i \end{aligned} \tag{2.38b}$$

where

$$\mathbf{q}_r^i = [x_c^i \quad y_c^i \quad \theta^i]^T$$

and

$$\mathbf{M}_r^i = \begin{bmatrix} m^i & 0 & 0 \\ 0 & m^i & 0 \\ 0 & 0 & I^i \end{bmatrix}$$

The absolute velocities $\dot{x}_c^i$, $\dot{y}_c^i$, and $\dot{\theta}_c^i$ can be expressed in terms of the generalized velocities as

$$\dot{x}_c^i = \sum_{j=1}^{n} \frac{\partial x_c^i}{\partial q_j}\dot{q}_j$$

$$\dot{y}_c^i = \sum_{j=1}^{n} \frac{\partial y_c^i}{\partial q_j}\dot{q}_j$$

$$\dot{\theta}^i = \sum_{j=1}^{n} \frac{\partial \theta^i}{\partial q_j}\dot{q}_j$$

That is,

$$\dot{\mathbf{q}}_r^i = \mathbf{B}^i\dot{\mathbf{q}} \tag{2.38c}$$

where the matrix $\mathbf{B}^i$ is a $3 \times n$ matrix defined as

$$\mathbf{B}^i = \begin{bmatrix} \dfrac{\partial x_c^i}{\partial q_1} & \dfrac{\partial x_c^i}{\partial q_2} & \cdots & \dfrac{\partial x_c^i}{\partial q_n} \\ \dfrac{\partial y_c^i}{\partial q_1} & \dfrac{\partial y_c^i}{\partial q_2} & \cdots & \dfrac{\partial y_c^i}{\partial q_n} \\ \dfrac{\partial \theta^i}{\partial q_1} & \dfrac{\partial \theta^i}{\partial q_2} & \cdots & \dfrac{\partial \theta^i}{\partial q_n} \end{bmatrix} \tag{2.38d}$$

Substituting Eq. 38c into Eq. 38b leads to an expression for the kinetic energy in terms of the generalized velocities as

$$T^i = \tfrac{1}{2}\dot{\mathbf{q}}^{\mathrm{T}}\mathbf{B}^{i\mathrm{T}}\mathbf{M}_{\mathrm{r}}^i\mathbf{B}^i\dot{\mathbf{q}}$$

which can also be written as

$$T^i = \tfrac{1}{2}\dot{\mathbf{q}}^{\mathrm{T}}\mathbf{M}^i\dot{\mathbf{q}} \tag{2.39a}$$

where $\mathbf{M}^i$ is the mass matrix of the rigid body defined as

$$\mathbf{M}^i = \mathbf{B}^{i\mathrm{T}}\mathbf{M}_{\mathrm{r}}^i\mathbf{B}^i \tag{2.39b}$$

The kinetic energy of a system of rigid bodies is

$$T = \sum_{i=1}^{n_b} T^i = \frac{1}{2}\sum_{i=1}^{n_b} \dot{\mathbf{q}}^{\mathrm{T}}\mathbf{M}^i\dot{\mathbf{q}}$$

where n_b is the total number of bodies in the system.

The system kinetic energy can also be written as

$$T = \tfrac{1}{2}\dot{\mathbf{q}}^{\mathrm{T}}\left[\sum_{i=1}^{n_b} \mathbf{M}^i\right]\dot{\mathbf{q}} = \tfrac{1}{2}\dot{\mathbf{q}}^{\mathrm{T}}\mathbf{M}\dot{\mathbf{q}}$$

where $\mathbf{M}$ is the system mass matrix defined as

$$\mathbf{M} = \sum_{i=1}^{n_b} \mathbf{M}^i$$

Illustrative Example Consider the system shown in Fig. 7a, in which AB is a uniform rod that has mass m, mass moment of inertia I, and length l. The kinetic energy of the rod can be defined using Eq. 38b. However, the rod has

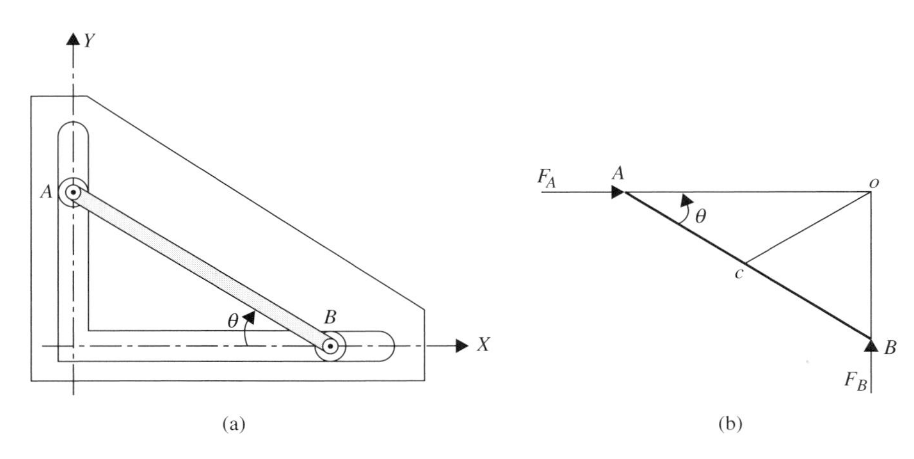

FIG. 2.7. Generalized mass.

only one degree of freedom since point A slides in the vertical direction only, while point B slides in the horizontal direction only. Because of these two restrictions on the motion of the rod, one has the following kinematic relationships:

$$x_A = x_c - \frac{l}{2}\cos\theta = 0$$

$$y_B = y_c - \frac{l}{2}\sin\theta = 0$$

where x_A and y_B are the horizontal and vertical coordinates of points A and B, respectively. The preceding two equations, which show that the center of mass of the rod follows a circular trajectory which has a radius $l/2$, lead to

$$\begin{bmatrix} \dot{x}_c \\ \dot{y}_c \end{bmatrix} = \frac{\dot{\theta} l}{2}\begin{bmatrix} -\sin\theta \\ \cos\theta \end{bmatrix}$$

The absolute velocities of the rod can then be expressed in terms of the independent angular velocity as

$$\begin{bmatrix} \dot{x}_c \\ \dot{y}_c \\ \dot{\theta} \end{bmatrix} = \dot{\theta}\begin{bmatrix} -\frac{l}{2}\sin\theta \\ \frac{l}{2}\cos\theta \\ 1 \end{bmatrix} = \mathbf{B}\dot{\theta}$$

where $\mathbf{B}$ is a 3×1 matrix which is recognized as

$$\mathbf{B} = \begin{bmatrix} -\frac{l}{2}\sin\theta \\ \frac{l}{2}\cos\theta \\ 1 \end{bmatrix}$$

Using an equation similar to Eq. 39a, the kinetic energy of the rod can be expressed in terms of the independent angular velocity as

$$T = \frac{1}{2}\mathbf{B}^{\mathrm{T}}\mathbf{M}_r\mathbf{B}\dot{\theta}^2 = \left(I + \frac{ml^2}{4}\right)\dot{\theta}^2 = \frac{l}{2}I_o\dot{\theta}^2$$

where I_o is the mass moment of inertia about point o shown in Fig. 7b. In this figure, point o is the point of intersection of the lines of action of the reaction forces F_A and F_B. The preceding equation shows that the generalized mass matrix associated with the independent coordinate of the rod reduces to a scalar defined by the mass moment of inertia about point o. Note that by using simple trigonometric identities, one can show that the length oc is $l/2$.

Definitions In both cases of systems of particles and rigid bodies, the kinetic energy is a quadratic function in the system velocities, or more precisely, a homogeneous, second degree *quadratic form* in the velocities $\dot{q}_1, \dot{q}_2, \ldots, \dot{q}_n$. The mass matrix $\mathbf{M}$ is said to be the *matrix* of *the quadratic form.* A quadratic form is said to be *positive* (*negative*) if it is equal or greater (less) than zero for all the real values of its variables. A positive (negative) quadratic form is said to be *positive* (*negative*) *definite* if it is equal to zero only when all its variables are equal to zero. A positive (negative) quadratic form which is equal to zero for some nonzero values of its variables is said to be *positive* (*negative*) *semidefinite.* A quadratic form which can be positive or negative for the real values of its variables is said to be *indefinite.* The matrix of quadratic form is said to be *positive* (*negative*) *definite*, *positive* (*negative*) *semidefinite*, or *indefinite* according to the nature of the quadratic form.

Since there can be no motion of particles and rigid bodies while the kinetic energy remains equal to zero, the kinetic energy is always a positive definite quadratic form. Consequently, the mass matrix is always positive definite. Therefore, the mass matrix is nonsingular, its determinant has always nonzero value and its inverse exists.

Vector Form of Lagrange's Equation Using the quadratic form of Eq. 36, Lagrange's equation can be expressed in a vector form. First, we introduce the following notation:

$$\frac{\partial T}{\partial \dot{\mathbf{q}}} = T_{\dot{\mathbf{q}}} = \dot{\mathbf{q}}^{\mathrm{T}}\mathbf{M} \tag{2.40}$$

$$\frac{\partial T}{\partial \mathbf{q}} = T_{\mathbf{q}} \tag{2.41}$$

where vector subscript implies differentiation with respect to this vector. Using this notation, one can then write Lagrange's equation in the following vector form:

$$\frac{d}{dt}(T_{\dot{\mathbf{q}}})^{\mathrm{T}} - (T_{\mathbf{q}})^{\mathrm{T}} = \mathbf{Q} \tag{2.42}$$

or

$$\frac{d}{dt}(\mathbf{M}\dot{\mathbf{q}}) - (T_{\mathbf{q}})^{\mathrm{T}} = \mathbf{Q} \tag{2.43}$$

where $\mathbf{Q}$ is the vector of generalized forces associated with the generalized coordinates and defined as

$$\mathbf{Q} = [Q_1 \quad Q_2 \quad \cdots \quad Q_n]^{\mathrm{T}} \tag{2.44}$$

Observe that Lagrange's equation can also be written compactly as

$$\mathbf{Q}_i = \mathbf{Q}$$

where $\mathbf{Q}_i$ is the vector of the *generalized inertia forces* expressed in terms of the kinetic energy as

$$\mathbf{Q}_i = \frac{d}{dt}(\mathbf{M}\dot{\mathbf{q}}) - (T_{\mathbf{q}})^{\mathrm{T}}$$

Example 2.8

Using Lagrange's equation, derive the differential equation of motion of the system shown in Fig. 8.

Solution. The position vectors of the two masses are given by

$$\mathbf{r}^1 = \begin{bmatrix} x_1 \\ y_1 \end{bmatrix} = \begin{bmatrix} l_1 \sin\theta_1 \\ -l_1 \cos\theta_1 \end{bmatrix}$$

$$\mathbf{r}^2 = \begin{bmatrix} x_2 \\ y_2 \end{bmatrix} = \begin{bmatrix} l_1 \sin\theta_1 + l_2 \sin\theta_2 \\ -l_1 \cos\theta_1 - l_2 \cos\theta_2 \end{bmatrix}$$

The velocity vectors are given by

$$\dot{\mathbf{r}}^1 = \begin{bmatrix} \dot{x}_1 \\ \dot{y}_1 \end{bmatrix} = \begin{bmatrix} \dot{\theta}_1 l_1 \cos\theta_1 \\ \dot{\theta}_1 l_1 \sin\theta_1 \end{bmatrix} = \begin{bmatrix} l_1 \cos\theta_1 & 0 \\ l_1 \sin\theta_1 & 0 \end{bmatrix} \begin{bmatrix} \dot{\theta}_1 \\ \dot{\theta}_2 \end{bmatrix} = \mathbf{B}^1 \dot{\mathbf{q}}$$

$$\dot{\mathbf{r}}^2 = \begin{bmatrix} \dot{x}_2 \\ \dot{y}_2 \end{bmatrix} = \begin{bmatrix} \dot{\theta}_1 l_1 \cos\theta_1 + \dot{\theta}_2 l_2 \cos\theta_2 \\ \dot{\theta}_1 l_1 \sin\theta_1 + \dot{\theta}_2 l_2 \sin\theta_2 \end{bmatrix} = \begin{bmatrix} l_1 \cos\theta_1 & l_2 \cos\theta_2 \\ l_1 \sin\theta_1 & l_2 \sin\theta_2 \end{bmatrix} \begin{bmatrix} \dot{\theta}_1 \\ \dot{\theta}_2 \end{bmatrix}$$

$$= \mathbf{B}^2 \dot{\mathbf{q}}$$

where

$$\mathbf{q} = \begin{bmatrix} q_1 \\ q_2 \end{bmatrix} = \begin{bmatrix} \theta_1 \\ \theta_2 \end{bmatrix}$$

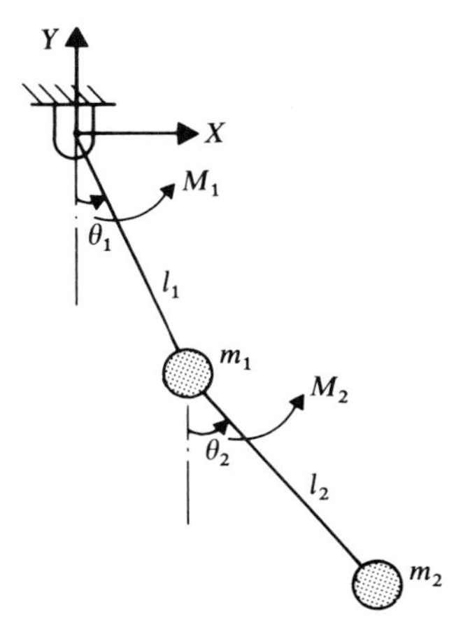

FIG. 2.8. Double pendulum.

and the coefficient matrices $\mathbf{B}^1$ and $\mathbf{B}^2$ are given by

$$\mathbf{B}^1 = \begin{bmatrix} l_1 \cos\theta_1 & 0 \\ l_1 \sin\theta_1 & 0 \end{bmatrix}$$

$$\mathbf{B}^2 = \begin{bmatrix} l_1 \cos\theta_1 & l_2 \cos\theta_2 \\ l_1 \sin\theta_1 & l_2 \sin\theta_2 \end{bmatrix}$$

The kinetic energy of the system is

$$\begin{aligned} T &= \sum_{i=1}^{2} T^i = \frac{1}{2}\sum_{i=1}^{2} m^i \dot{\mathbf{r}}^{i\mathrm{T}}\dot{\mathbf{r}}^i \\ &= \tfrac{1}{2}m_1\dot{\mathbf{r}}^{1\mathrm{T}}\dot{\mathbf{r}}^1 + \tfrac{1}{2}m_2\dot{\mathbf{r}}^{2\mathrm{T}}\dot{\mathbf{r}}^2 \\ &= \tfrac{1}{2}\dot{\mathbf{q}}^{\mathrm{T}}m_1\mathbf{B}^{1\mathrm{T}}\mathbf{B}^1\dot{\mathbf{q}} + \tfrac{1}{2}\dot{\mathbf{q}}^{\mathrm{T}}m_2\mathbf{B}^{2\mathrm{T}}\mathbf{B}^2\dot{\mathbf{q}} \\ &= \tfrac{1}{2}\dot{\mathbf{q}}^{\mathrm{T}}[m_1\mathbf{B}^{1\mathrm{T}}\mathbf{B}^1 + m_2\mathbf{B}^{2\mathrm{T}}\mathbf{B}^2]\dot{\mathbf{q}} \\ &= \tfrac{1}{2}\dot{\mathbf{q}}^{\mathrm{T}}\mathbf{M}\dot{\mathbf{q}} \end{aligned}$$

where the mass matrix $\mathbf{M}$ is defined as

$$\begin{aligned} \mathbf{M} &= m_1\mathbf{B}^{1\mathrm{T}}\mathbf{B}^1 + m_2\mathbf{B}^{2\mathrm{T}}\mathbf{B}^2 \\ &= m_1\begin{bmatrix} l_1\cos\theta_1 & l_1\sin\theta_1 \\ 0 & 0 \end{bmatrix}\begin{bmatrix} l_1\cos\theta_1 & 0 \\ l_1\sin\theta_1 & 0 \end{bmatrix} \\ &\quad + m_2\begin{bmatrix} l_1\cos\theta_1 & l_1\sin\theta_1 \\ l_2\cos\theta_2 & l_2\sin\theta_2 \end{bmatrix}\begin{bmatrix} l_1\cos\theta_1 & l_2\cos\theta_2 \\ l_1\sin\theta_1 & l_2\sin\theta_2 \end{bmatrix} \\ &= \begin{bmatrix} m_1 l_1^2 & 0 \\ 0 & 0 \end{bmatrix} + \begin{bmatrix} m_2 l_1^2 & m_2 l_1 l_2 \cos(\theta_1 - \theta_2) \\ m_2 l_1 l_2 \cos(\theta_1 - \theta_2) & m_2 l_2^2 \end{bmatrix} \\ &= \begin{bmatrix} (m_1 + m_2) l_1^2 & m_2 l_1 l_2 \cos(\theta_1 - \theta_2) \\ m_2 l_1 l_2 \cos(\theta_1 - \theta_2) & m_2 l_2^2 \end{bmatrix} \end{aligned}$$

The virtual work of the forces that act on the system is given by

$$\delta W = -m_1 g\delta y_1 - m_2 g\delta y_2 + M_1\delta\theta_1 + M_2\delta\theta_2$$

where the virtual displacements δy_1 and δy_2 are

$$\begin{aligned} \delta y_1 &= l_1 \sin\theta_1\delta\theta_1 \\ \delta y_2 &= l_1 \sin\theta_1\delta\theta_1 + l_2\sin\theta_2\delta\theta_2 \end{aligned}$$

That is,

$$\begin{aligned} \delta W &= -m_1 g l_1 \sin\theta_1\delta\theta_1 - m_2 g(l_1\sin\theta_1\delta\theta_1 + l_2\sin\theta_2\delta\theta_2) \\ &\quad + M_1\delta\theta_1 + M_2\delta\theta_2 \\ &= (M_1 - m_1 g l_1 \sin\theta_1 - m_2 g l_1 \sin\theta_1)\delta\theta_1 + (M_2 - m_2 g l_2 \sin\theta_2)\delta\theta_2 \end{aligned}$$

which can be written as

$$\delta W = Q_{\theta 1}\delta\theta_1 + Q_{\theta 2}\delta\theta_2 = \mathbf{Q}^{\mathrm{T}}\delta\mathbf{q}$$

where $Q_{\theta 1}$ and $Q_{\theta 2}$ are, respectively, the generalized forces associated with the coordinates θ_1 and θ_2 and are given by

$$\mathbf{Q} = \begin{bmatrix} Q_{\theta 1} \\ Q_{\theta 2} \end{bmatrix} = \begin{bmatrix} M_1 - m_1 g l_1 \sin\theta_1 - m_2 g l_1 \sin\theta_1 \\ M_2 - m_2 g l_2 \sin\theta_2 \end{bmatrix}$$

Having determined the kinetic energy and the generalized forces, one can use Lagrange's equation. To this end, we evaluate the following:

$$\frac{d}{dt}\left(\frac{\partial T}{\partial \dot{\mathbf{q}}}\right) = \frac{d}{dt}(T_{\dot{\mathbf{q}}}) = \frac{d}{dt}(\dot{\mathbf{q}}^{\mathrm{T}}\mathbf{M}) = \ddot{\mathbf{q}}^{\mathrm{T}}\mathbf{M} + \dot{\mathbf{q}}^{\mathrm{T}}\dot{\mathbf{M}}$$

That is,

$$\frac{d}{dt}(T_{\dot{\mathbf{q}}})^{\mathrm{T}} = \mathbf{M}\ddot{\mathbf{q}} + \dot{\mathbf{M}}\dot{\mathbf{q}}$$

where $\dot{\mathbf{M}}$ is the matrix

$$\dot{\mathbf{M}} = \begin{bmatrix} 0 & -m_2 l_1 l_2(\dot\theta_1 - \dot\theta_2)\sin(\theta_1 - \theta_2) \\ -m_2 l_1 l_2(\dot\theta_1 - \dot\theta_2)\sin(\theta_1 - \theta_2) & 0 \end{bmatrix}$$

Therefore,

$$\frac{d}{dt}\left(\frac{\partial T}{\partial \dot{\mathbf{q}}}\right)^{\mathrm{T}} = \begin{bmatrix} (m_1 + m_2) l_1^2 & m_2 l_1 l_2 \cos(\theta_1 - \theta_2) \\ m_2 l_1 l_2 \cos(\theta_1 - \theta_2) & m_2 l_2^2 \end{bmatrix} \begin{bmatrix} \ddot\theta_1 \\ \ddot\theta_2 \end{bmatrix} - \begin{bmatrix} m_2 l_1 l_2 \dot\theta_2(\dot\theta_1 - \dot\theta_2)\sin(\theta_1 - \theta_2) \\ m_2 l_1 l_2 \dot\theta_1(\dot\theta_1 - \dot\theta_2)\sin(\theta_1 - \theta_2) \end{bmatrix}$$

One can also show that

$$\left(\frac{\partial T}{\partial \mathbf{q}}\right)^{\mathrm{T}} = (T_{\mathbf{q}})^{\mathrm{T}} = \begin{bmatrix} -m_2 l_1 l_2 \dot\theta_1 \dot\theta_2 \sin(\theta_1 - \theta_2) \\ m_2 l_1 l_2 \dot\theta_1 \dot\theta_2 \sin(\theta_1 - \theta_2) \end{bmatrix}$$

Substituting into Lagrange's equation of motion given by Eq. 42 or 43, one obtains the differential equations of motion of this system.

2.5 STRAIN ENERGY

The generalized elastic forces of the springs, which were obtained in the preceding sections using the virtual work, can also be obtained using the strain energy expression. For example, in the single degree of freedom mass–spring system shown in Fig. 3, the strain energy due to the deformation of the spring is given by

$$U = \tfrac{1}{2} k x^2 \tag{2.45}$$

By using this expression for the strain energy, the generalized force of the spring can simply be defined as

$$Q_s = -\frac{\partial U}{\partial x} = -kx \tag{2.46}$$

For this system, one can write Lagrange's equation of motion in the following form:

$$\frac{d}{dt}\left(\frac{\partial T}{\partial \dot{x}}\right) - \left(\frac{\partial T}{\partial x}\right) = \bar{Q} + Q_s \tag{2.47}$$

where $\bar{Q}$ is the vector of all generalized forces excluding the spring force. Using Eq. 46, Eq. 47 can be written as

$$\frac{d}{dt}\left(\frac{\partial T}{\partial \dot{x}}\right) - \left(\frac{\partial T}{\partial x}\right) + \left(\frac{\partial U}{\partial x}\right) = \bar{Q} \tag{2.48}$$

In general, one may write Lagrange's equation in the case of n generalized coordinates in the following alternate form

$$\frac{d}{dt}\left(\frac{\partial T}{\partial \dot{q}_j}\right) - \frac{\partial T}{\partial q_j} + \frac{\partial U}{\partial q_j} = \bar{Q}_j, \qquad j = 1, 2, \ldots, n \tag{2.49}$$

where $U = U(q_1, q_2, \ldots, q_n)$ is the strain energy function that depends on the system generalized coordinates. Keeping in mind that the differentiation of the scalar strain energy function with respect to the generalized coordinate vector $\mathbf{q} = [q_1 \ q_2 \ q_3 \ \cdots \ q_n]^{\mathrm{T}}$ leads to a row vector, that is,

$$\frac{\partial U}{\partial \mathbf{q}} = U_{\mathbf{q}} = \begin{bmatrix} \dfrac{\partial U}{\partial q_1} & \dfrac{\partial U}{\partial q_2} & \cdots & \dfrac{\partial U}{\partial q_n} \end{bmatrix}, \tag{2.50}$$

Eq. 49 can be written using vector notation as

$$\frac{d}{dt}\left(\frac{\partial T}{\partial \dot{\mathbf{q}}}\right)^{\mathrm{T}} - \left(\frac{\partial T}{\partial \mathbf{q}}\right)^{\mathrm{T}} + \left(\frac{\partial U}{\partial \mathbf{q}}\right)^{\mathrm{T}} = \bar{\mathbf{Q}} \tag{2.51}$$

where $\bar{\mathbf{Q}} = [\bar{Q}_1 \ \bar{Q}_2 \ \cdots \ \bar{Q}_n]^{\mathrm{T}}$ is the vector of all generalized forces excluding the elastic forces which are accounted for using the strain energy function U.

We have previously shown that the kinetic energy function T can be expressed as a quadratic form in the vector of system generalized velocities. One can also show that the strain energy function U of linear systems can be expressed as a quadratic form in the vector of generalized coordinates. In fact U can be written as

$$U = \tfrac{1}{2}\mathbf{q}^{\mathrm{T}}\mathbf{K}\mathbf{q} \tag{2.52}$$

where $\mathbf{K}$ is the symmetric stiffness matrix of the system given by

$$\mathbf{K} = \begin{bmatrix} k_{11} & & & \\ k_{21} & k_{22} & \text{symmetric} & \\ \vdots & \vdots & \ddots & \\ k_{n1} & k_{n2} & \cdots & k_{nn} \end{bmatrix} \tag{2.53}$$

Consequently, the term $\partial U/\partial \mathbf{q}$ in Lagrange's equation of Eq. 51 is the following row vector:

$$\frac{\partial U}{\partial \mathbf{q}} = \mathbf{q}^{\mathrm{T}}\mathbf{K} = \left[\sum_{j=1}^{n} k_{j1} q_j \quad \sum_{j=1}^{n} k_{j2} q_j \quad \cdots \quad \sum_{j=1}^{n} k_{jn} q_j\right]$$

Since the stiffness matrix is symmetric, that is, $\mathbf{K} = \mathbf{K}^{\mathrm{T}}$, one has

$$\left(\frac{\partial U}{\partial \mathbf{q}}\right)^{\mathrm{T}} = \mathbf{Kq} = \begin{bmatrix} \sum_{j=1}^{n} k_{1j} q_j \\ \sum_{j=1}^{n} k_{2j} q_j \\ \vdots \\ \sum_{j=1}^{n} k_{nj} q_j \end{bmatrix} \tag{2.54}$$

where $k_{ij} = k_{ji}$. It is important, however, to emphasize that, while the kinetic energy is always a positive-definite quadratic form in the velocities, the strain energy can be a positive-semidefinite quadratic form in the coordinates. This situation occurs when the system has rigid-body degrees of freedom. The motion of such a system can occur without any change in the potential energy. The stiffness matrix of such a system is always singular and the system is said to be *semidefinite*. The dynamics of the semidefinite systems will be discussed in more detail in the following chapter.

Example 2.9

Using the strain energy function, determine the stiffness matrix of the two degree of freedom mass–spring system shown in Fig. 4.

Solution. The strain energy of the spring k_1 is

$$U_1 = \tfrac{1}{2}k_1 x_1^2 = \tfrac{1}{2}k_1 q_1^2$$

The strain energy of the spring k_2 is

$$U_2 = \tfrac{1}{2}k_2(x_2 - x_1)^2 = \tfrac{1}{2}k_2(q_2 - q_1)^2$$

The total strain energy of the system is

$$\begin{aligned} U &= U_1 + U_2 = \tfrac{1}{2}k_1 q_1^2 + \tfrac{1}{2}k_2(q_2 - q_1)^2 \\ &= \tfrac{1}{2}(k_1 + k_2)q_1^2 - k_2 q_1 q_2 + \tfrac{1}{2}k_2 q_2^2 \end{aligned}$$

which is a quadratic form in the generalized coordinates q_1 and q_2 and can be written as

$$\begin{aligned} U &= \tfrac{1}{2}[q_1 \quad q_2]\begin{bmatrix} (k_1 + k_2) & -k_2 \\ -k_2 & k_2 \end{bmatrix}\begin{bmatrix} q_1 \\ q_2 \end{bmatrix} \\ &= \tfrac{1}{2}\mathbf{q}^{\mathrm{T}}\mathbf{Kq} \end{aligned}$$

where **K** is the stiffness matrix defined as

$$\mathbf{K} = \begin{bmatrix} k_1 + k_2 & -k_2 \\ -k_2 & k_2 \end{bmatrix}$$

which is the same stiffness matrix obtained in Example 6 using the virtual work.

2.6 HAMILTON'S PRINCIPLE

Another alternate approach for deriving the differential equations of motion from scalar energy quantities is *Hamilton's principle* which states that

$$\delta \int_{t_1}^{t_2} (T - V)\, dt + \int_{t_1}^{t_2} \delta W_{nc}\, dt = 0 \tag{2.55}$$

where T is the system kinetic energy, V is the system potential energy, and δW_{nc} is the virtual work of the nonconservative forces. Hamilton's principle is sometimes written in the following form:

$$\delta \int_{t_1}^{t_2} L\, dt + \int_{t_1}^{t_2} \delta W_{nc}\, dt = 0 \tag{2.56}$$

where L is called the *Lagrangian* and defined as

$$L = T - V \tag{2.57}$$

Hamilton's principle of Eq. 55 or its alternate form of Eq. 56 states that the variation of the kinetic and potential energy plus the line integral of the virtual work done by the nonconservative forces during any time interval between t_1 and t_2 must be equal to zero.

The potential energy V is defined as the strain energy minus the work done by the conservative forces, that is,

$$V = U - W_c \tag{2.58}$$

where W_c is the work done by the conservative forces. Excluding the forces that contribute to the strain energy U, one may define the virtual work of all other forces, conservative and nonconservative, as

$$\delta W = \delta W_c + \delta W_{nc} \tag{2.59}$$

Substituting Eqs. 58 and 59 into Eq. 55 leads to the following alternate form of Hamilton's principle:

$$\delta \int_{t_1}^{t_2} (T - U)\, dt + \int_{t_1}^{t_2} \delta W\, dt = 0 \tag{2.60}$$

When applying Hamilton's principle it is assumed that the system coordinates are specified at the two end points t_1 and t_2, that is,

$$\delta \mathbf{q}(t_1) = \delta \mathbf{q}(t_2) = \mathbf{0} \tag{2.61}$$

or

$$
\begin{aligned}
\delta q_1(t_1) &= \delta q_1(t_2) = 0 \\
\delta q_2(t_1) &= \delta q_2(t_2) = 0 \\
&\vdots \\
\delta q_n(t_1) &= \delta q_n(t_2) = 0
\end{aligned}
\tag{2.62}
$$

Relationship Between Hamilton's Principle and Lagrange's Equation The use of Hamilton's principle is equivalent to the application of Lagrange's equation. In fact Hamilton's principle can be used to derive Lagrange's equation of motion. To this end, we write

$$\delta \int_{t_1}^{t_2} T\,dt = \int_{t_1}^{t_2} \delta T\,dt = \int_{t_1}^{t_2} \left(\frac{\partial T}{\partial \mathbf{q}} \delta \mathbf{q} + \frac{\partial T}{\partial \dot{\mathbf{q}}} \delta \dot{\mathbf{q}} \right) dt \tag{2.63}$$

Using integration by parts, the second term in the preceding equation yields

$$\int_{t_1}^{t_2} \left(\frac{\partial T}{\partial \dot{\mathbf{q}}} \right) \delta \dot{\mathbf{q}}\,dt = \left(\frac{\partial T}{\partial \dot{\mathbf{q}}} \right) \delta \mathbf{q} \Big|_{t_1}^{t_2} - \int_{t_1}^{t_2} \frac{d}{dt} \left(\frac{\partial T}{\partial \dot{\mathbf{q}}} \right) \delta \mathbf{q}\,dt$$

Using the assumption of Eq. 61, the preceding equation reduces to

$$\int_{t_1}^{t_2} \left(\frac{\partial T}{\partial \dot{\mathbf{q}}} \right) \delta \dot{\mathbf{q}}\,dt = -\int_{t_1}^{t_2} \frac{d}{dt} \left(\frac{\partial T}{\partial \dot{\mathbf{q}}} \right) \delta \mathbf{q}\,dt \tag{2.64}$$

Substituting Eq. 64 into Eq. 63 yields

$$\delta \int_{t_1}^{t_2} T\,dt = \int_{t_1}^{t_2} \left[\frac{\partial T}{\partial \mathbf{q}} - \frac{d}{dt} \left(\frac{\partial T}{\partial \dot{\mathbf{q}}} \right) \right] \delta \mathbf{q}\,dt \tag{2.65}$$

Similarly,

$$\delta \int_{t_1}^{t_2} U\,dt = \int_{t_1}^{t_2} \left(\frac{\partial U}{\partial \mathbf{q}} \right) \delta \mathbf{q}\,dt \tag{2.66}$$

$$\int_{t_1}^{t_2} \delta W\,dt = \int_{t_1}^{t_2} \bar{\mathbf{Q}}^{\mathrm{T}} \delta \mathbf{q}\,dt \tag{2.67}$$

where $\bar{\mathbf{Q}}$ is the vector of generalized forces.

Substituting Eqs. 65–67 into Hamilton's principle of Eq. 60 yields

$$\int_{t_1}^{t_2} \left[\frac{\partial T}{\partial \mathbf{q}} - \frac{d}{dt} \left(\frac{\partial T}{\partial \dot{\mathbf{q}}} \right) - \frac{\partial U}{\partial \mathbf{q}} + \bar{\mathbf{Q}}^{\mathrm{T}} \right] \delta \mathbf{q}\,dt = 0 \tag{2.68}$$

If the coordinates $q_1, q_2, \ldots, q_n$ are independent, the integrand in Eq. 68 must be identically zero. This leads, after rearranging terms, to the following familiar form of Lagrange's equation:

$$\frac{d}{dt} \left(\frac{\partial T}{\partial \dot{\mathbf{q}}} \right)^{\mathrm{T}} - \left(\frac{\partial T}{\partial \mathbf{q}} \right)^{\mathrm{T}} + \left(\frac{\partial U}{\partial \mathbf{q}} \right)^{\mathrm{T}} = \bar{\mathbf{Q}} \tag{2.69}$$

Example 2.10

Use Hamilton's principle to derive the differential equation of motion of the single degree of freedom system of Example 2.

Solution. The system kinetic energy is

$$T = \tfrac{1}{2}m\dot{x}^2$$

The system strain energy is

$$U = \tfrac{1}{2}kx^2$$

The virtual work of all forces, excluding the spring force, that act on this system is

$$\delta W = F(t)\delta x - c\dot{x}\delta x$$

Using Hamilton's principle of Eq. 60, one has

$$\delta \int_{t_1}^{t_2} (T - U)\, dt + \int_{t_1}^{t_2} \delta W\, dt = 0$$

where

$$\delta \int_{t_1}^{t_2} (T - U)\, dt = \delta \int_{t_1}^{t_2} (\tfrac{1}{2}m\dot{x}^2 - \tfrac{1}{2}kx^2)\, dt$$

$$= \int_{t_1}^{t_2} (m\dot{x}\delta\dot{x} - kx\delta x)\, dt$$

which, by integrating by parts the first term, leads to

$$\delta \int_{t_1}^{t_2} (T - U)\, dt = m\dot{x}\delta x \Big|_{t_1}^{t_2} - \int_{t_1}^{t_2} (m\ddot{x}\delta x + kx\delta x)\, dt$$

Substituting this expression and the expression of the virtual work into Hamilton's principle and using Eq. 61 lead to

$$\int_{t_1}^{t_2} (m\ddot{x} + kx - F(t) + c\dot{x})\delta x\, dt = 0$$

Since δx is an independent coordinate, the above equation leads to

$$m\ddot{x} + c\dot{x} + kx = F(t)$$

which is the same equation of motion obtained in Example 5.

Example 2.11

Use Hamilton's principle to derive the differential equations of motion of the system shown in Fig. 9.

Solution. The kinetic and strain energies of the system are given by

$$T = \tfrac{1}{2}m_1\dot{x}_1^2 + \tfrac{1}{2}m_2\dot{x}_2^2$$

$$U = \tfrac{1}{2}k_1x_1^2 + \tfrac{1}{2}k_2(x_2 - x_1)^2$$

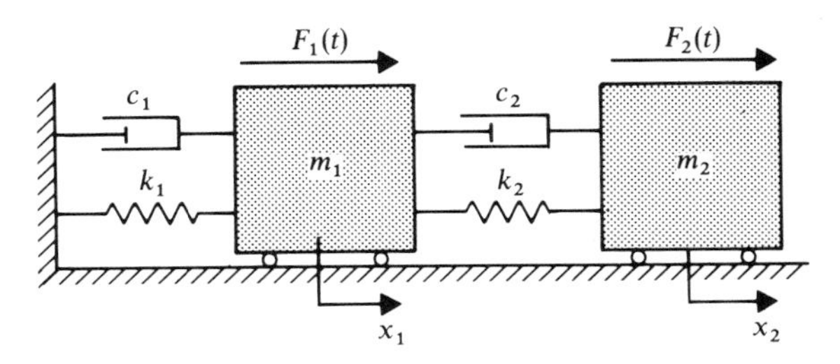

FIG. 2.9. Two degree of freedom system.

The virtual work of the damping and external forces is given by

$$\delta W = -c_1\dot{x}_1\delta x_1 - c_2(\dot{x}_2 - \dot{x}_1)\delta(x_2 - x_1) + F_1\delta x_1 + F_2\delta x_2$$

The variation of the time integral of $(T - U)$ can be obtained as

$$\delta\int_{t_1}^{t_2}(T - U)\,dt = \delta\int_{t_1}^{t_2}\left[\tfrac{1}{2}m_1\dot{x}_1^2 + \tfrac{1}{2}m_2\dot{x}_2^2 - \tfrac{1}{2}k_1x_1^2 - \tfrac{1}{2}k_2(x_2 - x_1)^2\right]dt$$

$$= \int_{t_1}^{t_2}\left[m_1\dot{x}_1\delta\dot{x}_1 + m_2\dot{x}_2\delta\dot{x}_2 - k_1x_1\delta x_1 - k_2(x_2 - x_1)\delta(x_2 - x_1)\right]dt$$

By integrating by parts and using the assumption of Hamilton's principle that the displacements are specified at the end points t_1 and t_2, one obtains

$$\delta\int_{t_1}^{t_2}(T - U)\,dt = \int_{t_1}^{t_2}[-m_1\ddot{x}_1\delta x_1 - m_2\ddot{x}_2\delta x_2 - k_1x_1\delta x_1 - k_2(x_2 - x_1)\delta(x_2 - x_1)]\,dt$$

By rearranging the terms, one obtains

$$\delta\int_{t_1}^{t_2}(T - U)\,dt = -\int_{t_1}^{t_2}\{[m_1\ddot{x}_1 + (k_1 + k_2)x_1 - k_2x_2]\delta x_1 + [m_2\ddot{x}_2 + k_2x_2 - k_2x_1]\delta x_2\}\,dt$$

Using this equation and the expression of the virtual work δW, Hamilton's principle yields

$$\delta\int_{t_1}^{t_2}(T - U)\,dt + \int_{t_1}^{t_2}\delta W\,dt = 0$$

That is,

$$-\int_{t_1}^{t_2}\{[m_1\ddot{x}_1 + (k_1 + k_2)x_1 - k_2x_2]\delta x_1 + [m_2\ddot{x}_2 + k_2x_2 - k_2x_1]\delta x_2\}\,dt$$

$$+\int_{t_1}^{t_2}[-c_1\dot{x}_1\delta x_1 - c_2(\dot{x}_2 - \dot{x}_1)\delta(x_2 - x_1) + F_1\delta x_1 + F_2\delta x_2]\,dt = 0$$

which yields, after rearranging terms,

$$\int_{t_1}^{t_2} [m_1\ddot{x}_1 + (c_1 + c_2)\dot{x}_1 - c_2\dot{x}_2 + (k_1 + k_2)x_1 - k_2x_2 - F_1]\delta x_1\, dt$$

$$+ \int_{t_1}^{t_2} [m_2\ddot{x}_2 + c_2\dot{x}_2 - c_2\dot{x}_1 + k_2x_2 - k_2x_1 - F_2]\delta x_2\, dt = 0$$

Since x_1 and x_2 are assumed to be independent coordinates, the preceding equation leads to the following two differential equations:

$$m_1\ddot{x}_1 + (c_1 + c_2)\dot{x}_1 - c_2\dot{x}_2 + (k_1 + k_2)x_1 - k_2x_2 = F_1$$

$$m_2\ddot{x}_2 + c_2\dot{x}_2 - c_2\dot{x}_1 + k_2x_2 - k_2x_1 = F_2$$

2.7 CONSERVATION THEOREMS

In this section, a special case of the preceding development is considered. This is the case of conservative systems in which all the forces acting on the system can be derived from the potential function V. In this case

$$\delta W_{nc} = 0 \tag{2.70}$$

Substituting this equation into Hamilton's principle of Eq. 55, one obtains

$$\delta \int_{t_1}^{t_2} (T - V)\, dt = 0 \tag{2.71}$$

or equivalently

$$\delta \int_{t_1}^{t_2} L\, dt = 0 \tag{2.72}$$

where $L = T - V$ is the Lagrangian of the system.

Clearly, for a conservative system Lagrange's equation reduces to

$$\frac{d}{dt}\left(\frac{\partial L}{\partial \dot{q}_j}\right) - \frac{\partial L}{\partial q_j} = 0, \qquad j = 1, 2, \ldots, n \tag{2.73}$$

or

$$\frac{d}{dt}\left(\frac{\partial L}{\partial \dot{q}_j}\right) = \frac{\partial L}{\partial q_j}, \qquad j = 1, 2, \ldots, n \tag{2.74}$$

Observe that the total time derivative of the Lagrangian L is given by

$$\frac{dL}{dt} = \sum_{j=1}^{n} \left(\frac{\partial L}{\partial \dot{q}_j}\ddot{q}_j + \frac{\partial L}{\partial q_j}\dot{q}_j\right) \tag{2.75}$$

Substituting Eq. 74 into Eq. 75, one obtains

$$\frac{dL}{dt} = \sum_{j=1}^{n} \left[\frac{\partial L}{\partial \dot{q}_j} \ddot{q}_j + \frac{d}{dt}\left(\frac{\partial L}{\partial \dot{q}_j}\right) \dot{q}_j \right]$$
$$= \sum_{j=1}^{n} \frac{d}{dt}\left(\frac{\partial L}{\partial \dot{q}_j} \dot{q}_j\right) \tag{2.76}$$

This equation can be rewritten as

$$\frac{dL}{dt} - \frac{d}{dt} \sum_{j=1}^{n} \frac{\partial L}{\partial \dot{q}_j} \dot{q}_j = 0$$

or equivalently

$$\frac{d}{dt}\left(L - \sum_{j=1}^{n} \frac{\partial L}{\partial \dot{q}_j} \dot{q}_j\right) = 0 \tag{2.77}$$

Since the potential energy V is independent of the velocity, one has

$$\frac{\partial L}{\partial \dot{q}_j} = \frac{\partial T}{\partial \dot{q}_j} \tag{2.78}$$

Substituting this equation into Eq. 77, one gets

$$\frac{d}{dt}\left(L - \sum_{j=1}^{n} \frac{\partial T}{\partial \dot{q}_j} \dot{q}_j\right) = 0 \tag{2.79}$$

That is,

$$L - \sum_{j=1}^{n} \frac{\partial T}{\partial \dot{q}_j} \dot{q}_j = -H \tag{2.80}$$

where H is a constant called the *Hamiltonian*. Equation 80 is valid for any conservative single or multi-degree of freedom system. The Hamiltonian H can be written in a more convenient form using the following identity:

$$\sum_{j=1}^{n} \frac{\partial T}{\partial \dot{q}_j} \dot{q}_j = 2T \tag{2.81}$$

Substituting this equation into Eq. 80 and using the fact that $L = T - V$, one obtains

$$H = -(T - V - 2T) = T + V \tag{2.82}$$

which implies that the Hamiltonian is equal to the total energy of the system. Since, for conservative systems, the Hamiltonian H is constant, one has

$$\frac{dH}{dt} = \frac{d}{dt}(T + V) = 0 \tag{2.83}$$

This equation can be used to develop the dynamic equations of motion of conservative single and multi-degree of freedom systems, as demonstrated by the following examples.

Example 2.12

Use the method of conservation of energy to derive the equation of free vibration of the single degree of freedom system of Example 2 assuming that the damping coefficient is equal to zero.

Solution. The Hamiltonian of the system is given by

$$H = T + V = \tfrac{1}{2}m\dot{x}^2 + \tfrac{1}{2}kx^2$$

Applying Eq. 83, one obtains

$$\frac{dH}{dt} = m\dot{x}\ddot{x} + kx\dot{x} = 0$$

This equation can be rewritten as

$$(m\ddot{x} + kx)\dot{x} = 0$$

Since $\dot{x}$ is not equal to zero for all values of t, one must have

$$m\ddot{x} + kx = 0$$

which is the equation of undamped free vibration of the single degree of freedom system.

Example 2.13

Use the method of conservation of energy to derive the differential equations of the free undamped vibration of the system in Example 9.

Solution. The kinetic energy of this system is

$$T = \tfrac{1}{2}m_1\dot{x}_1^2 + \tfrac{1}{2}m_2\dot{x}_2^2$$

and the strain energy is given by

$$U = V = \tfrac{1}{2}k_1x_1^2 + \tfrac{1}{2}k_2(x_2 - x_1)^2$$

The Hamiltonian of the system is

$$H = T + V = \tfrac{1}{2}m_1\dot{x}_1^2 + \tfrac{1}{2}m_2\dot{x}_2^2 + \tfrac{1}{2}k_1x_1^2 + \tfrac{1}{2}k_2(x_2 - x_1)^2$$

The application of Eq. 83 leads to

$$\frac{dH}{dt} = m_1\dot{x}_1\ddot{x}_1 + m_2\dot{x}_2\ddot{x}_2 + k_1x_1\dot{x}_1 + k_2(x_2 - x_1)(\dot{x}_2 - \dot{x}_1)$$

$$= 0$$

This equation can be rewritten as

$$[m_1\ddot{x}_1 + k_1x_1 - k_2(x_2 - x_1)]\dot{x}_1 + [m_2\ddot{x}_2 + k_2(x_2 - x_1)]\dot{x}_2 = 0$$

Since $\dot{x}_1$ and $\dot{x}_2$ are independent velocities, their coefficients in the above equation must be equal to zero. This leads to the following two differential equations of motion for the undamped two degree of freedom system:

$$m_1 \ddot{x}_1 + (k_1 + k_2)x_1 - k_2 x_2 = 0$$
$$m_2 \ddot{x}_2 + k_2 x_2 - k_1 x_1 = 0$$

Nonlinear Systems Conservation theorems can also be applied to nonlinear constrained multi-body systems. In order to demonstrate that, we consider the multi-body system shown in Fig. 10, which was considered in the preceding chapter. The system consists of two homogeneous circular cylinders, each of mass m and centroidal mass moment of inertia I, and a connecting rod AB of mass m_r and length l. The cylinders, which have radius r, are assumed to roll without slipping. Because of the rolling conditions, the velocities of the centers of mass of the cylinders are equal and given by

$$\mathbf{v}_O = \mathbf{v}_C = [-r\dot{\theta} \quad 0]^{\mathrm{T}}$$

The velocity of the center of mass of the connecting rod is equal to the absolute velocities of points A and B since the angular velocity of this rod is equal to zero. These velocities are defined as

$$\mathbf{v}_r = \mathbf{v}_B = \mathbf{v}_A = \mathbf{v}_O + \mathbf{v}_{AO} = \mathbf{v}_O + \boldsymbol{\omega} \times \mathbf{r}_{AO} = \begin{bmatrix} -(r - a\cos\theta)\dot{\theta} \\ a\dot{\theta}\sin\theta \end{bmatrix}$$

where $\mathbf{v}_{AO}$ is the velocity of point A with respect to point O, $\mathbf{v}_r$ is the absolute velocity of the center of mass of the connecting rod, $\boldsymbol{\omega}$ is the angular velocity vector of the cylinders, and $\mathbf{r}_{AO}$ is the position vector of point A with respect to point O defined as

$$\mathbf{r}_{AO} = [a\sin\theta \quad -a\cos\theta]^{\mathrm{T}}$$

The kinetic and potential energies of the system are

$$T = 2\left(\frac{1}{2}m\mathbf{v}_O^{\mathrm{T}}\mathbf{v}_O + \frac{1}{2}I\dot{\theta}^2\right) + \frac{1}{2}m_r\mathbf{v}_r^{\mathrm{T}}\mathbf{v}_r$$

$$V = m_r g a(1 - \cos\theta)$$

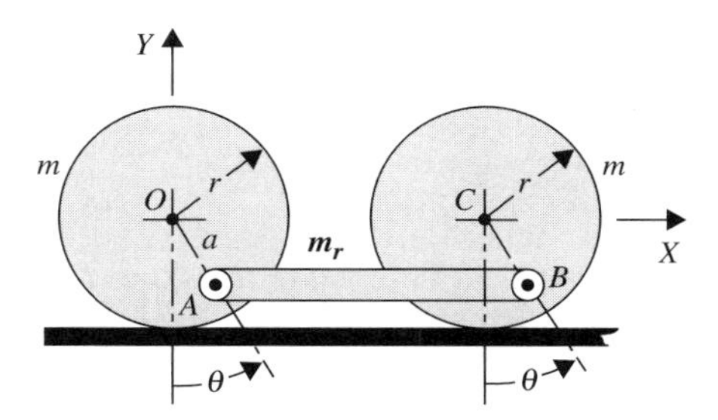

FIG. 2.10. Multi-body system.

Using the velocity expressions, the Hamiltonian can be written as

$$H = T + V = \left\{ I + mr^2 + \frac{1}{2} m_r (r^2 + a^2 - 2\,ar \cos\theta) \right\} \dot{\theta}^2 + m_r g a (1 - \cos\theta)$$

It follows that

$$\frac{dH}{dt} = 2\left\{ I + mr^2 + \frac{1}{2} m_r (r^2 + a^2 - 2ra \cos\theta) \right\} \ddot{\theta}\dot{\theta}$$
$$+ m_r r a \dot{\theta}^3 \sin\theta + m_r g a \dot{\theta} \sin\theta = 0$$

Since for a cylinder $I = mr^2/2$, the preceding equation leads to

$$2\left\{ \frac{3}{2} mr^2 + \frac{1}{2} m_r (r^2 + a^2 - 2ra \cos\theta) \right\} \ddot{\theta} + m_r r a \dot{\theta}^2 \sin\theta + m_r g a \sin\theta = 0$$

which is the same nonlinear differential equation obtained in the preceding chapter by applying D'Alembert's principle.

2.8 CONCLUDING REMARKS

In this chapter, a brief introduction to the subject of Lagrangian dynamics is presented, and the important concepts of the generalized coordinates, virtual work, and the generalized forces are introduced. These concepts are then used to derive Lagrange's equation from Newton's second law. The final form of Lagrange's equation is presented in terms of scalar energy quantities such as the strain and kinetic energies. The quadratic forms of the kinetic and strain energies are also examined and used to define the system mass and stiffness matrices. Hamilton's principle which represents an alternate approach for deriving the dynamic equations of motion using scalar quantities was also discussed. It was also shown that Lagrange's equation or Hamilton's principle can be used to develop conservation theorems that can be used to derive the differential equations of motion of conservative systems.

The development of Lagrange's equation using the vector equation of Newton's second law given by Eq. 15, consists of several steps. At the end of each intermediate step, the resulting equations can be used to derive the dynamic equations of the mechanical system. Equation 16, for example, is the well-known *principle of virtual work in dynamics*. This important equation can be written as

$$\delta W_i = \delta W_e \tag{2.84}$$

where δW_i is the virtual work of the inertia forces and δW_e is the virtual work of the applied forces. The virtual work of the workless constraint forces is equal to zero and, consequently, such forces do not appear in Eq. 84. Equation 84, even though it is an intermediate step in deriving Lagrange's equation, represents a powerful tool which can be used to develop the equations of motion of mechanical systems. The use of Eq. 84 can be demonstrated by

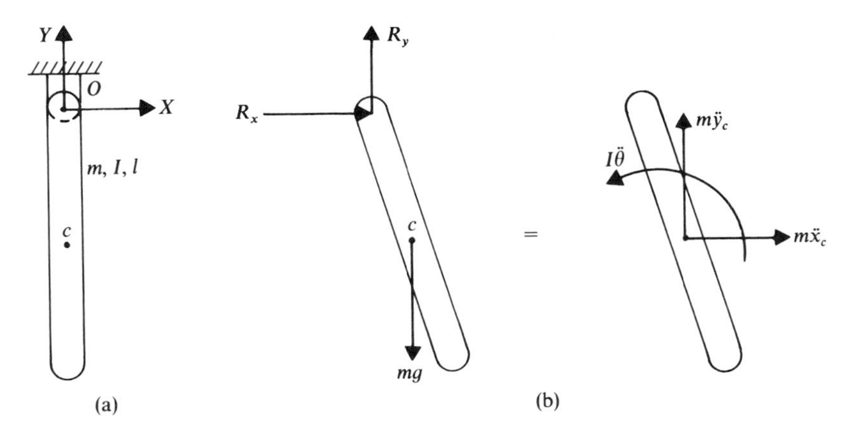

FIG. 2.11. Application of the principle of virtual work in dynamics.

considering the simple pendulum shown in Fig. 11(a). The free-body diagram as well as the inertia-force diagram are shown in Fig. 11(b). The position of the center of mass of the rod can be expressed in terms of the degree of freedom of the system θ as

$$x_c = \frac{l}{2} \sin \theta, \qquad y_c = -\frac{l}{2} \cos \theta$$

It follows that

$$\delta x_c = \frac{l}{2} \cos \theta \delta\theta, \qquad \delta y_c = \frac{l}{2} \sin \theta \delta\theta$$

$$\dot{x}_c = \frac{l\dot{\theta}}{2} \cos \theta, \qquad \dot{y}_c = \frac{l\dot{\theta}}{2} \sin \theta$$

The components of the acceleration of the center of mass are

$$\ddot{x}_c = \frac{l\ddot{\theta}}{2} \cos \theta - \frac{l}{2}\dot{\theta}^2 \sin \theta$$

$$\ddot{y}_c = \frac{l\ddot{\theta}}{2} \sin \theta + \frac{l}{2}\dot{\theta}^2 \cos \theta$$

The virtual work of the inertia forces and moments can then be written as

$$\begin{aligned} \delta W_i &= m\ddot{x}_c \delta x_c + m\ddot{y}_c \delta y_c + I\ddot{\theta}\delta\theta \\ &= m\left(\frac{l}{2}\ddot{\theta} \cos \theta - \frac{l}{2}\dot{\theta}^2 \sin \theta\right)\left(\frac{l}{2} \cos \theta\delta\theta\right) \\ &\quad + m\left(\frac{l}{2}\ddot{\theta} \sin \theta + \frac{l}{2}\dot{\theta}^2 \cos \theta\right)\left(\frac{l}{2} \sin \theta\delta\theta\right) \\ &\quad + I\ddot{\theta}\delta\theta \end{aligned}$$

This equation can be simplified and rewritten as

$$\delta W_i = m\left(\frac{l}{2}\right)^2 \ddot{\theta}(\cos^2\theta + \sin^2\theta)\delta\theta + I\ddot{\theta}\delta\theta$$
$$= \left[m\left(\frac{l}{2}\right)^2 + I\right]\ddot{\theta}\delta\theta$$

The virtual work of the external forces is given by

$$\delta W_e = -mg\delta y_c = -mg\frac{l}{2}\sin\theta\delta\theta$$

Note that the virtual work of the reactions R_x and R_y is equal to zero since the virtual change in the position vector of point O is equal to zero. Therefore, the application of Eq. 84 leads to

$$\left[m\left(\frac{l}{2}\right)^2 + I\right]\ddot{\theta}\delta\theta = -mg\frac{l}{2}\sin\theta\delta\theta$$

which leads to

$$I_O\ddot{\theta} + mg\frac{l}{2}\sin\theta = 0 \tag{2.85}$$

where

$$I_O = I + m\left(\frac{l}{2}\right)^2 \tag{2.86}$$

Equation 85 can also be obtained by the direct application of Newton's second law or Lagrange's equation. It is derived in this section using the principle of virtual work in dynamics. Equation 86 is the *parallel axis theorem* which states that the moment of inertia about an arbitrary axis is equal to the moment of inertia about a parallel axis passing through the center of mass plus the total mass of the body multiplied by the square of the distance between the two axes. Equation 86 can also be derived using the kinetic energy of the system. In the example shown in Fig. 11, the kinetic energy is

$$T = \tfrac{1}{2}m\dot{x}_c^2 + \tfrac{1}{2}m\dot{y}_c^2 + \tfrac{1}{2}I\dot{\theta}^2 \tag{2.87}$$

If $\dot{x}_c$ and $\dot{y}_c$ are expressed in terms of the generalized velocities, Eq. 87 yields

$$T = \tfrac{1}{2}m\left(\frac{l\dot{\theta}}{2}\cos\theta\right)^2 + \tfrac{1}{2}m\left(\frac{l\dot{\theta}}{2}\sin\theta\right)^2 + \tfrac{1}{2}I\dot{\theta}^2$$
$$= \tfrac{1}{2}m\dot{\theta}^2\left(\frac{l}{2}\right)^2[\cos^2\theta + \sin^2\theta] + \tfrac{1}{2}I\dot{\theta}^2$$

Since $\cos^2\theta + \sin^2\theta = 1$, the preceding equation leads to

$$T = \tfrac{1}{2}\left[m\left(\frac{l}{2}\right)^2 + I\right]\dot{\theta}^2$$
$$= \tfrac{1}{2}I_O\dot{\theta}^2$$

where I_O is defined by Eq. 86.

Problems

2.1. The single degree of freedom system shown in Fig. P1 consists of a mass m, a damper c, and two springs k_1 and k_2. Develop expressions for the kinetic and strain energies and the virtual work of the damping force. Use these energy and work expressions to develop the equations of free vibration of the system.

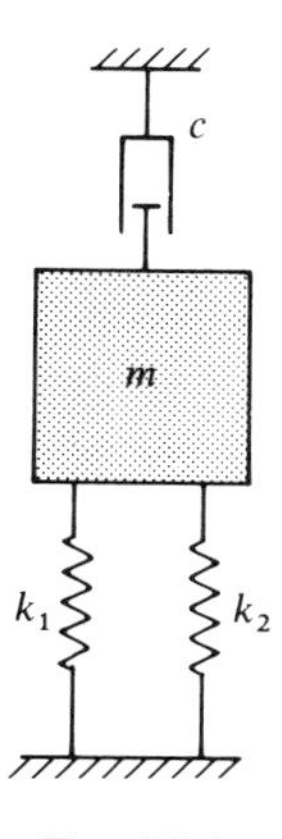

FIG. P2.1

2.2. In the case of free vibration of the system shown in Fig. P1, show that the time rate of change of the sum of the kinetic and strain energies is equal to the negative of the damping coefficient multiplied by the square of the velocity, that is,

$$\frac{d}{dt}(T + U) = -c\dot{x}^2$$

where x is the displacement of the mass.

2.3. Derive the differential equation of motion of the system shown in Fig. P1 by using Newton's second law, and also by using Lagrange's equation.

2.4. Show that the kinetic energy of a system of n particles can be written as

$$T = \frac{l}{2}mv_c^2 + \sum_{i=1}^{n}\frac{l}{2}m_i v_{ic}^2$$

where m is the total mass of the system of particles, v_c is the velocity of the center of mass, m_i is the mass of the particle i, and v_{ic} is the velocity of the particle i with respect to the center of mass.

2.5. In the case of a rigid body, derive the *principle of work and energy* which states that the change in the kinetic energy of a rigid body is equal to the work of the forces acting on the body.

2.6. Use the principle of work and energy obtained in Problem 5 to derive the equation of motion of the system shown in Fig. P1. Compare the use of this principle with the result presented in Problem 2.

2.7. Obtain expressions for the kinetic and potential energies and the virtual work for the system shown in Fig. P2 in the following two cases:

(a) small angular rotations,
(b) finite angular rotations.

Derive the equations of motion in the two cases using the principle of virtual work in dynamics.

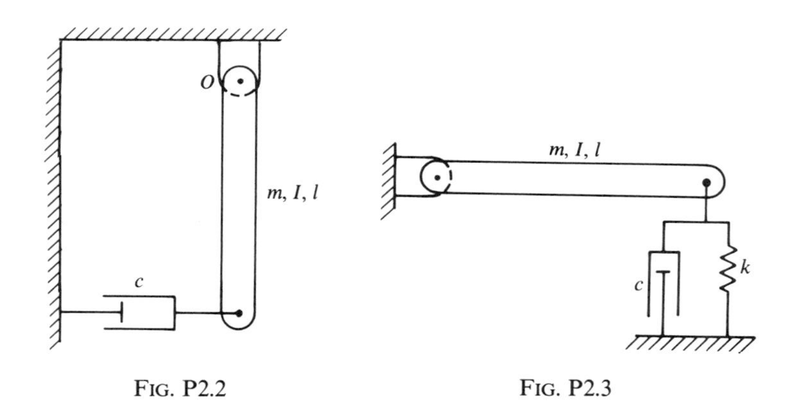

FIG. P2.2 FIG. P2.3

2.8. Obtain the differential equation of free vibration of the single degree of freedom system shown in Fig. P3. Assume small oscillations.

2.9. In Problem 8, determine the time rate of change of the sum of the kinetic and strain energies.

2.10. The system shown in Fig. P4 consists of a rigid massless bar of length $l_1 + l_2$, and two masses m_1 and m_2 which are rigidly attached to the bar. Obtain the differential equation of free vibration of this system by using Lagrange's equation. Use the assumption of small oscillations.

2.11. The virtual work of the inertia forces of a rigid body can be written as

$$\delta W_i = \int_V \rho \ddot{\mathbf{r}}^{\mathrm{T}} \delta \mathbf{r} \, dV$$

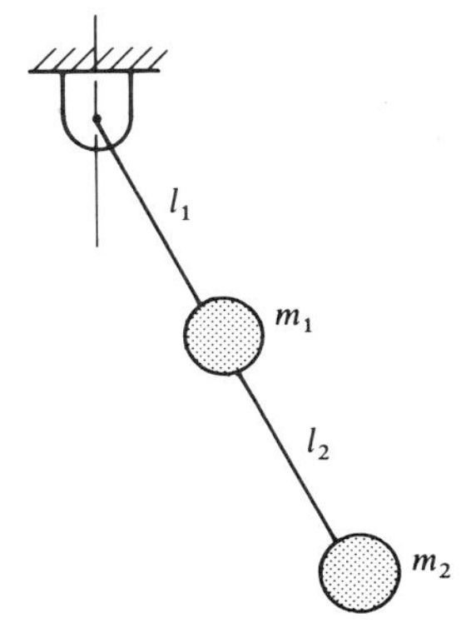

FIG. P2.4

where V is the volume of the body, ρ is the mass density, and $\mathbf{r}$ is the global position vector of an arbitrary point on the rigid body. Using the preceding equation, determine the virtual work of the inertia forces of the rigid body in terms of the acceleration of the center of mass, and the angular acceleration of the body. Assume the case of planar motion.

2.12. Repeat Problem 11 in the case of three-dimensional motion.

2.13. For the system shown in Fig. P5, determine the mass and stiffness matrices using, respectively, the kinetic and strain energies. Use the energy expressions and Lagrange's equation to develop the matrix equation of motion of this system.

2.14. Using the assumption of small oscillations, obtain the differential equations of motion of the double pendulum shown in Fig. P6 using Lagrange's equation.

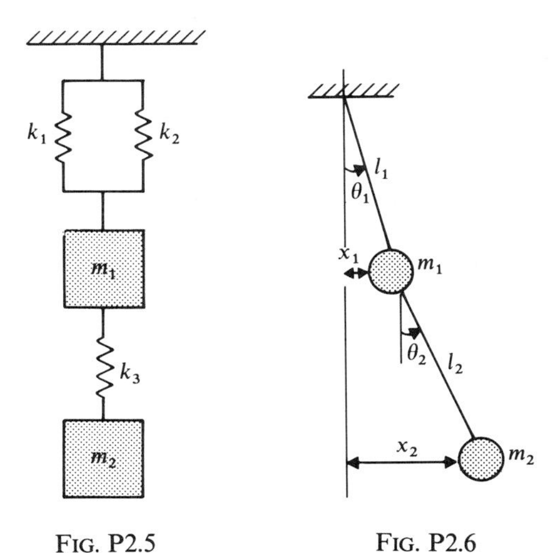

FIG. P2.5 FIG. P2.6

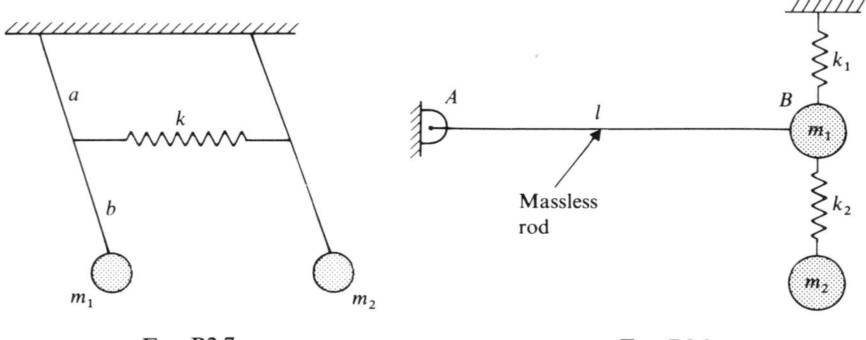

FIG. P2.7

FIG. P2.8

2.15. Derive the differential equations of motion of the two degree of freedom system shown in Fig. P7 using Lagrange's equation. Assume small oscillations.

2.16. In the system shown in Fig. P8, AB is a massless rod which pivots freely about a pin connection at A. If the generalized coordinates are selected to be the angular orientation of the rod θ and the displacement x_2 of the mass m_2, obtain the differential equations of this system using Lagrange's equation. Use the assumption of small oscillations.

2.17. By using Lagrange's equation, derive the equations of motion of the two degree of freedom system shown in Fig. P9.

2.18. In the system shown in Fig. P10, AB is a rigid bar which pivots freely about a pin connection at A. At equilibrium the bar AB is in a horizontal position. Derive the differential equation of motion of this system using Lagrange's equation.

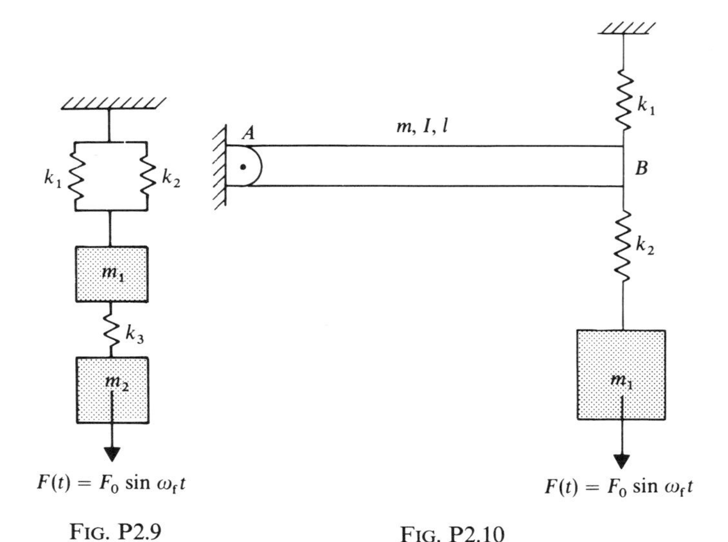

FIG. P2.9

FIG. P2.10

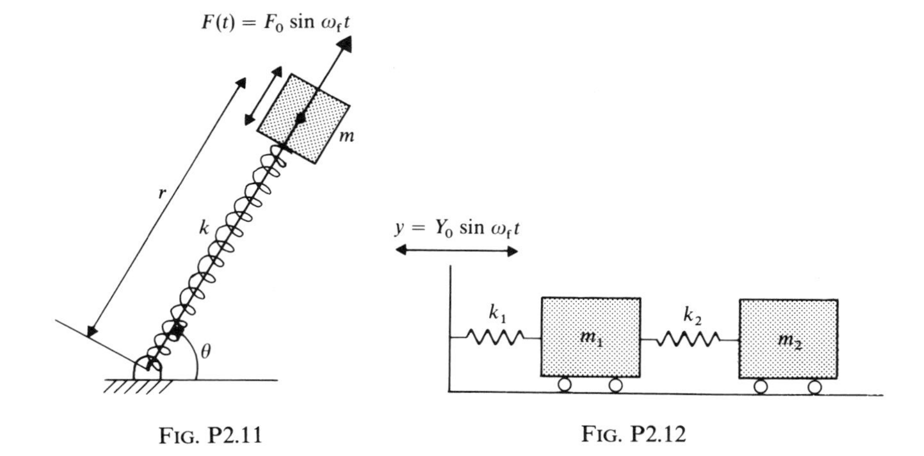

FIG. P2.11

FIG. P2.12

2.19. Using Hamilton's principle, obtain the differential equations of motion of the system shown in Fig. P9.

2.20. Using Hamilton's principle, obtain the differential equations of motion of the system shown in Fig. P11.

2.21. Obtain the differential equations of motion of the oscillatory system shown in Fig. P12 using Lagrange's equation.

2.22. Use Hamilton's principle to derive the equations of motion of the system shown in Fig. P12.

2.23. Use the principle of virtual work in dynamics to derive the equations of motion of the system shown in Fig. P9.

2.24. Use the principle of virtual work in dynamics to derive the nonlinear equation of motion of the system shown in Fig. P13. The system consists of two cylinders, each of which has mass m, and a connecting rod that has mass m_r and length l. The cylinders are assumed to roll without slipping.

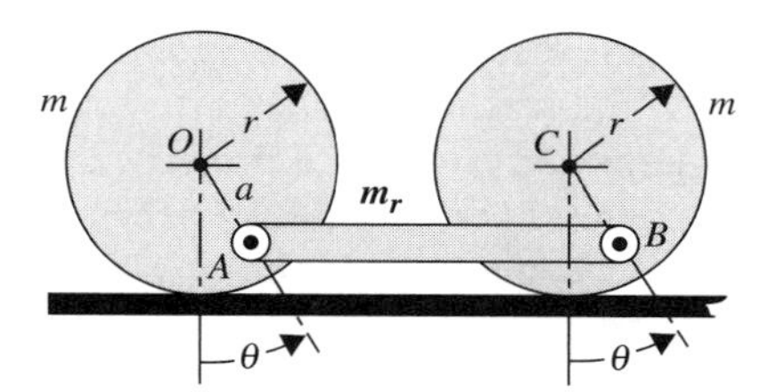

FIG. P2.13

3
Multi-Degree of Freedom Systems

The methods of vibration analysis of single degree of freedom systems can be generalized and extended to study systems with an arbitrary finite number of degrees of freedom. Mechanical systems in general consist of structural elements which have distributed mass and elasticity. In many cases, these systems can be represented by equivalent systems which consist of some elements which are bulky solids which can be treated as rigid elements with specified inertia properties while the other elements are elastic elements which have negligible inertia effects. In fact, the single degree of freedom systems discussed in the preceding chapters are examples of these equivalent models which are called *lumped mass systems*.

We have shown in the preceding chapters that a single degree of freedom system exhibits motion governed by one second-order ordinary differential equation while a two degree of freedom system exhibits motion governed by two second-order ordinary differential equations. It is expected, therefore, that a system having n degrees of freedom exhibits motion which is governed by a set of n simultaneous second-order differential equations. An example of these systems is shown in Fig. 1.

In Section 1 of this chapter, the general form of the second-order ordinary differential equations of motion that govern the vibration of multi-degree of freedom systems is presented, and the use of these equations is demonstrated by several applications. In Section 2, the undamped free vibration of multi-degree of freedom systems is discussed and it is shown that a system with n degrees of freedom has n natural frequencies. Methods for determining the mode shapes of the undamped systems are presented and the orthogonality of these mode shapes is discussed in Section 3. In this section, we also discuss the use of the modal transformation to obtain n uncoupled second-order ordinary differential equations of motion in terms of the modal coordinates. Sections 4 and 5 are devoted, respectively, to the analysis of semidefinite systems, and the conservation of energy in the case of undamped free vibration. In Section 6, the forced vibration of the undamped multi-degree of freedom systems is discussed. The solution of the equations of motion of

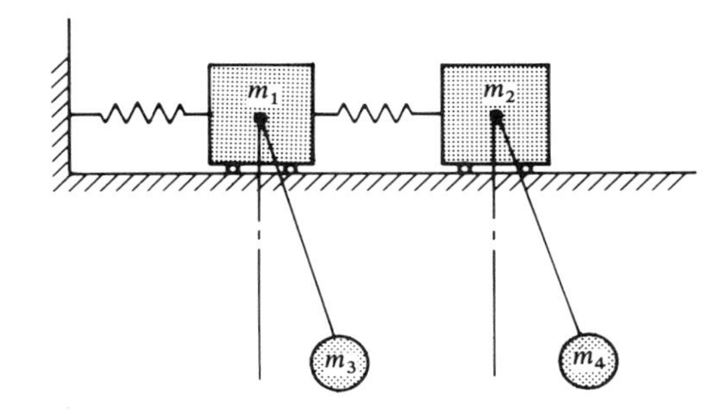

FIG. 3.1. Example of multidegree of freedom systems.

viscously damped multi-degree of freedom systems is obtained in Section 7 using proportional damping, while the case of general viscous damping is covered in Section 8. In Section 9, the *modal truncation* technique frequently used in order to reduce the number of degrees of freedom is discussed, and Sections 10 and 11 are devoted to the classical computer and numerical methods used in the vibration analysis of multi-degree of freedom systems.

3.1 EQUATIONS OF MOTION

Using matrix notation, the general form of the matrix equation of motion of the multi-degree of freedom system is given by

$$\mathbf{M}\ddot{\mathbf{q}} + \mathbf{C}\dot{\mathbf{q}} + \mathbf{K}\mathbf{q} = \mathbf{F} \tag{3.1}$$

where **M**, **C**, and **K** are, respectively, the mass, damping, and stiffness matrices, **q** is the vector of coordinates, and **F** is the vector of forces that act on the multi-degree of freedom system. If the system has n degrees of freedom, the vectors **q** is given by

$$\mathbf{q} = [q_1 \quad q_2 \quad \cdots \quad q_n]^{\mathrm{T}} \tag{3.2}$$

and the mass, damping, and stiffness matrices, and the vector **F** are given in a more explicit form as

$$\mathbf{M} = \begin{bmatrix} m_{11} & m_{12} & \cdots & m_{1n} \\ m_{21} & m_{22} & \cdots & m_{2n} \\ \vdots & \vdots & \ddots & \vdots \\ m_{n1} & m_{n2} & \cdots & m_{nn} \end{bmatrix} \tag{3.3}$$

$$\mathbf{C} = \begin{bmatrix} c_{11} & c_{12} & \cdots & c_{1n} \\ c_{21} & c_{22} & \cdots & c_{2n} \\ \vdots & \vdots & \ddots & \vdots \\ c_{n1} & c_{n2} & \cdots & c_{nn} \end{bmatrix} \tag{3.4}$$

$$\mathbf{K} = \begin{bmatrix} k_{11} & k_{12} & \cdots & k_{1n} \\ k_{21} & k_{22} & \cdots & k_{2n} \\ \vdots & \vdots & \ddots & \vdots \\ k_{n1} & k_{n2} & \cdots & k_{nn} \end{bmatrix} \tag{3.5}$$

and

$$\mathbf{F} = [F_1 \quad F_2 \quad \cdots \quad F_n]^{\mathrm{T}} \tag{3.6}$$

where m_{ij}, c_{ij}, and k_{ij}, $i, j = 1, 2, \ldots, n$, are, respectively, the mass, damping, and stiffness coefficients.

Special Cases Equation 1 is the general form of the matrix equation that governs the *forced damped vibration* of the multi-degree of freedom system. The equations of motion of the *free damped vibration* can be obtained from Eq. 1 by letting $\mathbf{F} = \mathbf{0}$, that is,

$$\mathbf{M}\ddot{\mathbf{q}} + \mathbf{C}\dot{\mathbf{q}} + \mathbf{K}\mathbf{q} = \mathbf{0} \tag{3.7a}$$

which can be written in a more explicit form as

$$\begin{bmatrix} m_{11} & m_{12} & \cdots & m_{1n} \\ m_{21} & m_{22} & \cdots & m_{2n} \\ \vdots & \vdots & \ddots & \vdots \\ m_{n1} & m_{n2} & \cdots & m_{nn} \end{bmatrix} \begin{bmatrix} \ddot{q}_1 \\ \ddot{q}_2 \\ \vdots \\ \ddot{q}_n \end{bmatrix} + \begin{bmatrix} c_{11} & c_{12} & \cdots & c_{1n} \\ c_{21} & c_{22} & \cdots & c_{2n} \\ \vdots & \vdots & \ddots & \vdots \\ c_{n1} & c_{n2} & \cdots & c_{nn} \end{bmatrix} \begin{bmatrix} \dot{q}_1 \\ \dot{q}_2 \\ \vdots \\ \dot{q}_n \end{bmatrix}$$
$$+ \begin{bmatrix} k_{11} & k_{12} & \cdots & k_{1n} \\ k_{21} & k_{22} & \cdots & k_{2n} \\ \vdots & \vdots & \ddots & \vdots \\ k_{n1} & k_{n2} & \cdots & k_{nn} \end{bmatrix} \begin{bmatrix} q_1 \\ q_2 \\ \vdots \\ q_n \end{bmatrix} = \begin{bmatrix} 0 \\ 0 \\ \vdots \\ 0 \end{bmatrix} \tag{3.7b}$$

Furthermore, if we assume the case of *free undamped vibration*, Eq. 7 reduces to

$$\mathbf{M}\ddot{\mathbf{q}} + \mathbf{K}\mathbf{q} = \mathbf{0} \tag{3.8a}$$

or

$$\begin{bmatrix} m_{11} & m_{12} & \cdots & m_{1n} \\ m_{21} & m_{22} & \cdots & m_{2n} \\ \vdots & \vdots & \ddots & \vdots \\ m_{n1} & m_{n2} & \cdots & m_{nn} \end{bmatrix} \begin{bmatrix} \ddot{q}_1 \\ \ddot{q}_2 \\ \vdots \\ \ddot{q}_n \end{bmatrix} + \begin{bmatrix} k_{11} & k_{12} & \cdots & k_{1n} \\ k_{21} & k_{22} & \cdots & k_{2n} \\ \vdots & \vdots & \ddots & \vdots \\ k_{n1} & k_{n2} & \cdots & k_{nn} \end{bmatrix} \begin{bmatrix} q_1 \\ q_2 \\ \vdots \\ q_n \end{bmatrix} = \begin{bmatrix} 0 \\ 0 \\ \vdots \\ 0 \end{bmatrix} \tag{3.8b}$$

Similarly, the case of undamped forced vibration can be described by the following matrix differential equation

$$\mathbf{M}\ddot{\mathbf{q}} + \mathbf{K}\mathbf{q} = \mathbf{F} \tag{3.9a}$$

which can also be written in a more explicit form as

$$\begin{bmatrix} m_{11} & m_{12} & \cdots & m_{1n} \\ m_{21} & m_{22} & \cdots & m_{2n} \\ \vdots & \vdots & \ddots & \vdots \\ m_{n1} & m_{n2} & \cdots & m_{nn} \end{bmatrix} \begin{bmatrix} \ddot{q}_1 \\ \ddot{q}_2 \\ \vdots \\ \ddot{q}_n \end{bmatrix} + \begin{bmatrix} k_{11} & k_{12} & \cdots & k_{1n} \\ k_{21} & k_{22} & \cdots & k_{2n} \\ \vdots & \vdots & \ddots & \vdots \\ k_{n1} & k_{n2} & \cdots & k_{nn} \end{bmatrix} \begin{bmatrix} q_1 \\ q_2 \\ \vdots \\ q_n \end{bmatrix} = \begin{bmatrix} F_1 \\ F_2 \\ \vdots \\ F_n \end{bmatrix} \tag{3.9b}$$

The use of Newton's second law and Lagrange's equation in deriving the differential equations of motion of several multi-degree of freedom systems is demonstrated by the following examples.

Rectilinear Motion Figure 2(a) shows a multi-degree of freedom system which consists of n masses connected by springs and dampers. The system has n degrees of freedom denoted as $x_1, x_2, \ldots, x_n$, which can be written in a vector form as

$$\mathbf{x} = [x_1 \quad x_2 \quad \cdots \quad x_n]^{\mathrm{T}} \tag{3.10}$$

Let $\mathbf{F}$ be the vector of forces that act on the masses, that is,

$$\mathbf{F} = [F_1 \quad F_2 \quad \cdots \quad F_n]^{\mathrm{T}} \tag{3.11}$$

The free-body diagrams of the masses are shown in Fig. 2(b). By direct application of Newton's second law, it can be verified that the equations of motion of the masses are given by

$$m_1\ddot{x}_1 = -k_1x_1 + k_2(x_2 - x_1) - c_1\dot{x}_1 + c_2(\dot{x}_2 - \dot{x}_1) + F_1$$

$$m_2\ddot{x}_2 = -k_2(x_2 - x_1) + k_3(x_3 - x_2) - c_2(\dot{x}_2 - \dot{x}_1) + c_3(\dot{x}_3 - \dot{x}_2) + F_2$$

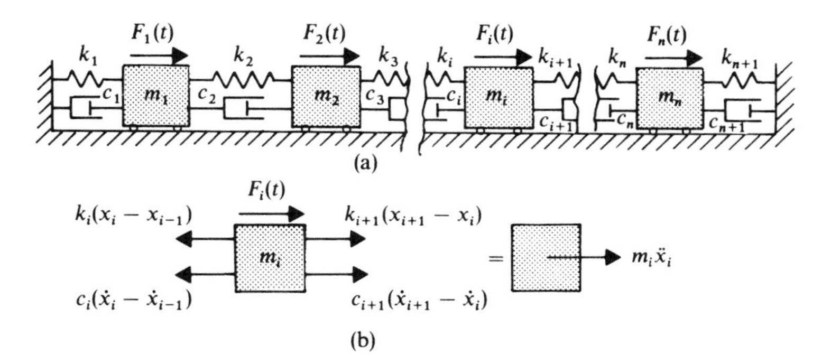

FIG. 3.2. Multidegree of freedom rectilinear system.

$$m_i\ddot{x}_i = -k_i(x_i - x_{i-1}) + k_{i+1}(x_{i+1} - x_i) - c_i(\dot{x}_i - \dot{x}_{i-1})$$
$$+ c_{i+1}(\dot{x}_{i+1} - \dot{x}_i) + F_i$$
$$\vdots$$
$$m_n\ddot{x}_n = -k_n(x_n - x_{n-1}) - k_{n+1}x_n - c_n(\dot{x}_n - \dot{x}_{n-1}) - c_{n+1}\dot{x}_n + F_n$$

which can be rearranged and written as

$$\left.\begin{aligned}
m_1\ddot{x}_1 + (c_1 + c_2)\dot{x}_1 - c_2\dot{x}_2 + (k_1 + k_2)x_1 - k_2x_2 &= F_1 \\
+ (c_2 + c_3)\dot{x}_2 - c_2\dot{x}_1 - c_3\dot{x}_3 + (k_2 + k_3)x_2 - k_2x_1 - k_3x_3 &= F_2 \\
&\vdots \\
m_i\lambda \qquad \dot{x}_i - c_i\dot{x}_{i-1} - c_{i+1}\dot{x}_{i+1} + (k_i + k_{i+1})x_i - k_ix_{i-1} - k_{i+1}x_{i+1} &= F_i \\
&\vdots \\
m_n\ddot{x}_n + (c_n + c_{n+1})\dot{x}_n - c_n\dot{x}_{n-1} + (k_n + k_{n+1})x_n - k_nx_{n-1} &= F_n
\end{aligned}\right\} \tag{3.12}$$

These equations can be written in a matrix form as

$$\mathbf{M}\ddot{\mathbf{x}} + \mathbf{C}\dot{\mathbf{x}} + \mathbf{K}\mathbf{x} = \mathbf{F} \tag{3.13}$$

where the vectors $\mathbf{x}$ and $\mathbf{F}$ are defined by Eqs. 10 and 11, and

$$\mathbf{M} = \begin{bmatrix} m_{11} & m_{12} & \cdots & m_{1n} \\ m_{21} & m_{22} & \cdots & m_{2n} \\ \vdots & \vdots & \ddots & \vdots \\ m_{n1} & m_{n2} & \cdots & m_{nn} \end{bmatrix} = \begin{bmatrix} m_1 & 0 & 0 & \cdots & 0 \\ 0 & m_2 & 0 & \cdots & 0 \\ \vdots & \vdots & \vdots & \ddots & \vdots \\ 0 & 0 & 0 & \cdots & m_n \end{bmatrix} \tag{3.14}$$

$$\mathbf{K} = \begin{bmatrix} k_{11} & k_{12} & k_{13} & \cdots & k_{1n} \\ k_{21} & k_{22} & k_{23} & \cdots & k_{2n} \\ \vdots & \vdots & \vdots & \ddots & \vdots \\ k_{n1} & k_{n2} & k_{n3} & \cdots & k_{nn} \end{bmatrix}$$
$$= \begin{bmatrix} k_1 + k_2 & -k_2 & 0 & \cdots & 0 \\ -k_2 & k_2 + k_3 & -k_3 & \cdots & 0 \\ \vdots & \vdots & \vdots & \ddots & \vdots \\ 0 & 0 & 0 & \cdots & k_n + k_{n+1} \end{bmatrix} \tag{3.15}$$

$$\mathbf{C} = \begin{bmatrix} c_{11} & c_{12} & c_{13} & \cdots & c_{1n} \\ c_{21} & c_{22} & c_{23} & \cdots & c_{2n} \\ \vdots & \vdots & \vdots & \ddots & \vdots \\ c_{n1} & c_{n2} & c_{n3} & \cdots & c_{nn} \end{bmatrix}$$
$$= \begin{bmatrix} c_1 + c_2 & -c_2 & 0 & \cdots & 0 \\ -c_2 & c_2 + c_3 & -c_3 & \cdots & 0 \\ \vdots & \vdots & \vdots & \ddots & \vdots \\ 0 & 0 & 0 & \cdots & c_n + c_{n+1} \end{bmatrix} \tag{3.16}$$

The differential equations of motion of Eq. 13 obtained from the application of Newton's second law can also be obtained using Lagrange's equation. To this end, the system kinetic energy, strain energy, and the virtual work of nonconservative forces are defined as

$$T = \tfrac{1}{2}m_1\dot{x}_1^2 + \tfrac{1}{2}m_2\dot{x}_2^2 + \cdots + \tfrac{1}{2}m_i\dot{x}_i^2 + \cdots + \tfrac{1}{2}m_n\dot{x}_n^2 \tag{3.17}$$

$$U = \tfrac{1}{2}k_1x_1^2 + \tfrac{1}{2}k_2(x_2 - x_1)^2 + \cdots + \tfrac{1}{2}k_i(x_i - x_{i-1})^2 + \cdots + \tfrac{1}{2}k_{n+1}x_n^2 \tag{3.18}$$

$$\begin{aligned}\delta W = {} & F_1\delta x_1 + F_2\delta x_2 + \cdots + F_i\delta x_i + \cdots + F_n\delta x_n - c_1\dot{x}_1\delta x_1 \\ & - c_2(\dot{x}_2 - \dot{x}_1)\delta(x_2 - x_1) - \cdots - c_i(\dot{x}_i - \dot{x}_{i-1})\delta(x_i - x_{i-1}) \\ & - \cdots - c_{n+1}\dot{x}_n\delta\dot{x}_n\end{aligned} \tag{3.19}$$

For this multi-degree of freedom system, Lagrange's equation can be stated as

$$\frac{d}{dt}\left(\frac{\partial T}{\partial \dot{x}_i}\right) - \frac{\partial T}{\partial x_i} + \frac{\partial U}{\partial x_i} = Q_i, \qquad i = 1, 2, \ldots, n \tag{3.20}$$

where Q_i is the generalized force associated with the coordinate x_i. In order to determine Q_i we rewrite Eq. 19 in the following form:

$$\begin{aligned}\delta W = {} & [F_1 - (c_1 + c_2)\dot{x}_1 + c_2\dot{x}_2]\delta x_1 + [F_2 - (c_2 + c_3)\dot{x}_2 + c_2\dot{x}_1 \\ & + c_3\dot{x}_3]\delta x_2 + \cdots + [F_i - (c_i + c_{i+1})\dot{x}_i + c_i\dot{x}_{i-1} + c_{i+1}\dot{x}_{i+1}]\delta x_i + \cdots \\ & + [F_n - (c_n + c_{n+1})\dot{x}_n + c_n\dot{x}_{n-1}]\delta x_n\end{aligned} \tag{3.21}$$

from which we can define the following:

$$\left.\begin{aligned} Q_1 &= F_1 - (c_1 + c_2)\dot{x}_1 + c_2\dot{x}_2 \\ Q_2 &= F_2 - (c_2 + c_3)\dot{x}_2 + c_2\dot{x}_1 + c_3\dot{x}_3 \\ &\ \ \vdots \\ Q_i &= F_i - (c_i + c_{i+1})\dot{x}_i + c_i\dot{x}_{i-1} + c_{i+1}\dot{x}_{i+1} \\ &\ \ \vdots \\ Q_n &= F_n - (c_n + c_{n+1})\dot{x}_n + c_n\dot{x}_{n-1} \end{aligned}\right\} \tag{3.22}$$

One can also verify that

$$\frac{d}{dt}\left(\frac{\partial T}{\partial \dot{x}_i}\right) - \frac{\partial T}{\partial x_i} = m_i\ddot{x}_i, \qquad i = 1, 2, \ldots, n \tag{3.23}$$

and

$$\left.\begin{aligned}
\frac{\partial U}{\partial x_1} &= (k_1 + k_2)x_1 - k_2x_2 \\
\frac{\partial U}{\partial x_2} &= (k_2 + k_3)x_2 - k_2x_1 - k_3x_3 \\
&\vdots \\
\frac{\partial U}{\partial x_i} &= (k_i + k_{i+1})x_i - k_ix_{i-1} - k_{i+1}x_{i+1} \\
&\vdots \\
\frac{\partial U}{\partial x_n} &= (k_n + k_{n+1})x_n - k_nx_{n-1}
\end{aligned}\right\} \tag{3.24}$$

Substituting Eqs. 22–24 into Eq. 20 leads to the following set of n second-order differential equations:

$$\left.\begin{aligned}
m_1\ddot{x}_1 + (k_1 + k_2)x_1 - k_2x_2 &= F_1 - (c_1 + c_2)\dot{x}_1 + c_2\dot{x}_2 \\
m_2\ddot{x}_2 + (k_2 + k_3)x_2 - k_2x_1 - k_3x_3 &= F_2 - (c_2 + c_3)\dot{x}_2 + c_2\dot{x}_1 + c_3\dot{x}_3 \\
&\vdots \\
m_i\ddot{x}_i + (k_i + k_{i+1})x_i - k_ix_{i-1} - k_{i+1}x_{i+1} &= F_i - (c_i + c_{i+1})\dot{x}_i + c_i\dot{x}_{i-1} \\
&\quad + c_{i+1}\dot{x}_{i+1} \\
&\vdots \\
m_n\ddot{x}_n + (k_n + k_{n+1})x_n - k_nx_{n-1} &= F_n - (c_n + c_{n+1})\dot{x}_n + c_n\dot{x}_{n-1}
\end{aligned}\right\} \tag{3.25}$$

which are the same set of equations given by Eq. 12.

Angular Oscillations Figure 3(a) shows a set of n masses which are rigidly connected to massless rods which have length l. The rods are pinned at their ends as shown in the figure. Let $T = [T_1 \;\; T_2 \;\; \cdots \;\; T_n]^{\mathrm{T}}$ be the vector of external torques that act on the rods. By using the free-body diagram shown in Fig. 3(b) and applying Newton's second law or D'Alembert's principle with the assumption of small angular oscillations, one can verify that the differential equations of motion of the masses are given by

$$\begin{aligned}
m_1l^2\ddot{\theta}_1 &= -m_1gl\theta_1 - k_1l^2\theta_1 + k_2l^2(\theta_2 - \theta_1) - c_1l^2\dot{\theta}_1 + c_2l^2(\dot{\theta}_2 - \dot{\theta}_1) + T_1 \\
m_2l^2\ddot{\theta}_2 &= -m_2gl\theta_2 - k_2l^2(\theta_2 - \theta_1) + k_3l^2(\theta_3 - \theta_2) - c_2l^2(\dot{\theta}_2 - \dot{\theta}_1) \\
&\quad + c_3l^2(\dot{\theta}_3 - \dot{\theta}_2) + T_2 \\
&\vdots
\end{aligned}$$

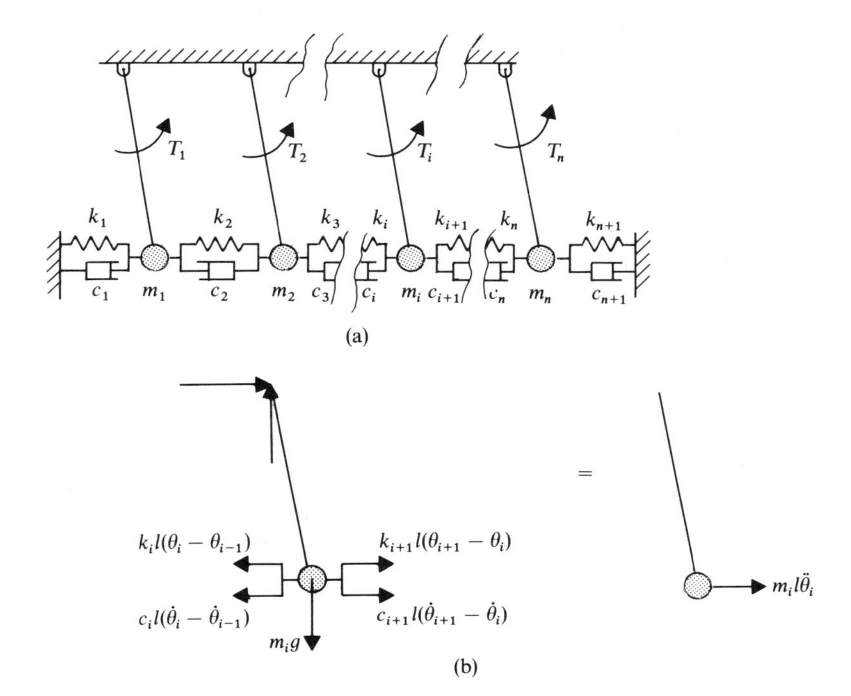

FIG. 3.3. Angular oscillations.

$$m_i l^2 \ddot{\theta}_i = -m_i g l \theta_i - k_i l^2(\theta_i - \theta_{i-1}) + k_{i+1} l^2(\theta_{i+1} - \theta_i) - c_i l^2(\dot{\theta}_i - \dot{\theta}_{i-1}) + c_{i+1} l^2(\dot{\theta}_{i+1} - \dot{\theta}_i) + T_i$$

$$\vdots$$

$$m_n l^2 \ddot{\theta}_n = -m_n g l \theta_n - k_n l^2(\theta_n - \theta_{n-1}) - k_{n+1} l^2 \theta_n - c_n l^2(\dot{\theta}_n - \dot{\theta}_{n-1}) - c_{n+1} l^2 \dot{\theta}_n + T_n$$

These equations can be rearranged and written as

$$m_1 l^2 \ddot{\theta}_1 + (c_1 + c_2) l^2 \dot{\theta}_1 - c_2 l^2 \dot{\theta}_2 + [(k_1 + k_2) l^2 + m_1 g l]\theta_1 - k_2 l^2 \theta_2 = T_1$$

$$m_2 l^2 \ddot{\theta}_2 + (c_2 + c_3) l^2 \dot{\theta}_2 - c_2 l^2 \dot{\theta}_1 - c_3 l^2 \dot{\theta}_3 + [(k_2 + k_3) l^2 + m_2 g l]\theta_2 - k_2 l^2 \theta_1 - k_3 l^2 \theta_3 = T_2$$

$$\vdots$$

$$m_i l^2 \ddot{\theta}_i + (c_i + c_{i+1}) l^2 \dot{\theta}_i - c_i l^2 \dot{\theta}_{i-1} - c_{i+1} l^2 \dot{\theta}_{i+1} + [(k_i + k_{i+1}) l^2 + m_i g l]\theta_i - k_i l^2 \theta_{i-1} - k_{i+1} l^2 \theta_{i+1} = T_i$$

$$\vdots$$

$$m_n l^2 \ddot{\theta}_n + (c_n + c_{n+1}) l^2 \dot{\theta}_n - c_n l^2 \dot{\theta}_{n-1} + [(k_n + k_{n+1}) l^2 + m_n g l]\theta_n - k_n l^2 \theta_{n-1} = T_n$$

which can be written in a matrix form as

$$\mathbf{M}\ddot{\boldsymbol{\theta}} + \mathbf{C}\dot{\boldsymbol{\theta}} + \mathbf{K}\boldsymbol{\theta} = \mathbf{T}$$

where $\boldsymbol{\theta}$ is the vector of angular oscillations given by

$$\boldsymbol{\theta} = [\theta_1 \quad \theta_2 \quad \cdots \quad \theta_n]^{\mathrm{T}}$$

and the mass, damping, and stiffness matrices, $\mathbf{M}$, $\mathbf{C}$, and $\mathbf{K}$ are given by

$$\mathbf{M} = \begin{bmatrix} m_1 l^2 & 0 & 0 & \cdots & 0 \\ 0 & m_2 l^2 & 0 & \cdots & 0 \\ 0 & 0 & m_3 l^2 & \cdots & 0 \\ \vdots & \vdots & \vdots & \ddots & \vdots \\ 0 & 0 & 0 & \cdots & m_n l^2 \end{bmatrix}$$

$$\mathbf{C} = \begin{bmatrix} (c_1 + c_2)l^2 & -c_2 l^2 & 0 & \cdots & 0 \\ -c_2 l^2 & (c_2 + c_3)l^2 & -c_3 l^2 & \cdots & 0 \\ 0 & -c_3 l^2 & (c_3 + c_4)l^2 & \cdots & 0 \\ \vdots & \vdots & \vdots & \ddots & \vdots \\ 0 & 0 & 0 & \cdots & (c_n + c_{n+1})l^2 \end{bmatrix}$$

$$\mathbf{K} = \begin{bmatrix} (k_1 + k_2)l^2 + m_1 gl & -k_2 l^2 & 0 & \cdots & 0 \\ -k_2 l^2 & [(k_2 + k_3)l^2 + m_2 gl] & -k_3 l^2 & \cdots & 0 \\ 0 & -k_3 l^2 & [(k_3 + k_4)l^2 + m_3 gl] & \cdots & 0 \\ \vdots & \vdots & \vdots & \ddots & \vdots \\ 0 & 0 & 0 & \cdots & [(k_n + k_{n+1})l^2 + m_n gl] \end{bmatrix}$$

In the case of angular oscillations, appropriate units must be used for the mass, damping, and stiffness coefficients. For example, the units for the mass coefficients must be kilogram · square meters (kg · m^2), for the damping coefficients newton · meters · seconds (N · m · s), and for the stiffness coefficients newton · meters (N · m).

The differential equations of motion of the system shown in Fig. 3(a) can also be obtained by applying Lagrange's equation. The system kinetic and strain energies in the case of small oscillations are defined as

$$\begin{aligned} T &= \tfrac{1}{2} m_1 l^2 \dot{\theta}_1^2 + \tfrac{1}{2} m_2 l^2 \dot{\theta}_2^2 + \cdots + \tfrac{1}{2} m_i l^2 \dot{\theta}_i^2 + \cdots + \tfrac{1}{2} m_n l^2 \dot{\theta}_n^2 \\ &= \frac{1}{2} \sum_{j=1}^{n} m_j l^2 \dot{\theta}_j^2 \end{aligned}$$

$$\begin{aligned} U &= \tfrac{1}{2} k_1 l^2 \theta_1^2 + \tfrac{1}{2} k_2 l^2 (\theta_2 - \theta_1)^2 + \cdots + \tfrac{1}{2} k_i l^2 (\theta_i - \theta_{i-1})^2 + \cdots \\ &\quad + \tfrac{1}{2} k_n l^2 (\theta_n - \theta_{n-1})^2 + \tfrac{1}{2} k_{n+1} l^2 \theta_n^2 \end{aligned}$$

The virtual work of the gravity and nonconservative forces acting on the system is given by

$$\begin{aligned}\delta W = {} & T_1\delta\theta_1 + T_2\delta\theta_2 + \cdots + T_i\delta\theta_i + \cdots + T_n\delta\theta_n - c_1 l^2\dot{\theta}_1\delta\theta_1 \\ & - c_2 l^2(\dot{\theta}_2 - \dot{\theta}_1)\delta(\theta_2 - \theta_1) - \cdots - c_i l^2(\dot{\theta}_i - \dot{\theta}_{i-1})\delta(\theta_i - \theta_{i-1}) - \cdots \\ & - c_n l^2(\dot{\theta}_n - \dot{\theta}_{n-1})\delta(\dot{\theta}_n - \dot{\theta}_{n-1}) - c_{n+1} l^2\dot{\theta}_n\delta\dot{\theta}_n - \sum_{i=1}^{n} m_i g l\theta_i\delta\theta_i\end{aligned}$$

It is left to the reader to show that the use of the kinetic energy, strain energy, and virtual work given above in Lagrange's equation leads to the same differential equations of motion obtained by applying Newton's second law.

Torsional Oscillations Figure 4(a) shows a set of n disks connected by massless shafts. The disk i in the system has mass moment of inertia I_i and is subjected to an external torque T_i. Figure 4(b) shows the free-body diagrams of the disks. The application of Euler equations leads to the following n differential equations of motion:

$$\begin{aligned}I_1\ddot{\theta}_1 &= -k_1\theta_1 + k_2(\theta_2 - \theta_1) + T_1 \\ I_2\ddot{\theta}_2 &= -k_2(\theta_2 - \theta_1) + k_3(\theta_3 - \theta_2) + T_2 \\ &\vdots \\ I_i\ddot{\theta}_i &= -k_i(\theta_i - \theta_{i-1}) + k_{i+1}(\theta_{i+1} - \theta_i) + T_i \\ &\vdots \\ I_n\ddot{\theta}_n &= -k_n(\theta_n - \theta_{n-1}) - k_{n+1}\theta_n + T_n\end{aligned}$$

where θ_i is the torsional oscillation of the disk i and k_i is the torsional stiffness of the massless shaft i defined as

$$k_i = \frac{G_i J_i}{l_i}$$

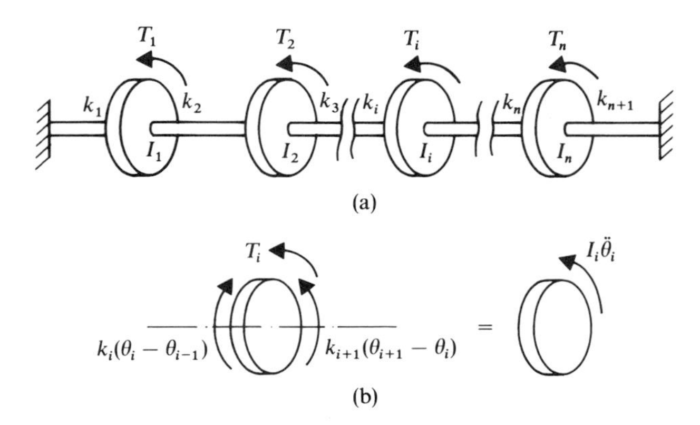

FIG. 3.4. Torsional oscillations.

in which G_i, J_i, and l_i are, respectively, the modulus of rigidity, the second moment of area, and the length of the shaft i. The preceding n differential equations can be written in the following matrix form

$$\mathbf{M}\ddot{\boldsymbol{\theta}} + \mathbf{K}\boldsymbol{\theta} = \mathbf{T}$$

where

$$\boldsymbol{\theta} = [\theta_1 \quad \theta_2 \quad \cdots \quad \theta_n]^{\mathrm{T}}$$

$$\mathbf{T} = [T_1 \quad T_2 \quad \cdots \quad T_n]^{\mathrm{T}}$$

$$\mathbf{M} = \begin{bmatrix} I_1 & 0 & 0 & \cdots & 0 \\ 0 & I_2 & 0 & \cdots & 0 \\ \vdots & \vdots & \vdots & \ddots & \vdots \\ 0 & 0 & 0 & \cdots & I_n \end{bmatrix}$$

and

$$\mathbf{K} = \begin{bmatrix} k_1 + k_2 & -k_2 & 0 & \cdots & 0 \\ -k_2 & k_2 + k_3 & -k_3 & \cdots & 0 \\ 0 & -k_3 & k_3 + k_4 & \cdots & 0 \\ \vdots & \vdots & \vdots & \ddots & \vdots \\ 0 & 0 & 0 & \cdots & k_n + k_{n+1} \end{bmatrix}$$

The differential equations of motion for the torsional system shown in Fig. 4(a) can also be obtained by using Lagrange's equations. For this system, the kinetic energy, strain energy, and virtual work are defined as

$$T = \tfrac{1}{2} I_1 \dot{\theta}_1^2 + \tfrac{1}{2} I_2 \dot{\theta}_2^2 + \cdots + \tfrac{1}{2} I_i \dot{\theta}_i^2 + \cdots + \tfrac{1}{2} I_n \dot{\theta}_n^2$$

$$U = \tfrac{1}{2} k_1 \theta_1^2 + \tfrac{1}{2} k_2 (\theta_2 - \theta_1)^2 + \cdots + \tfrac{1}{2} k_i (\theta_{i+1} - \theta_i)^2 + \cdots + \tfrac{1}{2} k_{n+1} \theta_n^2$$

$$\delta W = T_1 \delta\theta_1 + T_2 \delta\theta_2 + \cdots + T_i \delta\theta_i + \cdots + T_n \delta\theta_n$$

These scalar energy and work expressions can be used with Lagrange's equation in order to obtain the n second-order differential equations of motion of the n-degree of freedom torsional system shown in Fig. 4(a).

Example 3.1

The torsional system shown in Fig. 5 consists of three disks which have mass moment of inertia, $I_1 = 2.0 \times 10^3$ kg·m², $I_2 = 3.0 \times 10^3$ kg·m², and $I_3 = 4.0 \times 10^3$ kg·m². The stiffness coefficients of the shafts connecting these disks are $k_1 = 12 \times 10^5$ N·m, $k_2 = 24 \times 10^5$ N·m, and $k_3 = 36 \times 10^5$ N·m. The matrix equation

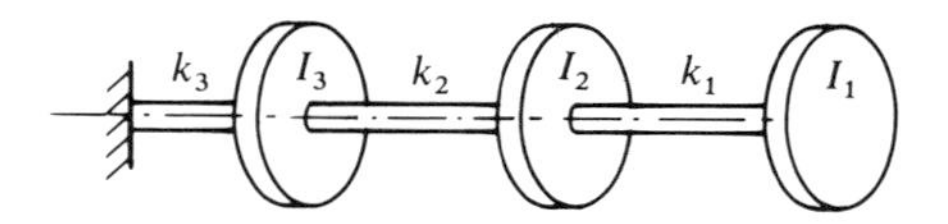

FIG. 3.5. Three degree of freedom torsional system.

of motion of the free vibration of this system is given by

$$\mathbf{M}\ddot{\boldsymbol{\theta}} + \mathbf{K}\boldsymbol{\theta} = \mathbf{0}$$

where, in this example, **M**, **K**, and $\boldsymbol{\theta}$ are given by

$$\mathbf{M} = \begin{bmatrix} I_1 & 0 & 0 \\ 0 & I_2 & 0 \\ 0 & 0 & I_3 \end{bmatrix} = 2 \times 10^3 \begin{bmatrix} 1.0 & 0 & 0 \\ 0 & 1.5 & 0 \\ 0 & 0 & 2 \end{bmatrix} \text{kg}\cdot\text{m}^2$$

$$\mathbf{K} = \begin{bmatrix} k_1 & -k_1 & 0 \\ -k_1 & k_1 + k_2 & -k_2 \\ 0 & -k_2 & k_2 + k_3 \end{bmatrix} = 12 \times 10^5 \begin{bmatrix} 1 & -1 & 0 \\ -1 & 3 & -2 \\ 0 & -2 & 5 \end{bmatrix} \text{N}\cdot\text{m}$$

$$\boldsymbol{\theta} = [\theta_1 \quad \theta_2 \quad \theta_3]^{\text{T}}$$

in which θ_1, θ_2, and θ_3 are the torsional oscillations of the disks.

Formulation of the Stiffness and Flexibility Matrices In this section, the stiffness matrix was obtained as the result of application of Newton's second law or Lagrange's equation. There are other techniques which can also be used to formulate the stiffness matrix. One of these methods utilizes the definition of the stiffness coefficients. In order to demonstrate the use of this method, let us consider the case of static analysis, in which the equations derived in this chapter reduce to

$$\mathbf{Kq} = \mathbf{Q}$$

where **K** is the stiffness matrix, **q** is the vector of generalized coordinates, and **Q** is the vector of generalized forces.

The preceding equation can also be written as

$$\sum_{j=1}^{n} k_{ij} q_j = Q_i$$

This equation can be written explicitly as

$$k_{i1}q_1 + k_{i2}q_2 + \cdots + k_{ii}q_i + \cdots + k_{in}q_n = Q_i$$

where k_{ij}, $i, j = 1, 2, \ldots, n$, are the elements of the stiffness matrix. Note that the stiffness coefficient k_{ii} is the force resulting from a unit displacement of the coordinate i, while holding all other coordinates equal to zero. The stiffness coefficient k_{ij} $(i \neq j)$, on the other hand, is the force associated with the coordinate i as the result of a unit displacement of the coordinate j while holding all other coordinates equal to zero. Since the stiffness matrix is symmetric, it is also clear that k_{ij} is the force associated with the coordinate j as the result of a unit displacement of the coordinate i. It is left to the reader to try to use this approach to develop the stiffness matrices of the systems presented in this section.

The *flexibility matrix* is defined to be the inverse of the stiffness matrix, we,

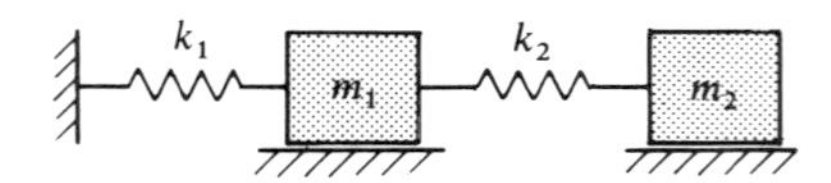

FIG. 3.6. Evaluation of the stiffness and flexibility coefficients.

therefore, have the following equation for the static equilibrium:

$$\mathbf{q} = \mathbf{K}^{-1}\mathbf{Q}$$

where $\mathbf{K}^{-1}$ is the flexibility matrix. The preceding equation yields

$$q_i = \sum_{j=1}^{n} f_{ij}Q_j = f_{i1}Q_1 + f_{i2}Q_2 + \cdots + f_{ii}Q_i + \cdots + f_{in}Q_n$$

where f_{ij} are the *flexibility coefficients*. Note that the flexibility coefficient f_{ii} is equal to the displacement of the coordinate i as the result of the application of a unit load Q_i while Q_j $(j \neq i)$ are all zeros. The flexibility coefficient f_{ij} is the displacement of the coordinate j as the result of the application of a unit load $Q_i = 1$ associated with the coordinate i while Q_j $(j \neq i)$ are all equal to zero. The flexibility matrix is symmetric and consequently f_{ij} is also equal to the displacement of the coordinate i as the result of the application of a unit load Q_j while all other forces are equal to zero. The flexibility matrix can be found using the unit load approach or by the direct inversion of the stiffness matrix.

We demonstrate the use of the methods discussed in this section for formulating the stiffness and the flexibility matrices using the system shown in Fig. 6. If the mass m_1 is given a unit displacement while holding the mass m_2 fixed, the force acting on the mass m_1 is given by

$$k_{11} = k_1 + k_2$$

while the force acting on the mass m_2 as the result of a unit displacement of the mass m_1 is

$$k_{12} = -k_2$$

Similarly, if the mass m_2 is given a unit displacement while holding the mass m_1 fixed, the force acting on the mass m_1 as the result of this displacement is

$$k_{21} = -k_2$$

and the force acting on m_2 is

$$k_{22} = k_2$$

Therefore, the stiffness matrix $\mathbf{K}$ is given by

$$\mathbf{K} = \begin{bmatrix} k_1 + k_2 & -k_2 \\ -k_2 & k_2 \end{bmatrix}$$

In order to determine the elements of the flexibility matrix, a unit load

$Q_1 = 1$ is first applied statically at the mass m_1, while the external force acting on the mass m_2 is assumed to be equal to zero. In this case, the displacements of the two masses are equal and are given by

$$f_{11} = f_{21} = \frac{1}{k_1}$$

In order to determine f_{12} and f_{22}, a unit load $Q_2 = 1$ is applied statically to the mass m_2 and Q_1 is assumed to be zero. Using the concept of equivalent springs (Shabana, 1996), it is clear that

$$f_{22} = \frac{k_1 + k_2}{k_1 k_2}, \qquad f_{12} = \frac{1}{k_1}$$

Therefore, the flexibility matrix $\mathbf{K}^{-1}$ is given by

$$\mathbf{K}^{-1} = \begin{bmatrix} 1/k_1 & 1/k_1 \\ 1/k_1 & (k_1 + k_2)/k_1 k_2 \end{bmatrix}$$

Note that $\mathbf{K}\mathbf{K}^{-1} = \mathbf{K}^{-1}\mathbf{K} = \mathbf{I}$, where $\mathbf{I}$ is the 2×2 identity matrix.

3.2 UNDAMPED FREE VIBRATION

In the case of undamped free vibration of multi-degree of freedom systems, Eq. 1 reduces to

$$\mathbf{M}\ddot{\mathbf{q}} + \mathbf{K}\mathbf{q} = \mathbf{0} \tag{3.26}$$

In like manner to the case of single degree of freedom systems, we assume a solution in the form

$$\mathbf{q} = \mathbf{A} \sin(\omega t + \phi) \tag{3.27}$$

where $\mathbf{A}$ is the vector of amplitudes, ω is the frequency, and ϕ is the phase angle. Differentiation of Eq. 27 twice with respect to time leads to

$$\ddot{\mathbf{q}} = -\omega^2 \mathbf{A} \sin(\omega t + \phi) \tag{3.28}$$

Substituting Eqs. 27 and 28 into Eq. 26 leads to

$$-\omega^2 \mathbf{M}\mathbf{A} \sin(\omega t + \phi) + \mathbf{K}\mathbf{A} \sin(\omega t + \phi) = \mathbf{0}$$

which leads to

$$\mathbf{K}\mathbf{A} - \omega^2 \mathbf{M}\mathbf{A} = \mathbf{0} \tag{3.29}$$

Natural Frequencies Equation 29 which is sometimes called the *standard eigenvalue problem* can be considered as a system of homogeneous equations in the vector of unknown amplitudes $\mathbf{A}$. This equation can be written in the following form:

$$[\mathbf{K} - \omega^2 \mathbf{M}]\mathbf{A} = \mathbf{0} \tag{3.30}$$

This equation has a nontrivial solution if and only if the coefficient matrix is singular, that is,

$$|\mathbf{K} - \omega^2 \mathbf{M}| = 0 \tag{3.31}$$

This equation is called the *characteristic equation* and can be written in a more explicit form as

$$\begin{vmatrix} k_{11} - \omega^2 m_{11} & k_{12} - \omega^2 m_{12} & k_{13} - \omega^2 m_{13} & \cdots & k_{1n} - \omega^2 m_{1n} \\ k_{21} - \omega^2 m_{21} & k_{22} - \omega^2 m_{22} & k_{23} - \omega^2 m_{23} & \cdots & k_{2n} - \omega^2 m_{2n} \\ k_{31} - \omega^2 m_{31} & k_{32} - \omega^2 m_{32} & k_{33} - \omega^2 m_{33} & \cdots & k_{3n} - \omega^2 m_{3n} \\ \vdots & \vdots & \vdots & \ddots & \vdots \\ k_{n1} - \omega^2 m_{n1} & k_{n2} - \omega^2 m_{n2} & k_{n3} - \omega^2 m_{n3} & \cdots & k_{nn} - \omega^2 m_{nn} \end{vmatrix} = 0 \tag{3.32}$$

The above equation leads to a polynomial of order n in ω^2, and as such, the term of highest order in this polynomial is $(\omega^2)^n$. The roots of this polynomial denoted as $\omega_1^2, \omega_2^2, \ldots, \omega_n^2$ are called the *characteristic values* or the *eigenvalues*. If the mass matrix $\mathbf{M}$ is positive definite and the stiffness matrix $\mathbf{K}$ is either positive definite or positive semidefinite, the characteristic values $\omega_1^2, \omega_2^2, \ldots, \omega_n^2$ are real nonnegative numbers. The square roots of these numbers, $\omega_1, \omega_2, \ldots, \omega_n$, are called the *natural frequencies* of the undamped multi-degree of freedom system. Thus, a system with n degrees of freedom has n natural frequencies.

Mode Shapes Associated with each characteristic value ω_i, there is an n-dimensional vector called the *characteristic vector* or the *eigenvector* $\mathbf{A}_i$ which can be obtained by using Eq. 30 as follows:

$$[\mathbf{K} - \omega_i^2 \mathbf{M}]\mathbf{A}_i = \mathbf{0} \tag{3.33}$$

This is a system of homogeneous algebraic equations with a singular coefficient matrix since ω_i^2 is one of the roots of the polynomial resulting from Eq. 32. Therefore, Eq. 33 has a nontrivial solution which defines the eigenvector $\mathbf{A}_i$ to within an arbitrary constant. The eigenvector (amplitude) $\mathbf{A}_i$ is sometimes referred to as the ith *mode shape*, *normal mode*, or *principal mode* of vibration.

As a generalization to the procedure used for the case of the two degree of freedom systems (Shabana, 1996), we may write the general solution in the case of undamped free vibration of the multi-degree of freedom system as

$$\begin{aligned} \mathbf{q} &= \alpha_1 \mathbf{A}_1 \sin(\omega_1 t + \phi_1) + \alpha_2 \mathbf{A}_2 \sin(\omega_2 t + \phi_2) + \cdots + \alpha_n \mathbf{A}_n \sin(\omega_n t + \phi_n) \\ &= \sum_{i=1}^{n} \alpha_i \mathbf{A}_i \sin(\omega_i t + \phi_i) \end{aligned} \tag{3.34}$$

where α_i and ϕ_i, $i = 1, 2, \ldots, n$, are $2n$ arbitrary constants which can be determined from the initial conditions.

Initial Conditions By differentiating Eq. 34 with respect to time, one obtains

$$\dot{\mathbf{q}} = \sum_{i=1}^{n} \alpha_i \omega_i \mathbf{A}_i \cos(\omega_i t + \phi_i) \tag{3.35}$$

Let $\mathbf{q}_0$ and $\dot{\mathbf{q}}_0$ be, respectively, the vectors of initial displacements and velocities. Substituting these initial conditons into Eqs. 34 and 35, one obtains

$$\mathbf{q}_0 = \sum_{i=1}^{n} \alpha_i \mathbf{A}_i \sin \phi_i \tag{3.36}$$

$$\dot{\mathbf{q}}_0 = \sum_{i=1}^{n} \alpha_i \omega_i \mathbf{A}_i \cos \phi_i \tag{3.37}$$

We may define the following two constants:

$$a_i = \alpha_i \sin \phi_i, \qquad i = 1, 2, \ldots, n \tag{3.38}$$

$$b_i = \alpha_i \cos \phi_i, \qquad i = 1, 2, \ldots, n \tag{3.39}$$

In terms of these constants, Eqs. 36 and 37 can be written as

$$\mathbf{q}_0 = \sum_{i=1}^{n} a_i \mathbf{A}_i \tag{3.40}$$

$$\dot{\mathbf{q}}_0 = \sum_{i=1}^{n} b_i \omega_i \mathbf{A}_i \tag{3.41}$$

which can be written in a matrix form as

$$\mathbf{\Phi a} = \mathbf{q}_0 \tag{3.42}$$

$$\mathbf{\Phi \omega b} = \dot{\mathbf{q}}_0 \tag{3.43}$$

where $\mathbf{a} = [a_1 \;\; a_2 \;\; \cdots \;\; a_n]^{\mathrm{T}}$, $\mathbf{b} = [b_1 \;\; b_2 \;\; \cdots \;\; b_n]^{\mathrm{T}}$, and $\boldsymbol{\omega}$ and $\mathbf{\Phi}$ are the matrices

$$\boldsymbol{\omega} = \begin{bmatrix} \omega_1 & 0 & 0 & \cdots & 0 \\ 0 & \omega_2 & 0 & \cdots & 0 \\ 0 & 0 & \omega_3 & \cdots & 0 \\ \vdots & \vdots & \vdots & \ddots & \vdots \\ 0 & 0 & 0 & \cdots & \omega_n \end{bmatrix} \tag{3.44}$$

$$\begin{aligned} \mathbf{\Phi} &= [\mathbf{A}_1 \quad \mathbf{A}_2 \quad \cdots \quad \mathbf{A}_n] \\ &= \begin{bmatrix} A_{11} & A_{21} & \cdots & A_{n1} \\ A_{12} & A_{22} & \cdots & A_{n2} \\ \vdots & \vdots & \ddots & \vdots \\ A_{1n} & A_{2n} & \cdots & A_{nn} \end{bmatrix} \end{aligned} \tag{3.45}$$

The matrix $\mathbf{\Phi}$, whose columns are the eigenvectors, is called the *modal matrix*.

Equations 42 and 43 represent a system of $2n$ linear algebraic equations that can be solved for the vectors of unknowns $\mathbf{a}$ and $\mathbf{b}$. If the number of these equations is large, numerical techniques can be used on digital computers to solve this system of equations. Once the vectors of the $2n$ unknowns $\mathbf{a} = [a_1 \ a_2 \ \cdots \ a_n]^{\mathrm{T}}$ and $\mathbf{b} = [b_1 \ b_2 \ \cdots \ b_n]^{\mathrm{T}}$ are determined the constants α_i and ϕ_i of Eqs. 38 and 39 can be calculated from the following relationships:

$$\alpha_i = \sqrt{a_i^2 + b_i^2}, \qquad i = 1, 2, \ldots, n \tag{3.46}$$

$$\phi_i = \tan^{-1}\frac{a_i}{b_i}, \qquad i = 1, 2, \ldots, n \tag{3.47}$$

These constants can be substituted into Eqs. 34 and 35 in order to determine the displacements and velocities as functions of time.

Example 3.2

In the torsional system of Example 1, the symmetric mass and stiffness matrices were given by

$$\mathbf{M} = \begin{bmatrix} I_1 & 0 & 0 \\ 0 & I_2 & 0 \\ 0 & 0 & I_3 \end{bmatrix} = 2 \times 10^3 \begin{bmatrix} 1 & 0 & 0 \\ 0 & 1.5 & 0 \\ 0 & 0 & 2 \end{bmatrix} \mathrm{kg \cdot m^2}$$

$$\mathbf{K} = \begin{bmatrix} k_1 & -k_1 & 0 \\ -k_1 & k_1 + k_2 & -k_2 \\ 0 & -k_2 & k_2 + k_3 \end{bmatrix} = 12 \times 10^5 \begin{bmatrix} 1 & -1 & 0 \\ -1 & 3 & -2 \\ 0 & -2 & 5 \end{bmatrix} \mathrm{N \cdot m}$$

and the governing equation of motion of the free vibration is

$$\mathbf{M}\ddot{\boldsymbol{\theta}} + \mathbf{K}\boldsymbol{\theta} = \mathbf{0}$$

We assume a solution in the form

$$\boldsymbol{\theta} = \mathbf{A} \sin(\omega t + \phi)$$

By substituting this assumed solution into the differential equation, one obtains

$$[\mathbf{K} - \omega^2\mathbf{M}]\mathbf{A} = \mathbf{0}$$

Substituting for the mass and stiffness matrices, we get

$$\left[6 \times 10^5 \begin{bmatrix} 1 & -1 & 0 \\ -1 & 3 & -2 \\ 0 & -2 & 5 \end{bmatrix} - \omega^2 \times 10^3 \begin{bmatrix} 1 & 0 & 0 \\ 0 & 1.5 & 0 \\ 0 & 0 & 2 \end{bmatrix}\right] \mathbf{A} = \mathbf{0}$$

which can be rewritten as

$$\left[\begin{bmatrix} 1 & -1 & 0 \\ -1 & 3 & -2 \\ 0 & -2 & 5 \end{bmatrix} - \beta \begin{bmatrix} 1 & 0 & 0 \\ 0 & 1.5 & 0 \\ 0 & 0 & 2 \end{bmatrix}\right] \mathbf{A} = \mathbf{0}$$

This system has a nontrivial solution if the determinant of the coefficient matrix is equal to zero. This leads to the following characteristic equation:

$$\beta^3 - 5.5\beta^2 + 7.5\beta - 2 = 0$$

which has the following roots:

$$\beta_1 = 0.3516, \qquad \beta_2 = 1.606, \qquad \beta_3 = 3.542$$

Since $\omega^2 = 600\beta$, the natural frequencies associated with the roots β_1, β_2, and β_3 are given, respectively, by

$$\omega_1 = 14.52 \text{ rad/s}, \qquad \omega_2 = 31.05 \text{ rad/s}, \quad \text{and} \quad \omega_3 = 46.1 \text{ rad/s}$$

For a given root β_i, $i = 1, 2, 3$, the mode shapes can be determined using the equation

$$\begin{bmatrix} 1-\beta_i & -1 & 0 \\ -1 & 3-1.5\beta_i & -2 \\ 0 & -2 & 5-2\beta_i \end{bmatrix} \begin{bmatrix} A_{i1} \\ A_{i2} \\ A_{i3} \end{bmatrix} = \begin{bmatrix} 0 \\ 0 \\ 0 \end{bmatrix}$$

which, by partitioning the coefficient matrix, leads to

$$\begin{bmatrix} -1 \\ 0 \end{bmatrix} A_{i1} + \begin{bmatrix} (3-1.5\beta_i) & -2 \\ -2 & 5-2\beta_i \end{bmatrix} \begin{bmatrix} A_{i2} \\ A_{i3} \end{bmatrix} = \begin{bmatrix} 0 \\ 0 \end{bmatrix}$$

or

$$\begin{bmatrix} A_{i2} \\ A_{i3} \end{bmatrix} = \frac{1}{3\beta_i^2 - 13.5\beta_i + 11} \begin{bmatrix} 5-2\beta_i \\ 2 \end{bmatrix} A_{i1}$$

Using the values obtained previously for β_i, $i = 1, 2, 3$, we have

$$\begin{bmatrix} A_{12} \\ A_{13} \end{bmatrix} = \begin{bmatrix} 0.649 \\ 0.302 \end{bmatrix} A_{11} \qquad \text{for} \quad \omega_1 = 14.52 \text{ rad/s}$$

$$\begin{bmatrix} A_{22} \\ A_{23} \end{bmatrix} = \begin{bmatrix} -0.607 \\ -0.679 \end{bmatrix} A_{21} \qquad \text{for } \omega_2 = 31.05 \text{ rad/s}$$

$$\begin{bmatrix} A_{32} \\ A_{33} \end{bmatrix} = \begin{bmatrix} -2.54 \\ 2.438 \end{bmatrix} A_{31} \qquad \text{for} \quad \omega_3 = 46.1 \text{ rad/s}$$

Since the mode shapes are determined to within an arbitrary constant, we may assume $A_{i1} = 1$, for $i = 1, 2, 3$. This leads to the following mode shapes:

$$\mathbf{A}_1 = \begin{bmatrix} A_{11} \\ A_{12} \\ A_{13} \end{bmatrix} = \begin{bmatrix} 1 \\ 0.649 \\ 0.302 \end{bmatrix}$$

$$\mathbf{A}_2 = \begin{bmatrix} A_{21} \\ A_{22} \\ A_{23} \end{bmatrix} = \begin{bmatrix} 1 \\ -0.607 \\ -0.679 \end{bmatrix}$$

$$\mathbf{A}_3 = \begin{bmatrix} A_{31} \\ A_{32} \\ A_{33} \end{bmatrix} = \begin{bmatrix} 1 \\ -2.54 \\ 2.438 \end{bmatrix}$$

The modal matrix $\mathbf{\Phi}$ is then defined as

$$\mathbf{\Phi} = \begin{bmatrix} 1 & 1 & 1 \\ 0.649 & -0.607 & -2.54 \\ 0.302 & -0.679 & 2.438 \end{bmatrix}$$

Figure 7 shows the modes of vibration of the system.

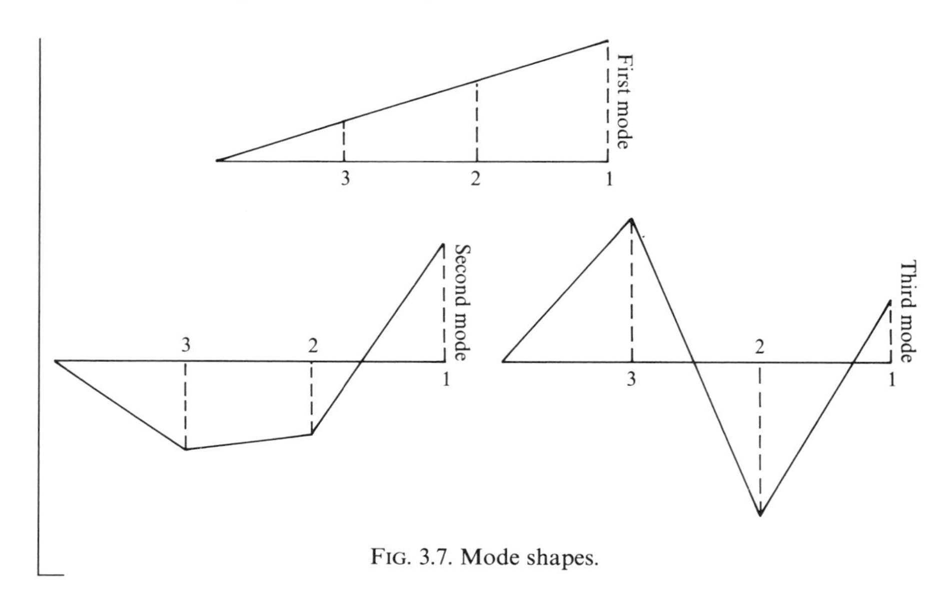

FIG. 3.7. Mode shapes.

3.3 ORTHOGONALITY OF THE MODE SHAPES

An important property of the mode shapes is the orthogonality. This property guarantees the existence of a unique solution to the system of algebraic equations given by Eqs. 42 and 43. In this section, we discuss this important property which can be used to obtain a set of n decoupled differential equations of motion for the multi-degree of freedom systems. These decoupled equations are expressed in terms of a new set of coordinates called *modal coordinates* defined using the modal matrix which is sometimes referred to as the *modal transformation.*

For the ith and jth natural frequencies ω_i and ω_j and the ith and jth mode shapes $\mathbf{A}_i$ and $\mathbf{A}_j$, Eq. 33 can be written as

$$\mathbf{K}\mathbf{A}_i = \omega_i^2 \mathbf{M}\mathbf{A}_i \tag{3.48}$$

$$\mathbf{K}\mathbf{A}_j = \omega_j^2 \mathbf{M}\mathbf{A}_j \tag{3.49}$$

Premultiplying Eq. 48 by the transpose of the vector $\mathbf{A}_j$ leads to

$$\mathbf{A}_j^{\mathrm{T}}\mathbf{K}\mathbf{A}_i = \omega_i^2 \mathbf{A}_j^{\mathrm{T}}\mathbf{M}\mathbf{A}_i \tag{3.50}$$

Taking the transpose of Eq. 49 and postmultiplying the resulting equation by the vector $\mathbf{A}_i$, we obtain

$$\mathbf{A}_j^{\mathrm{T}}\mathbf{K}^{\mathrm{T}}\mathbf{A}_i = \omega_j^2 \mathbf{A}_j^{\mathrm{T}}\mathbf{M}^{\mathrm{T}}\mathbf{A}_i$$

Since the mass and stiffness matrices are symmetric, the preceding equation leads to

$$\mathbf{A}_j^{\mathrm{T}}\mathbf{K}\mathbf{A}_i = \omega_j^2 \mathbf{A}_j^{\mathrm{T}}\mathbf{M}\mathbf{A}_i \tag{3.51}$$

Subtracting Eq. 51 from Eq. 50 yields

$$(\omega_i^2 - \omega_j^2)\mathbf{A}_j^{\mathrm{T}}\mathbf{M}\mathbf{A}_i = 0 \tag{3.52}$$

If ω_i^2 and ω_j^2 are distinct eigenvalues, that is, $\omega_i^2 \neq \omega_j^2$, and $\mathbf{M}$ is a positive-definite matrix, we conclude from Eq. 52 that

$$\begin{aligned} \mathbf{A}_j^{\mathrm{T}}\mathbf{M}\mathbf{A}_i &= 0 \qquad \text{for } i \neq j \\ &\neq 0 \qquad \text{for } i = j \end{aligned} \tag{3.53}$$

That is, the eigenvectors associated with distinct eigenvalues are orthogonal with respect to the mass matrix. The positive definiteness of the mass matrix guarantees that $\mathbf{A}_i^{\mathrm{T}}\mathbf{M}\mathbf{A}_i$ is not equal to zero for the nonzero vector $\mathbf{A}_i$. Therefore, we may write Eq. 52 as

$$\begin{aligned} \mathbf{A}_j^{\mathrm{T}}\mathbf{M}\mathbf{A}_i &= 0 \qquad \text{for } i \neq j \\ &= m_i \qquad \text{for } i = j \end{aligned} \tag{3.54}$$

where m_i is a real positive scalar.

The mode shapes are also orthogonal with respect to the stiffness matrix. This can be proved by writing Eqs. 48 and 49 in the following alternate form:

$$\mathbf{M}\mathbf{A}_i = \frac{1}{\omega_i^2}\mathbf{K}\mathbf{A}_i \tag{3.55}$$

$$\mathbf{M}\mathbf{A}_j = \frac{1}{\omega_j^2}\mathbf{K}\mathbf{A}_j \tag{3.56}$$

Following a procedure similar to the one used to prove the orthogonality of the mode shapes with respect to the mass matrix, one can easily verify that

$$\begin{aligned} \mathbf{A}_j^{\mathrm{T}}\mathbf{K}\mathbf{A}_i &= 0 \qquad \text{for } i \neq j \\ &= k_i \qquad \text{for } i = j \end{aligned} \tag{3.57}$$

where k_i is a nonnegative scalar.

Linear Independence of the Mode Shapes The orthogonality with respect to the mass or stiffness matrix can be used to show that the mode shapes of a multi-degree of freedom system are linearly independent, and as a consequence, any of these mode shapes cannot be written as a linear combination of the others. The linear independence is an important property which assures us that the modal matrix of Eq. 45 has a full rank and accordingly a solution of Eqs. 42 and 43 does exist.

The set of vectors $\mathbf{A}_1, \mathbf{A}_2, \ldots, \mathbf{A}_n$ are said to be linearly independent, if the following relationship,

$$\beta_1\mathbf{A}_1 + \beta_2\mathbf{A}_2 + \cdots + \beta_i\mathbf{A}_i + \cdots + \beta_n\mathbf{A}_n = \mathbf{0} \tag{3.58}$$

where β_i, $i = 1, 2, \ldots, n$, are scalars, holds only when the scalars $\beta_1, \beta_2, \ldots, \beta_n$ are all identically equal to zero. In order to prove that this is indeed the case, we premultiply Eq. 58 by $\mathbf{A}_i^{\mathrm{T}}\mathbf{M}$ and use the orthogonality conditions of

Eq. 54 to obtain

$$\beta_i \mathbf{A}_i^{\mathrm{T}} \mathbf{M} \mathbf{A}_i = \beta_i m_i = 0$$

Since m_i is not equal to zero, we conclude that

$$\beta_i = 0$$

for any $i = 1, 2, \ldots, n$. That is, the mode shapes of the multi-degree of freedom system are indeed linearly independent. Accordingly, the modal matrix $\mathbf{\Phi}$ of Eq. 45 has a full rank and is thus nonsingular.

In proving the orthogonality conditions, it was assumed that the eigenvalues ω_i^2 and ω_j^2 are distinct. In some engineering applications, however, the case of *repeated roots*, in which two or more eigenvalues are equal, may be encountered. For a general eigenvalue problem, the eigenvectors associated with repeated roots may or may not be linearly independent. Let ω_r^2 be an eigenvalue with multiplicity s, that is, $\omega_r^2, \omega_{r+1}^2, \ldots, \omega_{r+s-1}^2$ are equal eigenvalues. If the rank of the matrix $[\mathbf{K} - \omega_r^2 \mathbf{M}]$ is equal to $n - s$ where n is the number of the system coordinates, it can be proved that the system of equations

$$[\mathbf{K} - \omega_r^2 \mathbf{M}]\mathbf{A}_r = \mathbf{0}$$

has s nontrivial linearly independent eigenvectors $\mathbf{A}_r, \mathbf{A}_{r+1}, \ldots, \mathbf{A}_{r+s-1}$. If the rank of the matrix $[\mathbf{K} - \omega_r^2 \mathbf{M}]$ is greater than $n - s$, the vectors $\mathbf{A}_r, \mathbf{A}_{r+1}, \ldots,$ $\mathbf{A}_{r+s-1}$ associated with the repeated eigenvalues are not totally independent, and in this case, the number of linearly independent eigenvectors is less than the number of the system coordinates. However, if the mass matrix $\mathbf{M}$ and the stiffness matrix $\mathbf{K}$ are real symmetric, it can be shown (Wylie and Barrett, 1982) that the eigenvectors associated with repeated eigenvalues are linearly independent.

Modal Transformation The orthogonality of the mode shapes can be used to obtain a set of n uncoupled second-order differential equations in terms of a new set of coordinates called *modal coordinates*. Each of the resulting equations is similar to the equation of the single degree of freedom system.

For convenience, we reproduce the following matrix equation of motion of the undamped free vibration of the multi-degree of freedom system:

$$\mathbf{M}\ddot{\mathbf{q}} + \mathbf{K}\mathbf{q} = \mathbf{0} \tag{3.59}$$

We now make the following coordinate transformation:

$$\mathbf{q} = \mathbf{\Phi}\mathbf{P}$$

$$= \begin{bmatrix} \phi_{11} & \phi_{12} & \phi_{13} & \cdots & \phi_{1n} \\ \phi_{21} & \phi_{22} & \phi_{23} & \cdots & \phi_{2n} \\ \vdots & & & & \vdots \\ \phi_{n1} & \phi_{n2} & \phi_{n3} & \cdots & \phi_{nn} \end{bmatrix} \begin{bmatrix} P_1 \\ P_2 \\ \vdots \\ P_n \end{bmatrix} \tag{3.60a}$$

or equivalently,

$$q_j = \sum_{i=1}^{n} \phi_{ji} P_i \tag{3.60b}$$

where $\mathbf{\Phi}$ is the modal transformation defined by Eq. 45, and $\mathbf{P}$ is the vector of *modal coordinates*. Differentiating Eq. 60 twice with respect to time and substituting into Eq. 59 yields

$$\mathbf{M\Phi\ddot{P}} + \mathbf{K\Phi P} = \mathbf{0} \tag{3.61}$$

Premultiplying this equation by $\mathbf{\Phi}^{\mathrm{T}}$ yields

$$\mathbf{\Phi}^{\mathrm{T}}\mathbf{M\Phi\ddot{P}} + \mathbf{\Phi}^{\mathrm{T}}\mathbf{K\Phi P} = \mathbf{0} \tag{3.62}$$

Using the orthogonality of the mode shapes with respect to the mass and stiffness matrices (Eqs. 54 and 57), the above equation yields

$$\mathbf{M}_p\mathbf{\ddot{P}} + \mathbf{K}_p\mathbf{P} = \mathbf{0} \tag{3.63}$$

where

$$\mathbf{M}_p = \mathbf{\Phi}^{\mathrm{T}}\mathbf{M\Phi} = \begin{bmatrix} \mathbf{A}_1^{\mathrm{T}}\mathbf{M}\mathbf{A}_1 & 0 & 0 & \cdots & 0 \\ 0 & \mathbf{A}_2^{\mathrm{T}}\mathbf{M}\mathbf{A}_2 & 0 & \cdots & 0 \\ 0 & 0 & \mathbf{A}_3^{\mathrm{T}}\mathbf{M}\mathbf{A}_3 & \cdots & 0 \\ \vdots & \vdots & \vdots & \ddots & \vdots \\ 0 & 0 & 0 & \cdots & \mathbf{A}_n^{\mathrm{T}}\mathbf{M}\mathbf{A}_n \end{bmatrix}$$

$$= \begin{bmatrix} m_1 & 0 & 0 & \cdots & 0 \\ 0 & m_2 & 0 & \cdots & 0 \\ 0 & 0 & m_3 & \cdots & 0 \\ \vdots & \vdots & \vdots & \ddots & \vdots \\ 0 & 0 & 0 & \cdots & m_n \end{bmatrix} \tag{3.64}$$

and

$$\mathbf{K}_p = \mathbf{\Phi}^{\mathrm{T}}\mathbf{K\Phi} = \begin{bmatrix} \mathbf{A}_1^{\mathrm{T}}\mathbf{K}\mathbf{A}_1 & 0 & 0 & \cdots & 0 \\ 0 & \mathbf{A}_2^{\mathrm{T}}\mathbf{K}\mathbf{A}_2 & 0 & \cdots & 0 \\ 0 & 0 & \mathbf{A}_3^{\mathrm{T}}\mathbf{K}\mathbf{A}_3 & \cdots & 0 \\ \vdots & \vdots & \vdots & \ddots & \vdots \\ 0 & 0 & 0 & \cdots & \mathbf{A}_n^{\mathrm{T}}\mathbf{K}\mathbf{A}_n \end{bmatrix}$$

$$= \begin{bmatrix} k_1 & 0 & 0 & \cdots & 0 \\ 0 & k_2 & 0 & \cdots & 0 \\ 0 & 0 & k_3 & \cdots & 0 \\ \vdots & \vdots & \vdots & \ddots & \vdots \\ 0 & 0 & 0 & \cdots & k_n \end{bmatrix} \tag{3.65}$$

Note that $\mathbf{M}_p$ and $\mathbf{K}_p$ are diagonal matrices. The matrix $\mathbf{M}_p$ is called the *modal mass matrix* and the matrix $\mathbf{K}_p$ is called the *modal stiffness matrix*. The scalars m_i and k_i defined by Eqs. 54 and 57 are called, respectively, the *modal mass* and *stiffness coefficients*.

Owing to the fact that the modal mass and stiffness matrices $\mathbf{M}_p$ and $\mathbf{K}_p$ are diagonal, the modal coordinates in Eq. 63 are not coupled. That is, Eq. 63 contains n uncoupled second-order ordinary differential equations which can be written as

$$m_i\ddot{P}_i + k_iP_i = 0, \qquad i = 1, 2, \ldots, n \tag{3.66}$$

This equation is similar to the equation of motion of the undamped single degree of freedom system. Furthermore, it is clear from Eq. 48 that the natural frequency ω_i can be obtained using the equation

$$\omega_i^2 = \frac{\mathbf{A}_i^{\mathrm{T}}\mathbf{K}\mathbf{A}_i}{\mathbf{A}_i^{\mathrm{T}}\mathbf{M}\mathbf{A}_i} \tag{3.67}$$

or equivalently

$$\omega_i^2 = \frac{k_i}{m_i} \tag{3.68}$$

That is, the square of the natural frequency of a mode is equal to the modal stiffness divided by the modal mass; a relationship which is similar to the definition used in the case of single degree of freedom systems.

The solutions of the differential equations of Eq. 66 take the form

$$P_i = C_i \sin(\omega_i t + \psi_i), \qquad i = 1, 2, \ldots, n \tag{3.69}$$

in which C_i is the amplitude and ψ_i is the phase angle. The constants C_i and ψ_i can be determined from the initial conditions on the modal coordinates. These initial conditions can be obtained using the transformation of Eq. 60 as

$$\mathbf{P}_0 = \mathbf{\Phi}^{-1}\mathbf{q}_0 \tag{3.70}$$

where $\mathbf{P}_0$ is the vector of the initial values of the modal coordinates, $\mathbf{q}_0$ is the vector of the given initial displacements, and $\mathbf{\Phi}^{-1}$ is the inverse of the modal matrix. Similarly,

$$\dot{\mathbf{P}}_0 = \mathbf{\Phi}^{-1}\dot{\mathbf{q}}_0 \tag{3.71}$$

Thus, the initial modal coordinates and velocities can be determined once the initial displacements and velocities are given. The constants C_i and ψ_i can be expressed in terms of the initial modal coordinates and velocities as

$$C_i = \sqrt{\left(\frac{\dot{P}_{i0}}{\omega_i}\right)^2 + (P_{i0})^2} \tag{3.72}$$

$$\psi_i = \tan^{-1}\frac{\omega_i P_{i0}}{\dot{P}_{i0}} \tag{3.73}$$

The jth physical coordinate can then be determined using Eqs. 60 and 69 as

$$\begin{aligned} q_j &= \sum_{i=1}^{n} \phi_{ji} P_i = \sum_{i=1}^{n} \phi_{ji} C_i \sin(\omega_i t + \psi_i) \\ &= \sum_{i=1}^{n} C_i \mathbf{A}_i \sin(\omega_i t + \psi_i) \end{aligned} \tag{3.74}$$

which is the same solution obtained in the preceding section and defined by Eq. 34, in which the coefficients of the mode shapes are the modal coordinates defined by Eq. 69.

Inverse of the Modal Matrix In Eqs. 70 and 71, the initial modal coordinates and velocities are expressed in terms of the inverse of the modal matrix. The orthogonality of the modal matrix with respect to the mass or stiffness matrices can be used to define the inverse of this matrix. From Eq. 64 we have

$$\mathbf{\Phi}^{\mathrm{T}}\mathbf{M}\mathbf{\Phi} = \mathbf{M}_p$$

where $\mathbf{M}_p$ is a diagonal matrix whose inverse can be easily obtained. Pre-multiplying the preceding equation by the inverse of the matrix $\mathbf{M}_p$, one obtains

$$\mathbf{M}_p^{-1}\mathbf{\Phi}^{\mathrm{T}}\mathbf{M}\mathbf{\Phi} = \mathbf{I}$$

where $\mathbf{I}$ is the $n \times n$ identity matrix. By postmultiplying the above equation by $\mathbf{\Phi}^{-1}$, we get

$$\mathbf{M}_p^{-1}\mathbf{\Phi}^{\mathrm{T}}\mathbf{M}\mathbf{\Phi}\mathbf{\Phi}^{-1} = \mathbf{\Phi}^{-1}$$

or

$$\mathbf{\Phi}^{-1} = \mathbf{M}_p^{-1}\mathbf{\Phi}^{\mathrm{T}}\mathbf{M}$$

This equation is useful in defining the initial modal coordinates and velocities given in Eqs. 70 and 71.

Normalized Mode Shapes In many situations, it is desirable to normalize the mode shapes with respect to the mass matrix or with respect to the stiffness matrix. In order to normalize the mode shapes with respect to the mass matrix, we divide each mode shape by the square root of the corresponding modal mass coefficient, that is,

$$\mathbf{A}_{im} = \frac{\mathbf{A}_i}{\sqrt{m_i}} = \frac{\mathbf{A}_i}{\sqrt{\mathbf{A}_i^{\mathrm{T}}\mathbf{M}\mathbf{A}_i}}, \qquad i = 1, 2, \ldots, n$$

where $\mathbf{A}_{im}$ is the mode shape i normalized with respect to the mass matrix. It is clear that

$$\mathbf{A}_{im}^{\mathrm{T}}\mathbf{M}\mathbf{A}_{im} = \frac{1}{m_i}\mathbf{A}_i^{\mathrm{T}}\mathbf{M}\mathbf{A}_i = 1$$

or

$$\mathbf{\Phi}_{\mathrm{m}}^{\mathrm{T}}\mathbf{M}\mathbf{\Phi}_{\mathrm{m}} = \mathbf{I}$$

where $\mathbf{I}$ is an $n \times n$ identity matrix, and $\mathbf{\Phi}_{\mathrm{m}}$ is the modal matrix whose columns are the mode shapes which are normalized with respect to the mass matrix. We also observe in this case the following

$$\mathbf{A}_{im}^{\mathrm{T}}\mathbf{K}\mathbf{A}_{im} = \frac{1}{m_i}\mathbf{A}_i^{\mathrm{T}}\mathbf{K}\mathbf{A}_i = \frac{k_i}{m_i} = \omega_i^2$$

It follows that

$$\mathbf{\Phi}_{\mathrm{m}}^{\mathrm{T}}\mathbf{K}\mathbf{\Phi}_{\mathrm{m}} = \boldsymbol{\omega}_{\mathrm{m}}$$

where $\boldsymbol{\omega}_{\mathrm{m}}$ is a diagonal matrix whose diagonal elements are the square of the system natural frequencies. This matrix is defined as

$$\boldsymbol{\omega}_{\mathrm{m}} = \begin{bmatrix} \omega_1^2 & 0 & 0 & \cdots & 0 \\ 0 & \omega_2^2 & 0 & \cdots & 0 \\ 0 & 0 & \omega_3^2 & \cdots & 0 \\ \vdots & \vdots & \vdots & \ddots & \vdots \\ 0 & 0 & 0 & \cdots & \omega_n^2 \end{bmatrix}$$

The mode shapes can also be normalized with respect to the stiffness matrix. To this end, we divide each mode shape by the square root of the corresponding modal stiffness coefficient, that is,

$$\mathbf{A}_{is} = \frac{\mathbf{A}_i}{\sqrt{k_i}} = \frac{\mathbf{A}_i}{\sqrt{\mathbf{A}_i^{\mathrm{T}}\mathbf{K}\mathbf{A}_i}}, \qquad i = 1, 2, \ldots, n$$

where $\mathbf{A}_{is}$ is the ith mode shape normalized with respect to the stiffness matrix. Clearly, in this case we have

$$\mathbf{A}_{is}^{\mathrm{T}}\mathbf{K}\mathbf{A}_{is} = \frac{1}{k_i}\mathbf{A}_i^{\mathrm{T}}\mathbf{K}\mathbf{A}_i = 1$$

$$\mathbf{A}_{is}^{\mathrm{T}}\mathbf{M}\mathbf{A}_{is} = \frac{1}{k_i}\mathbf{A}_i^{\mathrm{T}}\mathbf{M}\mathbf{A}_i = \frac{m_i}{k_i} = \frac{1}{\omega_i^2}$$

and as a consequence,

$$\boldsymbol{\Phi}_{\mathrm{s}}^{\mathrm{T}}\mathbf{K}\boldsymbol{\Phi}_{\mathrm{s}} = \mathbf{I}$$

$$\boldsymbol{\Phi}_{\mathrm{s}}^{\mathrm{T}}\mathbf{M}\boldsymbol{\Phi}_{\mathrm{s}} = \boldsymbol{\omega}_{\mathrm{s}}$$

where $\boldsymbol{\Phi}_{\mathrm{s}}$ is the modal matrix whose columns are the mode shapes normalized with respect to the stiffness matrix, and $\boldsymbol{\omega}_{\mathrm{s}}$ is the matrix

$$\boldsymbol{\omega}_{\mathrm{s}} = \begin{bmatrix} 1/\omega_1^2 & 0 & 0 & \cdots & 0 \\ 0 & 1/\omega_2^2 & 0 & \cdots & 0 \\ 0 & 0 & 1/\omega_3^2 & \cdots & 0 \\ \vdots & \vdots & \vdots & \ddots & \vdots \\ 0 & 0 & 0 & \cdots & 1/\omega_n^2 \end{bmatrix}$$

The matrix $\boldsymbol{\Phi}_{\mathrm{m}}$ is said to be *orthonormal with respect to the mass matrix*, while the matrix $\boldsymbol{\Phi}_{\mathrm{s}}$ is said to be *orthonormal with respect to the stiffness matrix.*

Example 3.3

In Example 2, the mass and stiffness matrices were given by

$$\mathbf{M} = 2 \times 10^3 \begin{bmatrix} 1 & 0 & 0 \\ 0 & 1.5 & 0 \\ 0 & 0 & 2 \end{bmatrix} \mathrm{kg \cdot m^2}$$

$$\mathbf{K} = 12 \times 10^5 \begin{bmatrix} 1 & -1 & 0 \\ -1 & 3 & -2 \\ 0 & -2 & 5 \end{bmatrix} \mathrm{N \cdot m}$$

The natural frequencies were found to be $\omega_1 = 14.52$ rad/s, $\omega_2 = 31.05$ rad/s, and $\omega_3 = 46.1$ rad/s and the modal matrix

$$\boldsymbol{\Phi} = \begin{bmatrix} 1.000 & 1.000 & 1.000 \\ 0.649 & -0.607 & -2.541 \\ 0.302 & -0.679 & 2.438 \end{bmatrix}$$

This modal matrix is orthogonal with respect to the mass and stiffness matrices. One can show the following:

$$\mathbf{M}_p = \boldsymbol{\Phi}^{\mathrm{T}}\mathbf{M}\boldsymbol{\Phi} = 2 \times 10^3 \begin{bmatrix} 1.814 & 0 & 0 \\ 0 & 2.475 & 0 \\ 0 & 0 & 22.573 \end{bmatrix} = \begin{bmatrix} m_1 & 0 & 0 \\ 0 & m_2 & 0 \\ 0 & 0 & m_3 \end{bmatrix}$$

$$\mathbf{K}_p = \boldsymbol{\Phi}^{\mathrm{T}}\mathbf{K}\boldsymbol{\Phi} = 2 \times 10^5 \begin{bmatrix} 3.828 & 0 & 0 \\ 0 & 23.86 & 0 \\ 0 & 0 & 497.7 \end{bmatrix}$$

$$= \begin{bmatrix} k_1 & 0 & 0 \\ 0 & k_2 & 0 \\ 0 & 0 & k_3 \end{bmatrix} = \begin{bmatrix} \omega_1^2 m_1 & 0 & 0 \\ 0 & \omega_2^2 m_2 & 0 \\ 0 & 0 & \omega_3^2 m_3 \end{bmatrix}$$

where m_i and k_i are, respectively, the modal mass and stiffness coefficients.

The modal matrix can be normalized with respect to the mass matrix $\mathbf{M}$ or with respect to the stiffness matrix $\mathbf{K}$. For example, in order to make the modal matrix orthonormal with respect to the stiffness matrix, for each mode shape $\mathbf{A}_i$ we use the following equation:

$$\mathbf{A}_{is} = \frac{\mathbf{A}_i}{\sqrt{\mathbf{A}_i^{\mathrm{T}}\mathbf{K}\mathbf{A}_i}}$$

It follows that

$$\mathbf{A}_{1s} = \frac{\mathbf{A}_1}{\sqrt{\mathbf{A}_1^{\mathrm{T}}\mathbf{K}\mathbf{A}_1}} = \frac{1}{8.7497 \times 10^2} \begin{bmatrix} 1.000 \\ 0.649 \\ 0.302 \end{bmatrix} = 10^{-2} \begin{bmatrix} 0.1143 \\ 0.0742 \\ 0.0345 \end{bmatrix}$$

$$\mathbf{A}_{2s} = \frac{\mathbf{A}_2}{\sqrt{\mathbf{A}_2^{\mathrm{T}}\mathbf{K}\mathbf{A}_2}} = \frac{1}{21.845 \times 10^2} \begin{bmatrix} 1.000 \\ -0.607 \\ -0.679 \end{bmatrix} = 10^{-2} \begin{bmatrix} 0.0458 \\ -0.0278 \\ -0.0311 \end{bmatrix}$$

$$\mathbf{A}_{3s} = \frac{\mathbf{A}_3}{\sqrt{\mathbf{A}_3^{\mathrm{T}}\mathbf{K}\mathbf{A}_3}} = \frac{1}{97.949 \times 10^2} \begin{bmatrix} 1.000 \\ -2.541 \\ 2.438 \end{bmatrix} = 10^{-2} \begin{bmatrix} 0.0102 \\ -0.0259 \\ 0.0249 \end{bmatrix}$$

The new modal matrix which is orthonormal with respect to the stiffness matrix is then given by

$$\boldsymbol{\Phi}_s = 10^{-2} \begin{bmatrix} 0.1143 & 0.0458 & 0.0102 \\ 0.0742 & -0.0278 & -0.0259 \\ 0.0345 & -0.0311 & 0.0249 \end{bmatrix}$$

One can show that

$$\mathbf{\Phi}_s^T \mathbf{K} \mathbf{\Phi}_s = \mathbf{I}$$

where $\mathbf{I}$ is the 3×3 identity matrix, that is, the modal stiffness coefficients are all equal to one. One can also show that

$$\mathbf{\Phi}_s^T \mathbf{M} \mathbf{\Phi}_s = 10^{-3} \begin{bmatrix} 4.743 & 0 & 0 \\ 0 & 1.0372 & 0 \\ 0 & 0 & 0.4705 \end{bmatrix} = \begin{bmatrix} 1/\omega_1^2 & 0 & 0 \\ 0 & 1/\omega_2^2 & 0 \\ 0 & 0 & 1/\omega_3^2 \end{bmatrix}$$

Similarly, if the modal matrix is made orthonormal with respect to the mass matrix, one has

$$\mathbf{\Phi}_m^T \mathbf{M} \mathbf{\Phi}_m = \mathbf{I}$$

and

$$\mathbf{\Phi}_m^T \mathbf{K} \mathbf{\Phi}_m = 10^3 \begin{bmatrix} 0.2108 & 0 & 0 \\ 0 & 0.964 & 0 \\ 0 & 0 & 2.1254 \end{bmatrix} = \begin{bmatrix} \omega_1^2 & 0 & 0 \\ 0 & \omega_2^2 & 0 \\ 0 & 0 & \omega_3^2 \end{bmatrix}$$

That is, the modal mass coefficient $m_i = 1$ for $i = 1, 2, 3$ and the modal stiffness coefficient $k_i = \omega_i^2$, $i = 1, 2, 3$.

Example 3.4

In this example, we consider the free vibration of the system in the preceding example as the result of the initial conditions

$$\mathbf{\theta}_0 = \begin{bmatrix} 0.1 \\ 0.05 \\ 0.01 \end{bmatrix} \text{rad} \quad \text{and} \quad \dot{\mathbf{\theta}}_0 = \begin{bmatrix} 10 \\ 15 \\ 20 \end{bmatrix} \text{rad/s}$$

The equations of motion of this system in terms of the modal coordinates are given by Eq. 66. The solution of these equations are given by Eq. 69 and the arbitrary constants are defined by Eqs. 72 and 73. The initial modal coordinates and velocities can be obtained using Eqs. 70 and 71. It is required, however, in these equations to evaluate the inverse of the modal matrix. We have previously shown that

$$\mathbf{\Phi}^{-1} = \mathbf{M}_p^{-1} \mathbf{\Phi}^T \mathbf{M}$$

By using this equation and the results of the preceding example, the inverse of the modal matrix $\mathbf{\Phi}^{-1}$ may be written as

$$\mathbf{\Phi}^{-1} = \begin{bmatrix} 0.2755 & 0 & 0 \\ 0 & 0.202 & 0 \\ 0 & 0 & 0.02215 \end{bmatrix} \begin{bmatrix} 1.000 & 0.649 & 0.302 \\ 1.000 & -0.607 & -0.679 \\ 1.000 & -2.541 & 2.438 \end{bmatrix} \begin{bmatrix} 2.0 & 0 & 0 \\ 0 & 3.0 & 0 \\ 0 & 0 & 4.0 \end{bmatrix}$$

$$= \begin{bmatrix} 0.551 & 0.537 & 0.333 \\ 0.404 & -0.368 & -0.549 \\ 0.044 & -0.169 & 0.216 \end{bmatrix}$$

The initial modal coordinates are given by

$$\mathbf{P}_0 = \mathbf{\Phi}^{-1}\mathbf{\theta}_0 = \begin{bmatrix} 0.551 & 0.537 & 0.333 \\ 0.404 & -0.368 & -0.549 \\ 0.044 & -0.169 & 0.216 \end{bmatrix} \begin{bmatrix} 0.1 \\ 0.05 \\ 0.01 \end{bmatrix} = \begin{bmatrix} 0.085 \\ 0.017 \\ -0.002 \end{bmatrix} \text{rad}$$

and the initial modal velocities are

$$\dot{\mathbf{P}}_0 = \mathbf{\Phi}^{-1}\dot{\mathbf{\theta}}_0 = \begin{bmatrix} 0.551 & 0.537 & 0.333 \\ 0.404 & -0.368 & -0.549 \\ 0.044 & -0.169 & 0.216 \end{bmatrix} \begin{bmatrix} 10 \\ 15 \\ 20 \end{bmatrix} = \begin{bmatrix} 20.225 \\ -12.460 \\ 2.228 \end{bmatrix} \text{rad/s}$$

Using the initial modal coordinates and velocities, the modal coordinates can be defined by using Eq. 69, or its equivalent form given by

$$P_i = P_{i0} \cos \omega_i t + \frac{\dot{P}_{i0}}{\omega_i} \sin \omega_i t, \qquad i = 1, 2, 3$$

which yields

$$\mathbf{P} = \begin{bmatrix} P_1 \\ P_2 \\ P_3 \end{bmatrix} = \begin{bmatrix} 0.085 \cos 14.52t + 1.393 \sin 14.52t \\ 0.017 \cos 31.05t - 0.401 \sin 31.05t \\ -0.002 \cos 46.1t + 0.048 \sin 46.1t \end{bmatrix}$$

The physical coordinates $\mathbf{\theta}$ can then be obtained using the relationship

$$\mathbf{\theta} = \mathbf{\Phi P}$$

Special Case As was pointed out earlier, in order to determine the initial conditions of the modal coordinates $\mathbf{P}_0$ and $\dot{\mathbf{P}}_0$, it is not necessary to obtain the inverse of the modal matrix shown in Eqs. 70 and 71. If the number of degrees of freedom is large, an LU factorization of the modal matrix can be used to solve for the initial modal coordinates and velocities. The orthogonality of the mode shapes with respect to the mass or stiffness matrices can also be used as an alternative to find the inverse of the modal matrix. In this later case, as shown previously, the inverse of the modal matrix can be expressed in terms of the inverse of the diagonal modal mass matrix as

$$\mathbf{\Phi}^{-1} = \mathbf{M}_p^{-1}\mathbf{\Phi}^{\mathrm{T}}\mathbf{M}$$

where $\mathbf{M}_p$ is the diagonal matrix defined in Eq. 64. Therefore, the initial conditions associated with the modal coordinates can be obtained using Eqs. 70 and 71. These two equations yield

$$\mathbf{M}_p\mathbf{P}_0 = \mathbf{\Phi}^{\mathrm{T}}\mathbf{M}\mathbf{q}_0$$

$$\mathbf{M}_p\dot{\mathbf{P}}_0 = \mathbf{\Phi}^{\mathrm{T}}\mathbf{M}\dot{\mathbf{q}}_0$$

Using these equations, one can show that if the vectors of initial coordinates $\mathbf{q}_0$ and initial velocities $\dot{\mathbf{q}}_0$ are proportional to a given mode shape, then the multi-degree of freedom system will oscillate in this mode of vibration, a case which is equivalent to the vibration of a single degree of freedom system. In order to demonstrate this, let us consider the special case in which $\mathbf{q}_0$ and $\dot{\mathbf{q}}_0$ are proportional to the mode shape i. One can then write $\mathbf{q}_0$ and $\dot{\mathbf{q}}_0$ as

$$\mathbf{q}_0 = d_1 \mathbf{A}_i$$
$$\dot{\mathbf{q}}_0 = d_2 \mathbf{A}_i$$

where d_1 and d_2 are proportionality constants. Therefore, the initial conditions associated with the modal coordinates can be expressed as

$$\mathbf{M}_p \mathbf{P}_0 = d_1 \mathbf{\Phi}^{\mathrm{T}} \mathbf{M} \mathbf{A}_i$$
$$\mathbf{M}_p \dot{\mathbf{P}}_0 = d_2 \mathbf{\Phi}^{\mathrm{T}} \mathbf{M} \mathbf{A}_i$$

Because of the orthogonality of the mode shapes with respect to the mass matrix, one can easily verify that the preceding equations yield

$$m_j P_{j0} = \begin{cases} d_1 m_j, & j = i \\ 0, & j \neq i \end{cases}$$

and

$$m_j \dot{P}_{j0} = \begin{cases} d_2 m_j, & j = i \\ 0, & j \neq i \end{cases}$$

which yields the following:

$$P_{j0} = \begin{cases} d_1, & j = i \\ 0, & j \neq i \end{cases}$$

$$\dot{P}_{j0} = \begin{cases} d_2, & j = i \\ 0, & j \neq i \end{cases}$$

which implies that among the equations

$$m_j \ddot{P}_j + k_j P_j = 0, \qquad j = 1, 2, \ldots, n$$

only one equation which is associated with mode i has a nontrivial solution and the vibration of the multi-degree of freedom system is indeed equivalent to the vibration of a single degree of freedom system. Therefore, every mode shape can be excited independently. Similarly, one can show that if the vector of initial coordinates or velocities is proportional to a linear combination of m mode shapes where $m < n$, only m mode shapes are excited and only m equations associated with m modal coordinates have nontrivial solutions. In this special case, the motion of the n-degree of freedom system is equivalent to the motion of an m-degree of freedom system.

3.4 RIGID-BODY MODES

There are situations in which one or more of the eigenvalues may be equal to zero. This is the case of a *semidefinite system*, in which the eigenvector associated with the zero eigenvalue corresponds to a rigid-body mode of vibration.

In a rigid body mode, the system can move as a rigid body without deformation in the elastic elements. For a semidefinite system, the stiffness matrix is positive semidefinite and, consequently, its determinant is equal to zero. By using Eq. 68, one may rewrite Eq. 66 for mode k as

$$\ddot{P}_k + \omega_k^2 P_k = 0$$

If $\omega_k = 0$, which is the case of a *rigid-body mode*, the preceding equation leads to

$$\ddot{P}_k = 0$$

which can be integrated to obtain the modal coordinate as

$$P_k = \dot{P}_{k0} t + P_{k0}$$

where P_{k0} and $\dot{P}_{k0}$ are, respectively, the initial modal coordinates and velocities associated with the rigid-body mode.

Observe that for the rigid body mode k, Eq. 33 reduces to

$$\mathbf{K}\mathbf{A}_k = \mathbf{0}$$

Since $\mathbf{K}$ is a singular matrix, the preceding system of equations has a nontrivial solution, and there exists a nonzero vector $\mathbf{A}_k$ such that

$$\mathbf{A}_k^{\mathrm{T}} \mathbf{K} \mathbf{A}_k = 0$$

As a consequence, there is no change in the potential energy of the system as the result of the rigid-body motion. Observe also that the rank of the stiffness matrix must be equal to the number of degrees of freedom minus the number of the rigid-body modes of the system. Consequently, the equation $\mathbf{K}\mathbf{A}_k = \mathbf{0}$ has as many independent nontrivial solutions as the number of the rigid-body modes of the system. In general, the degree of singularity or the rank deficiency of the matrix $\mathbf{K}$ defines the number of dependent equations in the system $\mathbf{K}\mathbf{A}_k = \mathbf{0}$. The number of the remaining independent equations is less than the number of the elements of the vector $\mathbf{A}_k$, and as such, these independent equations can be solved for only a number of unknown elements equal to the rank of the matrix. The other elements can be treated as independent variables that can be varied arbitrarily in order to define a number of independent solutions equal to $(n - r)$, where n and r are, respectively, the dimension and rank of the stiffness matrix.

It is important to emphasize at this point that while $\mathbf{A}_k^{\mathrm{T}} \mathbf{K} \mathbf{A}_k = 0$ for the rigid-body mode k, $\mathbf{A}_k^{\mathrm{T}} \mathbf{M} \mathbf{A}_k$ is not equal to zero. However, the equation

$$\omega_k^2 \mathbf{A}_k^{\mathrm{T}} \mathbf{M} \mathbf{A}_k = \mathbf{A}_k^{\mathrm{T}} \mathbf{K} \mathbf{A}_k$$

is satisfied since $\omega_k = 0$. Nonrigid-body modes are sometimes referred to as *deformation modes*. The complete solution of the equations of free vibration can thus be expressed as a linear combination of the rigid-body and deformation modes.

For the purpose of demonstration, consider the four degree of freedom system shown in Fig. 8. The equation of motion of this system can be developed using the *principle of virtual work in dynamics* discussed in the preceding chapter. For this system, the virtual work of the inertia and applied forces

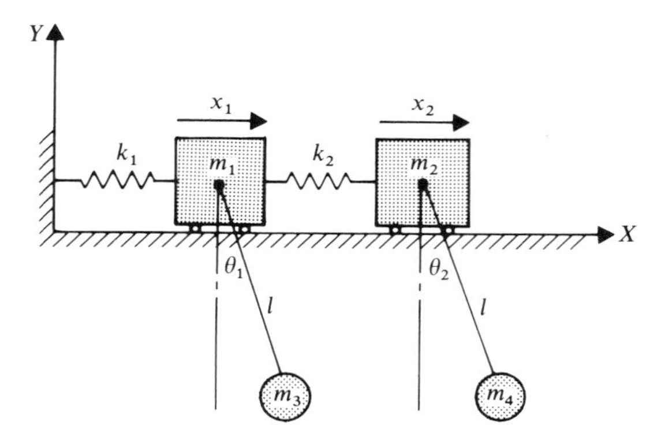

FIG. 3.8. Semidefinite system.

can be written in the case of small oscillations as

$$\delta W_i = m_1 \ddot{x}_1 \, \delta x_1 + m_2 \ddot{x}_2 \, \delta x_2 + m_3 \ddot{x}_3 \, \delta x_3 + m_4 \ddot{x}_4 \, \delta x_4$$

$$\delta W_e = -k_1 x_1 \, \delta x_1 - k_2 (x_2 - x_1) \, \delta(x_2 - x_1) - m_3 g \, \delta y_3 - m_4 g \, \delta y_4$$

where x_3, x_4, y_3, and y_4 in the case of small oscillations are given by

$$x_3 = x_1 + l \sin \theta_1 \approx x_1 + l\theta_1$$

$$x_4 = x_2 + l \sin \theta_2 \approx x_2 + l\theta_2$$

$$y_3 = -l \cos \theta_1$$

$$y_4 = -l \cos \theta_2$$

The coordinates $\ddot{x}_3$ and $\ddot{x}_4$ in the case of small angular oscillations are

$$\ddot{x}_3 = \ddot{x}_1 + l\ddot{\theta}_1$$

$$\ddot{x}_4 = \ddot{x}_2 + l\ddot{\theta}_2$$

Therefore, the virtual work of the inertia forces is given by

$$\begin{aligned} \delta W_i &= m_1 \ddot{x}_1 \, \delta x_1 + m_2 \ddot{x}_2 \, \delta x_2 + m_3 (\ddot{x}_1 + l\ddot{\theta}_1) \, \delta(x_1 + l\theta_1) \\ &\quad + m_4 (\ddot{x}_2 + l\ddot{\theta}_2) \, \delta(x_2 + l\theta_2) \\ &= [(m_1 + m_3)\ddot{x}_1 + m_3 l\ddot{\theta}_1] \, \delta x_1 + [(m_2 + m_4)\ddot{x}_2 + m_4 l\ddot{\theta}_2] \, \delta x_2 \\ &\quad + m_3 l(\ddot{x}_1 + l\ddot{\theta}_1) \, \delta\theta_1 + m_4 l(\ddot{x}_2 + l\ddot{\theta}_2) \, \delta\theta_2 \end{aligned}$$

which can be written in a matrix form as

$$\delta W_i = [\delta x_1 \quad \delta x_2 \quad \delta\theta_1 \quad \delta\theta_2] \begin{bmatrix} (m_1 + m_3) & 0 & m_3 l & 0 \\ 0 & (m_2 + m_4) & 0 & m_4 l \\ m_3 l & 0 & m_3 l^2 & 0 \\ 0 & m_4 l & 0 & m_4 l^2 \end{bmatrix} \begin{bmatrix} \ddot{x}_1 \\ \ddot{x}_2 \\ \ddot{\theta}_1 \\ \ddot{\theta}_2 \end{bmatrix}$$

$$= \delta \mathbf{q}^{\mathrm{T}} \mathbf{Q}_i$$

where $\mathbf{q} = [x_1 \;\; x_2 \;\; \theta_1 \;\; \theta_2]^{\mathrm{T}}$, and $\mathbf{Q}_{\mathrm{i}}$ is the vector of generalized inertia forces given by

$$\mathbf{Q}_{\mathrm{i}} = \begin{bmatrix} (m_1 + m_3) & 0 & m_3 l & 0 \\ 0 & (m_2 + m_4) & 0 & m_4 l \\ m_3 l & 0 & m_3 l^2 & 0 \\ 0 & m_4 l & 0 & m_4 l^2 \end{bmatrix} \begin{bmatrix} \ddot{x}_1 \\ \ddot{x}_2 \\ \ddot{\theta}_1 \\ \ddot{\theta}_2 \end{bmatrix}$$

The virtual work of the generalized applied forces is

$$\begin{aligned} \delta W_{\mathrm{e}} &= -k_1 x_1\, \delta x_1 - k_2(x_2 - x_1)\, \delta(x_2 - x_1) - m_3 g l \theta_1\, \delta\theta_1 - m_4 g l \theta_2\, \delta\theta_2 \\ &= -[(k_1 + k_2)x_1 - k_2 x_2]\, \delta x_1 - (k_2 x_2 - k_2 x_1)\, \delta x_2 \\ &\quad - m_3 g l \theta_1\, \delta\theta_1 - m_4 g l \theta_2\, \delta\theta_2 \end{aligned}$$

which can be written in a matrix form as

$$\begin{aligned} \delta W_{\mathrm{e}} &= -[\delta x_1 \quad \delta x_2 \quad \delta\theta_1 \quad \delta\theta_2] \begin{bmatrix} (k_1 + k_2) & -k_2 & 0 & 0 \\ -k_2 & k_2 & 0 & 0 \\ 0 & 0 & m_3 g l & 0 \\ 0 & 0 & 0 & m_4 g l \end{bmatrix} \begin{bmatrix} x_1 \\ x_2 \\ \theta_1 \\ \theta_2 \end{bmatrix} \\ &= \delta\mathbf{q}^{\mathrm{T}}\mathbf{Q}_{\mathrm{e}} \end{aligned}$$

where $\mathbf{Q}_{\mathrm{e}}$ is the vector of generalized applied forces defined as

$$\mathbf{Q}_{\mathrm{e}} = \begin{bmatrix} (k_1 + k_2) & -k_2 & 0 & 0 \\ -k_2 & k_2 & 0 & 0 \\ 0 & 0 & m_3 g l & 0 \\ 0 & 0 & 0 & m_4 g l \end{bmatrix} \begin{bmatrix} x_1 \\ x_2 \\ \theta_1 \\ \theta_2 \end{bmatrix}$$

The principle of virtual work in dynamics states that

$$\delta W_{\mathrm{i}} = \delta W_{\mathrm{e}}$$

Since the coordinates x_1, x_2, θ_1, and θ_2 are independent, one has as the result of the application of the principle of virtual work in dynamics

$$\mathbf{Q}_{\mathrm{i}} = \mathbf{Q}_{\mathrm{e}}$$

which leads to the differential equation

$$\mathbf{M}\ddot{\mathbf{q}} + \mathbf{K}\mathbf{q} = \mathbf{0}$$

where the mass matrix $\mathbf{M}$ and the stiffness matrix $\mathbf{K}$ are

$$\mathbf{M} = \begin{bmatrix} (m_1 + m_3) & 0 & m_3 l & 0 \\ 0 & (m_2 + m_4) & 0 & m_4 l \\ m_3 l & 0 & m_3 l^2 & 0 \\ 0 & m_4 l & 0 & m_4 l^2 \end{bmatrix}$$

$$\mathbf{K} = \begin{bmatrix} (k_1 + k_2) & -k_2 & 0 & 0 \\ -k_2 & k_2 & 0 & 0 \\ 0 & 0 & m_3 gl & 0 \\ 0 & 0 & 0 & m_4 gl \end{bmatrix}$$

In this system, there are inertia coupling terms between the translational and rotational coordinates, while there are no elastic coupling terms. Note that the stiffness matrix is positive definite and it is nonsingular. Therefore, the system shown in Fig. 8 has no rigid-body modes. Any possible motion leads to a change in the system potential energy which in this case is a positive-definite quadratic from.

Let us consider now the case in which the effect of gravity is not considered. In this case, the stiffness matrix becomes

$$\mathbf{K} = \begin{bmatrix} (k_1 + k_2) & -k_2 & 0 & 0 \\ -k_2 & k_2 & 0 & 0 \\ 0 & 0 & 0 & 0 \\ 0 & 0 & 0 & 0 \end{bmatrix}$$

The rank of this stiffness matrix is two since there are only two linearly independent rows. Therefore, the stiffness matrix is singular, its determinant is equal to zero, and the potential energy becomes a positive semidefinite quadratic form. That is, there exist possible configurations of the system, different from the zero configuration, such that the system potential energy is equal to zero. In fact, for this system, as the result of neglecting the effect of gravity, there are two rigid-body modes. The natural frequency associated with each of these modes can be determined using the characteristic equation and is precisely equal to zero. The associated mode shape can be determined using the equation

$$[\mathbf{K} - \omega_i^2 \mathbf{M}]\mathbf{A}_i = \mathbf{0}$$

Since $\omega_1 = \omega_2 = 0$, the above equation for $i = 1, 2$ reduces to

$$\mathbf{K}\mathbf{A}_i = \mathbf{0}, \qquad i = 1, 2$$

Recall that the rank of the 4×4 stiffness matrix is two. One is then guaranteed that the preceding matrix equation has two linearly independent solutions. These two linearly independent solutions can be written as

$$\mathbf{A}_1 = \alpha_1 \begin{bmatrix} 0 \\ 0 \\ 1 \\ 0 \end{bmatrix}, \qquad \mathbf{A}_2 = \alpha_2 \begin{bmatrix} 0 \\ 0 \\ 0 \\ 1 \end{bmatrix}$$

where α_1 and α_2 are arbitrary nonzero constants. The physical interpretation of these two solutions is that when the gravity effect is neglected, the rods connected to the masses m_1 and m_2 can have independent rigid-body rotations without any change in the system potential energy. That is

$$\mathbf{A}_i^{\mathrm{T}}\mathbf{K}\mathbf{A}_i = 0, \qquad i = 1, 2$$

Note also that a linear combination of the vectors $\mathbf{A}_1$ and $\mathbf{A}_2$ can still be selected as an eigenvector, for example, one may choose $\mathbf{A}_1$ and $\mathbf{A}_2$ as

$$\mathbf{A}_1 = \alpha_1 \begin{bmatrix} 0 \\ 0 \\ 1 \\ 0 \end{bmatrix}, \qquad \mathbf{A}_2 = \begin{bmatrix} 0 \\ 0 \\ \alpha_2 \\ \alpha_3 \end{bmatrix}$$

with α_1, α_2, and α_3 as constants. These two vectors can also be used as the rigid-body modes since they are linearly independent and satisfy the equation $\mathbf{K}\mathbf{A}_i = \mathbf{0}$, $i = 1, 2$. Observe that the condition $\mathbf{A}_i^{\mathrm{T}}\mathbf{K}\mathbf{A}_i = 0$ remains valid, while $\mathbf{A}_i^{\mathrm{T}}\mathbf{M}\mathbf{A}_i \neq 0$ ($i = 1, 2$) since the kinetic energy is always positive definite. In fact, for the two rigid-body modes $\mathbf{A}_1$ and $\mathbf{A}_2$, one can show that

$$\sum_{i=1}^{2} \mathbf{A}_i^{\mathrm{T}}\mathbf{M}\mathbf{A}_i = \alpha_1^2 m_3 l^2 + \alpha_2^2 m_4 l^2$$

which is proportional to the rotational kinetic energy of the two rods, and α_1 and α_2 are constants. Observe also that the orthogonality conditions

$$\left.\begin{aligned} \mathbf{A}_i^{\mathrm{T}}\mathbf{M}\mathbf{A}_j = 0, \\ \mathbf{A}_i^{\mathrm{T}}\mathbf{K}\mathbf{A}_j = 0, \end{aligned}\right\}, \qquad i \neq j, \qquad i, j = 1, 2, 3, 4$$

are still in effect.

Example 3.5

We consider the system of Example 1, when $k_3 = 0$. This is the case in which the shaft is free to rotate about its own axis. In this case, the stiffness matrix is given by

$$\mathbf{K} = \begin{bmatrix} k_1 & -k_1 & 0 \\ -k_1 & k_1 + k_2 & -k_2 \\ 0 & -k_2 & k_2 \end{bmatrix} = 12 \times 10^5 \begin{bmatrix} 1 & -1 & 0 \\ -1 & 3 & -2 \\ 0 & -2 & 2 \end{bmatrix} \mathrm{N \cdot m}$$

which is singular since the sum of the second and third rows is the negative of the first row. Consequently, the rows of the stiffness matrix are not linearly independent. The rank of this matrix is equal to two and, therefore, the system has one rigid body mode. The mass matrix, however, remains the same and is given by

$$\mathbf{M} = \begin{bmatrix} I_1 & 0 & 0 \\ 0 & I_2 & 0 \\ 0 & 0 & I_3 \end{bmatrix} = 2 \times 10^3 \begin{bmatrix} 1 & 0 & 0 \\ 0 & 1.5 & 0 \\ 0 & 0 & 2 \end{bmatrix} \mathrm{kg \cdot m^2}$$

The matrix equation of the free vibration of this system is given by

$$\mathbf{M}\ddot{\boldsymbol{\theta}} + \mathbf{K}\boldsymbol{\theta} = \mathbf{0}$$

By assuming a solution in the form of Eq. 27, we obtain the following standard eigenvalue problem:

$$[\mathbf{K} - \omega^2\mathbf{M}]\mathbf{A} = \mathbf{0}$$

in which the determinant of the coefficient matrix can be written as

$$\begin{vmatrix} 1-\beta & -1 & 0 \\ -1 & 3-1.5\beta & -2 \\ 0 & -2 & 2-2\beta \end{vmatrix} = 0$$

where $\beta = \omega^2/600$. Therefore, the characteristic equation can be written in terms of β as

$$\beta(1-\beta)(\beta-3) = 0$$

which has the following roots

$$\beta_1 = 0, \qquad \beta_2 = 1, \quad \text{and} \quad \beta_3 = 3$$

The corresponding natural frequencies are

$$\omega_1 = 0, \qquad \omega_2 = 24.495 \text{ rad/s}, \quad \text{and} \quad \omega_3 = 42.426 \text{ rad/s}$$

The first natural frequency ω_1 corresponds to a rigid-body mode. In order to obtain the mode shapes, we use the following equation:

$$\begin{bmatrix} 1-\beta_i & -1 & 0 \\ -1 & 3-1.5\beta_i & -2 \\ 0 & -2 & 2(1-\beta_i) \end{bmatrix} \begin{bmatrix} A_{i1} \\ A_{i2} \\ A_{i3} \end{bmatrix} = \begin{bmatrix} 0 \\ 0 \\ 0 \end{bmatrix}, \qquad i = 1, 2, 3$$

which, by partitioning of the coefficient matrix, leads to the following equation:

$$\begin{bmatrix} -1 \\ 0 \end{bmatrix} A_{i1} + \begin{bmatrix} 3-1.5\beta_i & -2 \\ -2 & 2(1-\beta_i) \end{bmatrix} \begin{bmatrix} A_{i2} \\ A_{i3} \end{bmatrix} = \begin{bmatrix} 0 \\ 0 \end{bmatrix}$$

or

$$\begin{bmatrix} 3-1.5\beta_i & -2 \\ -2 & 2(1-\beta_i) \end{bmatrix} \begin{bmatrix} A_{i2} \\ A_{i3} \end{bmatrix} = \begin{bmatrix} 1 \\ 0 \end{bmatrix} A_{i1}$$

That is,

$$\begin{bmatrix} A_{i2} \\ A_{i3} \end{bmatrix} = \frac{1}{2(1-\beta_i)(3-1.5\beta_i) - 4} \begin{bmatrix} 2(1-\beta_i) & 2 \\ 2 & (3-1.5\beta_i) \end{bmatrix} \begin{bmatrix} 1 \\ 0 \end{bmatrix} A_{i1}$$

For the rigid-body mode, $\beta_1 = 0$ and, accordingly,

$$\begin{bmatrix} A_{12} \\ A_{13} \end{bmatrix} = \frac{1}{2} \begin{bmatrix} 2 & 2 \\ 2 & 3 \end{bmatrix} \begin{bmatrix} 1 \\ 0 \end{bmatrix} A_{11} = \begin{bmatrix} 1 \\ 1 \end{bmatrix} A_{11}$$

In this mode of vibration, all the disks have the same amplitude and the system moves as one rigid body. For the second and third natural frequencies, one has the following deformation modes

$$\begin{bmatrix} A_{22} \\ A_{23} \end{bmatrix} = \frac{-1}{4} \begin{bmatrix} 0 & 2 \\ 2 & 1.5 \end{bmatrix} \begin{bmatrix} 1 \\ 0 \end{bmatrix} A_{21} = \begin{bmatrix} 0 \\ -0.5 \end{bmatrix} A_{21}$$

$$\begin{bmatrix} A_{32} \\ A_{33} \end{bmatrix} = \frac{-1}{2} \begin{bmatrix} -4 & 2 \\ 2 & -1.5 \end{bmatrix} \begin{bmatrix} 1 \\ 0 \end{bmatrix} A_{31} = \begin{bmatrix} -2.0 \\ 1.0 \end{bmatrix} A_{31}$$

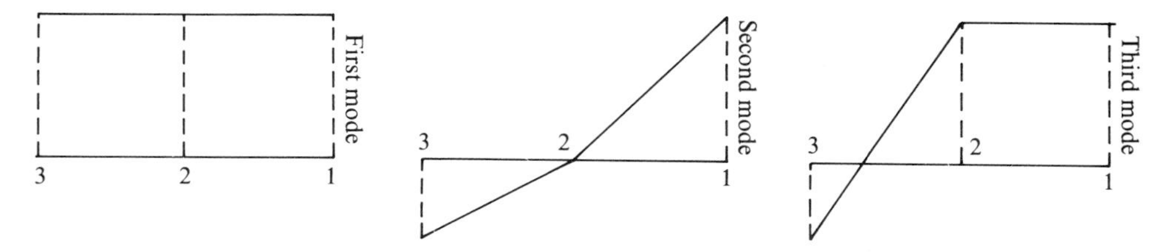

FIG. 3.9. Modes of vibration of the semidefinite system.

We may assume that $A_{i1} = 1$ for $i = 1, 2, 3$. The resulting modal matrix is given by

$$\mathbf{\Phi} = \begin{bmatrix} 1.0 & 1.0 & 1.0 \\ 1.0 & 0.0 & -2.0 \\ 1.0 & -0.5 & 1.0 \end{bmatrix}$$

Figure 9 shows the modes of vibration of the system examined in this example. Observe that

$$\mathbf{A}_1^{\mathrm{T}}\mathbf{K}\mathbf{A}_1 = 12 \times 10^5 [1.0 \quad 1.0 \quad 1.0] \begin{bmatrix} 1 & -1 & 0 \\ -1 & 3 & -2 \\ 0 & -2 & 2 \end{bmatrix} \begin{bmatrix} 1.0 \\ 1.0 \\ 1.0 \end{bmatrix} = 0$$

while

$$\mathbf{A}_1^{\mathrm{T}}\mathbf{M}\mathbf{A}_1 = 2 \times 10^3 [1.0 \quad 1.0 \quad 1.0] \begin{bmatrix} 1 & 0 & 0 \\ 0 & 1.5 & 0 \\ 0 & 0 & 2 \end{bmatrix} \begin{bmatrix} 1.0 \\ 1.0 \\ 1.0 \end{bmatrix} = 9 \times 10^3$$

that is, the modal stiffness coefficient associated with the rigid-body mode is equal to zero, while the modal mass coefficient is not equal to zero. Therefore, if the system vibrates freely in its rigid-body mode, the strain energy is equal to zero while the kinetic energy is not equal to zero. Observe also that $\mathbf{A}_1^{\mathrm{T}}\mathbf{M}\mathbf{A}_k = 0$, and $\mathbf{A}_1^{\mathrm{T}}\mathbf{K}\mathbf{A}_k = 0$, for $k = 2, 3$.

3.5 CONSERVATION OF ENERGY

Using the modal expansion, it can be shown that in the case of the undamped free vibration of the multi-degree of freedom systems, the sum of the kinetic and potential energies is a constant of motion. To this end, we write the kinetic and potential energies of the multi-degree of freedom system as

$$T = \tfrac{1}{2}\dot{\mathbf{q}}^{\mathrm{T}}\mathbf{M}\dot{\mathbf{q}}$$

$$U = \tfrac{1}{2}\mathbf{q}^{\mathrm{T}}\mathbf{K}\mathbf{q}$$

By using the modal transformation, the kinetic and potential energies can be expressed in terms of the modal variables as

$$T = \tfrac{1}{2}\dot{\mathbf{P}}^{\mathrm{T}}\mathbf{\Phi}^{\mathrm{T}}\mathbf{M}\mathbf{\Phi}\dot{\mathbf{P}}$$

$$= \tfrac{1}{2}\dot{\mathbf{P}}^{\mathrm{T}}\mathbf{M}_P\dot{\mathbf{P}}$$

$$U = \tfrac{1}{2}\mathbf{P}^{\mathrm{T}}\mathbf{\Phi}^{\mathrm{T}}\mathbf{K}\mathbf{\Phi}\mathbf{P}$$

$$= \tfrac{1}{2}\mathbf{P}^{\mathrm{T}}\mathbf{K}_P\mathbf{P}$$

where $\mathbf{M}_P$ and $\mathbf{K}_P$ are, respectively, the diagonal modal mass and stiffness matrices, and $\mathbf{P}$ is the vector of modal coordinates. In terms of the modal mass and stiffness coefficients m_i and k_i, the kinetic and potential energies can be written as

$$T = \frac{1}{2}\sum_{i=1}^{n} m_i \dot{P}_i^2$$

$$U = \frac{1}{2}\sum_{i=1}^{n} k_i P_i^2$$

In the case of free vibration, the modal coordinates and velocities are harmonic functions given by

$$P_i = C_i \sin(\omega_i t + \psi_i)$$

$$\dot{P}_i = C_i\omega_i \cos(\omega_i t + \psi_i)$$

Substituting these two equations into the expressions of the kinetic and potential energies, one obtains

$$T = \frac{1}{2}\sum_{i=1}^{n} m_i C_i^2 \omega_i^2 \cos^2(\omega_i t + \psi_i)$$

$$U = \frac{1}{2}\sum_{i=1}^{n} k_i C_i^2 \sin^2(\omega_i t + \psi_i)$$

Adding these two equations and keeping in mind that $k_i = m_i\omega_i^2$, one obtains

$$T + U = \frac{1}{2}\sum_{i=1}^{n} k_i C_i^2 [\cos^2(\omega_i t + \psi_i) + \sin^2(\omega_i t + \psi_i)]$$

which can simply be written as

$$T + U = \frac{1}{2}\sum_{i=1}^{n} k_i C_i^2$$

or, alternatively,

$$T + U = \frac{1}{2}\sum_{i=1}^{n} m_i \omega_i^2 C_i^2$$

That is, the sum of the kinetic and potential energies of the conservative multi-degree of freedom system is constant.

Rayleigh Quotient In the preceding sections, it was shown that the vibration of the multi-degree of freedom system can be represented as a linear combination of its mode shapes. However, a multi-degree of freedom system can vibrate at one of its mode shapes only if this mode is the only mode that is excited. Let us assume that the system is made to vibrate freely at its ith mode shape. In this case, the sum of the kinetic and potential energies of the system can be written as

$$T + U = \tfrac{1}{2}k_i C_i^2 = \tfrac{1}{2}m_i \omega_i^2 C_i^2$$

which implies that

$$\omega_i^2 = \frac{k_i}{m_i} = \frac{\mathbf{A}_i^{\mathrm{T}}\mathbf{K}\mathbf{A}_i}{\mathbf{A}_i^{\mathrm{T}}\mathbf{M}\mathbf{A}_i}$$

That is, the square of the natural frequency associated with the mode i can be written in a form of quotient where the numerator is proportional to the potential energy, while the denominator is proportional to the kinetic energy. This quotient is called the *Rayleigh quotient.*

If the vibration of the multi-degree of freedom system is described by an arbitrary vector $\mathbf{q}$, the Rayleigh quotient can be written as

$$R(\mathbf{q}) = \frac{\mathbf{q}^{\mathrm{T}}\mathbf{K}\mathbf{q}}{\mathbf{q}^{\mathrm{T}}\mathbf{M}\mathbf{q}}$$

If $\mathbf{q}$ is parallel to any of the eigenvectors, the Rayleigh quotient is precisely the eigenvalue associated with this eigenvector. In the general case, however, $\mathbf{q}$ can be expressed as a linear combination of the linearly independent mode shapes as

$$\mathbf{q} = \sum_{i=1}^{n} \alpha_i \mathbf{A}_i$$

If we assume that the mode shapes are made orthonormal with respect to the mass matrix, that is

$$\mathbf{A}_i^{\mathrm{T}}\mathbf{M}\mathbf{A}_i = 1,$$

the Rayleigh quotient can be written as

$$R(\mathbf{q}) = \frac{\mathbf{q}^{\mathrm{T}}\mathbf{K}\mathbf{q}}{\mathbf{q}^{\mathrm{T}}\mathbf{M}\mathbf{q}} = \frac{\sum_{i=1}^{n} \alpha_i^2 \mathbf{A}_i^{\mathrm{T}}\mathbf{K}\mathbf{A}_i}{\sum_{i=1}^{n} \alpha_i^2 \mathbf{A}_i^{\mathrm{T}}\mathbf{M}\mathbf{A}_i}$$

$$= \frac{\sum_{i=1}^{n} \alpha_i^2 \omega_i^2}{\sum_{i=1}^{n} \alpha_i^2}$$

Since $\omega_1 \leq \omega_2 \leq \omega_3 \cdots \leq \omega_n$, one has

$$\frac{\sum_{i=1}^{m} \alpha_i^2 \omega_i^2}{\sum_{i=1}^{m} \alpha_i^2} = \frac{\alpha_1^2 \omega_1^2 + \alpha_2^2 \omega_2^2 + \cdots + \alpha_m^2 \omega_m^2}{\alpha_1^2 + \alpha_2^2 + \cdots + \alpha_m^2}$$

$$\geq \frac{\omega_1^2(\alpha_1^2 + \alpha_2^2 + \cdots + \alpha_m^2)}{\alpha_1^2 + \alpha_2^2 + \cdots + \alpha_m^2} = \omega_1^2$$

for any integer m. It follows that the minimum value of the Rayleigh quotient is the square of the natural frequency associated with the fundamental mode of vibration. If, however, the shape of the ith mode can be assumed, the Rayleigh quotient can be used to provide an estimate of the square of the natural frequency corresponding to this mode. In practical applications, this is seldom done for high-frequency modes, since the shape of these modes cannot be assumed with good accuracy. Therefore, the Rayleigh quotient is often used to obtain an estimate for the fundamental frequency of the system, since in many applications the shape of the fundamental mode can be assumed with sufficient accuracy.

In order to demonstrate the use of the Rayleigh quotient to predict the fundamental frequency, the system of Example 1 is considered. The mass and stiffness matrices of this three degree of freedom system were found to be

$$\mathbf{M} = 2 \times 10^3 \begin{bmatrix} 1 & 0 & 0 \\ 0 & 1.5 & 0 \\ 0 & 0 & 2 \end{bmatrix} \mathrm{kg \cdot m^2}$$

$$\mathbf{K} = 12 \times 10^5 \begin{bmatrix} 1 & -1 & 0 \\ -1 & 3 & -2 \\ 0 & -2 & 5 \end{bmatrix} \mathrm{N \cdot m}$$

Let us assume that the deformation of the system in its first mode can be described by the harmonic function $\sin(\pi x/2l)$. Using this function and assuming that the distances between the disks are equal, one has an estimate for the elements of the first mode shape as

$$A_{11} = \sin\frac{\pi l}{2l} = 1.00$$

$$A_{12} = \sin\frac{\pi(2l/3)}{2l} = \sin\frac{\pi}{3} = 0.8660$$

$$A_{13} = \sin\frac{\pi(l/3)}{2l} = \sin\frac{\pi}{6} = 0.5$$

Therefore, a rough estimate of the first mode shape is

$$\mathbf{A}_1 = \begin{bmatrix} 1.00 \\ 0.8660 \\ 0.50 \end{bmatrix}$$

In this case, one has

$$\mathbf{A}_1^{\mathrm{T}}\mathbf{K}\mathbf{A}_1 = 12 \times 10^5 [1 \quad 0.8660 \quad 0.5] \begin{bmatrix} 1 & -1 & 0 \\ -1 & 3 & -2 \\ 0 & -2 & 5 \end{bmatrix} \begin{bmatrix} 1.00 \\ 0.8660 \\ 0.5 \end{bmatrix}$$

$$= 12.43 \times 10^5$$

$$\mathbf{A}_1^{\mathrm{T}}\mathbf{M}\mathbf{A}_1 = 2 \times 10^3[1 \quad 0.8660 \quad 0.5]\begin{bmatrix} 1 & 0 & 0 \\ 0 & 1.5 & 0 \\ 0 & 0 & 2 \end{bmatrix}\begin{bmatrix} 1.00 \\ 0.8660 \\ 0.5 \end{bmatrix}$$

$$= 5.2499 \times 10^3$$

Therefore, an estimate for the fundamental frequency using the Rayleigh quotient can be obtained as

$$\omega_1^2 \approx \frac{\mathbf{A}_1^{\mathrm{T}}\mathbf{K}\mathbf{A}_1}{\mathbf{A}_1^{\mathrm{T}}\mathbf{M}\mathbf{A}_1} = \frac{12.43 \times 10^5}{5.2499 \times 10^3} = 2.3677 \times 10^2$$

which yields $\omega_1 = 15.387$ rad/s. The exact fundamental frequency evaluated in Example 2 is 14.52 rad/s. That is, with a rough estimate for the first mode shape, the error in the obtained fundamental frequency using the Rayleigh quotient is less than 6 percent.

3.6 FORCED VIBRATION OF THE UNDAMPED SYSTEMS

The general matrix equation which governs the forced vibration of the undamped multi-degree of freedom systems is given by

$$\mathbf{M}\ddot{\mathbf{q}} + \mathbf{K}\mathbf{q} = \mathbf{F} \tag{3.75}$$

where **M** and **K** are the symmetric mass and stiffness matrices of the system, **q** is the vector of the displacements, and **F** is the vector of the forcing functions. Equation 75 is a set of n coupled differential equations which must be solved simultaneously. When the number of degrees of freedom is very large, determining the solution of these equations becomes a difficult task. An elegant approach to overcome this problem is to use the modal superposition technique discussed in the preceding sections. In this technique, the vector of displacements is expressed in terms of the modal coordinates using the modal transformation matrix. The displacement vector, in this case, is written as a linear combination of the mode shapes. By utilizing the orthogonality of the mode shapes with respect to the mass and stiffness matrices, n uncoupled differential equations of motion can be obtained. These equations can be solved using techniques that are similar to the ones used in the solution of the equations of motion of single degree of freedom systems.

By using the mode superposition technique, the displacement vector **q** is expressed as

$$\mathbf{q} = \mathbf{\Phi}\mathbf{P} \tag{3.76}$$

where **P** is the vector of modal coordinates, and $\mathbf{\Phi}$ is the modal transformation matrix whose columns are the mode shapes of the multi-degree of freedom systems. Differentiation of Eq. 76 twice with respect to time yields

$$\ddot{\mathbf{q}} = \mathbf{\Phi}\ddot{\mathbf{P}} \tag{3.77}$$

Substituting Eqs. 76 and 77 into Eq. 75 yields

$$\mathbf{M}\mathbf{\Phi}\ddot{\mathbf{P}} + \mathbf{K}\mathbf{\Phi}\mathbf{P} = \mathbf{F}$$

Premultiplying this equation by the transpose of the modal transformation matrix, one obtains

$$\mathbf{\Phi}^{\mathrm{T}}\mathbf{M}\mathbf{\Phi}\ddot{\mathbf{P}} + \mathbf{\Phi}^{\mathrm{T}}\mathbf{K}\mathbf{\Phi}\mathbf{P} = \mathbf{\Phi}^{\mathrm{T}}\mathbf{F} \tag{3.78}$$

which can be rewritten as

$$\mathbf{M}_p\ddot{\mathbf{P}} + \mathbf{K}_p\mathbf{P} = \mathbf{Q} \tag{3.79}$$

where $\mathbf{M}_p$, $\mathbf{K}_p$, and $\mathbf{Q}$ are, respectively, the modal mass and stiffness matrices, and the vector of modal forcing functions defined as

$$\mathbf{M}_p = \mathbf{\Phi}^{\mathrm{T}}\mathbf{M}\mathbf{\Phi} \tag{3.80}$$

$$\mathbf{K}_p = \mathbf{\Phi}^{\mathrm{T}}\mathbf{K}\mathbf{\Phi} \tag{3.81}$$

$$\mathbf{Q} = \mathbf{\Phi}^{\mathrm{T}}\mathbf{F} \tag{3.82}$$

By using the fact that $\mathbf{M}_p$ and $\mathbf{K}_p$ are diagonal matrices and by using the definition of the elements of $\mathbf{M}_p$ and $\mathbf{K}_p$ given by Eqs. 64 and 65, Eq. 79 can be written in the following alternate form:

$$m_i\ddot{P}_i + k_iP_i = Q_i, \qquad i = 1, 2, \ldots, n \tag{3.83}$$

where Q_i is the ith element in the vector $\mathbf{Q}$.

The differential equations given by Eq. 83 have the same form as the differential equations that govern the vibration of the undamped single degree of freedom systems. As discussed in Chapter 1, the solution of the differential equations given by Eq. 83 can be expressed as

$$\begin{aligned} P_i(t) &= P_{i0}\cos\omega_i t + \frac{\dot{P}_{i0}}{\omega_i}\sin\omega_i t \\ &\quad + \frac{1}{m_i\omega_i}\int_0^t Q_i(\tau)\sin\omega_i(t-\tau)\,d\tau, \qquad i = 1, 2, \ldots, n \end{aligned} \tag{3.84}$$

where P_{i0} and $\dot{P}_{i0}$ are the initial conditions associated with the ith modal coordinate. Having determined the modal coordinates from Eq. 84, the vector of displacements $\mathbf{q}$ can be obtained using Eq. 76.

Example 3.6

Find the response of the system of Example 1 to the step load given by

$$\mathbf{F}(t) = 10^4\begin{bmatrix} 1 \\ 2 \\ 4 \end{bmatrix} \mathrm{N\cdot m}$$

Assume that the system is to start from rest and neglect the damping effect.

Solution. In order to be able to use Eq. 79, we first evaluate the vector of modal forces $\mathbf{Q}$ defined by Eq. 82 as

$$\mathbf{Q} = \mathbf{\Phi}^{\mathrm{T}}\mathbf{F}$$

Using the modal matrix $\mathbf{\Phi}$ obtained in Example 2, the vector $\mathbf{Q}$ is given by

$$\mathbf{Q} = \mathbf{\Phi}^{\mathrm{T}}\mathbf{F} = \begin{bmatrix} 1.0 & 0.649 & 0.302 \\ 1.0 & -0.607 & -0.679 \\ 1.0 & -2.54 & 2.438 \end{bmatrix} \begin{bmatrix} 1 \\ 2 \\ 4 \end{bmatrix} \times 10^4$$

$$= 10^4 \begin{bmatrix} 3.506 \\ -2.930 \\ 5.670 \end{bmatrix}$$

The equations of motion can then be written in terms of the modal coordinates as

$$m_i \ddot{P}_i + k_i P_i = Q_i, \qquad i = 1, 2, 3$$

where m_i and k_i, $i = 1, 2, 3$, are, respectively, the modal mass and stiffness coefficients defined in Example 3. Since Q_i is constant, one can verify that the solutions of the above equations, in the case of zero initial conditions, are given by

$$P_i = \frac{Q_i}{k_i}(1 - \cos \omega_i t)$$

Using the results of Examples, 2 and 3, we have

$$\mathbf{P} = \begin{bmatrix} P_1 \\ P_2 \\ P_3 \end{bmatrix} = \begin{bmatrix} 0.0463(1 - \cos 14.52t) \\ -0.0062(1 - \cos 31.05t) \\ 0.0006(1 - \cos 46.1t) \end{bmatrix}$$

The vector of torsional oscillations $\mathbf{\theta}$ is given by

$$\mathbf{\theta} = \mathbf{\Phi}\mathbf{P} = \begin{bmatrix} 1.0 & 1.0 & 1.0 \\ 0.649 & -0.607 & -2.54 \\ 0.302 & -0.679 & 2.438 \end{bmatrix} \begin{bmatrix} P_1 \\ P_2 \\ P_3 \end{bmatrix}$$

$$= \begin{bmatrix} P_1 + P_2 + P_3 \\ 0.649P_1 - 0.607P_2 - 2.54P_3 \\ 0.302P_1 - 0.679P_2 + 2.432P_3 \end{bmatrix}$$

3.7 VISCOUSLY DAMPED SYSTEMS

We considered thus far the free and forced vibrations of undamped multi-degree of freedom systems, and it was demonstrated that, for the undamped system, the eigenvalues and eigenvectors are real. This is not, however, the case when damping is included in the mathematical model, since for a viscously damped multi-degree of freedom system, the eigenvalues and eigenvectors can be complex numbers. Even though in many applications damping is small, its effect on the system stability and system response in the resonance region may be significant.

The general matrix equation that governs the vibration of the viscously damped multi-degree of freedom system is given by

$$\mathbf{M}\ddot{\mathbf{q}} + \mathbf{C}\dot{\mathbf{q}} + \mathbf{K}\mathbf{q} = \mathbf{F} \tag{3.85}$$

in which $\mathbf{C}$ is the damping matrix defined as

$$\mathbf{C} = \begin{bmatrix} c_{11} & c_{12} & c_{13} & \cdots & c_{1n} \\ c_{21} & c_{22} & c_{23} & \cdots & c_{2n} \\ c_{31} & c_{32} & c_{33} & \cdots & c_{3n} \\ \vdots & \vdots & \vdots & \ddots & \vdots \\ c_{n1} & c_{n2} & c_{n3} & \cdots & c_{nn} \end{bmatrix} \tag{3.86}$$

In general, $\mathbf{\Phi}^{\mathrm{T}}\mathbf{C}\mathbf{\Phi}$ is not a diagonal matrix, and as a consequence, the use of the modal coordinates does not lead to a system of uncoupled independent differential equations. In the remainder of this section, we will consider an important special case in which the damping matrix is proportional to the mass and stiffness matrices. In this special case, a system of uncoupled differential equations expressed in terms of the modal coordinates can be obtained. The case of a general damping matrix is discussed in the following section.

Proportional Damping If the damping matrix can be written as a linear combination of the mass and stiffness matrices, the matrix $\mathbf{C}$ takes the form

$$\mathbf{C} = \alpha\mathbf{M} + \beta\mathbf{K} \tag{3.87}$$

where α and β are constants. Substituting Eq. 87 into Eq. 85 yields

$$\mathbf{M}\ddot{\mathbf{q}} + (\alpha\mathbf{M} + \beta\mathbf{K})\dot{\mathbf{q}} + \mathbf{K}\mathbf{q} = \mathbf{F} \tag{3.88}$$

The vector of displacement $\mathbf{q}$ can be expressed in terms of the modal coordinates by using the mode shapes of the undamped system as

$$\mathbf{q} = \mathbf{\Phi}\mathbf{P} \tag{3.89}$$

where $\mathbf{\Phi}$ is the modal transformation matrix and $\mathbf{P}$ is the vector of modal coordinates. Substituting Eq. 89 into Eq. 88 yields

$$\mathbf{M}\mathbf{\Phi}\ddot{\mathbf{P}} + (\alpha\mathbf{M} + \beta\mathbf{K})\mathbf{\Phi}\dot{\mathbf{P}} + \mathbf{K}\mathbf{\Phi}\mathbf{P} = \mathbf{F} \tag{3.90}$$

Premultiplying this equation by $\mathbf{\Phi}^{\mathrm{T}}$, one obtains

$$\mathbf{\Phi}^{\mathrm{T}}\mathbf{M}\mathbf{\Phi}\ddot{\mathbf{P}} + \mathbf{\Phi}^{\mathrm{T}}(\alpha\mathbf{M} + \beta\mathbf{K})\mathbf{\Phi}\dot{\mathbf{P}} + \mathbf{\Phi}^{\mathrm{T}}\mathbf{K}\mathbf{\Phi}\mathbf{P} = \mathbf{\Phi}^{\mathrm{T}}\mathbf{F} \tag{3.91}$$

Using the orthogonality of the mode shapes with respect to the mass and stiffness matrices yields

$$\mathbf{M}_p\ddot{\mathbf{P}} + \mathbf{C}_p\dot{\mathbf{P}} + \mathbf{K}_p\mathbf{P} = \mathbf{Q} \tag{3.92}$$

in which

$$\mathbf{M}_p = \mathbf{\Phi}^{\mathrm{T}}\mathbf{M}\mathbf{\Phi} \tag{3.93}$$

$$\mathbf{K}_p = \mathbf{\Phi}^{\mathrm{T}}\mathbf{K}\mathbf{\Phi} \tag{3.94}$$

$$\mathbf{C}_p = \mathbf{\Phi}^{\mathrm{T}}(\alpha\mathbf{M} + \beta\mathbf{K})\mathbf{\Phi} = \alpha\mathbf{M}_p + \beta\mathbf{K}_p \tag{3.95}$$

$$\mathbf{Q} = \mathbf{\Phi}^{\mathrm{T}}\mathbf{F} \tag{3.96}$$

Because the matrix $\mathbf{C}$ is assumed to be a linear combination of the mass matrix $\mathbf{M}$ and the stiffness matrix $\mathbf{K}$, the *modal damping matrix* $\mathbf{C}_p$ is diagonal. The elements on the diagonal of this matrix are denoted as c_i, $i = 1, 2, \ldots, n$. The *modal damping coefficient* c_i can be expressed in terms of the modal mass and stiffness coefficients m_i and k_i, respectively, by using Eq. 95 as

$$c_i = \alpha m_i + \beta k_i \tag{3.97}$$

Equation 92 can be then written as n uncoupled differential equations as

$$m_i\ddot{P}_i + c_i\dot{P}_i + k_iP_i = Q_i, \qquad i = 1, 2, \ldots, n \tag{3.98}$$

These equations, which are in the same form as the differential equation of motion which governs the forced vibration of damped single degree of freedom systems, can be also written in the following form

$$\ddot{P}_i + 2\xi_i\omega_i\dot{P}_i + \omega_i^2 P_i = \frac{Q_i}{m_i} \tag{3.99}$$

where the *modal damping factor* ξ_i is defined as

$$\xi_i = \frac{c_i}{2m_i\omega_i} \tag{3.100}$$

As discussed in the case of the single degree of freedom systems, the solution of Eq. 99 will depend on the modal damping factor ξ_i. In the case of an underdamped system in which $\xi_i < 1$, the solution of Eq. 99 is given by

$$P_i(t) = e^{-\xi_i\omega_i t}\left[P_{i0}\cos\omega_{di}t + \frac{\dot{P}_{i0} + \xi_i\omega_i P_{i0}}{\omega_{di}}\sin\omega_{di}t\right] + \int_0^t Q_i(\tau)h_i(t-\tau)\,d\tau, \qquad i = 1, 2, \ldots, n \tag{3.101}$$

where P_{i0} and $\dot{P}_{i0}$ are, respectively, the initial modal coordinate and velocity associated with mode i and

$$\left.\begin{aligned} \omega_{\mathrm{d}i} &= \omega_i\sqrt{1-\xi_i^2} \\ h_i(t) &= \frac{e^{-\xi_i\omega_i t}}{m_i\omega_{\mathrm{d}i}}\sin\omega_{\mathrm{d}i}t \end{aligned}\right\} \tag{3.102}$$

in which $\omega_{\mathrm{d}i}$ is the *damped natural frequency* of mode i, and h_i is the *unit impulse response* associated with the ith mode of vibration.

Experimental Modal Analysis The concept of proportional damping discussed in this section can also be used in cases in which the damping is small. Despite the fact that in these cases the resulting modal damping matrix

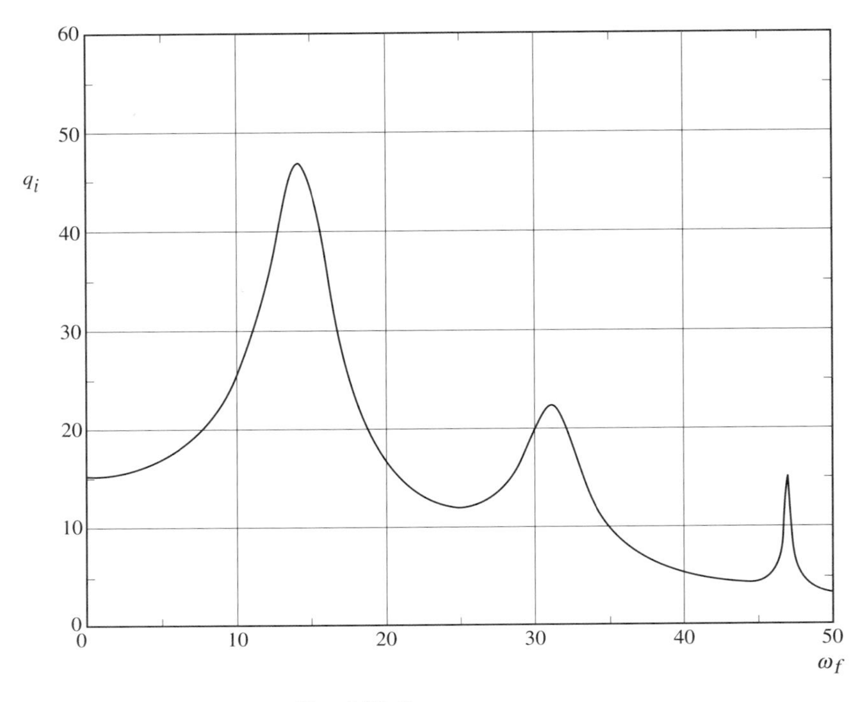

FIG. 3.10. Resonance curve.

is not, in general, diagonal, a reasonable approximation can be made by neglecting the off-diagonal terms. The matrix $\mathbf{C}_p$ can then be considered as a diagonal matrix and a set of independent uncoupled differential equations expressed in terms of the modal coordinates can be obtained. This approximation is acceptable in many engineering applications, since structural damping is usually small. Furthermore, there are several experimental teehniques which can be used to determine the natural frequencies, mode shapes, and modal damping coefficients. Figure 10 shows a resonance curve that can be obtained experimentally for a multi-degree of freedom system by exciting the system using a vibration exciter and measuring the coordinate q_i at a selected point. The resonance curve of a multi-degree of freedom system has a number of resonance regions equal to the number of the system degrees of freedom, which is equal to the number of the system natural frequencies. Each resonance region can be considered as a resonance curve of a single degree of freedom system, and, therefore, for a lightly damped system, the frequencies at which the peaks occur define approximate values for the system natural frequencies. The techniques described in the first volume of this book (Shabana, 1995) for the experimental determination of the damping factors of single degree of freedom systems then can be used to determine the damping factor ξ_i associated with the ith mode shape. Experimental modal analysis techniques also can be used to determine the mode shapes of the multi-degree of freedom systems. In order to demonstrate this, we rewrite Eq. 92 as

$$\mathbf{M}_p\ddot{\mathbf{P}} + \mathbf{C}_p\dot{\mathbf{p}} + \mathbf{K}_p\mathbf{P} = \mathbf{\Phi}^{\mathrm{T}}\mathbf{F}$$

This equation also can be written in the case of a harmonic excitation as

$$m_i\ddot{P}_i + c_i\dot{P}_i + k_iP_i = \mathbf{A}_i^{\mathrm{T}}\mathbf{F}_o \sin \omega_f t, \qquad i = 1, 2, \ldots, n$$

where $\mathbf{F}_o$ is the vector of force amplitudes. The preceding equation defines the steady state solution for the modal coordinate i as

$$P_i = \frac{\mathbf{A}_i^{\mathrm{T}}\mathbf{F}_o/k_i}{\sqrt{(1 - r_i^2)^2 + (2r_i\xi_i)^2}} \sin(\omega_f t - \psi_i)$$

where

$$r_i = \frac{\omega_f}{\omega_i}, \qquad \psi_i = \tan^{-1}\left(\frac{2r_i\xi_i}{1 - r_i^2}\right)$$

The vector of physical displacements can be obtained using the modal transformation as

$$\mathbf{q} = \mathbf{\Phi}\mathbf{P} = \sum_{i=1}^{n} \mathbf{A}_1P_1 + \mathbf{A}_2P_2 + \cdots + \mathbf{A}_nP_n$$

$$= \sum_{i=1}^{n} \frac{\mathbf{A}_i\mathbf{A}_i^{\mathrm{T}}}{k_i\sqrt{(1 - r_i^2)^2 + (2r_i\xi_i)^2}} \mathbf{F}_o \sin(\omega_f t - \psi_i)$$

In the resonance region of a mode i, the solution is dominated by this mode, and as a consequence, one has

$$\mathbf{q} = \frac{\mathbf{A}_i\mathbf{A}_i^{\mathrm{T}}}{m_i\sqrt{(\omega_i^2 - \omega_f^2)^2 + (2\xi_i\omega_i\omega_f)^2}} \mathbf{F}_o \sin(\omega_f t - \psi_i)$$

In the resonance region,

$$\frac{\mathbf{A}_i\mathbf{A}_i^{\mathrm{T}}}{\sqrt{(\omega_i^2 - \omega_f^2)^2 + (2\xi_i\omega_i\omega_f)^2}} \approx \frac{\mathbf{A}_i\mathbf{A}_i^{\mathrm{T}}}{2\xi_i\omega_i^2}$$

It follows that

$$\mathbf{q} = \frac{\mathbf{A}_i\mathbf{A}_i^{\mathrm{T}}}{2\xi_ik_i} \mathbf{F}_o \sin(\omega_f t - \psi_i) = \frac{1}{2\xi_ik_i} \mathbf{B}_i\mathbf{F}_o \sin(\omega_f t - \psi_i) \tag{3.103}$$

where

$$\mathbf{B}_i = \mathbf{A}_i\mathbf{A}_i^{\mathrm{T}}$$

The symmetric matrix $\mathbf{B}_i$ has n^2 elements, but only n of these elements are unique. There are two methods which can be used to determine the n unique elements of the matrix $\mathbf{B}_i$. In the first method, the excitation is made at n nodes, one at a time, and the measurement is taken at one node. The obtained n measurements define, to within an arbitrary constant, the row of the matrix $\mathbf{B}_i$ that corresponds to the measured coordinate. In the second method, the excitation is made at only one node, and all of the elements of the vector $\mathbf{q}$ are measured. In this case, the n measurements can be used to define, to within an arbitrary constant, the column of the matrix $\mathbf{B}_i$ that corresponds to the point of excitation.

If one row or one column of the matrix $\mathbf{B}_i$ is determined, one can use this row or column to determine the entire mode shape vector $\mathbf{A}_i$. For example, if the vector $\mathbf{A}_i = [a_1 \ a_2 \ \dots \ a_n]^{\mathrm{T}}$, the matrix $\mathbf{B}_i$ takes the following form:

$$\mathbf{B}_i = \begin{bmatrix} a_1^2 & a_1 a_2 & a_1 a_3 & \cdots & a_1 a_n \\ a_2 a_1 & a_2^2 & a_2 a_3 & \cdots & a_2 a_n \\ \vdots & \vdots & \vdots & \ddots & \vdots \\ a_n a_1 & a_n a_2 & a_n a_3 & \cdots & a_n^2 \end{bmatrix}$$

If, for instance, the excitation is made at the kth nodal point, the measurements of the n elements of the nodal displacement vector $\mathbf{q}$ can be used to define the kth column of the matrix $\mathbf{B}_i$ given by

$$[a_1 a_k \quad a_2 a_k \quad \dots \quad a_k^2 \quad \dots \quad a_n a_k]^{\mathrm{T}}$$

Knowing the numerical value $(a_k)^2$, one can determine a_k, and use other elements of the preceding vector to evaluate all the elements of the mode shape $\mathbf{A}_i$.

We also note that, because of the phase angle ψ_i, the phase between the force and the displacement must be determined. At resonance, $\psi_i = \pi/2$ or $-\pi/2$. This phase must be determined in order to correctly determine the sine of the elements of the matrix $\mathbf{B}_i$. As previously pointed out, the matrix $\mathbf{B}_i$ can be determined using Eq. 103 to within any arbitrary constant. Once the mode shapes are determined, the modal mass coefficients can be calculated. For example, if the elements of the vector $\mathbf{q}$ are measured, all of the other variables in Eq. 103 are known except the modal stiffness coefficient k_i, which can be calculated using any of the scalar equations of Eq. 103. Using this stiffness coefficient and the natural frequency of the mode i, the modal mass coefficient also can be determined. As we previously demonstrated, the mode shapes can be scaled such that either the modal mass or the modal stiffness coefficient is equal to one.

The procedure for the experimental determination of the mode shapes and the modal parameters is outlined in this section in order to demonstrate the basic concepts underlying an important practical tool widely used in industry for the vibration analysis of many mechanical systems. The actual implementation of this procedure requires a more detailed presentation (Brown et al., 1979; Ewins, 1984; Shabana, 1986) that includes elaborate statistical and error analysis. Figure 11 shows an example that demonstrates the wide use of experimental modal identification techniques in construction machine industry. Figure 12 shows some the mode shapes and frequencies that were determined experimentally for the hydraulic excavator shown in Fig. 11.

FIG. 3.11. Hydraulic excavator

3.8 GENERAL VISCOUS DAMPING

The assumption made in the preceding section that the damping matrix is proportional to the mass and stiffness matrices enables us to use the modal transformation matrix to decouple the equations of motion. The procedure outlined in the preceding section led to n uncoupled differential equations expressed in terms of the modal coordinates. Each of these equations is similar to the differential equation of motion of the viscously damped single degree of freedom system.

In this section, we discuss the case of a general viscous damping matrix which cannot be expressed as a linear combination of the mass and stiffness matrices. Before we start our discussion on the case of general damping matrix, we present some matrix results that will be used frequently in this chapter.

If $\mathbf{B}$ is a nonsingular matrix that has λ_i as an eigenvalue, then $1/\lambda_i$ is an eigenvalue of the matrix $\mathbf{B}^{-1}$ since

$$\mathbf{B}\mathbf{A}_i = \lambda_i \mathbf{A}_i$$

implies that

$$\mathbf{A}_i = \lambda_i \mathbf{B}^{-1}\mathbf{A}_i$$

or

$$\frac{1}{\lambda_i}\mathbf{A}_i = \mathbf{B}^{-1}\mathbf{A}_i$$

This equation shows that $1/\lambda_i$ is an eigenvalue of $\mathbf{B}^{-1}$ and $\mathbf{A}_i$ is the corresponding eigenvector. Furthermore, it can be shown that $\mathbf{A}_i$ is also an eigen-

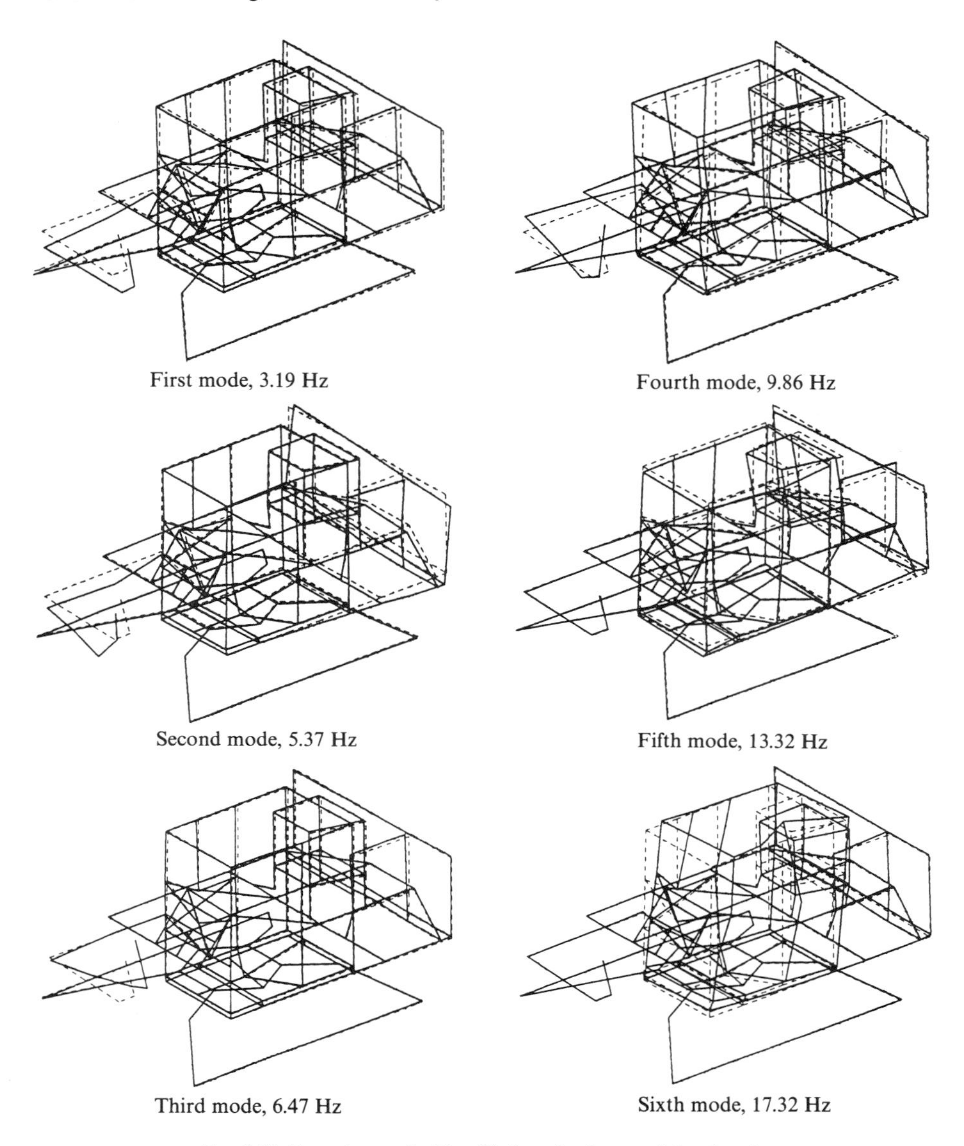

First mode, 3.19 Hz

Fourth mode, 9.86 Hz

Second mode, 5.37 Hz

Fifth mode, 13.32 Hz

Third mode, 6.47 Hz

Sixth mode, 17.32 Hz

FIG. 3.12. Experimentally identified mode shapes of the chassis

vector of the matrix $\mathbf{B}^2$. This can be demonstrated by premultiplying the eigenvalue problem by $\mathbf{B}$ to yield

$$\mathbf{B}^2\mathbf{A}_i = \lambda_i\mathbf{B}\mathbf{A}_i = \lambda_i^2\mathbf{A}_i$$

which implies that λ_i^2 is an eigenvalue of $\mathbf{B}^2$, and $\mathbf{A}_i$ is an eigenvector for both $\mathbf{B}$ and $\mathbf{B}^2$. Using a similar procedure, it can be shown that if λ_i is an eigenvalue of the matrix $\mathbf{B}$, then λ_i^k is an eigenvalue of $\mathbf{B}^k$, and the matrices $\mathbf{B}$ and $\mathbf{B}^k$ have the same eigenvector.

If $\mathbf{B}$ is a symmetric matrix, the eigenvectors of $\mathbf{B}$ are orthogonal. If $\mathbf{A}_i$ is normalized such that $\mathbf{A}_i^{\mathrm{T}}\mathbf{A}_i = 1$, then one has

$$\mathbf{A}_i^{\mathrm{T}}\mathbf{B}\mathbf{A}_i = \lambda_i, \qquad \mathbf{A}_i^{\mathrm{T}}\mathbf{B}^k\mathbf{A}_i = \lambda_i^k$$

It follows that if $\mathbf{\Phi}$ is the matrix of the eigenvectors of the symmetric matrix $\mathbf{B}$ and the matrix $\mathbf{B}^k$, then

$$\mathbf{\Phi}^{\mathrm{T}}\mathbf{B}\mathbf{\Phi} = \mathbf{\Lambda}, \qquad \mathbf{\Phi}^{\mathrm{T}}\mathbf{B}^k\mathbf{\Phi} = \mathbf{\Lambda}^k$$

where $\mathbf{\Lambda}$ is a diagonal matrix whose diagonal elements are the eigenvalues.

Exponential Matrix Forms If a is a scalar, the use of Taylor series shows that

$$e^a = 1 + a + \frac{a^2}{2!} + \frac{a^3}{3!} + \cdots$$

If $\mathbf{B}$ is a matrix, one also has

$$e^{\mathbf{B}t} = \mathbf{I} + \mathbf{B}t + \frac{(\mathbf{B}t)^2}{2!} + \frac{(\mathbf{B}t)^3}{3!} + \cdots$$

This is a convergent series that has the properties

$$e^{\mathbf{B}r}e^{\mathbf{B}t} = e^{\mathbf{B}(r+t)}$$

$$e^{\mathbf{B}t}e^{-\mathbf{B}t} = \mathbf{I}$$

$$\frac{d}{dt}(e^{\mathbf{B}t}) = \mathbf{B}e^{\mathbf{B}t}$$

It is then clear that if $\mathbf{B}$ is skew symmetric, then $e^{\mathbf{B}t}$ is an orthogonal matrix.

It was previously demonstrated in this section that if λ_i is an eigenvalue and $\mathbf{A}_i$ is the corresponding normalized eigenvector of the matrix $\mathbf{B}$, then λ_i^k and $\mathbf{A}_i$ are the eigenvalue and eigenvector of the matrix $\mathbf{B}^k$. Therefore, if $\mathbf{\Phi}$ is the matrix of the eigenvectors, one has

$$\mathbf{B}^k = \mathbf{\Phi}\mathbf{\Lambda}^k\mathbf{\Phi}^{-1}$$

where $\mathbf{\Lambda}$ is a diagonal matrix whose diagonal elements are the eigenvalues of the matrix $\mathbf{B}$. Using the preceding identity, the exponential form of the matrix $\mathbf{B}t$ can be written as

$$\begin{aligned} e^{\mathbf{B}t} &= \mathbf{I} + \mathbf{\Phi}\mathbf{\Lambda}\mathbf{\Phi}^{-1} + \frac{\mathbf{\Phi}\mathbf{\Lambda}^2\mathbf{\Phi}^{-1}t^2}{2} + \frac{\mathbf{\Phi}\mathbf{\Lambda}^3\mathbf{\Phi}^{-1}t^3}{3} + \cdots \\ &= \mathbf{\Phi}\left(\mathbf{I} + \mathbf{\Lambda}t + \frac{(\mathbf{\Lambda}t)^2}{2!} + \frac{(\mathbf{\Lambda}t)^3}{3!} + \cdots\right)\mathbf{\Phi}^{-1} \\ &= \mathbf{\Phi}e^{\mathbf{\Lambda}t}\mathbf{\Phi}^{-1} \end{aligned}$$

Note that the matrix $e^{\mathbf{B}t}$ is never singular since its inverse is defined by the matrix $e^{-\mathbf{B}t}$. This is also clear from the determinant

$$|e^{\mathbf{B}t}| = e^{\lambda_1 t} e^{\lambda_2 t} \cdots e^{\lambda_n t} = e^{tr(\mathbf{B}t)}$$

This determinant cannot be equal to zero.

As an example, we consider the matrix

$$\mathbf{B} = \begin{bmatrix} 4 & 1 & 2 \\ 1 & 0 & 0 \\ 2 & 0 & 0 \end{bmatrix}$$

which has the eigenvalues

$$\lambda_1 = 0, \qquad \lambda_2 = 5, \qquad \lambda_3 = -1$$

The corresponding orthonormal eigenvectors are

$$\mathbf{A}_1 = \frac{1}{\sqrt{5}} \begin{bmatrix} 0 \\ 2 \\ -1 \end{bmatrix}, \qquad \mathbf{A}_2 = \frac{1}{\sqrt{30}} \begin{bmatrix} 5 \\ 1 \\ 2 \end{bmatrix}, \qquad \mathbf{A}_3 = \frac{1}{\sqrt{6}} \begin{bmatrix} 1 \\ -1 \\ -2 \end{bmatrix}$$

The matrix $\mathbf{\Phi}$ whose columns are the orthonormal eigenvectors is

$$\mathbf{\Phi} = \begin{bmatrix} 0 & \frac{5}{\sqrt{30}} & \frac{1}{\sqrt{6}} \\ \frac{2}{\sqrt{5}} & \frac{1}{\sqrt{30}} & \frac{-1}{\sqrt{6}} \\ \frac{-1}{\sqrt{5}} & \frac{2}{\sqrt{30}} & \frac{-2}{\sqrt{6}} \end{bmatrix}$$

The matrix $e^{\mathbf{A}t}$ is

$$e^{\mathbf{A}t} = \begin{bmatrix} 1 & 0 & 0 \\ 0 & e^{5t} & 0 \\ 0 & 0 & e^{-t} \end{bmatrix}$$

It follows that

$$e^{\mathbf{B}t} = \mathbf{\Phi} e^{\mathbf{A}t} \mathbf{\Phi}^{-1} = \begin{bmatrix} \frac{5e^{5t} + e^{-t}}{6} & \frac{e^{5t} - e^{-t}}{6} & \frac{e^{5t} - e^{-t}}{3} \\ \frac{e^{5t} - e^{-t}}{6} & \frac{e^{5t} + 5e^{-t} + 24}{30} & \frac{e^{5t} + 5e^{-t} - 6}{15} \\ \frac{e^{5t} - e^{-t}}{3} & \frac{e^{5t} + 5e^{-t} - 6}{15} & \frac{2e^{5t} + 10e^{-t} + 3}{15} \end{bmatrix}$$

State Space Representation In the remainder of this section, we will present a technique for solving Eq. 85 in the case of general viscous damping matrix. To this end, we define the *state vector* $\mathbf{y}$ as

$$\mathbf{y} = \begin{bmatrix} \mathbf{q} \\ \dot{\mathbf{q}} \end{bmatrix} \tag{3.104}$$

Clearly, the state vector has dimension $2n$. Differentiation of this vector with respect to time leads to

$$\dot{\mathbf{y}} = \begin{bmatrix} \dot{\mathbf{q}} \\ \ddot{\mathbf{q}} \end{bmatrix} \tag{3.105}$$

Premultiplying Eq. 85 which governs the vibration of a viscously damped multi-degree of freedom system by the inverse of the mass matrix leads to

$$\ddot{\mathbf{q}} + \mathbf{M}^{-1}\mathbf{C}\dot{\mathbf{q}} + \mathbf{M}^{-1}\mathbf{K}\mathbf{q} = \mathbf{M}^{-1}\mathbf{F}$$

which can be written as

$$\ddot{\mathbf{q}} = -\mathbf{M}^{-1}\mathbf{C}\dot{\mathbf{q}} - \mathbf{M}^{-1}\mathbf{K}\mathbf{q} + \mathbf{M}^{-1}\mathbf{F}$$

This equation with the following identity

$$\dot{\mathbf{q}} = \mathbf{I}\dot{\mathbf{q}}$$

where $\mathbf{I}$ is an $n \times n$ identity matrix leads to

$$\dot{\mathbf{y}} = \begin{bmatrix} \dot{\mathbf{q}} \\ \ddot{\mathbf{q}} \end{bmatrix} = \begin{bmatrix} \mathbf{0} & \mathbf{I} \\ -\mathbf{M}^{-1}\mathbf{K} & -\mathbf{M}^{-1}\mathbf{C} \end{bmatrix} \begin{bmatrix} \mathbf{q} \\ \dot{\mathbf{q}} \end{bmatrix} + \begin{bmatrix} \mathbf{0} \\ \mathbf{M}^{-1}\mathbf{F} \end{bmatrix}$$

which can be written compactly as

$$\dot{\mathbf{y}} = \mathbf{B}\mathbf{y} + \mathbf{R}(t) \tag{3.106}$$

in which the state vector $\mathbf{y}$ is defined by Eq. 105 and the $2n \times 2n$ *matrix* $\mathbf{B}$ and the $2n$-dimensional vector $\mathbf{R}(t)$ are given, respectively, by

$$\mathbf{B} = \begin{bmatrix} \mathbf{0} & \mathbf{I} \\ -\mathbf{M}^{-1}\mathbf{K} & -\mathbf{M}^{-1}\mathbf{C} \end{bmatrix} \tag{3.107}$$

$$\mathbf{R}(t) = \begin{bmatrix} \mathbf{0} \\ \mathbf{M}^{-1}\mathbf{F} \end{bmatrix} \tag{3.108}$$

Free Vibration If the viscously damped multi-degree of freedom system is vibrating freely, the vector of forcing functions $\mathbf{F}$ is identically zero, and accordingly

$$\mathbf{R}(t) = \mathbf{0} \tag{3.109}$$

Equation 106 then reduces to

$$\dot{\mathbf{y}} = \mathbf{B}\mathbf{y} \tag{3.110}$$

To this equation, we assume a solution in the form

$$\mathbf{y} = \mathbf{A}e^{\mu t} \tag{3.111}$$

where in this case, the vector $\mathbf{A}$ has dimension $2n$ and μ is a constant. Differentiating Eq. 111 with respect to time leads to

$$\dot{\mathbf{y}} = \mu \mathbf{A} e^{\mu t} \tag{3.112}$$

Substituting Eqs. 111 and 112 into Eq. 110 yields

$$\mu \mathbf{A} e^{\mu t} = \mathbf{B} \mathbf{A} e^{\mu t}$$

which leads to

$$[\mathbf{B} - \mu \mathbf{I}]\mathbf{A} = \mathbf{0} \tag{3.113}$$

where $\mathbf{I}$ in this equation is a $2n \times 2n$ identity matrix. Equation 113 is an eigenvalue problem in which the coefficient matrix

$$\mathbf{H} = \mathbf{B} - \mu \mathbf{I} \tag{3.114}$$

is called the *characteristic matrix*.

Equation 113 has a nontrivial solution if and only if the determinant of the coefficient matrix is equal to zero, that is,

$$|\mathbf{H}| = |\mathbf{B} - \mu \mathbf{I}| = 0 \tag{3.115}$$

This equation defines a polynomial of order $2n$ in μ. The roots of this polynomial, which may be complex numbers, define the eigenvalues $\mu_1, \mu_2, \ldots, \mu_{2n}$. Associated with each eigenvalue μ_i, there exists an eigenvector $\mathbf{A}_i$ which can be determined to within an arbitrary constant using Eq. 113 as

$$\mathbf{B}\mathbf{A}_i = \mu_i \mathbf{A}_i \tag{3.116}$$

In this case, the modal matrix is defined as

$$\mathbf{\Phi} = [\mathbf{A}_1 \quad \mathbf{A}_2 \quad \cdots \quad \mathbf{A}_{2n}] \tag{3.117}$$

It can be verified that

$$\mathbf{\Phi}^{-1}\mathbf{B}\mathbf{\Phi} = \boldsymbol{\mu} \tag{3.118}$$

where $\mathbf{\Phi}$ is the $2n \times 2n$ modal matrix whose columns consist of the $2n$ eigenvectors $\mathbf{A}_i$, $i = 1, 2, \ldots, 2n$, and $\boldsymbol{\mu}$ is the diagonal matrix

$$\boldsymbol{\mu} = \begin{bmatrix} \mu_1 & 0 & 0 & \cdots & 0 \\ 0 & \mu_2 & 0 & \cdots & 0 \\ 0 & 0 & \mu_3 & \cdots & 0 \\ \vdots & \vdots & \vdots & & \vdots \\ 0 & 0 & 0 & \cdots & \mu_{2n} \end{bmatrix} \tag{3.119}$$

Recall that, for any integer r, the following identity holds:

$$\mathbf{\Phi}^{-1}\mathbf{B}^r\mathbf{\Phi} = \boldsymbol{\mu}^r$$

The modal transformation of Eq. 117 can be used to obtain $2n$ uncoupled first-order differential equations. To this end, we write

$$\mathbf{y} = \mathbf{\Phi}\mathbf{P} \tag{3.120}$$

and

$$\dot{\mathbf{y}} = \mathbf{\Phi}\dot{\mathbf{P}} \tag{3.121}$$

where $\mathbf{P}$ is a $2n$-dimensional vector. Substituting Eqs. 120 and 121 into Eq. 110 yields

$$\mathbf{\Phi}\dot{\mathbf{P}} = \mathbf{B}\mathbf{\Phi}\mathbf{P}$$

or

$$\dot{\mathbf{P}} = \mathbf{\Phi}^{-1}\mathbf{B}\mathbf{\Phi}\mathbf{P}$$

which upon using Eq. 118 can be written as

$$\dot{\mathbf{P}} = \boldsymbol{\mu}\mathbf{P} \tag{3.122}$$

This is a system of $2n$ uncoupled first-order differential equations which can be written as

$$\dot{P}_i = \mu_i P_i, \qquad i = 1, 2, \ldots, 2n \tag{3.123}$$

The solution of these homogeneous first-order differential equations is simple and can be obtained as follows. Equation 123 can be written as

$$\frac{dP_i}{dt} = \mu_i P_i$$

or

$$\frac{dP_i}{P_i} = \mu_i \, dt$$

which upon integration yields

$$\ln P_i = \mu_i t + C_0 \tag{3.124}$$

where C_0 is the constant of integration.

An alternate form of Eq. 124 is

$$P_i = e^{\mu_i t} e^{C_0}$$

Using the initial conditons, one can show that $e^{C_0} = P_{i0}$ and consequently,

$$P_i = P_{i0} e^{\mu_i t} \tag{3.125}$$

where P_{i0} is the initial condition associated with the coordinate P_i. These initial conditions can be obtained by using Eq. 120 as

$$\mathbf{P}_0 = \mathbf{\Phi}^{-1}\mathbf{y}_0 \tag{3.126}$$

where $\mathbf{y}_0$ is the state vector which contains the initial displacements and velocities. Having determined the $2n$ coordinates of Eq. 125, the displacements and velocities of the multi-degree of freedom system can be obtained using the transformation of Eq. 120.

Forced Vibration Equation 106 describes the forced vibration of viscously damped multi-degree of freedom systems. This equation with Eqs. 120 and 121 can be used to obtain $2n$ uncoupled differential equations. In order to demonstrate this, we substitute Eqs. 120 and 121 into Eq. 106. This yields

$$\boldsymbol{\Phi}\dot{\mathbf{P}} = \mathbf{B}\boldsymbol{\Phi}\mathbf{P} + \mathbf{R}(t)$$

or

$$\dot{\mathbf{P}} = \boldsymbol{\Phi}^{-1}\mathbf{B}\boldsymbol{\Phi}\mathbf{P} + \boldsymbol{\Phi}^{-1}\mathbf{R}(t)$$

which upon using Eq. 118 can be written as

$$\dot{\mathbf{P}} = \boldsymbol{\mu}\mathbf{P} + \mathbf{Q} \tag{3.127}$$

where the matrix $\boldsymbol{\mu}$ is defined by Eq. 119 and $\mathbf{Q} = [Q_1 \;\; Q_2 \;\; \cdots \;\; Q_{2n}]^{\mathrm{T}}$ is the vector

$$\mathbf{Q} = \boldsymbol{\Phi}^{-1}\mathbf{R} \tag{3.128}$$

Equation 127 is a set of $2n$ uncoupled first-order differential equations which can be written as

$$\dot{P}_i = \mu_i P_i + Q_i, \qquad i = 1, 2, \ldots, 2n \tag{3.129}$$

The solution of these equations is given by

$$P_i = e^{\mu_i t} P_{i0} + \int_0^t e^{\mu_i (t-\tau)} Q_i(\tau)\, d\tau, \qquad i = 1, 2, 3, \ldots, 2n \tag{3.130}$$

Define the diagonal matrix $\mathbf{T}_1$ as

$$\mathbf{T}_1(t) = \begin{bmatrix} e^{\mu_1 t} & 0 & 0 & \cdots & 0 \\ 0 & e^{\mu_2 t} & 0 & \cdots & 0 \\ 0 & 0 & e^{\mu_3 t} & \cdots & 0 \\ \vdots & \vdots & \vdots & & \vdots \\ 0 & 0 & 0 & \cdots & e^{\mu_{2n} t} \end{bmatrix} = \mathbf{e}^{\boldsymbol{\mu} t} \tag{3.131a}$$

and

$$\mathbf{T}_1(t, \tau) = \mathbf{e}^{\boldsymbol{\mu}(t-\tau)} \tag{3.131b}$$

or more generally

$$\mathbf{T}_1(t_1, t_2) = \mathbf{e}^{\boldsymbol{\mu}(t_1 - t_2)} \tag{3.131c}$$

The matrix $\mathbf{T}_1(t)$ can also be expressed in a series form as

$$\mathbf{T}_1(t) = \mathbf{e}^{\boldsymbol{\mu} t} = \mathbf{I} + t\boldsymbol{\mu} + \frac{t^2}{2!}\boldsymbol{\mu}^2 + \frac{t^3}{3!}\boldsymbol{\mu}^3 + \cdots \tag{3.131d}$$

In terms of the matrix $\mathbf{T}_1$, Eq. 130 can be written as

$$\mathbf{P} = \mathbf{T}_1(t)\mathbf{P}_0 + \int_0^t \mathbf{T}_1(t-\tau)\mathbf{Q}(\tau)\, d\tau \tag{3.132}$$

This equation is expressed in terms of the vector of modal coordinates $\mathbf{P}$. In order to obtain the space coordinates and velocities, we use Eqs. 120, 126, and 128 and multiply Eq. 132 by the modal matrix $\boldsymbol{\Phi}$ to obtain

$$\mathbf{y} = \boldsymbol{\Phi}\mathbf{P} = \boldsymbol{\Phi}\mathbf{T}_1\boldsymbol{\Phi}^{-1}\mathbf{y}_0 + \int_0^t \boldsymbol{\Phi}\mathbf{T}_1(t-\tau)\boldsymbol{\Phi}^{-1}\mathbf{R}(\tau)\, d\tau$$

which can be written compactly as

$$\mathbf{y} = \mathbf{T}(t)\mathbf{y}_0 + \int_0^t \mathbf{T}(t-\tau)\mathbf{R}(\tau)\, d\tau \tag{3.133}$$

where $\mathbf{T}$ is called the *state transition matrix* and is defined as

$$\mathbf{T} = \mathbf{\Phi}\mathbf{T}_1\mathbf{\Phi}^{-1} \tag{3.134}$$

The state transition matrix which plays a fundamental role in theory of linear dynamic systems satisfies the following identities:

$$\begin{aligned}
\mathbf{T}(t_1, t_1) &= \mathbf{I} = \mathbf{T}(0)\\
\mathbf{T}(t_3, t_1) &= \mathbf{T}(t_3, t_2)\mathbf{T}(t_2, t_1)\\
\mathbf{T}(t_1, t_2)\mathbf{T}(t_2, t_1) &= \mathbf{I}\\
[\mathbf{T}(t_1, t_2)]^{-1} &= \mathbf{T}(t_2, t_1)\\
\mathbf{T}(t_1)\mathbf{T}(t_2) &= \mathbf{T}(t_1 + t_2)
\end{aligned}$$

where $\mathbf{I}$ is the identity matrix.

Illustrative Example In order to demonstrate the use of the state transition matrix in the vibration analysis of mechanical systems, a simple viscously damped single degree of freedom system is considered. The equation of motion of such a system is given by

$$m\ddot{q} + c\dot{q} + kq = F(t)$$

where m, c, and k are, respectively, the mass, damping, and stiffness coefficients, $F(t)$ is the forcing function, and q is the degree of freedom of the system. The differential equation of the system can also be written in the following form

$$\begin{aligned}
\ddot{q} &= -\frac{k}{m}q - \frac{c}{m}\dot{q} + \frac{F(t)}{m}\\
&= -\omega^2 q - 2\xi\omega\dot{q} + \frac{F(t)}{m}
\end{aligned}$$

where ω is the natural frequency of the system and ξ is the damping factor, that is

$$\omega = \sqrt{\frac{k}{m}}, \qquad \xi = \frac{c}{2m\omega}$$

One can, therefore, define the state equations as

$$\dot{\mathbf{y}} = \begin{bmatrix} \dot{q} \\ \ddot{q} \end{bmatrix} = \begin{bmatrix} 0 & 1 \\ -\omega^2 & -2\xi\omega \end{bmatrix}\begin{bmatrix} q \\ \dot{q} \end{bmatrix} + \begin{bmatrix} 0 \\ F(t)/m \end{bmatrix}$$

From which the matrix $\mathbf{B}$ and the vector $\mathbf{R}(t)$ of Eq. 106 can be defined as

$$\mathbf{B} = \begin{bmatrix} 0 & 1 \\ -\omega^2 & -2\xi\omega \end{bmatrix}$$

$$\mathbf{R}(t) = \begin{bmatrix} 0 \\ F(t)/m \end{bmatrix}$$

The characteristic matrix $\mathbf{H}$ of Eq. 114 can then be defined for this system as

$$\mathbf{H} = \mathbf{B} - \mu\mathbf{I}$$
$$= \begin{bmatrix} -\mu & 1 \\ -\omega^2 & -2\xi\omega - \mu \end{bmatrix}$$

which defines the characteristic equation as

$$|\mathbf{H}| = \begin{vmatrix} -\mu & 1 \\ -\omega^2 & -2\xi\omega - \mu \end{vmatrix}$$
$$= \mu(2\xi\omega + \mu) + \omega^2 = 0$$

This is a quadratic equation in μ which can simply be written as

$$\mu^2 + 2\xi\omega\mu + \omega^2 = 0$$

The roots of this equation are

$$\mu_1 = -\xi\omega + \omega\sqrt{\xi^2 - 1}$$
$$\mu_2 = -\xi\omega - \omega\sqrt{\xi^2 - 1}$$

The eigenvector $\mathbf{A}_i$ associated with the eigenvalue μ_i can be obtained by using Eq. 113 or Eq. 116. In this case one has

$$[\mathbf{B} - \mu_i\mathbf{I}]\mathbf{A}_i = \begin{bmatrix} -\mu_i & 1 \\ -\omega^2 & -2\xi\omega - \mu_i \end{bmatrix} \begin{bmatrix} A_{i1} \\ A_{i2} \end{bmatrix} = 0$$

which yields

$$A_{i2} = \mu_i A_{i1}$$

The eigenvector $\mathbf{A}_i$ can then be written as

$$\mathbf{A}_i = \begin{bmatrix} 1 \\ \mu_i \end{bmatrix}, \qquad i = 1, 2$$

and the modal matrix $\mathbf{\Phi}$ of Eq. 117 is

$$\mathbf{\Phi} = [\mathbf{A}_1 \quad \mathbf{A}_2] = \begin{bmatrix} 1 & 1 \\ \mu_1 & \mu_2 \end{bmatrix}$$

It follows that $\mathbf{\Phi}^{-1}$ is the matrix

$$\mathbf{\Phi}^{-1} = \frac{1}{\mu_2 - \mu_1} \begin{bmatrix} \mu_2 & -1 \\ -\mu_1 & 1 \end{bmatrix}$$

which must satisfy the identity of Eq. 118. The state transition matrix $\mathbf{T}$ is given in this example by

$$\mathbf{T} = \mathbf{\Phi}\mathbf{T}_1\mathbf{\Phi}^{-1}$$
$$= \frac{1}{\mu_2 - \mu_1} \begin{bmatrix} 1 & 1 \\ \mu_1 & \mu_2 \end{bmatrix} \begin{bmatrix} e^{\mu_1 t} & 0 \\ 0 & e^{\mu_2 t} \end{bmatrix} \begin{bmatrix} \mu_2 & -1 \\ -\mu_1 & 1 \end{bmatrix}$$

$$= \frac{1}{\mu_2 - \mu_1}\begin{bmatrix} \mu_2 e^{\mu_1 t} - \mu_1 e^{\mu_2 t} & e^{\mu_2 t} - e^{\mu_1 t} \\ \mu_1 \mu_2 (e^{\mu_1 t} - e^{\mu_2 t}) & \mu_2 e^{\mu_2 t} - \mu_1 e^{\mu_1 t} \end{bmatrix}$$

It is left to the reader as an exercise to show that the use of Eq. 125 and 133 leads, respectively, to the free and forced vibration solution obtained in Chapter 1 in the case of the single degree of freedom systems.

3.9 APPROXIMATION AND NUMERICAL METHODS

We have shown in the preceding sections that the solution of the equations of motion of the multi-degree of freedom system can be expressed in terms of the modal coordinates. It was also shown that the generalized forces associated with the modal coordinates can be obtained by premultiplying the vector of externally applied forces by the transpose of the modal matrix, and as a result, the frequency contents in a generalized force associated with a certain mode depend on the frequency contents in the externally applied forces. Clearly, if the modal force has a frequency close to the natural frequency of the corresponding mode, this mode of vibration will be excited and this mode will have a significant effect on the dynamic response of the system. Quite often, the frequency content in a given forcing function has an upper limit and accordingly mode shapes which have frequencies higher than this upper limit are not excited, and their effects on the dynamic response of the system can be ignored. In this case, the multi-degree of freedom system can be represented by an equivalent system which has a smaller number of degrees of freedom determined by the number of the significant low-frequency modes of vibration. If the multi-degree of freedom system has only one significant mode of vibration, the system can be replaced by an equivalent system which has only one degree of freedom. Similarly, the multi-degree of freedom system can be represented by a two degree of freedom system if only two modes are significant. The procedure of eliminating insignificant modes is called *modal truncation.*

Consider a system which has n degrees of freedom and assume that only m modes of vibration are significant, where $m < n$. We have previously shown that the vibration of the n degree of freedom system is governed by n second-order ordinary differential equations which can be written in the following form:

$$\mathbf{M}\ddot{\mathbf{q}} + \mathbf{C}\dot{\mathbf{q}} + \mathbf{K}\mathbf{q} = \mathbf{F}(t) \tag{3.135}$$

where $\mathbf{M}$, $\mathbf{C}$, and $\mathbf{K}$ are, respectively, the $n \times n$ mass, damping, and stiffness matrices, and $\mathbf{q}$ and $\mathbf{F}$ are, respectively, the n-dimensional vectors of displacements and externally applied forces. Let $\mathbf{\Phi}_t$ be the modal matrix which contains only the m significant modes of vibration, that is, $\mathbf{\Phi}_t$ has n rows and m columns and accordingly it is not a square matrix. The elimination of insignificant mode shapes can be expressed mathematically as

$$\mathbf{q} = \mathbf{\Phi}_t \mathbf{P} \tag{3.136}$$

where the vector of modal coordinates $\mathbf{P}$ consists in this case of m elements. Substituting Eq. 136 into Eq. 135 and premultiplying by the transpose of the modal matrix $\mathbf{\Phi}_t$ one obtains

$$\mathbf{\Phi}_t^{\mathrm{T}}\mathbf{M}\mathbf{\Phi}_t\ddot{\mathbf{P}} + \mathbf{\Phi}_t^{\mathrm{T}}\mathbf{C}\mathbf{\Phi}_t\dot{\mathbf{P}} + \mathbf{\Phi}_t^{\mathrm{T}}\mathbf{K}\mathbf{\Phi}_t\mathbf{P} = \mathbf{\Phi}_t^{\mathrm{T}}\mathbf{F}(t) \tag{3.137}$$

Assuming the case of proportional damping, and utilizing the orthogonality of the mode shapes with respect to the mass and stiffness matrices, Eq. 137 can be rewritten as

$$\mathbf{M}_p\ddot{\mathbf{P}} + \mathbf{C}_p\dot{\mathbf{P}} + \mathbf{K}_p\mathbf{P} = \mathbf{Q} \tag{3.138}$$

where $\mathbf{M}_p$, $\mathbf{C}_p$, and $\mathbf{K}_p$ are square diagonal matrices with dimension $m \times m$, and $\mathbf{P}$ is the m-dimensional vector of the modal coordinates.

Equation 138 has m second-order uncoupled differential equations which can be solved for the m modal coordinates as discussed in the preceding sections. The n-dimensional vector of actual displacements can be recovered by using the transformation of Eq. 136.

Example 3.7

For the three degree of freedom torsional system discussed in Example 2, it was shown that the mass and stiffness matrices are given by

$$\mathbf{M} = 10^3 \begin{bmatrix} 2 & 0 & 0 \\ 0 & 3.0 & 0 \\ 0 & 0 & 4 \end{bmatrix} \mathrm{kg \cdot m^2}$$

$$\mathbf{K} = 12 \times 10^5 \begin{bmatrix} 1 & -1 & 0 \\ -1 & 3 & -2 \\ 0 & -2 & 5 \end{bmatrix} \mathrm{N \cdot m}$$

The modal matrix $\mathbf{\Phi}$ for this system is given by

$$\mathbf{\Phi} = \begin{bmatrix} 1.000 & 1.000 & 1.000 \\ 0.649 & -0.607 & -2.541 \\ 0.302 & -0.679 & 2.438 \end{bmatrix}$$

The equation of motion of the undamped system is

$$\mathbf{M}\ddot{\boldsymbol{\theta}} + \mathbf{K}\boldsymbol{\theta} = \mathbf{F}$$

where $\boldsymbol{\theta}$ is the vector of the torsional oscillations of the system. It was also shown that the natural frequencies of this system are $\omega_1 = 14.52$ rad/s, $\omega_2 = 31.05$ rad/s, and $\omega_3 = 46.1$ rad/s. If we assume that the frequency content in the forcing function $\mathbf{F}$ is not in the resonance region of the third mode, we may ignore this mode by deleting the third column of the modal transformation $\mathbf{\Phi}$. In this case, the reduced number of degrees of freedom m is equal to two and the new modal transformation $\mathbf{\Phi}_t$ is given by

$$\mathbf{\Phi}_t = \begin{bmatrix} 1.000 & 1.000 \\ 0.649 & -0.607 \\ 0.302 & -0.679 \end{bmatrix}$$

By using Eq. 136, the system equations of motion can be expressed in terms of the modal coordinates as

$$\mathbf{M}_p\ddot{\mathbf{P}} + \mathbf{K}_p\mathbf{P} = \mathbf{Q}$$

where $\mathbf{M}_p$ and $\mathbf{K}_p$ are 2×2 diagonal matrices defined as

$$\mathbf{M}_p = \mathbf{\Phi}_t^{\mathrm{T}}\mathbf{M}\mathbf{\Phi}_t = 10^3\begin{bmatrix} 3.628 & 0 \\ 0 & 4.95 \end{bmatrix} = \begin{bmatrix} m_1 & 0 \\ 0 & m_2 \end{bmatrix}$$

$$\mathbf{K}_p = \mathbf{\Phi}_t^{\mathrm{T}}\mathbf{K}\mathbf{\Phi}_t = 10^5\begin{bmatrix} 7.656 & 0 \\ 0 & 47.72 \end{bmatrix} = \begin{bmatrix} k_1 & 0 \\ 0 & k_2 \end{bmatrix}$$

where m_i and k_i, $i = 1, 2$, are the modal mass and stiffness coefficients. The generalized modal force vector $\mathbf{Q}$ is a two-dimensional vector defined by

$$\mathbf{Q} = \mathbf{\Phi}_t^{\mathrm{T}}\mathbf{F}$$

Similarly, if the upper limit on the frequency content in the forcing function is much below the second natural frequency of the system, we may replace the three degree of freedom system by an equivalent single degree of freedom system. This is the case in which we ignore the second and third modes of vibrations. In this case, the modal transformation is reduced to a column vector defined as

$$\mathbf{\Phi}_t = \begin{bmatrix} 1.000 \\ 0.649 \\ 0.302 \end{bmatrix}$$

where, in this case, $m = 1$. Using this modal transformation, the equation of motion of the equivalent single degree of freedom system is given by

$$m_1\ddot{P}_1 + k_1\ddot{P}_1 = Q_1$$

where $m_1 = 3.628 \times 10^3$, $k_1 = 7.656 \times 10^5$, and Q_1 is the scalar defined by

$$Q_1 = \mathbf{\Phi}_t^{\mathrm{T}}\mathbf{F}$$

Direct Numerical Integration The method of mode superposition discussed in this chapter is widely used in the vibration analysis of multi-degree of freedom systems. In addition to the physical insight gained from its use, the mode-superposition technique leads to a set of uncoupled differential equations which can be solved in a closed form in a manner similar to the equations that describe the vibration of single degree of freedom systems. Furthermore, if high-frequency modes of vibration do not significantly contribute to the solution of the vibration equations, modal truncation can be used to reduce the number of degrees of freedom as described in this section. In this case, one has to determine only the low-frequency modes of vibration.

In some other applications, however, such as in the case of impact loading, the contribution of the high-frequency modes of vibration is significant. If the number of degrees of freedom is very large, the use of the mode-superposition technique is not recommended since the solution of the eigenvalue problem for a large system becomes a difficult task. Another situation where the use of mode superposition is not recommended is in the cases where the vibration equations are nonlinear. As a simple example, consider the multi-degree of freedom system shown in Fig. 3 and discussed in Section 1. In deriving the equations of motion of this system, the assumption of small oscillations was used. This leads to a set of linear differential equations which can be solved using the techniques discussed in the preceding sections. If the displacements of the masses are large such that the assumption of small oscillations is no longer valid, the resulting differential equations of motion are no longer linear, and the mode shapes and natural frequencies of the system are no longer constant. Consequently, the use of the mode-superposition technique can be very expensive computationally.

An alternate approach for the use of the modal-expansion method is the direct numerical integration of the system differential equations.This subject was discussed in Chapter 5 of the first volume of this book (Shabana, 1996) and the method presented there for the analysis of single degree of freedom systems can be generalized to the case of multi-degree of freedom systems. To this end, we write the equations of motion of the linear or the nonlinear system in the following general form:

$$\mathbf{M}\ddot{\mathbf{q}} = \mathbf{f}_1(\mathbf{q}, \dot{\mathbf{q}}, t)$$

where $\mathbf{f}_1$ is a vector function which can be nonlinear in the coordinates, velocities, and time. In the case of linear systems, the vector function $\mathbf{f}_1$ is given by

$$\mathbf{f}_1(\mathbf{q}, \dot{\mathbf{q}}, t) = -\mathbf{C}\dot{\mathbf{q}} - \mathbf{K}\mathbf{q} + \mathbf{F}(t)$$

Assuming that the mass matrix is nonsingular, the preceding two equations can be used to define the accelerations as follows:

$$\ddot{\mathbf{q}} = \mathbf{M}^{-1}\mathbf{f}_1(\mathbf{q}, \dot{\mathbf{q}}, t)$$

One may define the vectors $\mathbf{y} = [\mathbf{y}_1 \;\; \mathbf{y}_2]^{\mathrm{T}}$ as

$$\mathbf{y} = \begin{bmatrix} \mathbf{y}_1 \\ \mathbf{y}_2 \end{bmatrix} = \begin{bmatrix} \mathbf{q} \\ \dot{\mathbf{q}} \end{bmatrix}$$

Using the preceding two equations, $\dot{\mathbf{y}}_2$ can be defined as

$$\begin{aligned} \dot{\mathbf{y}}_2 &= \mathbf{M}^{-1}\mathbf{f}_1(\mathbf{q}, \dot{\mathbf{q}}, t) = \mathbf{M}^{-1}\mathbf{f}_1(\mathbf{y}_1, \mathbf{y}_2, t) \\ &= \mathbf{M}^{-1}\mathbf{f}_1(\mathbf{y}, t) \end{aligned}$$

which can be written compactly as

$$\dot{\mathbf{y}}_2 = \mathbf{G}(\mathbf{y}, t) \tag{3.139}$$

where

$$\mathbf{G}(\mathbf{y}, t) = \mathbf{M}^{-1}\mathbf{f}_1(\mathbf{y}, t)$$

Equation 139 can be combined with the equation

$$\dot{\mathbf{y}}_1 = \mathbf{y}_2$$

to yield the following system of equations

$$\dot{\mathbf{y}} = \begin{bmatrix} \dot{\mathbf{y}}_1 \\ \dot{\mathbf{y}}_2 \end{bmatrix} = \begin{bmatrix} \mathbf{y}_2 \\ \mathbf{G}(\mathbf{y}, t) \end{bmatrix}$$

For a system of n degrees of freedom, the preceding equation has $2n$ first-order differential equations which can be integrated numerically using direct numerical-integration methods such as the Runge–Kutta method (Shabana, 1996).

3.10 MATRIX-ITERATION METHODS

The methods presented thus far for the vibration analysis of multi-degree of freedom systems require the evaluation of the mode shapes and natural frequencies which define the solution of the vibration equations. If the number of degrees of freedom is very large, the solution of the characteristic equation to determine the natural frequencies becomes more difficult and in these cases, the use of the numerical methods is recommended.

It was shown in the preceding sections that the equations of the free undamped vibration of the multi-degree of freedom system can be written as

$$\mathbf{M}\ddot{\mathbf{q}} + \mathbf{K}\mathbf{q} = \mathbf{0} \tag{3.140}$$

We assumed a solution for this equation in the form

$$\mathbf{q} = \mathbf{A}\sin(\omega t + \phi)$$

which upon substitution into Eq. 140 leads to

$$\omega^2\mathbf{M}\mathbf{A} = \mathbf{K}\mathbf{A} \tag{3.141}$$

If the mass matrix is assumed to be nonsingular, Eq. 141 can be written as

$$\omega^2\mathbf{A} = \mathbf{M}^{-1}\mathbf{K}\mathbf{A} \tag{3.142}$$

which can be written as

$$\lambda\mathbf{A} = \mathbf{H}_s\mathbf{A} \tag{3.143}$$

in which

$$\lambda = \omega^2 \tag{3.144a}$$

$$\mathbf{H}_s = \mathbf{M}^{-1}\mathbf{K} \tag{3.144b}$$

Equation 143 is referred to as the *stiffness formulation* of the eigenvalue problem of the multi-degree of freedom system. Note that in this formulation, the lowest eigenvalue corresponds to the lowest natural frequency in the system.

Alternatively, if the stiffness matrix of the system is nonsingular, Eq. 141 leads to

$$\omega^2 \mathbf{K}^{-1}\mathbf{M}\mathbf{A} = \mathbf{A}$$

which can be written as

$$\mu \mathbf{A} = \mathbf{H}_f \mathbf{A} \tag{3.145}$$

where

$$\mu = \frac{1}{\omega^2} \tag{3.146a}$$

$$\mathbf{H}_f = \mathbf{K}^{-1}\mathbf{M} \tag{3.146b}$$

Equation 145 is referred to as the *flexibility formulation* of the eigenvalue problem of the multi-degree of freedom system. In this formulation, the lowest eigenvalue corresponds to the highest natural frequency of the system.

In the following, we discuss some numerical methods which can be used to determine the eigenvalues and eigenvectors of the system using either the stiffness or the flexibility formulation.

Matrix-Iteration Method This method, which is an iterative procedure refered to as the *Stodla method*, can be used with either the stiffness or the flexibility formulation. When it is used with the stiffness formulation, the solution converges to the mode shape which corresponds to the highest eigenvalue, or equivalently the highest natural frequency. On the other hand, if the flexibility formulation is used the solution converges to the mode shape which corresponds to the highest eigenvalue which corresponds in this case to the lowest natural frequency.

In the remainder of this section, the matrix-iteration method is discussed using the flexibility formulation of Eq. 145. In this method, a trial shape associated with the lowest natural frequency is assumed. We denote this trial shape which should represent the best possible estimate of the first mode as $\mathbf{A}_{10}$, where the first subscript refers to the mode number and the second subscript refers to the iteration number. Since the mode shape can be determined to within an arbitrary constant, without any loss of generality, we assume that the first element in $\mathbf{A}_{10}$ is equal to one, that is,

$$\hat{\mathbf{A}}_{10} = \begin{bmatrix} 1 \\ (A_{10})_2 \\ (A_{10})_3 \\ \vdots \end{bmatrix} \tag{3.147}$$

In the matrix-iteration method the vector $\hat{\mathbf{A}}_{10}$ is substituted into the right-hand side of Eq. 145. This yields

$$\mathbf{A}_{11} = \mu_{11}\hat{\mathbf{A}}_{11} = \mathbf{H}_f \hat{\mathbf{A}}_{10} \tag{3.148}$$

Since the eigenvalue μ is not known at this stage, we select $\hat{\mathbf{A}}_{11}$ such that its first element is equal to one. This in turn will automatically define μ_{11}. If $\hat{\mathbf{A}}_{11}$ is the same as $\hat{\mathbf{A}}_{10}$, then $\hat{\mathbf{A}}_{10}$ is the true mode shape. If this is not the case, we substitute $\hat{\mathbf{A}}_{11}$ in the right-hand side of Eq. 145. This leads to

$$\mathbf{A}_{12} = \mu_{12}\hat{\mathbf{A}}_{12} = \mathbf{H}_f\hat{\mathbf{A}}_{11} = \frac{1}{\mu_{11}}\mathbf{H}_f^2\hat{\mathbf{A}}_{10} \tag{3.149}$$

If $\hat{\mathbf{A}}_{12}$ is the same or close enough to $\mathbf{A}_{11}$, convergence is achieved, otherwise $\hat{\mathbf{A}}_{12}$ is substituted into the right-hand side of Eq. 145 to determine $\mathbf{A}_{13}$. This process continues until $\hat{\mathbf{A}}_{1j}$ is close enough to $\hat{\mathbf{A}}_{1(j-1)}$ for some integer j. The vector $\hat{\mathbf{A}}_{1j}$ then defines the first mode shape and the scalar μ_{1j} defines the first (highest) eigenvalue. This eigenvalue corresponds to the lowest natural frequency which can be determined by using Eq. 146a as

$$\omega_1 = \sqrt{\frac{1}{\mu_{1j}}} \tag{3.150}$$

Clearly, the stiffness formulation can be used if the interest is in determining the higher mode of vibration.

Convergence It can be proved that the iterative procedure described in this section converges to the first mode shape of vibration of the system. Let $\mathbf{A}_1, \mathbf{A}_2, \ldots, \mathbf{A}_n$ be the linearly independent true mode shapes of the multi-degree of freedom system. Since these mode shapes are linearly independent, any other n-dimensional vector can be expressed as a linear combination of these mode shapes. Therefore, the initial guess $\mathbf{A}_{10}$ can be written in terms of these mode shapes as

$$\mathbf{A}_{10} = \alpha_1\mathbf{A}_1 + \alpha_2\mathbf{A}_2 + \cdots + \alpha_n\mathbf{A}_n = \sum_{k=1}^{n} \alpha_k\mathbf{A}_k \tag{3.151}$$

where $\alpha_1, \alpha_2, \ldots, \alpha_n$ are scalar coefficients. Since the mode shapes can be determined to within an arbitrary constant, the iteration procedure described in this section leads to the following relationships:

$$\mathbf{A}_{1j} = \mathbf{H}_f\mathbf{A}_{1(j-1)} = \mathbf{H}_f(\mathbf{H}_f\mathbf{A}_{1(j-2)}) = \mathbf{H}_f^j\mathbf{A}_{10} \tag{3.152}$$

Substituting Eq. 151 into Eq. 152, one obtains

$$\mathbf{A}_{1j} = \mathbf{H}_f^j \sum_{k=1}^{n} \alpha_k\mathbf{A}_k = \sum_{k=1}^{n} \alpha_k\mathbf{H}_f^j\mathbf{A}_k \tag{3.153}$$

It is clear from the eigenvalue problem of Eq. 145 that

$$\mathbf{H}_{\mathrm{f}}\mathbf{A}_k = \mu_k \mathbf{A}_k$$

If both sides of this equation are multiplied by $\mathbf{H}_{\mathrm{f}}$, one obtains

$$\mathbf{H}_{\mathrm{f}}^2\mathbf{A}_k = \mu_k \mathbf{H}_{\mathrm{f}}\mathbf{A}_k = \mu_k^2 \mathbf{A}_k$$

If we multiply both sides of this equation again by $\mathbf{H}_{\mathrm{f}}$, we get

$$\mathbf{H}_{\mathrm{f}}^3\mathbf{A}_k = \mu_k^2 \mathbf{H}_{\mathrm{f}}\mathbf{A}_k = \mu_k^3 \mathbf{A}_k$$

One, therefore, has the following general relationship:

$$\mathbf{H}_{\mathrm{f}}^j\mathbf{A}_k = \mu_k^j \mathbf{A}_k \tag{3.154}$$

Substituting this equation into Eq. 153 yields

$$\mathbf{A}_{1j} = \sum_{k=1}^{n} \alpha_k \mathbf{H}_{\mathrm{f}}^j \mathbf{A}_k = \sum_{k=1}^{n} \alpha_k \mu_k^j \mathbf{A}_k \tag{3.155}$$

in which

$$\mu_1 > \mu_2 > \mu_3 > \cdots > \mu_n$$

As the number of iterations j increases, the first term in the summation of Eq. 155 becomes larger in comparison with other terms. Consequently, $\mathbf{A}_{1j}$ will resemble the first true mode $\mathbf{A}_1$, that is,

$$\mathbf{A}_{1j} = \sum_{k=1}^{n} \alpha_k \mu_k^j \mathbf{A}_k \approx \alpha_1 \mu_1^j \mathbf{A}_1 \tag{3.156}$$

as j increases.

Example 3.8

For the torsional system of Example 1, the symmetric mass and stiffness matrices are given by

$$\mathbf{M} = 10^3 \begin{bmatrix} 2 & 0 & 0 \\ 0 & 3 & 0 \\ 0 & 0 & 4 \end{bmatrix}$$

$$\mathbf{K} = 12 \times 10^5 \begin{bmatrix} 1 & -1 & 0 \\ -1 & 3 & -2 \\ 0 & -2 & 5 \end{bmatrix}$$

In order to use the flexibility formulation, one has to find the inverse of the stiffness matrix as

$$\mathbf{K}^{-1} = \frac{1}{72 \times 10^5} \begin{bmatrix} 11 & 5 & 2 \\ 5 & 5 & 2 \\ 2 & 2 & 2 \end{bmatrix}$$

The matrix $\mathbf{H}_f$ of Eq. 145 is given by

$$\mathbf{H}_f = \mathbf{K}^{-1}\mathbf{M} = \frac{1}{3600}\begin{bmatrix} 11 & 7.5 & 4 \\ 5 & 7.5 & 4 \\ 2 & 3 & 4 \end{bmatrix}$$

Based on the initial guess, we assume the trial shape of the first mode as

$$\hat{\mathbf{A}}_{10} = \begin{bmatrix} 1 \\ 0.8 \\ 0.4 \end{bmatrix}$$

Substituting this trial shape into the right-hand side of Eq. 145, one obtains

$$\mathbf{A}_{11} = \mathbf{H}_f\hat{\mathbf{A}}_{10} = \frac{1}{3600}\begin{bmatrix} 11 & 7.5 & 4 \\ 5 & 7.5 & 4 \\ 2 & 3 & 4 \end{bmatrix}\begin{bmatrix} 1 \\ 0.8 \\ 0.4 \end{bmatrix} = \frac{1}{3600}\begin{bmatrix} 18.6 \\ 12.6 \\ 6 \end{bmatrix}$$

which can be used to define $\hat{\mathbf{A}}_{11}$, μ_{11}, and ω_{11} as

$$\hat{\mathbf{A}}_{11} = \begin{bmatrix} 1.00 \\ 0.6774 \\ 0.3226 \end{bmatrix}, \qquad \mu_{11} = \frac{18.6}{3600} = 5.1667 \times 10^{-3}, \qquad \omega_{11} = 13.912 \text{ rad/s}$$

Substituting the vector $\hat{\mathbf{A}}_{11}$ into the right-hand side of Eq. 145

$$\mathbf{A}_{12} = \mathbf{H}_f\hat{\mathbf{A}}_{11} = \frac{1}{3600}\begin{bmatrix} 11 & 7.5 & 4 \\ 5 & 7.5 & 4 \\ 2 & 3 & 4 \end{bmatrix}\begin{bmatrix} 1.00 \\ 0.6774 \\ 0.3226 \end{bmatrix} = \frac{1}{3600}\begin{bmatrix} 17.3709 \\ 11.3709 \\ 5.3226 \end{bmatrix}$$

which yields

$$\hat{\mathbf{A}}_{12} = \begin{bmatrix} 1.000 \\ 0.6546 \\ 0.3064 \end{bmatrix}, \qquad \mu_{12} = \frac{17.3709}{3600} = 4.8253 \times 10^{-3}, \qquad \omega_{12} = 14.3959 \text{ rad/s}$$

Using $\hat{\mathbf{A}}_{12}$ as the new trial vector, we obtain

$$\mathbf{A}_{13} = \mathbf{H}_f\hat{\mathbf{A}}_{12} = \frac{1}{3600}\begin{bmatrix} 11 & 7.5 & 4 \\ 5 & 7.5 & 4 \\ 2 & 3 & 4 \end{bmatrix}\begin{bmatrix} 1.000 \\ 0.6546 \\ 0.3064 \end{bmatrix} = \frac{1}{3600}\begin{bmatrix} 17.1351 \\ 11.1351 \\ 5.1894 \end{bmatrix}$$

which defines $\hat{\mathbf{A}}_{13}$, μ_{13}, and ω_{13} as

$$\hat{\mathbf{A}}_{13} = \begin{bmatrix} 1.000 \\ 0.6498 \\ 0.3028 \end{bmatrix}, \qquad \mu_{13} = \frac{17.1351}{3600} = 4.7598 \times 10^{-3}, \qquad \omega_{13} = 14.4947 \text{ rad/s}$$

Observe that the matrix-iteration method is converging to the solution obtained in Example 2. If we make another iteration, we have

$$\mathbf{A}_{14} = \mathbf{H}_f\hat{\mathbf{A}}_{13} = \frac{1}{3600}\begin{bmatrix} 11 & 7.5 & 4 \\ 5 & 7.5 & 4 \\ 2 & 3 & 4 \end{bmatrix}\begin{bmatrix} 1.000 \\ 0.6498 \\ 0.3028 \end{bmatrix} = \frac{1}{3600}\begin{bmatrix} 17.0847 \\ 11.0847 \\ 5.1606 \end{bmatrix}$$

which defines $\hat{\mathbf{A}}_{14}$, μ_{14}, and ω_{14} as

$$\hat{\mathbf{A}}_{14} = \begin{bmatrix} 1.000 \\ 0.649 \\ 0.302 \end{bmatrix}, \qquad \mu_{14} = \frac{17.0847}{3600} = 4.7458 \times 10^{-3}, \qquad \omega_{14} = 14.516 \text{ rad/s}$$

Analysis of Higher Modes In order to prove the convergence of the matrix-iteration method to the mode shape associated with the lowest natural frequency, we utilized the fact that our initial estimate for the trial shape can be expressed as a linear combination of the linearly independent true mode shapes. Clearly, the convergence will be faster if the trial shape is close to the shape of the first mode. Furthermore, the matrix-iteration method is more efficient if the natural frequencies are widely separated, since in our analysis of convergence we have made the argument that the jth power of the first eigenvalue is much higher than the jth power of the other eigenvalues.

In order to determine the second mode of vibration, we simply eliminate the dependence of our trial shape on the first mode of vibration. By using a similar argument to the one we previously made we can prove that this trial shape will converge to the second mode of vibration, since the jth power of μ_2, for large j, is assumed to be much higher than the jth power of $\mu_3, \mu_4, \ldots, \mu_n$.

Let us assume that the trial shape for the second mode $\mathbf{A}_{20}$ can be expressed as a linear combination of the true mode shapes, that is,

$$\mathbf{A}_{20} = \sum_{k=1}^{n} \alpha_k \mathbf{A}_k \tag{3.157}$$

where $\alpha_1, \alpha_2, \ldots, \alpha_n$ are constant scalars. In order to eliminate the dependence of the trial shape on the first mode of vibration, we use the orthogonality condition. Since the mode shapes are orthogonal with respect to the mass matrix, one has

$$\begin{aligned} \mathbf{A}_1^{\mathrm{T}}\mathbf{M}\mathbf{A}_{20} &= \mathbf{A}_1^{\mathrm{T}}\mathbf{M}\left[\sum_{k=1}^{n} \alpha_k \mathbf{A}_k\right] \\ &= \alpha_1 \mathbf{A}_1^{\mathrm{T}}\mathbf{M}\mathbf{A}_1 + \alpha_2 \mathbf{A}_1^{\mathrm{T}}\mathbf{M}\mathbf{A}_2 + \cdots + \alpha_n \mathbf{A}_1^{\mathrm{T}}\mathbf{M}\mathbf{A}_n \\ &= \alpha_1 \mathbf{A}_1^{\mathrm{T}}\mathbf{M}\mathbf{A}_1 \end{aligned}$$

If the trial shape is assumed to be independent of the first mode of vibration, the preceding equation reduces to

$$\mathbf{A}_1^{\mathrm{T}}\mathbf{M}\mathbf{A}_{20} = 0 \tag{3.158}$$

This is a scalar equation, which can be written as

$$b_1(A_{20})_1 + b_2(A_{20})_2 + \cdots + b_n(A_{20})_n = 0 \tag{3.159}$$

where $b_1, b_2, \ldots, b_n$ are constant scalars, and $(A_{20})_k$ is the kth element of the trial shape $\mathbf{A}_{20}$. The constant scalars $b_1, b_2, \ldots, b_n$ are assumed to be known at this stage, since the first mode is assumed to be known. These constants are simply given by

$$\mathbf{b}^{\mathrm{T}} = [b_1 \quad b_2 \quad \cdots \quad b_n] = \mathbf{A}_1^{\mathrm{T}}\mathbf{M} \tag{3.160}$$

Dividing Eq. 159 by b_1, one obtains

$$(A_{20})_1 = s_{12}(A_{20})_2 + s_{13}(A_{20})_3 + \cdots + s_{1n}(A_{20})_n \tag{3.161}$$

in which the first element in the trial mode shape is expressed as a linear combination of the other elements, and

$$s_{1k} = -\frac{b_k}{b_1}, \qquad k = 2, 3, \ldots, n \tag{3.162}$$

In addition to Eq. 161, one has the following simple equations:

$$\left.\begin{array}{l} (A_{20})_2 = (A_{20})_2 \\ (A_{20})_3 = \qquad (A_{20})_3 \\ \vdots \qquad\qquad\qquad \ddots \\ (A_{20})_n = \qquad\qquad (A_{20})_n \end{array}\right\} \tag{3.163}$$

Combining Eqs. 161 and 163 one obtains

$$\mathbf{A}_{20} = \mathbf{S}_1\mathbf{A}_{20} \tag{3.164}$$

where $\mathbf{S}_1$ is called the *sweeping matrix* of the first mode and is given by

$$\mathbf{S}_1 = \begin{bmatrix} 0 & s_{12} & s_{13} & \cdots & s_{1n} \\ 0 & 1 & 0 & \cdots & 0 \\ 0 & 0 & 1 & \cdots & 0 \\ \vdots & \vdots & \vdots & \ddots & \vdots \\ 0 & 0 & 0 & \cdots & 1 \end{bmatrix} \tag{3.165}$$

In Eq. 164, the dependence of the trial shape on the first mode is eliminated. When Eq. 164 is substituted in the right-hand side of Eq. 145, one obtains

$$\mathbf{A}_{21} = \mathbf{H}_{\mathrm{f}}\mathbf{S}_1\hat{\mathbf{A}}_{20}$$

This leads to the following iterative procedure:

$$\mathbf{A}_{2j} = \mathbf{H}_{\mathrm{f}}\mathbf{S}_1\hat{\mathbf{A}}_{2(j-1)} \tag{3.166}$$

This iterative procedure will converge to the second mode of vibration, thus defining μ_2, and $\mathbf{A}_2$. The scalar μ_2 can then be used to determine the second natural frequency.

In order to determine the third mode of vibration using the matrix-iteration method, the dependence of the trial shape on the first and second modes must be eliminated. To this end, we use the following two orthogonality conditions:

$$\mathbf{A}_1^{\mathrm{T}}\mathbf{M}\mathbf{A}_{30} = 0 \tag{3.167a}$$

$$\mathbf{A}_2^{\mathrm{T}}\mathbf{M}\mathbf{A}_{30} = 0 \tag{3.167b}$$

Using these two conditions, the first two elements in the trial-shape vector $\mathbf{A}_{30}$ can be expressed in terms of the other elements in a manner similar to Eq. 161.

This leads to the sweeping matrix $\mathbf{S}_2$ which has the first two columns equal to zero. That is, the form of this sweeping matrix is as follows:

$$\mathbf{S}_2 = \begin{bmatrix} 0 & 0 & \bar{s}_{13} & \bar{s}_{14} & \cdots & \bar{s}_{1n} \\ 0 & 0 & \bar{s}_{23} & \bar{s}_{24} & \cdots & \bar{s}_{2n} \\ 0 & 0 & 1 & 0 & \cdots & 0 \\ \vdots & \vdots & \vdots & \vdots & \ddots & \vdots \\ 0 & 0 & 0 & 0 & \cdots & 1 \end{bmatrix} \tag{3.168}$$

Observe that the sweeping matrix $\mathbf{S}_1$ of Eq. 165 has a column rank $n - 1$, and the sweeping matrix $\mathbf{S}_2$ of Eq. 168 has a column rank $n - 2$.

Using the sweeping matrix $\mathbf{S}_2$, one can define the trial shape vector which is independent of the first and second modes as

$$\mathbf{A}_{30} = \mathbf{S}_2\mathbf{A}_{30} \tag{3.169}$$

By substituting this equation into Eq. 145, one obtains

$$\mathbf{A}_{31} = \mathbf{H}_{\mathrm{f}}\mathbf{S}_2\hat{\mathbf{A}}_{30}$$

This leads to the iterative procedure, which can be expressed mathematically as

$$\mathbf{A}_{3j} = \mathbf{H}_{\mathrm{f}}\mathbf{S}_2\hat{\mathbf{A}}_{3(j-1)} \tag{3.170}$$

In order to determine an arbitrary mode i using the matrix-iteration method, one has to determine the sweeping matrix $\mathbf{S}_{i-1}$ using the following orthogonality relationships:

$$\mathbf{A}_1^{\mathrm{T}}\mathbf{M}\mathbf{A}_{i0} = 0$$

$$\mathbf{A}_2^{\mathrm{T}}\mathbf{M}\mathbf{A}_{i0} = 0$$

$$\vdots$$

$$\mathbf{A}_{i-1}^{\mathrm{T}}\mathbf{M}\mathbf{A}_{i0} = 0$$

By using the sweeping matrix $\mathbf{S}_{i-1}$, Eq. 170 can be generalized for the ith mode as

$$\mathbf{A}_{ij} = \mathbf{H}_{\mathrm{f}}\mathbf{S}_{i-1}\hat{\mathbf{A}}_{i(j-1)} \tag{3.171}$$

This iterative procedure converges to the ith mode since the dependence of the trial shape on mode $1, 2, \ldots, i - 1$ has been eliminated using the sweeping matrix $\mathbf{S}_{i-1}$.

Example 3.9

In Example 8, it was shown that the first mode of vibration of the torsional system of Example 1 is given by

$$\mathbf{A}_1 = \begin{bmatrix} 1.000 \\ 0.649 \\ 0.302 \end{bmatrix}$$

In order to determine the second mode, we use the orthogonality condition of Eq. 158

$$\mathbf{A}_1^{\mathrm{T}}\mathbf{M}\mathbf{A}_{20} = 0$$

The vector $\mathbf{b}$ of Eq. 160 is given by

$$\mathbf{b}^{\mathrm{T}} = \mathbf{A}_1^{\mathrm{T}}\mathbf{M} = [1.000 \quad 0.649 \quad 0.302]\begin{bmatrix} 2 & 0 & 0 \\ 0 & 3 & 0 \\ 0 & 0 & 4 \end{bmatrix} \times 10^3$$

$$= 10^3\,[2.000 \quad 1.947 \quad 1.208]$$

Therefore, s_{1k}, $k = 2, 3$, of Eq. 162 are defined as

$$[s_{12} \quad s_{13}] = [-0.9735 \quad -0.604]$$

The sweeping matrix $\mathbf{S}_1$ of Eq. 165 is given by

$$\mathbf{S}_1 = \begin{bmatrix} 0 & s_{12} & s_{13} \\ 0 & 1 & 0 \\ 0 & 0 & 1 \end{bmatrix} = \begin{bmatrix} 0 & -0.9735 & -0.604 \\ 0 & 1 & 0 \\ 0 & 0 & 1 \end{bmatrix}$$

Therefore,

$$\mathbf{H}_{\mathrm{f}}\mathbf{S}_1 = \frac{1}{3600}\begin{bmatrix} 11 & 7.5 & 4 \\ 5 & 7.5 & 4 \\ 2 & 3 & 4 \end{bmatrix}\begin{bmatrix} 0 & -0.9735 & -0.604 \\ 0 & 1 & 0 \\ 0 & 0 & 1 \end{bmatrix}$$

$$= \frac{1}{3600}\begin{bmatrix} 0 & -3.2085 & -2.644 \\ 0 & 2.6325 & 0.980 \\ 0 & 1.053 & 2.792 \end{bmatrix}$$

We may assume the following trial shape for the second mode:

$$\hat{\mathbf{A}}_{20} = \begin{bmatrix} 1.0 \\ -1.0 \\ -1.0 \end{bmatrix}$$

which leads to

$$\mathbf{A}_{21} = \mathbf{H}_{\mathrm{f}}\mathbf{S}_1\hat{\mathbf{A}}_{20} = \frac{1}{3600}\begin{bmatrix} 0 & -3.2085 & -2.644 \\ 0 & 2.6325 & 0.980 \\ 0 & 1.053 & 2.792 \end{bmatrix}\begin{bmatrix} 1.0 \\ -1.0 \\ -1.0 \end{bmatrix}$$

$$= \frac{1}{3600}\begin{bmatrix} 5.8525 \\ -3.6125 \\ -3.845 \end{bmatrix}$$

This vector can be used to define $\hat{\mathbf{A}}_{21}$, μ_{21}, and ω_{21} as

$$\hat{\mathbf{A}}_{21} = \begin{bmatrix} 1.000 \\ -0.61726 \\ -0.65698 \end{bmatrix}, \qquad \mu_{21} = \frac{5.8525}{3600} = 1.6257 \times 10^{-3}, \qquad \omega_{21} = 24.8016 \text{ rad/s}$$

For the second iteration, we substitute $\hat{\mathbf{A}}_{21}$ into the right-hand side of Eq. 166. This yields

$$\mathbf{A}_{22} = \mathbf{H}_f \mathbf{S}_1 \hat{\mathbf{A}}_{21} = \frac{1}{3600} \begin{bmatrix} 0 & -3.2085 & -2.644 \\ 0 & 2.6325 & 0.980 \\ 0 & 1.053 & 2.792 \end{bmatrix} \begin{bmatrix} 1.00 \\ -0.61726 \\ -0.65698 \end{bmatrix}$$

$$= \frac{1}{3600} \begin{bmatrix} 3.7175 \\ -2.2688 \\ -2.4843 \end{bmatrix}$$

which defines $\hat{\mathbf{A}}_{22}$, μ_{22}, and ω_{22} as

$$\hat{\mathbf{A}}_{22} = \begin{bmatrix} 1.000 \\ -0.6103 \\ -0.6683 \end{bmatrix}, \qquad \mu_{22} = \frac{3.7175}{3600} = 1.0326 \times 10^{-3}, \qquad \omega_{22} = 31.119 \text{ rad/s}$$

For the third iteration, we have

$$\mathbf{A}_{23} = \mathbf{H}_f \mathbf{S}_1 \hat{\mathbf{A}}_{22} = \frac{1}{3600} \begin{bmatrix} 0 & -3.2085 & -2.644 \\ 0 & 2.6325 & 0.980 \\ 0 & 1.053 & 2.792 \end{bmatrix} \begin{bmatrix} 1.000 \\ -0.6103 \\ -0.6683 \end{bmatrix}$$

$$= \frac{1}{3600} \begin{bmatrix} 3.725 \\ -2.2615 \\ -2.509 \end{bmatrix}$$

Using this vector, $\hat{\mathbf{A}}_{23}$, μ_{23}, and ω_{23} can be defined as

$$\hat{\mathbf{A}}_{23} = \begin{bmatrix} 1.000 \\ -0.607 \\ -0.673 \end{bmatrix}, \qquad \mu_{23} = \frac{3.725}{3600} = 1.034 \times 10^{-3}, \qquad \omega_{23} = 31.087 \text{ rad/s}$$

Another iteration shows that

$$\hat{\mathbf{A}}_{24} = \begin{bmatrix} 1.000 \\ -0.606 \\ -0.678 \end{bmatrix}, \qquad \omega_{24} = 31.05 \text{ rad/s}$$

Comparing these results with the result of Example 2 demonstrates the convergence of this method to the true second mode of vibration.

In order to use the matrix-iteration method to determine the third mode, we use the two orthogonality conditions of Eqs. 167a and 167b. This leads to the following sweeping matrix:

$$\mathbf{S}_2 = \begin{bmatrix} 0 & 0 & 0.4096 \\ 0 & 0 & -1.0412 \\ 0 & 0 & 1 \end{bmatrix}$$

Hence,

$$\mathbf{H}_f \mathbf{S}_2 = \frac{1}{3600} \begin{bmatrix} 11 & 7.5 & 4 \\ 5 & 7.5 & 4 \\ 2 & 3 & 4 \end{bmatrix} \begin{bmatrix} 0 & 0 & 0.4093 \\ 0 & 0 & -1.0415 \\ 0 & 0 & 1 \end{bmatrix}$$

$$= \frac{1}{3600}\begin{bmatrix} 0 & 0 & 0.69105 \\ 0 & 0 & -1.76475 \\ 0 & 0 & 1.6941 \end{bmatrix}$$

We may assume a trial shape for the third mode as

$$\hat{\mathbf{A}}_{30} = \begin{bmatrix} 1.00 \\ -1.00 \\ 1.00 \end{bmatrix}$$

The first iteration then yields

$$\mathbf{A}_{31} = \mathbf{H}_{\mathrm{f}}\mathbf{S}_2\mathbf{A}_{30} = \frac{1}{3600}\begin{bmatrix} 0 & 0 & 0.69105 \\ 0 & 0 & -1.76475 \\ 0 & 0 & 1.6941 \end{bmatrix}\begin{bmatrix} 1.00 \\ -1.00 \\ 1.00 \end{bmatrix}$$

$$= \frac{1}{3600}\begin{bmatrix} 0.69105 \\ -1.76475 \\ 1.6941 \end{bmatrix}$$

This vector can be used to define $\hat{\mathbf{A}}_{31}$, μ_{31}, and ω_{31} as

$$\hat{\mathbf{A}}_{31} = \begin{bmatrix} 1.000 \\ -2.5537 \\ 2.4515 \end{bmatrix}, \qquad \mu_{31} = \frac{0.69105}{3600} = 1.9196 \times 10^{-3}, \qquad \omega_{31} = 72.1766 \text{ rad/s}$$

The second iteration yields

$$\hat{\mathbf{A}}_{32} = \begin{bmatrix} 1.000 \\ -2.5537 \\ 2.4515 \end{bmatrix}, \qquad \mu_{32} = 4.7059 \times 10^{-4}, \qquad \omega_{32} = 46.0978 \text{ rad/s}$$

Rigid-Body Modes In a semidefinite system, the stiffness matrix is singular. In this case, the inverse of the stiffness matrix does not exist because of the rigid-body modes. The matrix equation of free undamped vibration of such systems can be written as

$$\mathbf{M}\ddot{\mathbf{q}} + \mathbf{K}\mathbf{q} = \mathbf{0} \tag{3.172}$$

Since the natural frequencies associated with the rigid-body modes are equal to zeros, these modes can be calculated or identified by inspection. These rigid-body modes will be denoted as $\mathbf{A}_{r1}, \mathbf{A}_{r2}, \ldots, \mathbf{A}_{rn_r}$, where n_r is the number of the rigid-body modes. The solution vector of Eq. 172 can be expressed using the mode superposition as

$$\mathbf{q} = \sum_{l=1}^{n_r} \alpha_l \mathbf{A}_{rl} + \sum_{k=1}^{n_d} \beta_k \mathbf{A}_k \tag{3.173}$$

where α_l and β_k are constants and n_d is the number of deformation modes. In order to use the flexibility formulation to determine the deformation modes, the rigid-body modes must be eliminated by writing the solution vector $\mathbf{q}$ in terms of the deformation modes only. Therefore, Eq. 173 reduces to

$$\mathbf{q} = \sum_{k=1}^{n_d} \beta_k \mathbf{A}_k \tag{3.174}$$

Using the orthogonality of the mode shapes and Eq. 174, we have the following relationships:

$$\left.\begin{aligned} \mathbf{A}_{r1}^{\mathrm{T}}\mathbf{M}\mathbf{q} &= 0 \\ \mathbf{A}_{r2}^{\mathrm{T}}\mathbf{M}\mathbf{q} &= 0 \\ &\vdots \\ \mathbf{A}_{rn_r}^{\mathrm{T}}\mathbf{M}\mathbf{q} &= 0 \end{aligned}\right\} \tag{3.175}$$

The scalar equations in Eq. 175 can be combined in one matrix equation, which can be written as

$$\mathbf{B}\mathbf{q} = \mathbf{0} \tag{3.176}$$

where $\mathbf{B}$ is the matrix

$$\mathbf{B} = \begin{bmatrix} \mathbf{A}_{r1}^{\mathrm{T}} \\ \mathbf{A}_{r2}^{\mathrm{T}} \\ \vdots \\ \mathbf{A}_{rn_r}^{\mathrm{T}} \end{bmatrix} \mathbf{M} \tag{3.177}$$

If the relationships of Eq. 175 are assumed to be linearly independent, the matrix $\mathbf{B}$ has a full row rank. Observe that the dimension of the matrix $\mathbf{B}$ is $n_r \times n$ where $n = n_r + n_d$ is the total number of generalized coordinates.

Equation 176 can then be used to write n_r coordinates in terms of the other n_d coordinates. In this case, Eq. 176 can be written in the following partitioned form:

$$[\mathbf{B}_d \quad \mathbf{B}_i] \begin{bmatrix} \mathbf{q}_d \\ \mathbf{q}_i \end{bmatrix} = \mathbf{0}$$

or

$$\mathbf{B}_d\mathbf{q}_d + \mathbf{B}_i\mathbf{q}_i = \mathbf{0} \tag{3.178}$$

where $\mathbf{B}_d$ is an $n_r \times n_r$ matrix and $\mathbf{B}_i$ is an $n_r \times n_d$ matrix. The partitioning in Eq. 178 is made in such a manner that the matrix $\mathbf{B}_d$ is a nonsingular matrix. Such a matrix exists since the matrix $\mathbf{B}$ is assumed to have a full row rank. The vector $\mathbf{q}_d$ of Eq. 178 can then be expressed in terms of the vector $\mathbf{q}_i$ as

$$\mathbf{q}_d = -\mathbf{B}_d^{-1}\mathbf{B}_i\mathbf{q}_i \tag{3.179}$$

Using Eq. 179, the vector $\mathbf{q}$ of the generalized coordinates can be written as

$$\mathbf{q} = \begin{bmatrix} \mathbf{q}_i \\ \mathbf{q}_d \end{bmatrix} = \begin{bmatrix} \mathbf{q}_i \\ -\mathbf{B}_d^{-1}\mathbf{B}_i\mathbf{q}_i \end{bmatrix} = \begin{bmatrix} \mathbf{I} \\ -\mathbf{B}_d^{-1}\mathbf{B}_i \end{bmatrix} \mathbf{q}_i$$

which can be written as

$$\mathbf{q} = \mathbf{B}_r\mathbf{q}_i \tag{3.180}$$

where $\mathbf{B}_r$ is the matrix defined as

$$\mathbf{B}_r = \begin{bmatrix} \mathbf{I} \\ -\mathbf{B}_d^{-1}\mathbf{B}_i \end{bmatrix}$$

Differentiating Eq. 180 twice with respect to time, one obtains

$$\ddot{\mathbf{q}} = \mathbf{B}_r\ddot{\mathbf{q}}_i \tag{3.181}$$

Substituting Eqs. 180 and 181 into Eq. 172 and premultiplying by the transpose of the matrix $\mathbf{B}_r$ one obtains

$$\mathbf{B}_r^{\mathrm{T}}\mathbf{M}\mathbf{B}_r\ddot{\mathbf{q}}_i + \mathbf{B}_r^{\mathrm{T}}\mathbf{K}\mathbf{B}_r\mathbf{q}_i = \mathbf{0}$$

which can be written as

$$\mathbf{M}_i\ddot{\mathbf{q}}_i + \mathbf{K}_i\mathbf{q}_i = \mathbf{0} \tag{3.182}$$

where

$$\mathbf{M}_i = \mathbf{B}_r^{\mathrm{T}}\mathbf{M}\mathbf{B}_r \tag{3.183}$$

$$\mathbf{K}_i = \mathbf{B}_r^{\mathrm{T}}\mathbf{K}\mathbf{B}_r \tag{3.184}$$

The stiffness matrix in Eq. 182 as defined by Eq. 184 is a nonsingular matrix since the rigid-body modes are eliminated. Therefore, the flexibility formulation as described earlier in this section can be used with the matrix-iteration method to determine the mode shapes numerically.

Example 3.10

In this example, the matrix-iteration method is used to determine the vibration modes of the torsional system of Example 5. The mass and stiffness matrices of this system were given by

$$\mathbf{M} = 10^3\begin{bmatrix} 2 & 0 & 0 \\ 0 & 3 & 0 \\ 0 & 0 & 4 \end{bmatrix}, \qquad \mathbf{K} = 12 \times 10^5\begin{bmatrix} 1 & -1 & 0 \\ -1 & 3 & -2 \\ 0 & -2 & 2 \end{bmatrix}$$

This system has one rigid-body mode which occurs when all the disks have the same displacement. Therefore, $n_r = 1$ and

$$\mathbf{A}_r = \begin{bmatrix} 1.00 \\ 1.00 \\ 1.00 \end{bmatrix}$$

The matrix $\mathbf{B}$ of Eq. 177 can then be determined as

$$\mathbf{B} = \mathbf{A}_r^{\mathrm{T}}\mathbf{M} = [1.000 \quad 1.000 \quad 1.000]\begin{bmatrix} 2 & 0 & 0 \\ 0 & 3 & 0 \\ 0 & 0 & 4 \end{bmatrix} \times 10^3$$

$$= 10^3[2.000 \quad 3.000 \quad 4.000]$$

Substituting this matrix into Eq. 176 with the vector $\mathbf{q} = [\theta_1 \quad \theta_2 \quad \theta_3]^{\mathrm{T}}$, one obtains

$$2\theta_1 + 3\theta_2 + 4\theta_3 = 0$$

or

$$[2 \quad 3 \quad 4]\begin{bmatrix}\theta_1\\ \theta_2\\ \theta_3\end{bmatrix} = 0$$

From which θ_1 can be expressed in terms of θ_2 and θ_3 as

$$\theta_1 = -1.5\theta_2 - 2\theta_3$$

Therefore, the vector $\mathbf{q}$ can be written in terms of θ_2 and θ_3 as

$$\mathbf{q} = \begin{bmatrix}\theta_1\\ \theta_2\\ \theta_3\end{bmatrix} = \begin{bmatrix}-1.5 & -2.00\\ 1 & 0\\ 0 & 1\end{bmatrix}\begin{bmatrix}\theta_2\\ \theta_3\end{bmatrix} \tag{3.185}$$

This defines the matrix $\mathbf{B}_r$ of Eq. 180 as

$$\mathbf{B}_r = \begin{bmatrix}-1.5 & -2.00\\ 1 & 0\\ 0 & 1\end{bmatrix}$$

The matrices $\mathbf{M}_i$ and $\mathbf{K}_i$ of Eq. 183 are given by

$$\mathbf{M}_i = \mathbf{B}_r^{\mathrm{T}}\mathbf{M}\mathbf{B}_r = 10^3\begin{bmatrix}-1.5 & 1 & 0\\ -2.00 & 0 & 1\end{bmatrix}\begin{bmatrix}2 & 0 & 0\\ 0 & 3 & 0\\ 0 & 0 & 4\end{bmatrix}\begin{bmatrix}-1.5 & -2.00\\ 1 & 0\\ 0 & 1\end{bmatrix}$$

$$= 10^3\begin{bmatrix}7.5 & 6.0\\ 6.0 & 12\end{bmatrix}$$

$$\mathbf{K}_i = \mathbf{B}_r^{\mathrm{T}}\mathbf{K}\mathbf{B}_r = 12 \times 10^5\begin{bmatrix}-1.5 & 1 & 0\\ -2.00 & 0 & 1\end{bmatrix}\begin{bmatrix}1 & -1 & 0\\ -1 & 3 & -2\\ 0 & -2 & 2\end{bmatrix}\begin{bmatrix}-1.5 & -2.00\\ 1.0 & 0.0\\ 0 & 1.0\end{bmatrix}$$

$$= 12 \times 10^5\begin{bmatrix}8.25 & 3.0\\ 3.0 & 6.0\end{bmatrix}$$

Using $\mathbf{M}_i$ and $\mathbf{K}_i$, the reduced system of equations can be written as

$$\mathbf{M}_i\ddot{\mathbf{q}}_i + \mathbf{K}_i\mathbf{q}_i = \mathbf{0}$$

or

$$10^3\begin{bmatrix}7.5 & 6.0\\ 6.0 & 12\end{bmatrix}\begin{bmatrix}\ddot{\theta}_2\\ \ddot{\theta}_3\end{bmatrix} + 12 \times 10^5\begin{bmatrix}8.25 & 3.0\\ 3.0 & 6.0\end{bmatrix}\begin{bmatrix}\theta_2\\ \theta_3\end{bmatrix} = \begin{bmatrix}0\\ 0\end{bmatrix}$$

This system has no rigid-body modes, and, therefore, other vibration modes can be determined using the analytical methods or the numerical procedure described in this section. Having determined the modes of θ_2 and θ_3, Eq. 180 or, equivalently, Eq. 185 can be used to determine the modes of θ_1, θ_2, and θ_3.

3.11 METHOD OF TRANSFER MATRICES

In the preceding section, a numerical procedure for determining the mode shapes and natural frequencies of the multi-degree of freedom systems was discussed. The convergence of this numerical procedure was proved and numerical examples that demonstrate its use were presented. The basic idea used in the matrix-iteration method or Stodla method, as it is sometimes referred to, is that an assumed mode shape is successively adjusted until the true mode shape is obtained. Once convergence is achieved the natural frequency associated with this mode can be evaluated. In this section, an alternative method which proceeds in the reverse sequence is discussed. That is, an initial assumption for the frequency is first made. By successive adjustments, the true frequency is obtained and the mode shape associated with this frequency is simultaneously evaluated. The procedure discussed in this section is suitable for the analysis of mechanical and structural *chain systems* whose elements are arranged along a basic axis.

In order to demonstrate the fundamental concept of the procedure, we first discuss *Holzer's method.* We then generalize this procedure by using a rectilinear mass–spring multi-degree of freedom system. The application of the method to other systems such as torsional systems are also discussed before we conclude this section.

Holzer's Method This method is basically a trial-and-error scheme for determining the natural frequencies of multi-degree of freedom systems. While the method has had its widest application in the analysis of multi-degree of freedom torsional systems, the same concept, can be applied to the analysis of other systems as demonstrated later in this section.

Figure 13 shows a torsional system which has n degrees of freedom. Recall that if this system vibrates at one of its natural frequencies, the shape of deformation will be the mode shape associated with this frequency. We have previously shown that in the case of free vibration, the equation of motion of the multi-degree of freedom system is given by

$$\mathbf{M}\ddot{\mathbf{q}} + \mathbf{K}\mathbf{q} = \mathbf{0}$$

where $\mathbf{M}$ and $\mathbf{K}$ are, respectively, the symmetric mass and stiffness matrices, and $\mathbf{q}$ is the vector of generalized coordinates. A solution of the equation of

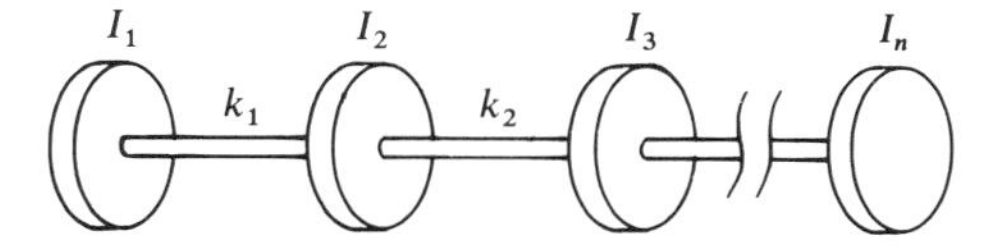

FIG. 3.13. Holzer's method.

motion is assumed in this case in the following form:

$$\mathbf{q} = \mathbf{A} \sin(\omega t + \psi)$$

where $\mathbf{A} = [A_1 \ A_2 \ \cdots \ A_n]^{\mathrm{T}}$ is the vector of amplitudes of the disks. Substituting this solution into the equation of motion yields the following relationships between the amplitudes of the disks:

$$[\mathbf{K} - \omega^2 \mathbf{M}]\mathbf{A} = \mathbf{0} \tag{3.186a}$$

where

$$\mathbf{M} = \begin{bmatrix} I_1 & 0 & 0 & \cdots & 0 \\ 0 & I_2 & 0 & \cdots & 0 \\ \vdots & \vdots & \vdots & \ddots & \vdots \\ 0 & 0 & 0 & \cdots & I_n \end{bmatrix}$$

and

$$\mathbf{K} = \begin{bmatrix} k_{11} & k_{12} & \cdots & k_{1n} \\ k_{21} & k_{22} & \cdots & k_{2n} \\ \vdots & \vdots & \ddots & \vdots \\ k_{n1} & k_{n2} & \cdots & K_{nn} \end{bmatrix}$$

$$= \begin{bmatrix} k_1 & -k_1 & 0 & \cdots & 0 \\ -k_1 & k_1 + k_2 & -k_2 & \cdots & 0 \\ \vdots & \vdots & \vdots & \ddots & \vdots \\ 0 & 0 & 0 & \cdots & k_{(n-1)} \end{bmatrix}$$

Therefore, Eq. 186a yields the following scalar equations:

$$\left.\begin{aligned} I_1\omega^2 A_1 + k_{12}(A_1 - A_2) &= 0 \\ I_2\omega^2 A_2 + k_{12}(A_2 - A_1) + k_{23}(A_2 - A_3) &= 0 \\ I_3\omega^2 A_3 + k_{23}(A_3 - A_2) + k_{34}(A_3 - A_4) &= 0 \\ &\vdots \\ I_n\omega^2 A_n + k_{(n-1)n}(A_n - A_{n-1}) &= 0 \end{aligned}\right\} \tag{3.186b}$$

where I_j denotes the mass moment of inertia of the disk j and k_{ij} is the ijth element of the stiffness matrix $\mathbf{K}$. Since there is no external force or couple acting on this system, adding the preceding equations yields

$$\sum_{j=1}^{n} I_j\omega^2 A_j = 0 \tag{3.187a}$$

Since the mode shape can be determined to within an arbitrary constant, without any loss of generality we assume that the amplitude of the first disk is equal to one. Assuming different values of ω, Eq. 186a or 186b can be solved for the amplitudes $A_2, A_3, \ldots, A_n$. For example, from the first equation in Eq.

186b, the amplitude A_2 can be written in terms of the amplitude A_1 as

$$A_2 = A_1 - \frac{I_1 \omega^2 A_1}{k_{12}}$$

Once A_2 is determined A_3 can be calculated by using the second equation in Eq. 186b as

$$A_3 = A_2 - \frac{k_{12}}{k_{23}}(A_2 - A_1) - \frac{I_2 \omega^2 A_2}{k_{23}}$$

Continuing in this manner, all the amplitudes corresponding to the assumed value of ω can be determined. The assumed value of ω and the calculated amplitudes can then be substituted into Eq. 187a. If this equation is satisfied, then ω is a root of this equation and accordingly the assumed ω is a natural frequency of the system and the corresponding amplitudes $A_1, A_2, \ldots, A_n$ represent the mode shape associated with this frequency. One, therefore, may write Eq. 187a as

$$\sum_{j=1}^{n} I_j \omega^2 A_j = f(\omega^2) \tag{3.187b}$$

One can then plot $f(\omega^2)$ as a function of ω^2 and the roots of this function which define the natural frequency of the system can easily be found.

The method of transfer matrices can be considered as a generalization of the Holzer's method. The application of the transfer-matrix method to the vibration analysis of rectilinear mass–spring systems as well as torsional systems is also discussed in this section.

Rectilinear Mass–Spring System Figure 14 shows a multi-degree of freedom mass–spring system. Let $\ddot{x}_i$ denote the acceleration of the mass i, F_i^r is the spring force acting on the mass from the right and F_i^l is the spring force action on the mass i from the left. The dynamic equilibrium condition for the mass i is given by

$$m_i \ddot{x}_i = F_i^r - F_i^l \tag{3.188}$$

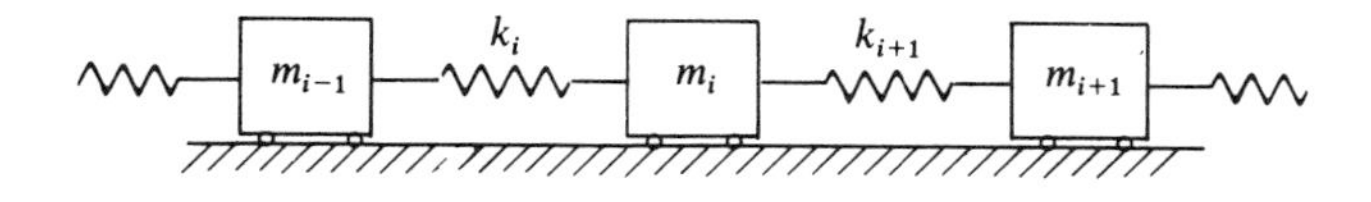

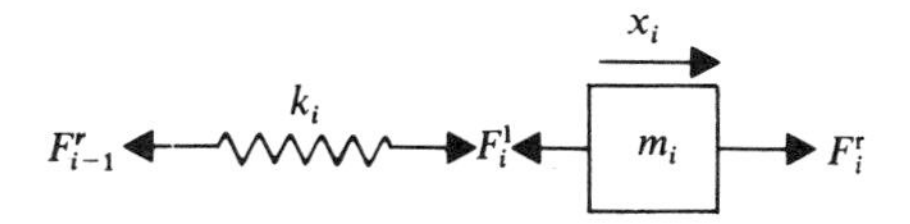

FIG. 3.14. Transfer matrix method.

If we assume harmonic oscillation, the displacement of the mass i can be written as

$$x_i = X_i \sin(\omega t + \phi) \tag{3.189a}$$

where X_i is the amplitude, ω is the frequency, and ϕ is the phase angle. Equation 189a implies that

$$\ddot{x}_i = -\omega^2 x_i \tag{3.189b}$$

Substituting this equation into Eq. 188 and rearranging the terms in this equation, one obtains

$$F_i^{\mathrm{r}} = -\omega^2 m_i x_i + F_i^{\mathrm{l}} \tag{3.190a}$$

Furthermore, for the point mass m_i we have

$$x_i^{\mathrm{r}} = x_i^{\mathrm{l}} = x_i \tag{3.190b}$$

Equations 190a and 190b can be combined into one matrix equation. This yields

$$\begin{bmatrix} x_i^{\mathrm{r}} \\ F_i^{\mathrm{r}} \end{bmatrix} = \begin{bmatrix} 1 & 0 \\ -\omega^2 m_i & 1 \end{bmatrix} \begin{bmatrix} x_i^{\mathrm{l}} \\ F_i^{\mathrm{l}} \end{bmatrix} \tag{3.191}$$

which can be written compactly as

$$\mathbf{s}_i^{\mathrm{r}} = \mathbf{P}_i \mathbf{s}_i^{\mathrm{l}} \tag{3.192}$$

where

$$\mathbf{s}_i^{\mathrm{r}} = \begin{bmatrix} x_i^{\mathrm{r}} \\ F_i^{\mathrm{r}} \end{bmatrix}, \qquad \mathbf{s}_i^{\mathrm{l}} = \begin{bmatrix} x_i^{\mathrm{l}} \\ F_i^{\mathrm{l}} \end{bmatrix}$$

and

$$\mathbf{P}_i = \begin{bmatrix} 1 & 0 \\ -\omega^2 m_i & 1 \end{bmatrix}$$

The vectors $\mathbf{s}_i^{\mathrm{r}}$ and $\mathbf{s}_i^{\mathrm{l}}$ are called the *state vectors*, and $\mathbf{P}_i$ is called the *point matrix*.

If we consider the equilibrium of the massless spring k_i, the following relationships can be easily verified:

$$x_i^{\mathrm{l}} = x_{i-1}^{\mathrm{r}} + \frac{1}{k_i} F_{i-1}^{\mathrm{r}}$$

$$F_i^{\mathrm{l}} = F_{i-1}^{\mathrm{r}}$$

These two equations can be combined in one matrix equation. This leads to

$$\begin{bmatrix} x_i^{\mathrm{l}} \\ F_i^{\mathrm{l}} \end{bmatrix} = \begin{bmatrix} 1 & 1/k_i \\ 0 & 1 \end{bmatrix} \begin{bmatrix} x_{i-1}^{\mathrm{r}} \\ F_{i-1}^{\mathrm{r}} \end{bmatrix}$$

which can be written as

$$\mathbf{s}_i^{\mathrm{l}} = \mathbf{H}_i \mathbf{s}_{i-1}^{\mathrm{r}} \tag{3.193}$$

where $\mathbf{H}_i$ is called the *field matrix*, and is defined as

$$\mathbf{H}_i = \begin{bmatrix} 1 & 1/k_i \\ 0 & 1 \end{bmatrix}$$

Substituting the vector $\mathbf{s}_i^l$ as defined by Eq. 193 into Eq. 192, one obtains

$$\mathbf{s}_i^r = \mathbf{P}_i \mathbf{H}_i \mathbf{s}_{i-1}^r \tag{3.194}$$

which can also be written as

$$\mathbf{s}_i^r = \mathbf{T}_i \mathbf{s}_{i-1}^r \tag{3.195}$$

where $\mathbf{T}_i$ is called the *transfer matrix* defined as

$$\begin{aligned} \mathbf{T}_i = \mathbf{P}_i \mathbf{H}_i &= \begin{bmatrix} 1 & 0 \\ -\omega^2 m_i & 1 \end{bmatrix} \begin{bmatrix} 1 & 1/k_i \\ 0 & 1 \end{bmatrix} \\ &= \begin{bmatrix} 1 & 1/k_i \\ -\omega^2 m_i & (1 - \omega^2 m_i/k_i) \end{bmatrix} \end{aligned} \tag{3.196}$$

Using a similar procedure, the state vector $\mathbf{s}_{i-1}^r$ can be expressed in terms of the state vector $\mathbf{s}_{i-2}^r$ as

$$\mathbf{s}_{i-1}^r = \mathbf{T}_{i-1} \mathbf{s}_{i-2}^r \tag{3.197}$$

which, upon substituting into Eq. 195, leads to

$$\mathbf{s}_i^r = \mathbf{T}_i \mathbf{T}_{i-1} \mathbf{s}_{i-2}^r$$

Continuing in this manner, one obtains

$$\mathbf{s}_i^r = \mathbf{T}_i \mathbf{T}_{i-1} \mathbf{T}_{i-2} \cdots \mathbf{T}_1 \mathbf{s}_0^r \tag{3.198}$$

where $\mathbf{s}_0^r$ is the state vector at the initial station. We may then write Eq. 198 in the simple form

$$\mathbf{s}_i^r = \mathbf{T}_{it} \mathbf{s}_0^r \tag{3.199}$$

where the matrix $\mathbf{T}_{it}$ is given by

$$\mathbf{T}_{it} = \mathbf{T}_i \mathbf{T}_{i-1} \mathbf{T}_{i-2} \cdots \mathbf{T}_1 \tag{3.200}$$

Equation 199, in which the state vector at station i is expressed in terms of the state vector of the initial station, represents two algebraic equations. Usually a force boundary condition or a displacement boundary condition would be known at each end of the system. When these boundary conditions are substituted into this equation, one obtains the characteristic equation which can be solved for the natural frequencies, which can then be used to determine the mode shapes of the system. This procedure represents one alternative for using Eq. 199. Another alternative which is more suited for a system with a large number of degrees of freedom is to use a numerical procedure based on Holzer's method. The natural frequency of the system and the unknown boundary condition at station 0 are assumed. The unknown values of the displacement and force at station 1 can then be determined by using the transfer matrix $\mathbf{T}_1$. This process can be continued until the other end of the system is reached, thus defining the boundary condition at this end. Since either the displacement or the force must be known at this end, the results obtained using the procedure of the transfer matrix can be checked and the error can be determined. If the error is not equal to zero or is relatively small, another value of the natural frequency is assumed and the entire procedure is repeated. This process is continued until the error becomes zero or relatively small.

Example 3.11

The use of the transfer-matrix method is demonstrated in this example by using the simple two degree of freedom system shown in Fig. 15. We assume for simplicity that $m_1 = m_2 = m$ and $k_1 = k_2 = k$. The boundary conditions at station 0 are $x = 0$, and $F = F_0$. The boundary conditions at mass 2 are $x = x_2$ and $F = 0$. By using Eq. 195 one has

$$\mathbf{s}_1^r = \mathbf{T}_1 \mathbf{s}_0^r$$

That is,

$$\begin{bmatrix} x_1^r \\ F_1^r \end{bmatrix} = \begin{bmatrix} 1 & 1/k \\ -\omega^2 m & (1 - \omega^2 m/k) \end{bmatrix} \begin{bmatrix} x_0^r \\ F_0^r \end{bmatrix} = \begin{bmatrix} 1 & 1/k \\ -\omega^2 m & (1 - \omega^2 m/k) \end{bmatrix} \begin{bmatrix} 0 \\ F_0 \end{bmatrix}$$

Similarly, we have

$$\mathbf{s}_2^r = \mathbf{T}_2 \mathbf{s}_1^r$$

which can be written in a more explicit form as

$$\begin{bmatrix} x_2^r \\ F_2^r \end{bmatrix} = \begin{bmatrix} x_2 \\ 0 \end{bmatrix} = \begin{bmatrix} 1 & 1/k \\ -\omega^2 m & (1 - \omega^2 m/k) \end{bmatrix} \begin{bmatrix} x_1^r \\ F_1^r \end{bmatrix}$$

which, upon substitution for x_1^r and F_1^r, yields

$$\begin{bmatrix} x_2 \\ 0 \end{bmatrix} = \begin{bmatrix} 1 & 1/k \\ -\omega^2 m & (1 - \omega^2 m/k) \end{bmatrix} \begin{bmatrix} 1 & 1/k \\ -\omega^2 m & (1 - \omega^2 m/k) \end{bmatrix} \begin{bmatrix} 0 \\ F_0 \end{bmatrix}$$

$$= \begin{bmatrix} (1 - \omega^2 m/k) & (1/k)(2 - \omega^2 m/k) \\ -\omega^2 m\left(2 - \dfrac{\omega^2 m}{k}\right) & -\omega^2 m/k + (1 - \omega^2 m/k)^2 \end{bmatrix} \begin{bmatrix} 0 \\ F_0 \end{bmatrix}$$

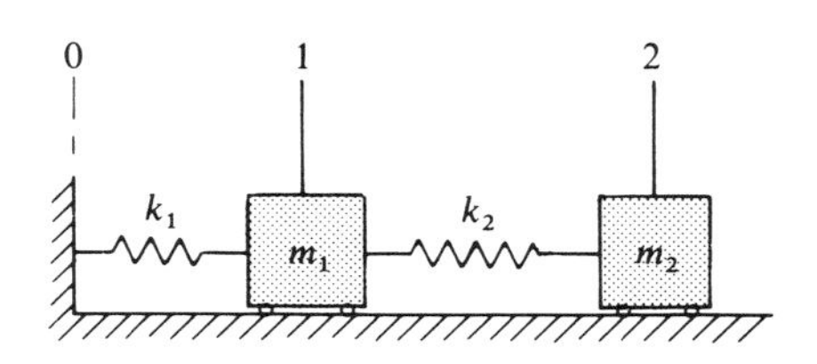

FIG. 3.15. A two degree of freedom system.

which yields the following two algebraic equations:

$$x_2 = \frac{F_0}{k}\left(2 - \frac{\omega^2 m}{k}\right)$$

$$0 = \left[\left(1 - \frac{\omega^2 m}{k}\right)^2 - \frac{\omega^2 m}{k}\right] F_0$$

The second algebraic equation yields

$$\left(1 - \frac{\omega^2 m}{k}\right)^2 - \frac{\omega^2 m}{k} = \left(\frac{\omega^2 m}{k}\right)^2 - 3\frac{\omega^2 m}{k} + 1 = 0$$

or

$$\omega^4 - 3\omega^2 \frac{k}{m} + \left(\frac{k}{m}\right)^2 = 0$$

Observe that this is the characteristic equation of the system which has the following roots:

$$\omega^2 = \frac{k}{2m}[3 \mp \sqrt{5}]$$

That is,

$$\omega_1^2 = 0.38197\frac{k}{m}, \qquad \omega_1 = 0.618\sqrt{\frac{k}{m}}$$

$$\omega_2^2 = 2.61803\frac{k}{m}, \qquad \omega_2 = 1.618\sqrt{\frac{k}{m}}$$

If we substitute these values for ω_1 and ω_2 into the first algebraic equation obtained from the application of the transfer-matrix procedure, we obtain for the first mode

$$x_2 = \frac{F_0}{k}\left(2 - \frac{\omega_1^2 m}{k}\right) = \frac{F_0}{k}(2 - 0.38197) = 1.61803\frac{F_0}{k}$$

and for the second mode

$$x_2 = \frac{F_0}{k}\left(2 - \frac{\omega_2^2 m}{k}\right) = \frac{F_0}{k}(2 - 2.61803) = -0.61803\frac{F_0}{k}$$

Since $x_1 = F_0/k$, the first and second mode shapes are given by

$$\mathbf{A}_1 = \begin{bmatrix} 1 \\ 1.61803 \end{bmatrix}, \qquad \mathbf{A}_2 = \begin{bmatrix} 1 \\ -0.61803 \end{bmatrix}$$

Torsional System The transfer-matrix method can also be used in the vibration analysis of torsional systems. The point, field, and transfer matrices can be obtained in a manner similar to the case of rectilinear mass–spring systems. Figure 16 shows a disk i, the left surface of which is connected to a shaft which has stiffness k_i, while the right surface is connected to a shaft which has stiffness k_{i+1}. Assuming harmonic oscillations, the equation of motion of the disk is given by

$$M_i^r = -\omega^2 I_i \theta_i + M_i^l \tag{3.201}$$

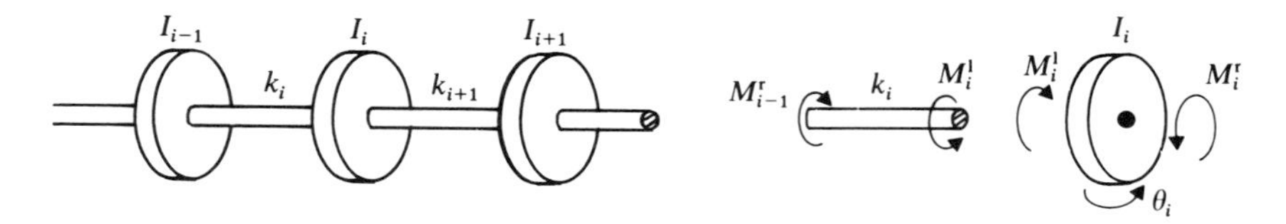

FIG. 3.16. Torsional system.

where M_i is the moment, I_i is the mass moment of inertia of the disk, and θ_i is the angular oscillation which can be used to define the equation

$$\theta_i = \theta^{\mathrm{r}}_i = \theta^{\mathrm{l}}_i \tag{3.202}$$

Equations 201 and 202 can be combined in order to obtain

$$\begin{bmatrix} \theta^{\mathrm{r}}_i \\ M^{\mathrm{r}}_i \end{bmatrix} = \begin{bmatrix} 1 & 0 \\ -\omega^2 I_i & 1 \end{bmatrix} \begin{bmatrix} \theta^{\mathrm{l}}_i \\ M^{\mathrm{l}}_i \end{bmatrix} \tag{3.203}$$

which is in the same form as Eq. 191 developed for the rectilinear system. It can, therefore, be written in the form of Eq. 192 as

$$\mathbf{s}^{\mathrm{r}}_i = \mathbf{P}_i \mathbf{s}^{\mathrm{l}}_i \tag{3.204}$$

where the state vectors $\mathbf{s}^{\mathrm{r}}_i$ and $\mathbf{s}^{\mathrm{l}}_i$ are defined by

$$\mathbf{s}^{\mathrm{r}}_i = \begin{bmatrix} \theta^{\mathrm{r}}_i \\ M^{\mathrm{r}}_i \end{bmatrix}, \qquad \mathbf{s}^{\mathrm{l}}_i = \begin{bmatrix} \theta^{\mathrm{l}}_i \\ M^{\mathrm{l}}_i \end{bmatrix}$$

and the point matrix $\mathbf{P}_i$ is given by

$$\mathbf{P}_i = \begin{bmatrix} 1 & 0 \\ -\omega^2 I_i & 1 \end{bmatrix}$$

If we consider the equilibrium of the shaft k_i which is assumed to be massless, we have

$$\theta^{\mathrm{l}}_i = \theta^{\mathrm{r}}_{i-1} + \frac{1}{k_i} M^{\mathrm{r}}_{i-1}$$

$$M^{\mathrm{l}}_i = M^{\mathrm{r}}_{i-1}$$

which can be combined in one matrix equation as

$$\begin{bmatrix} \theta^{\mathrm{l}}_i \\ M^{\mathrm{l}}_i \end{bmatrix} = \begin{bmatrix} 1 & 1/k_i \\ 0 & 1 \end{bmatrix} \begin{bmatrix} \theta^{\mathrm{r}}_{i-1} \\ M^{\mathrm{r}}_{i-1} \end{bmatrix} \tag{3.205}$$

where the field matrix $\mathbf{H}_i$ can be identified as

$$\mathbf{H}_i = \begin{bmatrix} 1 & 1/k_i \\ 0 & 1 \end{bmatrix}$$

Equation 205 can be substituted into Eq. 203. This yields

$$\mathbf{s}_i^{\mathrm{r}} = \mathbf{T}_i \mathbf{s}_{i-1}^{\mathrm{r}} \tag{3.206}$$

where $\mathbf{T}_i$ is the transfer matrix defined in this case as

$$\begin{aligned} \mathbf{T}_i = \mathbf{P}_i \mathbf{H}_i &= \begin{bmatrix} 1 & 0 \\ -\omega^2 I_i & 1 \end{bmatrix} \begin{bmatrix} 1 & 1/k_i \\ 0 & 1 \end{bmatrix} \\ &= \begin{bmatrix} 1 & 1/k_i \\ -\omega^2 I_i & (1 - \omega^2 I_i/k_i) \end{bmatrix} \end{aligned} \tag{3.207}$$

If Eq. 206 is applied to successive points of the torsional system one obtains equations similar to Eqs. 199 and 200 developed for the rectilinear mass–spring system.

Example 3.12

In order to demonstrate the use of the transfer-matrix method in the vibration analysis of torsional systems, we consider the simple two degree of freedom system shown in Fig. 17 which consists of two disks which have moments of inertia I_1 and I_2 and two shafts which have torsional stiffness coefficients k_1 and k_2. For simplicity, we assume that $I_1 = I_2 = I$ and $k_1 = k_2 = k$. The boundary conditions at station 0 is $\theta = 0$ and $M = M_0$, while the boundary conditions at station 2 are $\theta = \theta_2$ and $M = 0$. We, therefore, have

$$\begin{bmatrix} \theta_1^{\mathrm{r}} \\ M_1^{\mathrm{r}} \end{bmatrix} = \begin{bmatrix} 1 & 1/k \\ -\omega^2 I & (1 - \omega^2 I/k) \end{bmatrix} \begin{bmatrix} \theta_0^{\mathrm{r}} \\ M_0^{\mathrm{r}} \end{bmatrix} = \begin{bmatrix} 1 & 1/k \\ -\omega^2 I & (1 - \omega^2 I/k) \end{bmatrix} \begin{bmatrix} 0 \\ M_0 \end{bmatrix}$$

and

$$\begin{bmatrix} \theta_2^{\mathrm{r}} \\ M_2^{\mathrm{r}} \end{bmatrix} = \begin{bmatrix} \theta_2 \\ 0 \end{bmatrix} = \begin{bmatrix} 1 & 1/k \\ -\omega^2 I & (1 - \omega^2 I/k) \end{bmatrix} \begin{bmatrix} \theta_1^{\mathrm{r}} \\ M_1^{\mathrm{r}} \end{bmatrix}$$

These two matrix equations yield

$$\begin{aligned} \begin{bmatrix} \theta_2 \\ 0 \end{bmatrix} &= \begin{bmatrix} 1 & 1/k \\ -\omega^2 I & (1 - \omega^2 I/k) \end{bmatrix} \begin{bmatrix} 1 & 1/k \\ -\omega^2 I & (1 - \omega^2 I/k) \end{bmatrix} \begin{bmatrix} 0 \\ M_0 \end{bmatrix} \\ &= \begin{bmatrix} (1 - \omega^2 I/k) & \frac{1}{k}(2 - \omega^2 I/k) \\ -\omega^2 I(2 - \omega^2 I/k) & -\omega^2 I/k + (1 - \omega^2 I/k)^2 \end{bmatrix} \begin{bmatrix} 0 \\ M_0 \end{bmatrix} \end{aligned}$$

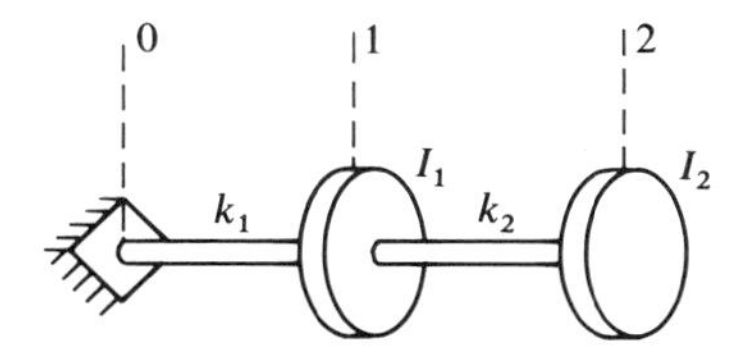

FIG. 3.17. Two degree of freedom system.

which yields the following two scalar equations:

$$\theta_2 = \frac{M_0}{k}\left(2 - \frac{\omega^2 I}{k}\right)$$

$$0 = \left[\left(1 - \frac{\omega^2 I}{k}\right)^2 - \frac{\omega^2 I}{k}\right] M_0$$

The second algebraic equation yields the characteristic equation of the system

$$\omega^4 - 3\omega^2 \frac{k}{I} + \left(\frac{k}{I}\right)^2 = 0$$

which defines the natural frequencies of the system ω_1 and ω_2 as

$$\omega_1 = 0.618\sqrt{\frac{k}{I}}, \qquad \omega_2 = 1.618\sqrt{\frac{k}{I}}$$

By using a similar procedure to the one described in the preceding example, one can show that the mode shapes of this system are given by

$$\mathbf{A}_1 = \begin{bmatrix} 1 \\ 1.61803 \end{bmatrix}, \qquad \mathbf{A}_2 = \begin{bmatrix} 1 \\ -0.61803 \end{bmatrix}$$

Problems

3.1. By using Newton's second law, obtain the differential equations of motion of the multi-degree of freedom system shown in Fig. P1.

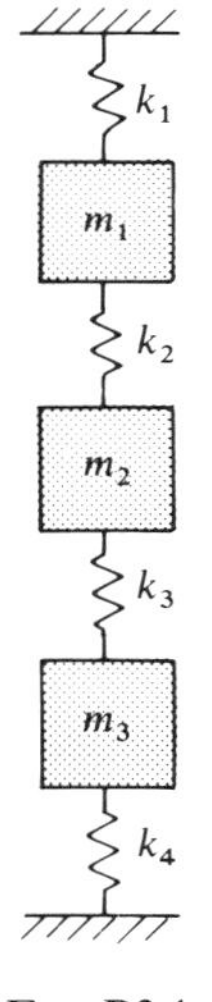

FIG. P3.1

3.2. By using Lagrange's equation, obtain the differential equations of motion of the system shown in Fig. P1.

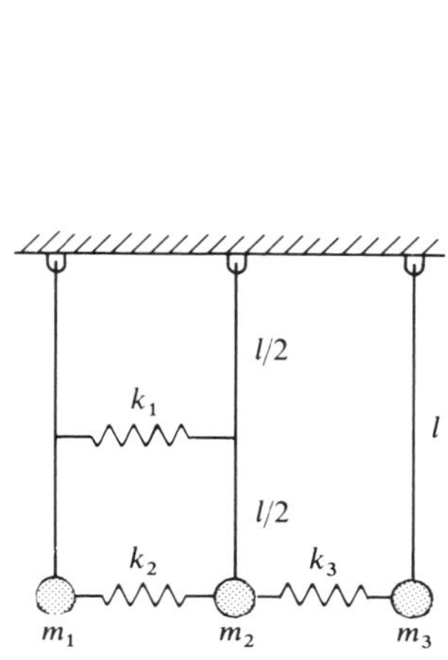

FIG. P3.2

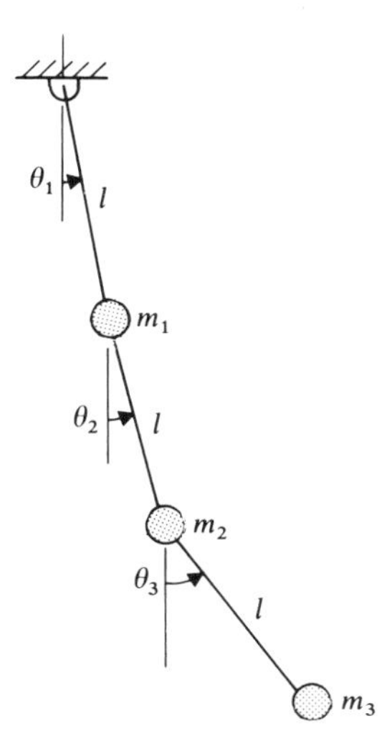

FIG. P3.3

3.3. Determine the natural frequencies and the mode shapes for Problem 1, taking $k_1 = k_2 = k_3 = k_4 = k = 1 \times 10^3$ N/m, and $m_1 = m_2 = m_3 = m = 2$ kg.

3.4. Repeat Problem 3, for the case in which $k_1 = k_2 = k_3 = k_4 = k$ and $m_1 = m$, $m_2 = 2m$, and $m_3 = m$.

3.5. Obtain the differential equations of free vibration of the three degree of freedom system shown in Fig. P2 by using Newton's second law. Assume small oscillations.

3.6. Use Lagrange's equation to derive the differential equations of free vibration of the system of Problem 5. Obtain also the natural frequencies and the mode shapes of vibration, assuming that $m_1 = m_2 = m_3 = 1$ kg, $l = 0.5$ m, and $k_1 = 0$, $k_2 = k_3 = 2 \times 10^3$ N/m.

3.7. Derive the differential equations of motion of the system shown in Fig. P3. Determine the natural frequencies and the mode shapes of vibration. Assume that $m_1 = m_2 = m_3 = 1$ kg, and $l = 0.5$ m. Assume small oscillations.

3.8. Assuming small oscillations, determine the natural frequencies and mode shapes of vibration of the system shown in Fig. P4. Assume that $m_1 = 2$ kg, $m_2 = m_3 = 0.5$ kg, $l_1 = l_2 = 0.5$ m, and $k_1 = 2 \times 10^3$ N/m.

3.9. Figure P5 depicts a shaft which supports three disks which have mass moments of inertia I_1, I_2, and I_3. The shaft, which is assumed to have negligible mass, has a circular cross section of diameter d and modulus of rigidity G. Obtain the differential equations that govern the torsional vibration of this system.

3.10. Obtain the natural frequencies and mode shapes of the torsional system of Problem 9 assuming that the shafts are made of steel and have equal diameters of 0.01 m, $I_1 = I_2 = I_3 = 1 \times 10^3$ kg·m^2 and $l_1 = l_2 = l_3 = 0.5$ m.

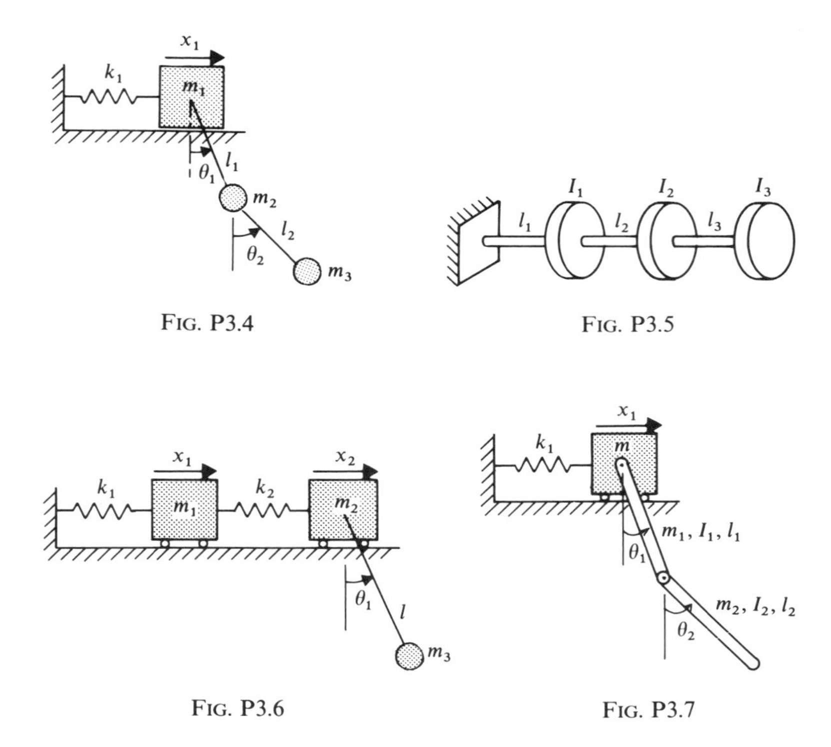

FIG. P3.4

FIG. P3.5

FIG. P3.6

FIG. P3.7

3.11. Assuming small oscillations, derive the differential equations of motion of the system shown in Fig. P6 by using Newton's second law. Obtain also the differential equations using Lagrange's equation of motion.

3.12. In Problem 11, if $m_1 = m_2 = m_3 = m = 1$ kg, and $k_1 = k_2 = k = 2 \times 10^3$ N/m, obtain the natural frequencies and mode shapes, assuming that the length of the rod is 0.5 m.

3.13. By using Newton's second law or Lagrange's equation, derive the differential equations of motion of the system shown in Fig. P7. Assume small oscillations.

3.14. In Problem 3, if $m = 3$ kg, $k = 1000$ N/m, and the initial displacements of the masses m_1, m_2, and m_3 are, respectively, 0.01, 0, and 0 m, determine the solution of the free vibration assuming that the masses have zero initial velocities.

3.15. For the system shown in Fig. P4, let $m_1 = m_2 = m_3 = 2$ kg, $l_1 = l_2 = 0.5$ m, and $k_1 = 1000$ N/m. The masses m_1, m_2, and m_3 are assumed to have zero initial displacements. Obtain the solution for the free vibration assuming that the rod l_2 has initial angular velocity of 3 rad/s.

3.16. Find the solution of the free vibration of the system of Problem 8 in terms of the initial displacements and velocities.

3.17. Obtain the solution for the free vibration of the system in Problem 12 in terms of the initial conditions.

3.18. Determine the modal mass and stiffness coefficients of the system of Problem 6.

3.19. Determine the modal mass and stiffness coefficients of the system of Problem 8.

3.20. Determine the modal mass and stiffness coefficients of the system of Problem 10.

3.21. Determine the modal mass and stiffness coefficients of the system of Problem 12.

3.22. In Problem 12, if the spring k_1 is removed, does the new system have a rigid-body mode? If the answer is yes, identify this rigid-body mode.

3.23. Show that, for an n-degree of freedom system, if the vectors of initial coordinates and velocities are proportional to a linear combination of m mode shapes, where $m < n$, then the motion of the n-degree of freedom system is equivalent to the motion of an m-degree of freedom system.

3.24. In Problem 12 determine the natural frequencies and mode shapes if the effect of gravity is neglected.

3.25. By inspection identify the rigid-body modes of the system shown in Fig. P7 if the effect of gravity is neglected.

3.26. Using Lagrange's equation or Newton's second law, obtain the differential equation of motion of the forced vibration for the system shown in Fig. P8. The moment M is given by $M = M_0 \sin \omega_f t$, where $M_0 = 2$ N·m, and $\omega_f = 5$ rad/s. Assume small oscillations, $m_1 = m_2 = m_3 = 1$ kg, $k_1 = k_2 = 10$ N/m, $c_1 = c_2 = 10$ N·s/m, and $l = 0.5$ m.

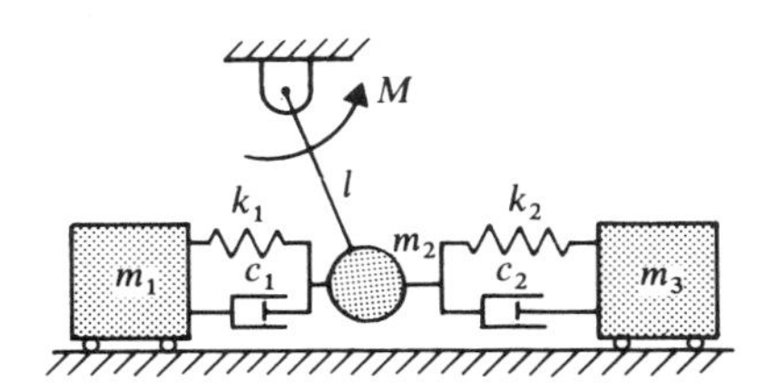

FIG. P3.8

3.27. In Problem 26, if $c_1 = c_2 = 0$, obtain the steady-state solution.

3.28. The system shown in Fig. P9 consists of masses which are connected by elastic elements and a damper. The system is subjected to support excitation as shown in the figure. Derive the differential equations of motion of this system.

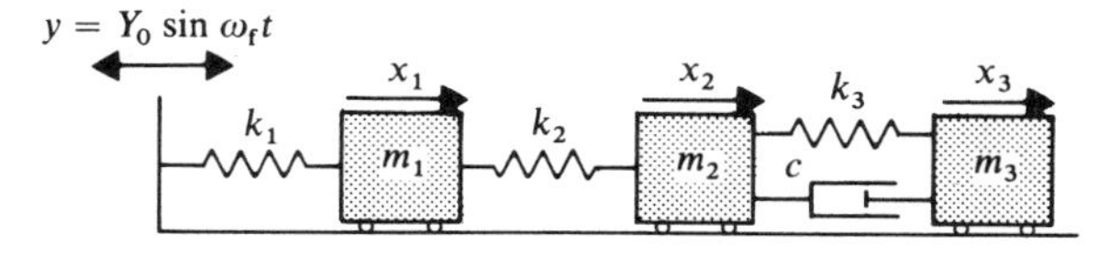

FIG. P3.9

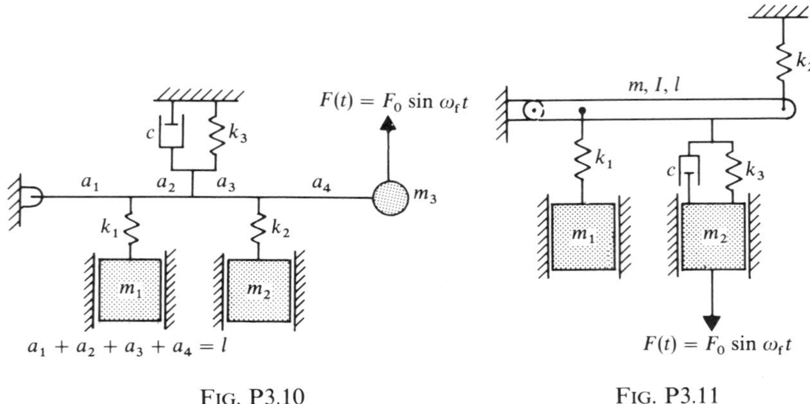

FIG. P3.10

FIG. P3.11

3.29. In Problem 28, obtain the steady-state solution for the undamped system. Assume $m_1 = m_2 = m_3 = 1$ kg, $k_1 = k_2 = k_3 = 1 \times 10^3$ N/m, $Y_0 = 0.01$ m, and $\omega_f = 4$ rad/s.

3.30. Assuming small oscillations, derive the differential equations of motion of the system shown in Fig. P10.

3.31. Obtain the differential equations of motion of the system shown in Fig. P11. Assume small oscillations.

3.32. Obtain the differential equations of motion of the system shown in Fig. P12. Assume small oscillations.

3.33. Use the state transition matrix to determine the solution of the free-vibration equation for the underdamped and overdamped single degree of freedom systems.

3.34. Use the state transition matrix to determine the solution of the equation of the forced vibration of the underdamped single degree of freedom systems.

3.35. Obtain the differential equations of motion of the vehicle system shown in Fig. P13. Assume small angular oscillations.

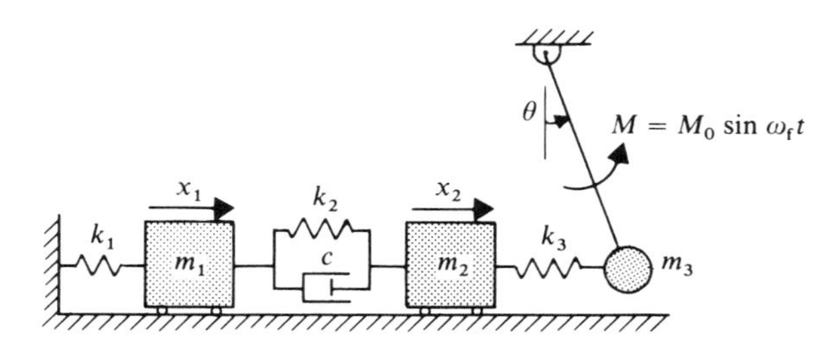

FIG. P3.12

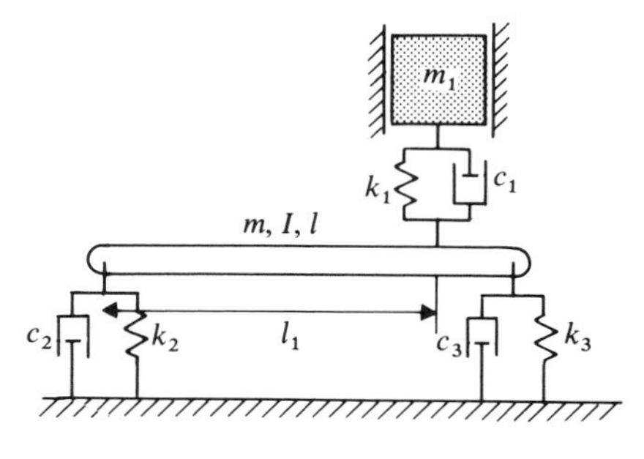

FIG. P3.13

3.36. If the damping coefficient of the system in Fig. P9 is equal to zero and if $m_1 = 1$ kg, $m_2 = 1.5$ kg, $m_3 = 2$ kg, $k_1 = 6 \times 10^5$ N/m, $k_2 = 8 \times 10^5$ N/m, $k_3 = 10 \times 10^5$ N/m, determine the natural frequencies and the mode shapes using the matrix-iteration method.

3.37. Show that the reduced system equations of motion obtained in Example 10 leads to the same natural frequencies and mode shapes determined in Example 5.

3.38. Use the matrix-iteration method to determine the natural frequencies and mode shapes of the system of Problem 3, assuming that $m = 2.5$ kg.

3.39. Use the matrix-iteration method to determine the natural frequencies and mode shapes of the system of Problem 8 assuming that $k_1 = 1.5 \times 10^3$ N/m.

3.40. Determine the natural frequencies and mode shapes of the system of Problem 27 using the matrix iteration method.

3.41. Determine the mode shapes and natural frequencies of the system of Problem 36 using the transfer-matrix method.

4
Vibration of Continuous Systems

Mechanical systems in general consist of structural components which have distributed mass and elasticity. Examples of these structural components are rods, beams, plates, and shells. Our study of vibration thus far has been limited to discrete systems which have a finite number of degrees of freedom. As has been shown in the preceding chapters, the vibration of mechanical systems with lumped masses and discrete elastic elements is governed by a set of second-order ordinary differential equations. Rods, beams, and other structural components on the other hand are considered as continuous systems which have an infinite number of degrees of freedom. The vibration of such systems is governed by partial differential equations which involve variables that depend on time as well as the spatial coordinates.

In this chapter, an introduction to the theory of vibration of continuous systems is presented. It is shown in the first two sections that the longitudinal and torsional vibration of rods can be described by second-order partial differential equations. Exact solutions for these equations are obtained using the method of separation of variables. In Section 3, the transverse vibrations of beams are examined and the fourth-order partial differential equation of motion that governs the transverse vibration is developed using the assumptions of the elementary beam theory. Solutions of the vibration equations are obtained for different boundary conditions. In Section 4, the orthogonality of the *eigenfunctions* (*mode shapes* or *principal modes*) is discussed for the cases of longitudinal, torsional, and transverse vibrations and modal parameters such as mass and stiffness coefficients are introduced. The material covered in this section is used to study the forced vibration of continuous systems in Section 5, where the solution of the vibration equations is expressed in term of the principal modes of vibration. Section 6 is devoted to the solution of the vibration equations subject to inhomogeneous boundary conditions, while in Section 7 the effect of damping on the vibration of viscoelastic materials is examined. In Section 8, the use of Lagrange's equation and the scalar energy quantities in deriving the differential equations of motion is demonstrated. The last three sections are devoted to the use of approximate methods in the analysis of continuous systems.

4.1 FREE LONGITUDINAL VIBRATIONS

In this section, we study the longitudinal vibration of prismatic rods such as the one shown in Fig. 1. The rod has length l and cross-sectional area A. The rod is assumed to be made of material which has a modulus of elasticity E and mass density ρ and is subjected to a distributed external force $F(x, t)$ per unit length. Figure 1 also shows the forces that act on an infinitesimal volume of length δx. The geometric center of this infinitesimal volume is located at a distance $x + \delta x/2$ from the end of the rod. Let P be the axial force that results from the vibration of the rod. The application of Newton's second law leads to the following condition for the dynamic equilibrium of the infinitesimal volume:

$$\rho A \frac{\partial^2 u}{\partial t^2} \delta x = P + \frac{\partial P}{\partial x} \delta x - P + F(x, t)\delta x$$

which can be simplified to yield

$$\rho A \frac{\partial^2 u}{\partial t^2} \delta x = \frac{\partial P}{\partial x} \delta x + F(x, t)\delta x$$

Dividing this equation by δx leads to

$$\rho A \frac{\partial^2 u}{\partial t^2} = \frac{\partial P}{\partial x} + F(x, t) \tag{4.1}$$

The force P can be expressed in terms of the axial stress σ as

$$P = A\sigma \tag{4.2}$$

The stress σ can be written in terms of the axial strain ε using Hooke's law as

$$\sigma = E\varepsilon, \tag{4.3}$$

while the strain displacement relationship is

$$\varepsilon = \frac{\partial u}{\partial x} \tag{4.4}$$

Substituting Eqs. 3 and 4 into Eq. 2, the axial force can be expressed in terms

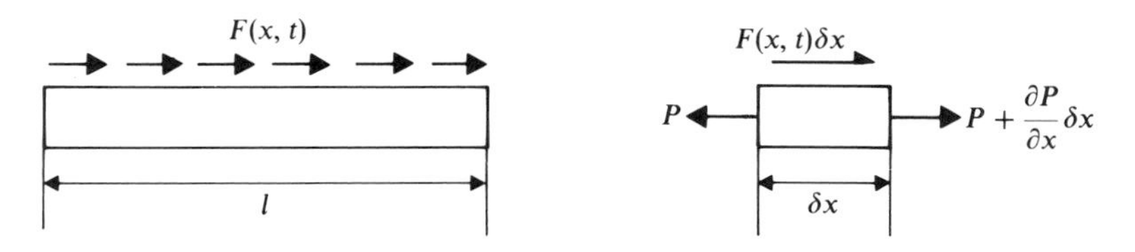

FIG. 4.1. Longitudinal vibration of rods.

of the longitudinal displacement as

$$P = EA\frac{\partial u}{\partial x} \tag{4.5}$$

Substituting Eq. 5 into Eq. 1 leads to

$$\rho A\frac{\partial^2 u}{\partial t^2} = \frac{\partial}{\partial x}\left(EA\frac{\partial u}{\partial x}\right) + F(x, t) \tag{4.6}$$

This is the partial differential equation that governs the forced longitudinal vibration of the rod.

Free Vibration The equation of free vibration can be obtained from Eq. 6 by letting $F(x, t) = 0$, that is,

$$\rho A\frac{\partial^2 u}{\partial t^2} = \frac{\partial}{\partial x}\left(EA\frac{\partial u}{\partial x}\right) \tag{4.7}$$

If the modulus of elasticity E and the cross-sectional area A are assumed to be constant, the partial differential equation of the longitudinal free vibration of the rod can be written as

$$\rho A\frac{\partial^2 u}{\partial t^2} = EA\frac{\partial^2 u}{\partial x^2} \tag{4.8}$$

or

$$\frac{\partial^2 u}{\partial t^2} = c^2\frac{\partial^2 u}{\partial x^2} \tag{4.9}$$

where c is a constant defined by

$$c = \sqrt{\frac{E}{\rho}} \tag{4.10}$$

Separation of Variables The general solution of Eq. 8 can be obtained using the method of the *separation of variables*. In this case we assume the solution in the form

$$u(x, t) = \phi(x)q(t) \tag{4.11}$$

where ϕ is a space-dependent function and q is a time-dependent function. The partial differentiation of Eq. 11 with respect to time and with respect to the spatial coordinate leads to

$$\frac{\partial^2 u}{\partial t^2} = \phi(x)\ddot{q}(t) \tag{4.12}$$

$$\frac{\partial^2 u}{\partial x^2} = \phi''(x)q(t) \tag{4.13}$$

where ($\cdot$) denotes differentiation with respect to time and ($'$) denotes differentiation with respect to the spatial coordinate x, that is,

$$\ddot{q}(t) = \frac{d^2q}{dt^2} \tag{4.14}$$

$$\phi''(x) = \frac{d^2\phi}{dx^2} \tag{4.15}$$

Substituting Eqs. 12 and 13 into Eq. 9 leads to

$$\phi\ddot{q} = c^2\phi''q \tag{4.16}$$

or

$$c^2\frac{\phi''}{\phi} = \frac{\ddot{q}}{q} \tag{4.17}$$

Since the left-hand side of this equation depends only on the spatial coordinate x and the right-hand side depends only on time, one concludes that Eq. 17 is satisfied only if both sides are equal to a constant, that is

$$c^2\frac{\phi''}{\phi} = \frac{\ddot{q}}{q} = -\omega^2 \tag{4.18}$$

where ω is a constant. A negative constant $-\omega^2$ was selected, since this choice leads to oscillatory motion. The choice of zero or a positive constant does not lead to vibratory motion and, therefore, it must be excluded. For example, one can show that if the constant is selected to be zero the solution increases linearly with time. While if a positive constant is selected, the solution contains two terms; one exponentially increasing function and the second is an exponentially decreasing function. This leads to an unstable solution which does not represent an oscillatory motion.

Equation 18 leads to the following two equations:

$$\phi'' + \left(\frac{\omega}{c}\right)^2\phi = 0 \tag{4.19}$$

$$\ddot{q} + \omega^2 q = 0 \tag{4.20}$$

The solution of these two equations is given by

$$\phi(x) = A_1 \sin\frac{\omega}{c}x + A_2 \cos\frac{\omega}{c}x \tag{4.21}$$

$$q(t) = B_1 \sin\omega t + B_2 \cos\omega t \tag{4.22}$$

By using Eq. 11, the longitudinal displacement $u(x, t)$ can then be written as

$$u(x, t) = \phi(x)q(t) = \left(A_1 \sin\frac{\omega}{c}x + A_2 \cos\frac{\omega}{c}x\right)(B_1 \sin\omega t + B_2 \cos\omega t) \tag{4.23a}$$

where A_1, A_2, B_1, B_2, and ω are arbitrary constants to be determined by using the boundary and initial conditions.

Elastic Waves Equation 23a can also be written in the following alternate form:

$$u(x, t) = A \sin\left(\frac{\omega}{c}x + \phi_1\right)\sin(\omega t + \phi_2)$$

where A, ϕ_1, and ϕ_2 are constants that can be expressed in terms of the constants A_1, A_2, B_1, and B_2. By using the trignometric identity

$$\sin\left(\frac{\omega}{c}x + \phi_1\right)\sin(\omega t + \phi_2)$$
$$= \frac{1}{2}\left[\cos\left(\frac{\omega}{c}x + \phi_1 - \omega t - \phi_2\right) - \cos\left(\frac{\omega}{c}x + \phi_1 + \omega t + \phi_2\right)\right],$$

the preceding equation for the displacement u can be written as

$$u(x, t) = \frac{A}{2}\cos\left(\frac{\omega}{c}x - \omega t + \phi_1 - \phi_2\right) - \frac{A}{2}\cos\left(\frac{\omega}{c}x + \omega t + \phi_1 + \phi_2\right)$$

which can be written as

$$u(x, t) = f_1(kx - \omega t + \phi_m) + f_2(kx + \omega t + \phi_p) \tag{4.23b}$$

where the constants ϕ_m, ϕ_p, and k are

$$\phi_m = \phi_1 - \phi_2$$
$$\phi_p = \phi_1 + \phi_2$$
$$k = \frac{\omega}{c}$$

and the functions f_1 and f_2 are defined as

$$f_1(kx - \omega t + \phi_m) = \frac{A}{2}\cos(kx - \omega t + \phi_m)$$

$$f_2(kx + \omega t + \phi_p) = -\frac{A}{2}\cos(kx + \omega t + \phi_p)$$

The function $f_1(kx - \omega t + \phi_m)$ represents a harmonic wave traveling in the positive x direction with a *wave velocity* $c = \omega/k$, while the function $f_2(kx + \omega t + \phi_p)$ represents another elastic harmonic wave traveling in the opposite direction with the same wave velocity c, where c is defined by Eq. 10. Therefore, Eq. 9 is often called the *wave equation* of the longitudinal vibration of the rod. Note that the wave velocity c is constant and depends on the material properties of the rod. The constant k which relates the wave velocity c to the frequency of the harmonic ω is called the *wave number*.

Boundary Conditions and the Orthogonality of the Eigenfunctions In order to demonstrate the procedure for determining the constants in Eq. 23a, we consider the example shown in Fig. 2 where the rod is fixed at one end and is free at the other end. The boundary condition at the fixed end is given by

$$u(0, t) = 0 \tag{4.24}$$

while at the free end, the stress must be equal to zero, that is,

$$\sigma(l, t) = E\varepsilon(l, t) = E\frac{\partial u(l, t)}{\partial x} = 0$$

which can be used to define the boundary condition at the free end as

$$\frac{\partial u(l, t)}{\partial x} = u'(l, t) = 0 \tag{4.25}$$

The boundary condition of Eq. 24 describes the state of displacement at the fixed end. This type of boundary condition is called *geometric boundary condition.* On the other hand, Eq. 25 describes the state of force or stress at the free end of the rod. This type of condition is often refered to as *natural boundary condition.* That is, the geometric boundary conditions describe the specified displacements and slopes, while the natural boundary conditions describe the specified forces and moments. Substituting Eqs. 24 and 25 into Eq. 23a results in

$$u(0, t) = \phi(0)q(t) = A_2 q(t) = 0$$

$$u'(l, t) = \phi'(l)q(t) = \frac{\omega}{c}\left(A_1 \cos\frac{\omega}{c}l - A_2 \sin\frac{\omega}{c}l\right)q(t) = 0$$

which lead to the following two conditions:

$$A_2 = 0 \tag{4.26}$$

$$A_1 \cos\frac{\omega l}{c} = 0 \tag{4.27}$$

For a nontrivial solution, Eqs. 26 and 27 lead, respectively, to

$$\phi(x) = A_1 \sin\frac{\omega}{c}x \tag{4.28}$$

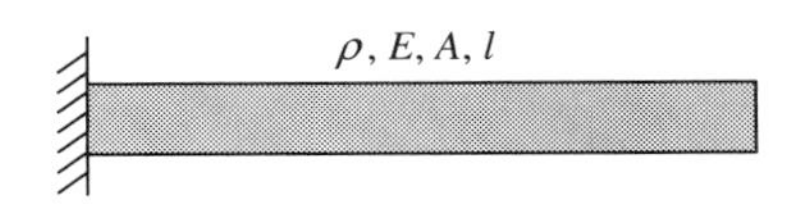

FIG. 4.2 Fixed-end conditions.

and

$$\cos \frac{\omega l}{c} = 0 \tag{4.29}$$

Equation 29 is called the *frequency* or the *characteristic equation.* The roots of this equation are given by

$$\frac{\omega l}{c} = \frac{\pi}{2}, \frac{3\pi}{2}, \frac{5\pi}{2}, \ldots, \frac{(2n-1)\pi}{2}, \ldots \tag{4.30}$$

This equation defines the *natural frequencies* of the rod as

$$\omega_j = \frac{(2j-1)\pi c}{2l}, \qquad j = 1, 2, 3, \ldots \tag{4.31}$$

Using the definition of the wave velocity c given by Eq. 10, the jth natural frequency ω_j can be defined as

$$\omega_j = \frac{(2j-1)\pi}{2l}\sqrt{\frac{E}{\rho}}, \qquad j = 1, 2, 3, \ldots \tag{4.32}$$

Thus, the continuous rod has an infinite number of natural frequencies. Corresponding to each of these natural frequencies, there is a *mode shape* or an *eigenfunction* ϕ_j defined by Eq. 28 as

$$\phi_j = A_{1j} \sin \frac{\omega_j}{c} x, \qquad j = 1, 2, 3, \ldots \tag{4.33}$$

where A_{1j}, $j = 1, 2, 3, \ldots$, are arbitrary constants. The eigenfunctions satisfy the following *orthogonality condition*:

$$\int_0^l \phi_i \phi_j \, dx = \begin{cases} 0 & \text{if} \quad i \neq j \\ h_i & \text{if} \quad i = j \end{cases} \tag{4.34}$$

where h_i is a constant. The longitudinal displacement $u(x, t)$ of the rod can then be expressed as

$$\begin{aligned} u(x,t) &= \sum_{j=1}^{\infty} \phi_j(x) q_j(t) \\ &= \sum_{j=1}^{\infty} (C_j \sin \omega_j t + D_j \cos \omega_j t) \sin \frac{\omega_j}{c} x \end{aligned} \tag{4.35}$$

where C_j and D_j are constants to be determined by using the initial conditions.

Initial Conditions Let us assume that the rod is subjected to the following initial conditions:

$$u(x, 0) = f(x) \tag{4.36}$$

$$\dot{u}(x, 0) = g(x) \tag{4.37}$$

Substituting these initial conditions into Eq. 35 leads to

$$u(x,0) = f(x) = \sum_{j=1}^{\infty} D_j \sin \frac{\omega_j}{c} x \tag{4.38}$$

$$\dot{u}(x,0) = g(x) = \sum_{j=1}^{\infty} \omega_j C_j \sin\frac{\omega_j}{c}x \tag{4.39}$$

In order to determine the constants D_j in Eq. 38, we multiply this equation by $\sin(\omega_j/c)x$ and integrate over the length of the rod. By using the orthogonality condition of Eq. 34, one obtains

$$D_j = \frac{2}{l}\int_0^l f(x)\sin\frac{\omega_j}{c}x\,dx, \qquad j = 1, 2, 3, \ldots \tag{4.40}$$

Similarly, in order to determine the constants C_j, we multiply Eq. 39 by $\sin(\omega_j x/c)$, integrate over the length of the rod, and use the orthogonality condition of Eq. 34. This leads to

$$C_j = \frac{2}{l\omega_j}\int_0^l g(x)\sin\frac{\omega_j}{c}x\,dx, \qquad j = 1, 2, 3, \ldots \tag{4.41}$$

Observe that the solution of the equation of the longitudinal vibration of the rod given by Eq. 38 is expressed as the sum of the modes of vibrations which in this case happen to be simple harmonic functions. The contribution of each of these modes to the solution will depend on the degree to which this particular mode is excited. As the result of a sudden application of a force, as in the case of impact loading, many of these modes will be excited and accordingly their contribution to the solution of the vibration equation is significant. It is important, however, to point out that it is possible for the rod to vibrate in only one of these modes of vibration. For instance, if the initial elastic deformation of the rod exactly coincides with one of these modes and the initial velocity is assumed to be zero, vibration will occur only in this mode of vibration and the system behaves as a single degree of freedom system. In order to demonstrate this, let us consider the special case in which the initial conditions are given by

$$u(x,0) = \sin\frac{\omega_k}{c}x$$

$$\dot{u}(x,0) = 0$$

Clearly, the initial displacement takes the shape of the kth mode. Substituting these initial conditions into Eqs. 40 and 41 and using the orthogonality of the mode shapes, one obtains

$$D_j = \begin{cases} 0 & \text{if } j \neq k \\ 1 & \text{if } j = k \end{cases}$$

$$C_j = 0, \qquad j = 1, 2, \ldots$$

That is, the solution of the equation of longitudinal vibration of Eq. 35 reduces in this special case to

$$u(x,t) = q_k(t)\sin\frac{\omega_k}{c}x = D_k\cos\omega_k t\sin\frac{\omega_k}{c}x$$

which demonstrates that the rod indeed vibrates in its kth mode due to the fact that the initial conditions in this special case cause only the kth mode of vibration to be excited. The continuous system in this special case behaves as a single degree of freedom system. This can be demonstrated by differentiating the displacement twice with respect to x and t, yielding, respectively,

$$\frac{\partial^2 u}{\partial x^2} = -\left(\frac{\omega_k}{c}\right)^2 q_k(t) \sin\frac{\omega_k}{c}x$$

$$\frac{\partial^2 u}{\partial t^2} = \ddot{q}_k(t) \sin\frac{\omega_k}{c}x$$

Substituting these two equations into the partial differential equation of Eq. 9, one obtains

$$\ddot{q}_k \sin\frac{\omega_k}{c}x = -\omega_k^2 q_k \sin\frac{\omega_k}{c}x$$

which leads to the following differential equation:

$$\ddot{q}_k + \omega_k^2 q_k = 0$$

This equation is in the same form as the equation which governs the free vibration of a single degree of freedom system. The frequency of oscillation in this case is ω_k.

If the initial conditions take a different form, the results will be different and other modes may be excited as well. For example, one can show that if the initial displacement of the system is a linear combination of several modes, only these modes will be excited. The solution can then be represented by a truncated series which has a finite number of terms instead of the infinite series of Eq. 35. In this case, the continuous system can be analyzed as a multi-degree of freedom system with the number of degrees of freedom being equal to the number of modes which are excited. In many applications in practice, this approach is being used in the study of vibration of continuous systems. Quite often, a finite number of modes are excited and the techniques used in the analysis of multi-degree of freedom systems are often used in the analysis of continuous systems as well. It is important, however, to point out that the accuracy of the reduced finite-dimensional multi-degree of freedom model in representing the actual infinite-dimensional continuous system will greatly depend on the judgment of which modes are significant. The analysis of the frequency content of the initial conditions and the forcing functions may be of great help in deciding which modes can be truncated.

Other Boundary Conditions From the analysis presented in this section it is clear that the mode shapes and the natural frequencies of the rod depend on the boundary conditions. Thus far we have considered only the case in which one end of the rod is fixed while the other end is free. Following a similar procedure to the one described in this section, the natural frequencies and mode shapes can be determined for other boundary conditons.

If the rod is assumed to be *free at both ends*, the boundary conditions are given by

$$\frac{\partial u(0,t)}{\partial x} = 0, \qquad \frac{\partial u(l,t)}{\partial x} = 0 \tag{4.42}$$

Using these boundary conditions and the separation of variables technique, one can show that the frequency equation is given by

$$\sin\frac{\omega l}{c} = 0 \tag{4.43}$$

which yields the natural frequencies of the longitudinal vibration of the rod with free ends as

$$\omega_j = \frac{j\pi c}{l} = \frac{j\pi}{l}\sqrt{\frac{E}{\rho}}, \qquad j = 1, 2, 3, \ldots \tag{4.44}$$

The associated eigenfunctions or mode shapes are given by

$$\phi_j = A_{2j}\cos\frac{\omega_j x}{c} \tag{4.45}$$

Another example is a rod with *both ends fixed.* The boundary conditions in this case are given by

$$u(0, t) = 0, \qquad u(l, t) = 0 \tag{4.46}$$

Using these boundary conditions and the separation of variables method, one can show that the frequency equation is given by

$$\sin\frac{\omega l}{c} = 0 \tag{4.47}$$

which yields the natural frequencies

$$\omega_j = \frac{j\pi c}{l} = \frac{j\pi}{l}\sqrt{\frac{E}{\rho}}, \qquad j = 1, 2, 3, \ldots \tag{4.48}$$

The mode shapes associated with these frequencies are

$$\phi_j = A_{1j}\sin\frac{j\pi x}{l}, \qquad j = 1, 2, 3, \ldots \tag{4.49}$$

Example 4.1

The system shown in Fig. 3 consists of a rigid mass m attached to a rod which has mass density ρ, length l, modulus of elasticity E, and cross-sectional area A. Determine the frequency equation and the mode shapes of the longitudinal vibration.

Solution. The end condition at the fixed end of the rod is given by

$$u(0, t) = 0 \tag{4.50}$$

The other end of the rod is attached to the mass m which exerts a force on the rod

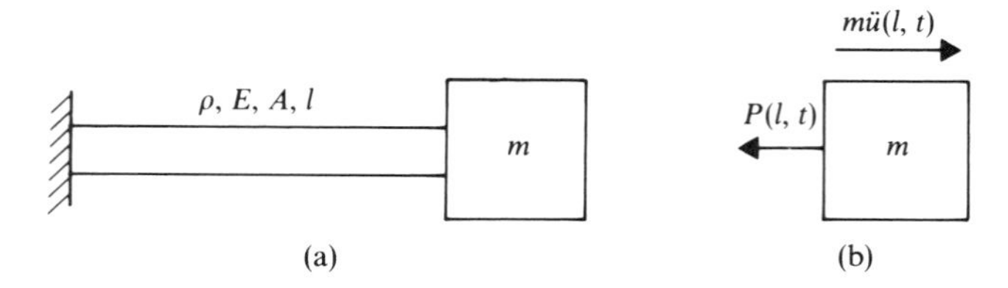

FIG. 4.3. Longitudinal vibration of a rod with a mass attached to its end.

because of the inertia effect. From the free-body digram shown in the figure, the second boundary condition is given by

$$m\frac{\partial^2 u(l,t)}{\partial t^2} = -P(l, t) = -\sigma(l, t)A = -EA\frac{\partial u(l,t)}{\partial x}$$

or

$$m\frac{\partial^2 u(l,t)}{\partial t^2} = -EA\frac{\partial u(l,t)}{\partial x} \tag{4.51}$$

The longitudinal vibration of the rod is governed by the equation

$$\begin{aligned} u(x,t) &= \phi(x)q(t) \\ &= \left(A_1 \sin\frac{\omega}{c}x + A_2 \cos\frac{\omega}{c}x\right)(B_1 \sin \omega t + B_2 \cos \omega t) \end{aligned} \tag{4.52}$$

Substituting Eq. 50 into Eq. 52 yields

$$A_2 = 0$$

Using this condition and Eq. 52, one gets

$$u(x, t) = \phi(x)q(t) = A_1 \sin\frac{\omega x}{c}(B_1 \sin \omega t + B_2 \cos \omega t)$$

which yields

$$\frac{\partial^2 u}{\partial t^2} = -\omega^2 A_1 \sin\frac{\omega x}{c}(B_1 \sin \omega t + B_2 \cos \omega t)$$

$$\frac{\partial u}{\partial x} = \frac{\omega}{c}A_1 \cos\frac{\omega x}{c}(B_1 \sin \omega t + B_2 \cos \omega t)$$

Therefore, the second boundary condition of Eq. 51 yields

$$-m\omega^2 A_1 \sin\frac{\omega l}{c}(B_1 \sin \omega t + B_2 \cos \omega t) = -EA\frac{\omega}{c}A_1 \cos\frac{\omega l}{c}(B_1 \sin \omega t + B_2 \cos \omega t)$$

This equation implies

$$\frac{m\omega c}{EA}\sin\frac{\omega l}{c} = \cos\frac{\omega l}{c}$$

or

$$\tan\frac{\omega l}{c} = \frac{EA}{m\omega c} \tag{4.53}$$

This is the frequency equation which upon multiplying both of its sides by $\omega l/c$, yields

$$\frac{\omega l}{c}\tan\frac{\omega l}{c} = \frac{\omega l E A}{m\omega c^2} = \frac{lA\rho}{m} = \frac{M}{m}$$

where M is the mass of the rod. The preceding equation can be written as

$$\gamma \tan \gamma = \mu \tag{4.54a}$$

where

$$\gamma = \frac{\omega l}{c}, \qquad \mu = \frac{M}{m}$$

Note that the frequency equation is a transcendental equation which has an infinite number of roots, and therefore, its solution defines an infinite number of natural frequencies. This equation can be expressed for each root as

$$\gamma_j \tan \gamma_j = \mu, \qquad j = 1, 2, 3, \ldots \tag{4.54b}$$

where

$$\gamma_j = \frac{\omega_j l}{c},$$

and the eigenfunction associated with the natural frequency ω_j is

$$\phi_j = A_{1j} \sin\frac{\omega_j x}{c}$$

Table 1 shows the first twenty roots of Eq. 54b for different values of the mass ratio μ.

TABLE 4.1. Roots of Eq. 54b

Mode number	γ				
	$1/\mu = 0$	$1/\mu = 0.01$	$1/\mu = 0.1$	$1/\mu = 1$	$1/\mu = 10$
1	1.5708	1.5552	1.4289	0.8603	0.3111
2	4.7124	4.6658	4.3058	3.4256	3.1731
3	7.8540	7.7764	7.2281	6.4372	6.2991
4	10.9956	10.8871	10.2003	9.5293	9.4354
5	14.1372	13.9981	13.2142	12.6453	12.5743
6	17.2788	17.1093	16.2594	15.7713	15.7143
7	20.4204	20.2208	19.3270	18.9024	18.8549
8	23.5619	23.3327	22.4108	22.0365	21.9957
9	26.7035	26.4451	25.5064	25.1725	25.1367
10	29.8451	29.5577	28.6106	28.3096	28.2779
11	32.9867	32.6710	31.7213	31.4477	31.4191
12	36.1283	35.7847	34.8371	34.5864	34.5604
13	39.2699	38.8989	37.9567	37.7256	37.7018
14	42.4115	42.0138	31.0795	40.8652	40.4832
15	45.5531	45.1292	44.2048	44.0050	43.9846
16	48.6947	48.2452	47.3321	47.1451	47.1260
17	51.8363	51.3618	50.4611	50.2854	50.2675
18	54.9779	54.4791	53.5916	53.4258	53.4089
19	58.1195	57.5969	56.7232	56.5663	56.5504
20	61.2611	60.7155	56.7232	59.7070	59.6919

Example 4.2
The system shown in Fig. 4 consists of a prismatic rod which has one end fixed and the other end attached to a spring with stiffness k as shown in the figure. The rod has length l, cross-sectional area A, mass density ρ, and modulus of elasticity E. Obtain the frequency equation and the eigenfunctions of this system.

Solution. As in the preceding example, the boundary condition at the fixed end is given by

$$u(0, t) = 0$$

At the other end, the axial force of the rod is equal in magnitude and opposite in direction to the spring force, that is,

$$P(l, t) = -ku(l, t)$$

Since $P = EAu'$, the preceding boundary condition is given by

$$EAu'(l, t) = -ku(l, t)$$

Using the technique of the separation of variables, the solution of the partial differential equation of the rod can be expressed as

$$\begin{aligned} u(x, t) &= \phi(x)q(t) \\ &= \left(A_1 \sin\frac{\omega}{c}x + A_2 \cos\frac{\omega}{c}x\right)(B_1 \sin \omega t + B_2 \cos \omega t) \end{aligned}$$

As in the preceding example, the boundary condition at the fixed end leads to

$$A_2 = 0$$

Therefore, the expression for the longitudinal displacement u may be simplified and written as

$$u = \phi(x)q(t) = A_1 \sin\frac{\omega}{c}x\,(B_1 \sin \omega t + B_2 \cos \omega t)$$

and

$$u' = \frac{\omega}{c}A_1 \cos\frac{\omega}{c}x\,(B_1 \sin \omega t + B_2 \cos \omega t)$$

By using the second boundary condition, the following equation is obtained:

$$EA\frac{\omega}{c}\cos\frac{\omega}{c}l = -k \sin\frac{\omega}{c}l$$

or

$$\tan\frac{\omega}{c}l = -\frac{EA\omega}{kc}$$

Multiplying both sides of this equation by $\omega l/c$ and using the definition of $c =$

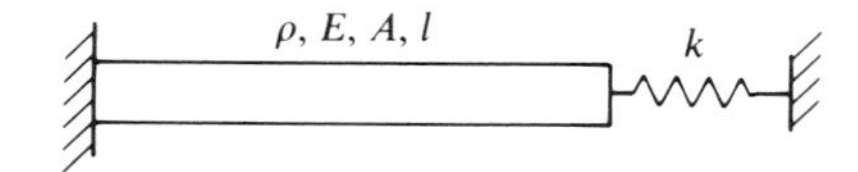

FIG. 4.4. Longitudinal vibration of a rod with one end attached to a spring.

$\sqrt{E/\rho}$, one obtains

$$\frac{\omega l}{c}\tan\frac{\omega l}{c} = -\frac{EA\omega^2 l}{kc^2} = -\frac{EA\omega^2 l\rho}{kE} = -\frac{\omega^2 M}{k}$$

where M is the mass of the rod. The above equation is the frequency equation which can be written as

$$\gamma \tan \gamma = -\frac{\omega^2 M}{k}$$

where

$$\gamma = \frac{\omega l}{c}$$

The roots of the frequency equation can be determined numerically and used to define the natural frequencies, ω_j, $j = 1, 2, 3, \ldots$. It is clear in this case that the eigenfunction associated with the jth natural frequency is

$$\phi_j = A_{1j} \sin\frac{\omega_j x}{c}$$

The frequency equations and mode shapes of the longitudinal vibrations obtained using different sets of boundary conditions are presented in Table 2.

4.2 FREE TORSIONAL VIBRATIONS

In this section, we consider the torsional vibration of straight circular shafts as the one shown in Fig. 5a. The shaft is assumed to have a length l and a cross section which has a polar moment of inertia I_p. The shaft is assumed to be made of material which has modulus of rigidity G and mass density ρ and is subjected to a distributed external torque defined per unit length by the function $T_e(x, t)$. In the analysis presented in this section, the cross sections of the shaft are assumed to remain in their planes and rotate as rigid surfaces about their centers. The equation of dynamic equilibrium of an infinitesimal volume of the shaft in torsional vibration is given by

$$\rho I_p \frac{\partial^2\theta}{\partial t^2}\delta x = T + \frac{\partial T}{\partial x}\delta x - T + T_e\delta x \tag{4.55}$$

where $\theta = \theta(x, t)$ is the angle of torsional oscillation of the infinitesimal volume about the axis of the shaft, T is the internal torque acting on the cross section at a distance x from the end of the shaft, and δx is the length of the infinitesimal volume. Equation 55 can be simplified and written as

$$\rho I_p \frac{\partial^2\theta}{\partial t^2} = \frac{\partial T}{\partial x} + T_e \tag{4.56}$$

From elementary strength of material theory, the internal torque T is proportional to the spatial derivative of the torsional oscillation θ. The relationship between the internal torque and the torsional oscillation is given in terms of the modulus of rigidity G as

$$T = GI_p \frac{\partial\theta}{\partial x} \tag{4.57}$$

TABLE 4.2. Frequency equations and mode shapes of the longitudinal vibrations $\left(c = \sqrt{\frac{E}{\rho}}, \gamma = \frac{\omega l}{c}, A_j = \text{constant}\right)$

Boundary conditions	Frequency equation	Mode shapes
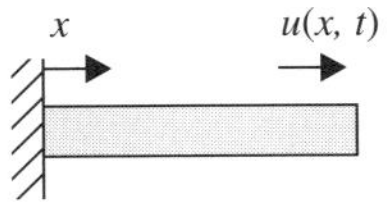 x, $u(x, t)$ Fixed-Free	$\cos \frac{\omega l}{c} = 0$ $\omega_j = \frac{(2j-1)\pi c}{2l}$	$\phi_j = A_j \sin \frac{\omega_j x}{c}$
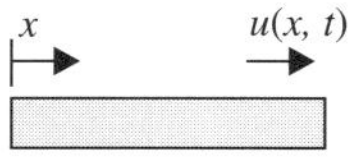 x, $u(x, t)$ Free-Free	$\sin \frac{\omega l}{c} = 0$ $\omega_j = \frac{j\pi c}{l}$	$\phi_j = A_j \cos \frac{\omega_j x}{c}$
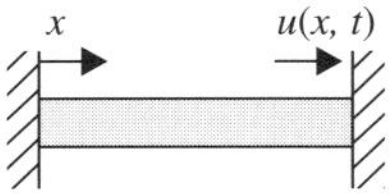 x, $u(x, t)$ Fixed-Fixed	$\sin \frac{\omega l}{c} = 0$ $\omega_j = \frac{j\pi c}{l}$	$\phi_j = A_j \sin \frac{\omega_j x}{c}$
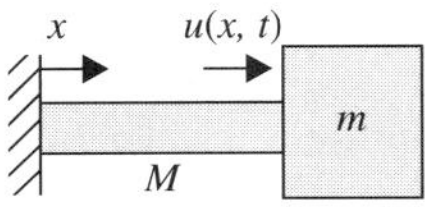 x, $u(x, t)$, m, M Fixed-Mass	$\gamma \tan \gamma = \mu$ where $\mu = \frac{M}{m}$	$\phi_j = A_j \sin \frac{\omega_j x}{c}$
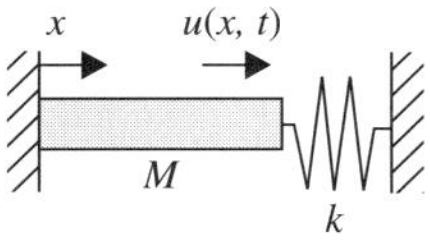 x, $u(x, t)$, M, k Fixed-Spring	$\gamma \tan \gamma = -\frac{\omega^2 M}{k}$	$\phi_j = A_j \sin \frac{\omega_j x}{c}$
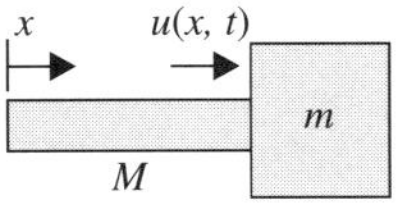 x, $u(x, t)$, m, M Free-Mass	$\gamma \cot \gamma = -\mu$ where $\mu = \frac{M}{m}$	$\phi_j = A_j \cos \frac{\omega_j x}{c}$
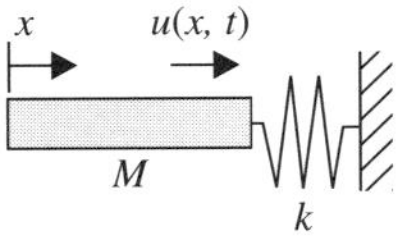 x, $u(x, t)$, M, k Free-Spring	$\gamma \cot \gamma = \frac{\omega^2 M}{k}$	$\phi_j = A_j \cos \frac{\omega_j x}{c}$

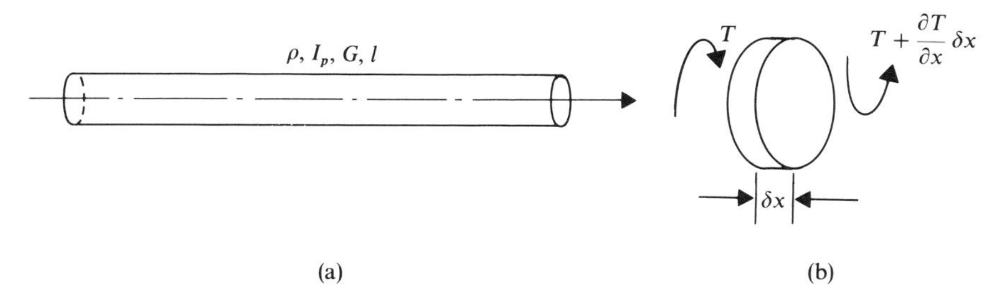

FIG. 4.5. Torsional vibration.

Substituting Eq. 57 into Eq. 56, one obtains

$$\rho I_p \frac{\partial^2 \theta}{\partial t^2} = \frac{\partial}{\partial x}\left(GI_p \frac{\partial \theta}{\partial x}\right) + T_e \tag{4.58}$$

If the shaft is assumed to have a constant cross-sectional area and a constant modulus of rigidity, Eq. 58 can be expressed as

$$\rho I_p \frac{\partial^2 \theta}{\partial t^2} = GI_p \frac{\partial^2 \theta}{\partial x^2} + T_e \tag{4.59}$$

Free Vibration The partial differential equation that governs the free torsional vibration of the shaft can be obtained from Eq. 59 if we let $T_e = 0$, that is,

$$\rho I_p \frac{\partial^2 \theta}{\partial t^2} = GI_p \frac{\partial^2 \theta}{\partial x^2} \tag{4.60}$$

which can be rewritten as

$$\frac{\partial^2 \theta}{\partial t^2} = c^2 \frac{\partial^2 \theta}{\partial x^2} \tag{4.61}$$

where c is a constant that depends on the inertia properties and the material of the shaft and is given by

$$c = \sqrt{\frac{G}{\rho}} \tag{4.62}$$

Equation 61 is in the same form as Eq. 9 that governs the longitudinal vibration of prismatic rods. Therefore, the same solution procedure described in the preceding section can be used to solve Eq. 61. Using the separation of variables technique, the torsional oscillation θ can be expressed as

$$\theta = \phi(x)q(t) \tag{4.63}$$

where $\phi(x)$ is a space-dependent function and $q(t)$ depends on time. Following the procedure described in the preceding section, one can show that

$$\phi(x) = A_1 \sin\frac{\omega x}{c} + A_2 \cos\frac{\omega x}{c} \tag{4.64}$$

$$q(t) = B_1 \sin \omega t + B_2 \cos \omega t \tag{4.65}$$

where A_1, A_2, B_1, B_2, and ω are arbitrary constants to be determined using the boundary and initial conditions. Substituting Eqs. 64 and 65 into Eq. 63 yields

$$\theta(x, t) = \left(A_1 \sin\frac{\omega x}{c} + A_2 \cos\frac{\omega x}{c}\right)(B_1 \sin \omega t + B_2 \cos \omega t) \quad (4.66)$$

Boundary Conditions Equation 66 is in the same form as Eq. 23a which describes the longitudinal vibration of prismatic rods. Therefore, one expects that the natural frequencies and mode shapes will also be in the same form. For example, in the case of a shaft with one end fixed and the other end free, the natural frequencies and eigenfunctions of torsional vibration are given by

$$\omega_j = \frac{(2j-1)\pi c}{2l} = \frac{(2j-1)\pi}{2l}\sqrt{\frac{G}{\rho}}, \qquad j = 1, 2, 3, \ldots \quad (4.67)$$

$$\phi_j = A_{1j} \sin\frac{\omega_j x}{c}, \qquad j = 1, 2, 3, \ldots \quad (4.68)$$

and the solution for the free torsional vibration is

$$\theta = \sum_{j=1}^{\infty} \sin\frac{\omega_j x}{c}(C_j \sin \omega_j t + D_j \cos \omega_j t) \quad (4.69)$$

where the constants C_j and D_j, $j = 1, 2, 3, \ldots$, can be determined by using the initial conditions as described in the preceding section.

In the case of a shaft with *free ends*, the natural frequencies and eigenfunctions are given by

$$\omega_j = \frac{j\pi c}{l} = \frac{j\pi}{l}\sqrt{\frac{G}{\rho}}, \qquad j = 1, 2, 3, \ldots \quad (4.70)$$

$$\phi_j = A_{2j} \cos\frac{\omega_j x}{c}, \qquad j = 1, 2, 3, \ldots \quad (4.71)$$

and the solution for the free torsional vibration of the shaft with free ends has the form

$$\theta = \sum_{j=1}^{\infty} \cos\frac{\omega_j x}{c}(C_j \sin \omega_j t + D_j \cos \omega_j t) \quad (4.72)$$

where C_j and D_j, $j = 1, 2, 3, \ldots$, are constants to be determined from the initial conditions.

Similarly one can show that the natural frequencies and mode shapes of the torsional vibration in the case of a shaft with *both ends fixed* are

$$\omega_j = \frac{j\pi c}{l} = \frac{j\pi}{l}\sqrt{\frac{G}{\rho}}, \qquad j = 1, 2, 3, \ldots \quad (4.73)$$

$$\phi_j = A_{1j} \sin\frac{\omega_j x}{c}, \qquad j = 1, 2, 3, \ldots \quad (4.74)$$

and the solution for the free torsional vibration is given by

$$\theta = \sum_{j=1}^{\infty} \sin\frac{\omega_j x}{c}(C_j \sin \omega_j t + D_j \cos \omega_j t) \tag{4.75}$$

As in the case of longitudinal vibrations discussed in the preceding section, the orthogonality of the mode shapes can be utilized in determining the arbitrary constants C_j and D_j using the initial conditions.

Example 4.3

The system shown in Fig. 6 consists of a disk with a mass moment of inertia I_d attached to a circular shaft which has mass density ρ, length l, cross-sectional polar moment of inertia I_p, and modulus of rigidity G. Obtain the frequency equation and the mode shapes of the torsional vibration.

Solution. The two boundary conditions in this case are given by

$$\theta(0, t) = 0$$

$$T(l, t) = GI_p\theta'(l, t) = -I_d\ddot{\theta}(l, t)$$

The solution for the free torsional vibration can be assumed in the form

$$\theta(x, t) = \phi(x)q(t)$$

$$= \left(A_1 \sin\frac{\omega x}{c} + A_2 \cos\frac{\omega x}{c}\right)(B_1 \sin \omega t + B_2 \cos \omega t)$$

Substituting the two boundary conditions into this solution and following the procedure described in Example 1, it can be shown that the frequency equation is given by

$$\gamma \tan \gamma = \mu$$

where

$$\gamma = \frac{\omega l}{c}, \qquad \mu = \frac{\rho I_p l}{I_d}, \qquad c = \sqrt{\frac{G}{\rho}}$$

This equation has an infinite number of roots γ_j, $j = 1, 2, 3, \ldots$, which can be used to define the natural frequencies ω_j as

$$\omega_j = \frac{\gamma_j c}{l}, \qquad j = 1, 2, 3, \ldots$$

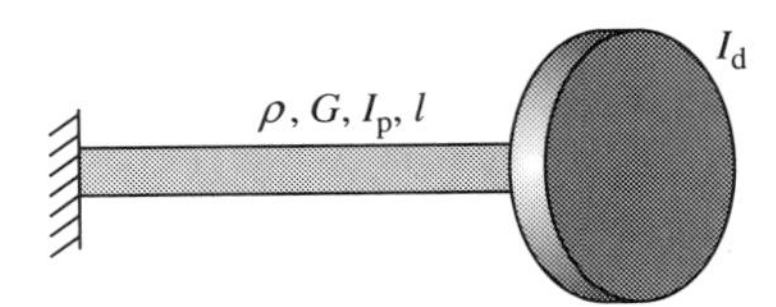

FIG. 4.6. Torsional vibration of a shaft with one end attached to a disk.

Example 4.4

The system shown in Fig. 7 consists of a straight cylindrical shaft which has one end fixed and the other end attached to a torsional spring with stiffness k_t as shown in the figure. The shaft has length l, mass density ρ, cross-sectional polar moment of inertia I_p, and modulus of rigidity G. Obtain the frequency equation and the eigenfunctions of this system.

Solution. The boundary conditions in this case are given by

$$\theta(0, t) = 0$$

$$T(l, t) = GI_p\theta'(l, t) = -k_t\theta(l, t)$$

The solution for the free torsional vibration is given by

$$\begin{aligned}\theta(x, t) &= \phi(x)q(t)\\ &= \left(A_1 \sin\frac{\omega x}{c} + A_2 \cos\frac{\omega x}{c}\right)(B_1 \sin \omega t + B_2 \cos \omega t)\end{aligned}$$

Substituting the boundary conditions into this solution and following the procedure described in Example 2, it can be shown that the frequency equation is given by

$$\gamma \tan \gamma = -\mu$$

where

$$\gamma = \frac{\omega l}{c}, \qquad \mu = \frac{\omega^2 \rho l I_p}{k_t}, \qquad c = \sqrt{\frac{G}{\rho}}$$

The roots $\gamma_j, j = 1, 2, 3, \ldots$, of the frequency equation can be obtained numerically. These roots define the natural frequencies $\omega_j, j = 1, 2, 3, \ldots$, as

$$\omega_j = \frac{\gamma_j c}{l}, \qquad j = 1, 2, 3, \ldots$$

The associated mode shapes are

$$\phi_j = A_{1j} \sin\frac{\omega_j x}{c}$$

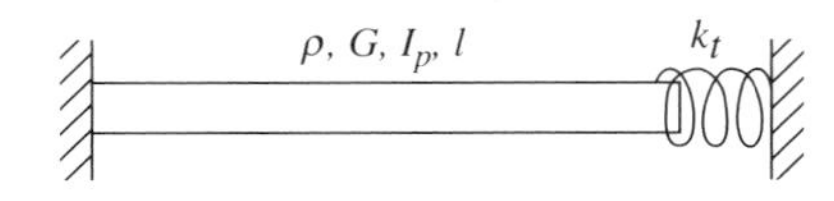

FIG. 4.7. Torsional vibration of a shaft with one end attached to a torsional spring.

Table 3 shows the frequency equations and mode shapes of the torsional vibration obtained using different sets of boundary conditions.

4.3 FREE TRANSVERSE VIBRATIONS OF BEAMS

It was shown in the preceding two sections that the differential equations that govern the longitudinal and torsional vibration of rods have the same form.

TABLE 4.3. Frequency equations and mode shapes of the torsional vibrations $\left(c = \sqrt{\frac{G}{\rho}}, \gamma = \frac{\omega l}{c}, A_j = \text{constant}\right)$

Boundary conditions	Frequency equation	Mode shapes
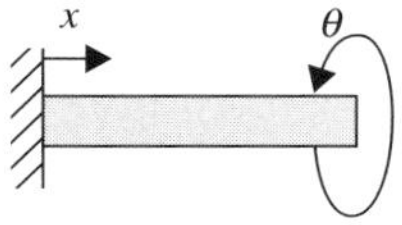 Fixed-Free	$\cos \frac{\omega l}{c} = 0$ $\omega_j = \frac{(2j-1)\pi c}{2l}$	$\phi_j = A_j \sin \frac{\omega_j x}{c}$
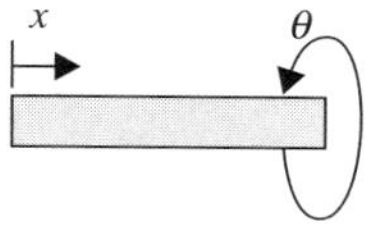 Free-Free	$\sin \frac{\omega l}{c} = 0$ $\omega_j = \frac{j\pi c}{l}$	$\phi_j = A_j \cos \frac{\omega_j x}{c}$
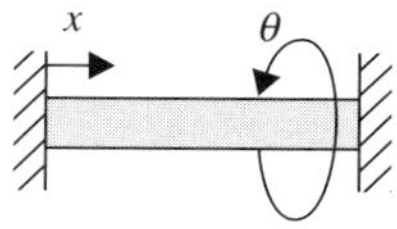 Fixed-Fixed	$\sin \frac{\omega l}{c} = 0$ $\omega_j = \frac{j\pi c}{l}$	$\phi_j = A_j \sin \frac{\omega_j x}{c}$
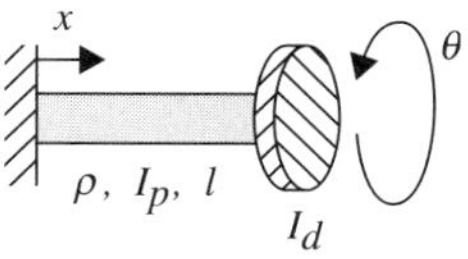 ρ, I_p, l; I_d Fixed-Inertial mass	$\gamma \tan \gamma = \mu$ where $\mu = \frac{\rho I_p l}{I_d}$	$\phi_j = A_j \sin \frac{\omega_j x}{c}$
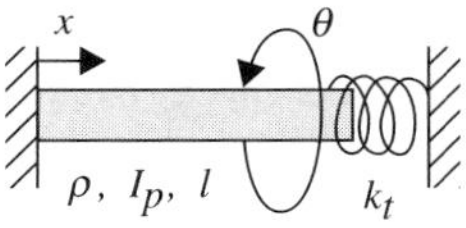 ρ, I_p, l; k_t Fixed-Spring	$\gamma \tan \gamma = -\mu$ where $\mu = \frac{\omega^2 \rho l I_p}{k_t}$	$\phi_j = A_j \sin \frac{\omega_j x}{c}$
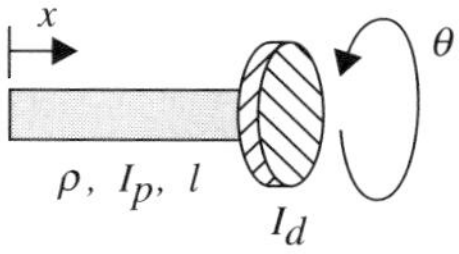 ρ, I_p, l; I_d Free-Inertial mass	$\gamma \cot \gamma = -\mu$ where $\mu = \frac{\rho I_p l}{I_d}$	$\phi_j = A_j \cos \frac{\omega_j x}{c}$
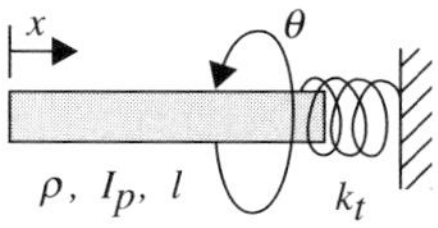 ρ, I_p, l; k_t Fixed-Spring	$\gamma \cot \gamma = \mu$ where $\mu = \frac{\omega^2 \rho l I_p}{k_t}$	$\phi_j = A_j \cos \frac{\omega_j x}{c}$

The theory of transverse vibration of beams is more difficult than that of the two types of vibration already considered in the preceding sections. In this section, the equations that govern the transverse vibration of beams are developed and methods for obtaining their solutions are discussed. There are several important applications of the theory of transverse vibration of beams; among these applications are the study of vibrations of rotating shafts and rotors and the transverse vibration of suspended cables.

Elementary Beam Theory In the elementary beam theory all stresses are assumed to be equal to zero except the normal stress σ which is assumed to vary linearly over the cross section with the y-coordinate of the beam as shown in Fig. 8. The normal stress σ can be written as

$$\sigma = ky \tag{4.76}$$

where k is constant, and $y = 0$ contains the neutral surface along which the normal stress σ is equal to zero. The assumption that all other stresses are equal to zero requires that the resultant of the internal forces be zero and that the moments of the internal forces about the neutral axis equal the bending moment M. Consequently,

$$\int_A \sigma\, dA = 0, \qquad \int_A y\sigma\, dA = M \tag{4.77}$$

where A is the cross-sectional area of the beam. Substituting Eq. 76 into Eq. 77 yields

$$k \int_A y\, dA = 0 \tag{4.78}$$

$$k \int_A y^2\, dA = M \tag{4.79}$$

Since k is a nonzero constant, the first equation implies that the neutral and centroidal axes of the cross section coincide. The second equation can be

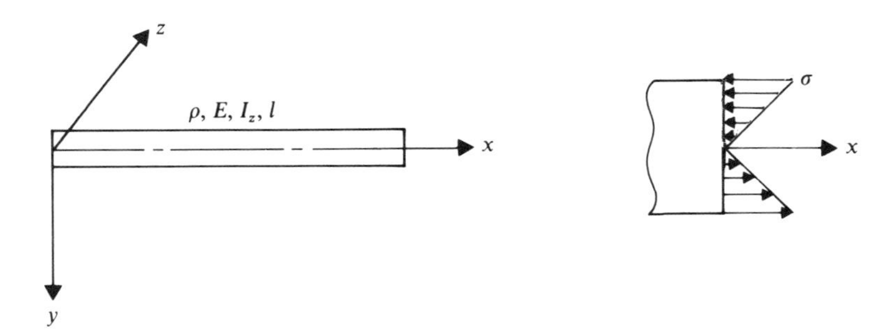

FIG. 4.8. Elementary beam theory.

used to define k as

$$k = \frac{M}{I_z} \tag{4.80}$$

where I_z is the moment of inertia of the cross section about the z-axis of the beam cross section, and is defined as

$$I_z = \int_A y^2 \, dA$$

Table 4 shows the area properties of selected shapes of the cross section. Substituting Eq. 80 into Eq. 76 yields

$$\sigma = \frac{My}{I_z} \tag{4.81}$$

Using Hooke's law, the strain ε is given by

$$\varepsilon = \frac{\sigma}{E} = \frac{My}{EI_z} \tag{4.82}$$

Let v denotes the transverse displacement of the beam. For small deformations $(dv/dx \ll 1)$, it can be shown that

$$\frac{1}{r} \approx \frac{d^2v}{dx^2} = -\frac{\varepsilon}{y} = -\frac{M}{EI_z} \tag{4.83}$$

where r is the radius of curvature of the beam. Equation 83 implies that

$$M = -EI_z v'' \tag{4.84}$$

This equation is known as the *Euler–Bernoulli law* of the elementary beam theory.

Partial Differential Equation In order to determine the differential equation for the transverse vibration of beams, we consider an infinitesimal volume at a distance x from the end of the beam as shown in Fig. 9. The length

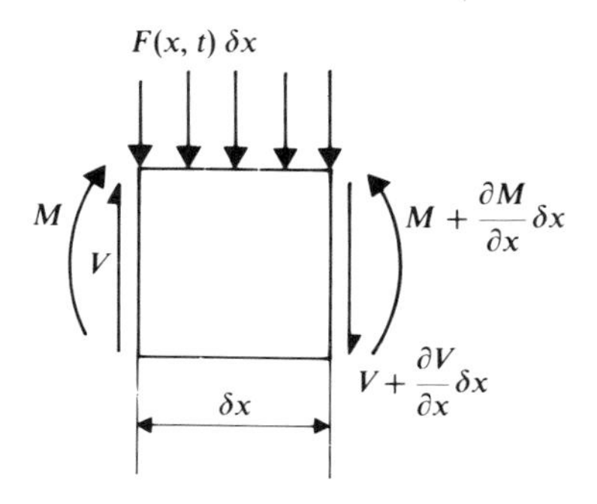

FIG. 4.9. Moments and shear forces.

TABLE 4.4. Moments of inertia of the cross sections

Shape	Area	Moment of inertia
Circle 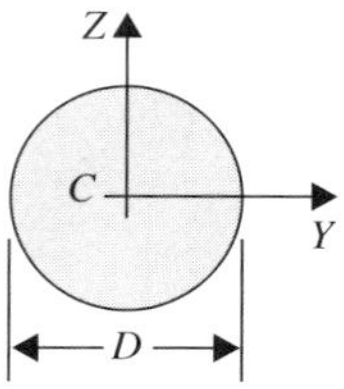	$A = \frac{\pi D^2}{4}$	$I_y = I_z = \frac{\pi D^4}{64}$
Rectangle 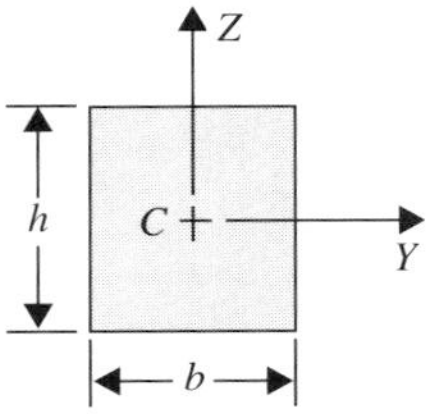	$A = bh$	$I_y = \frac{bh^3}{12}, I_2 = \frac{hb^3}{12}$
Triangle 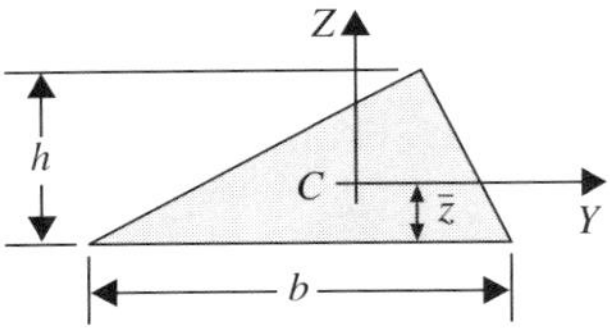	$A = \frac{bh}{2}$ $\bar{z} = \frac{h}{3}$	$I_y = \frac{bh^3}{36}$
Ellipse 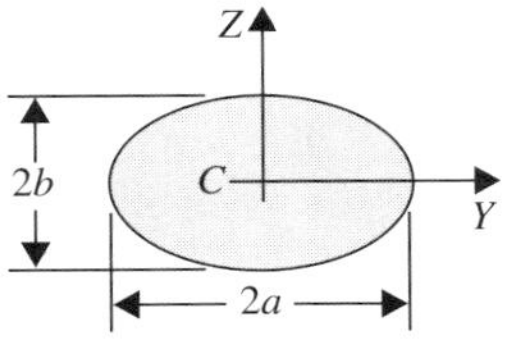	$A = \pi ab$	$I_y = \frac{\pi ab^3}{4}, I_z = \frac{\pi ba^3}{4}$
Circular Sector Area 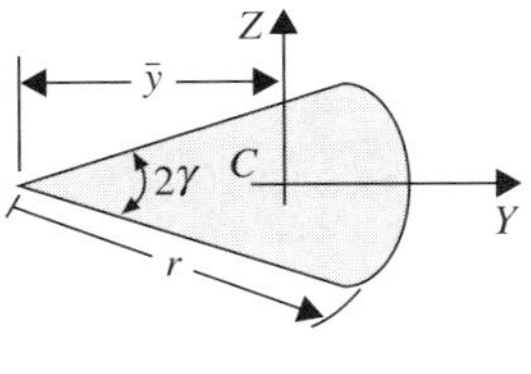	$A = \gamma r^2$ $\bar{y} = \frac{2r}{3}\frac{\sin\gamma}{\gamma}$	$I_y = \frac{r^4}{4}(\gamma - \frac{1}{2}\sin 2\gamma)$

of this infinitesimal volume is assumed to be δx. Let V and M be, respectively, the shear force and bending moment, and let $F(x, t)$ be the loading per unit length of the beam. Neglecting the rotary inertia, the sum of the moments about the left end of the section yields

$$M + \frac{\partial M}{\partial x}\delta x - M - V\delta x - \frac{\partial V}{\partial x}(\delta x)^2 - F(x, t)\frac{(\delta x)^2}{2} - \rho A\frac{\partial^2 v}{\partial t^2}\frac{(\delta x)^2}{2} = 0 \tag{4.85a}$$

Taking the limit as δx approaches zero, the preceding equation leads to

$$V = \frac{\partial M}{\partial x} \tag{4.85b}$$

The dynamic equilibrium condition for the transverse vibration of the beam is obtained by applying Newton's second law as

$$\rho A\delta x\frac{\partial^2 v}{\partial t^2} = V + \frac{\partial V}{\partial x}\delta x - V + F(x, t)\delta x \tag{4.86}$$

where ρ is the mass density and A is the cross-sectional area. Equation 86 can be rewritten after simplification as

$$\rho A\frac{\partial^2 v}{\partial t^2} = \frac{\partial V}{\partial x} + F(x, t) \tag{4.87}$$

Substituting Eq. 85b into Eq. 87 yields

$$\rho A\frac{\partial^2 v}{\partial t^2} = \frac{\partial^2 M}{\partial x^2} + F(x, t) \tag{4.88}$$

The moment M can be eliminated from this equation by using the moment displacement relationship. To this end, Eq. 84 is substituted into Eq. 88. This leads to

$$\rho A\frac{\partial^2 v}{\partial t^2} = -\frac{\partial^2}{\partial x^2}(EI_z v'') + F(x, t) \tag{4.89}$$

If E and I_z are assumed to be constant, Eq. 89 becomes

$$\rho A\frac{\partial^2 v}{\partial t^2} = -EI_z\frac{\partial^4 v}{\partial x^4} + F(x, t) \tag{4.90}$$

In the case of free vibration, $F(x, t) = 0$ and Eq. 90 reduces to

$$\frac{\partial^2 v}{\partial t^2} = -c^2\frac{\partial^4 v}{\partial x^4} \tag{4.91}$$

where c is a constant defined as

$$c = \sqrt{\frac{EI_z}{\rho A}} \tag{4.92}$$

Separation of Variables Equation 91 is a fourth-order partial differential equation that governs the free transverse vibration of the beam. The solution of this equation can be obtained by using the technique of the separation of variables. In this case, we assume a solution in the form

$$v = \phi(x)q(t) \tag{4.93}$$

where $\phi(x)$ is a space-dependent function, and $q(t)$ is a function that depends only on time. Equation 93 leads to

$$\frac{\partial^2 v}{\partial t^2} = \phi(x)\frac{d^2 q(t)}{dt^2} = \phi(x)\ddot{q}(t) \tag{4.94}$$

$$\frac{\partial^4 v}{\partial x^4} = \frac{d^4\phi(x)}{dx^4}q(t) = \phi^{iv}(x)q(t) \tag{4.95}$$

Substituting these equations into Eq. 91, one obtains

$$\phi(x)\ddot{q}(t) = -c^2\phi^{iv}(x)q(t)$$

which implies that

$$\frac{\ddot{q}(t)}{q(t)} = -c^2\frac{\phi^{iv}(x)}{\phi(x)} = -\omega^2 \tag{4.96}$$

where ω is a constant to be determined. Equation 96 leads to the following two equations:

$$\ddot{q} + \omega^2 q = 0 \tag{4.97}$$

$$\phi^{iv} - \left(\frac{\omega}{c}\right)^2 \phi = 0 \tag{4.98}$$

The solution of Eq. 97 is given by

$$q = B_1 \sin \omega t + B_2 \cos \omega t \tag{4.99}$$

For Eq. 98, we assume a solution in the form

$$\phi = Ae^{\lambda x}$$

Substituting this assumed solution into Eq. 98 yields

$$\left[\lambda^4 - \left(\frac{\omega}{c}\right)^2\right]Ae^{\lambda x} = 0$$

or

$$\lambda^4 - \left(\frac{\omega}{c}\right)^2 = 0$$

which can be written as

$$\lambda^4 - \eta^4 = 0 \tag{4.100}$$

where

$$\eta = \sqrt{\frac{\omega}{c}} \tag{4.101}$$

The roots of Eq. 100 are

$$\lambda_1 = \eta, \qquad \lambda_2 = -\eta, \qquad \lambda_3 = i\eta, \qquad \lambda_4 = -i\eta$$

where $i = \sqrt{-1}$. Therefore, the general solution of Eq. 98 can be written as

$$\phi(x) = A_1 e^{\eta x} + A_2 e^{-\eta x} + A_3 e^{i\eta x} + A_4 e^{-i\eta x} \tag{4.102}$$

which can be rewritten as

$$\begin{aligned}\phi(x) = {} & A_5 \frac{e^{\eta x} - e^{-\eta x}}{2} + A_6 \frac{e^{\eta x} + e^{-\eta x}}{2} \\ & + A_7(-i)\frac{e^{i\eta x} - e^{-i\eta x}}{2} + A_8 \frac{e^{i\eta x} + e^{-i\eta x}}{2}\end{aligned} \tag{4.103}$$

where

$$A_1 = \frac{A_5 + A_6}{2}, \qquad A_2 = \frac{A_6 - A_5}{2}$$

$$A_3 = \frac{A_8 - iA_7}{2}, \qquad A_4 = \frac{A_8 + iA_7}{2}$$

Equation 103 can then be rewritten, using Euler's formula of the complex variables, as

$$\phi(x) = A_5 \sinh \eta x + A_6 \cosh \eta x + A_7 \sin \eta x + A_8 \cos \eta x \tag{4.104}$$

Substituting Eqs. 99 and 104 into Eq. 93 yields

$$\begin{aligned}v(x, t) = {} & (A_5 \sinh \eta x + A_6 \cosh \eta x + A_7 \sin \eta x + A_8 \cos \eta x) \\ & \cdot (B_1 \sin \omega t + B_2 \cos \omega t)\end{aligned} \tag{4.105}$$

where ω is defined by Eq. 101 as

$$\omega = c\eta^2 \tag{4.106}$$

Boundary Conditions The natural frequencies of the beam, as well as the constants that appear in Eq. 105, depend on the boundary and initial conditions. For example, if the beam is *simply supported at both ends* the boundary conditions are

$$v(0, t) = 0, \qquad v''(0, t) = 0$$

$$v(l, t) = 0, \qquad v''(l, t) = 0$$

which imply that

$$\begin{aligned}\phi(0) = 0, \qquad \phi''(0) = 0 \\ \phi(l) = 0, \qquad \phi''(l) = 0\end{aligned} \tag{4.107}$$

It is clear that in this case, there are two geometric boundary conditions that specify the displacements, and there are two natural boundary conditions that specify the moments at the ends of the beam. Substituting these conditions into Eq. 104 yields

$$\left.\begin{aligned} A_6 + A_8 &= 0 \\ A_6 - A_8 &= 0 \\ A_5 \sinh \eta l + A_6 \cosh \eta l + A_7 \sin \eta l + A_8 \cos \eta l &= 0 \\ A_5 \sinh \eta l + A_6 \cosh \eta l - A_7 \sin \eta l - A_8 \cos \eta l &= 0 \end{aligned}\right\} \tag{4.108}$$

These equations are satisfied if $A_5 = A_6 = A_8 = 0$ and

$$A_7 \sin \eta l = 0 \tag{4.109}$$

The roots of Eq. 109 are

$$\eta l = j\pi, \qquad j = 1, 2, 3, \ldots \tag{4.110}$$

Therefore, the natural frequencies are given by

$$\omega_j = \frac{j^2\pi^2}{l^2} c = \frac{j^2\pi^2}{l^2} \sqrt{\frac{EI_z}{\rho A}}, \qquad j = 1, 2, 3, \ldots \tag{4.111}$$

and the corresponding modes of vibration are

$$\phi_j = A_{7j} \sin \eta_j x, \qquad j = 1, 2, 3, \ldots \tag{4.112}$$

The solution for the free vibration of the simply supported beam can then be written as

$$v(x, t) = \sum_{j=1}^{\infty} \phi_j q_j = \sum_{j=1}^{\infty} (C_j \sin \omega_j t + D_j \cos \omega_j t) \sin \eta_j x \tag{4.113}$$

The arbitrary constants C_j and D_j, $j = 1, 2, 3, \ldots$, can be determined using the initial conditions by the method described in Section 1 of this chapter.

In the case of a *cantilever beam*, the geometric boundary conditions at the fixed end are

$$v(0, t) = 0, \qquad v'(0, t) = 0$$

and the natural boundary conditions at the free end are

$$v''(l, t) = 0, \qquad v'''(l, t) = 0$$

These conditions imply that

$$\phi(0) = 0, \qquad \phi'(0) = 0$$
$$\phi''(l) = 0, \qquad \phi'''(l) = 0$$

Substituting these conditions into Eq. 104 yields the frequency equation

$$\cos \eta l \cosh \eta l = -1$$

The roots of this frequency equation can be determined numerically. The first six roots are (Timoshenko et al., 1974) $\eta_1 l = 1.875$, $\eta_2 l = 4.694$, $\eta_3 l = 7.855$, $\eta_4 l = 10.996$, $\eta_5 l = 14.137$, and $\eta_6 l = 17.279$. Approximate values of these roots can be calculated using the equation

$$\eta_j l \approx (j - \tfrac{1}{2})\pi$$

The fundamental natural frequencies of the system can be obtained using Eqs. 92 and 101 as

$$\omega_j = \eta_j^2 c = \eta_j^2 \sqrt{\frac{EI_z}{\rho A}}$$

The first six natural frequencies are

$$\omega_1 = 3.51563 \sqrt{\frac{EI_z}{ml^3}}, \qquad \omega_2 = 22.03364 \sqrt{\frac{EI_z}{ml^3}}$$

$$\omega_3 = 61.7010 \sqrt{\frac{EI_z}{ml^3}}, \qquad \omega_4 = 120.9120 \sqrt{\frac{EI_z}{ml^3}}$$

$$\omega_5 = 199.8548 \sqrt{\frac{EI_z}{ml^3}}, \qquad \omega_6 = 298.5638 \sqrt{\frac{EI_z}{ml^3}}$$

In this case it can be verified that the mode shapes are

$$\phi_j(x) = A_{6j}[\sin \eta_j x - \sinh \eta_j x + \bar{D}_j(\cos \eta_j x - \cosh \eta_j x)], \qquad j = 1, 2, \ldots$$

where A_{6j} is an arbitrary constant and

$$\bar{D}_j = \frac{\cos \eta_j l + \cosh \eta_j l}{\sin \eta_j l - \sinh \eta_j l}$$

Example 4.5

Neglecting the effect of rotary inertia, find the first four natural frequencies of the transverse vibration of a beam with free ends. The beam has length l, cross-sectional area A, moment of inertia I_z, mass density ρ, and modulus of elasticity E.

Solution. At the free end of a beam, the bending moment and shear force are equal to zero. Therefore, the boundary conditions in this example are given by

$$\phi''(0) = 0, \qquad \phi'''(0) = 0, \qquad \phi''(l) = 0, \qquad \phi'''(l) = 0$$

Substituting the first two boundary conditions into Eq. 104 yields

$$\phi''(0) = \eta^2 A_6 - \eta^2 A_8 = 0$$

$$\phi'''(0) = \eta^3 A_5 - \eta^3 A_7 = 0$$

which imply $A_6 = A_8$ and $A_5 = A_7$. Hence

$$\phi(x) = A_5(\sinh \eta x + \sin \eta x) + A_6(\cosh \eta x + \cos \eta x)$$

Substituting the third and fourth boundary conditions into this equation, one obtains

$$A_5(\sinh \eta l - \sin \eta l) + A_6(\cosh \eta l - \cos \eta l) = 0$$

$$A_5(\cosh \eta l - \cos \eta l) + A_6(\sinh \eta l + \sin \eta l) = 0$$

That is,

$$\begin{bmatrix} \sinh \eta l - \sin \eta l & \cosh \eta l - \cos \eta l \\ \cosh \eta l - \cos \eta l & \sinh \eta l + \sin \eta l \end{bmatrix} \begin{bmatrix} A_5 \\ A_6 \end{bmatrix} = \begin{bmatrix} 0 \\ 0 \end{bmatrix}$$

This system of homogeneous algebraic equations in the unknowns A_5 and A_6 has a nontrivial solution if and only if the determinant of the coefficient matrix is equal to zero. This leads to the following frequency equation:

$$(\sinh^2 \eta l - \sin^2 \eta l) - (\cosh \eta l - \cos \eta l)^2 = 0$$

Since

$$\cosh^2 \eta l - \sinh^2 \eta l = 1 \quad \text{and} \quad \cos^2 \eta l + \sin^2 \eta l = 1,$$

the frequency equation can be simplified to

$$\cos \eta l \cosh \eta l = 1$$

This equation is satisfied if $\eta l = 0$. Therefore, the first natural frequency is

$$\omega_1 = 0$$

This frequency is associated with a rigid-body motion of the beam. Thus the associated mode shape is given by

$$\phi_1(x) = A_{51} + A_{61}x$$

One can show that the second, third, and fourth natural frequencies are defined by the equations

$$\omega_2 = \eta^2 c = \eta^2 l^2 \frac{c}{l^2} = (4.730)^2 \frac{c}{l^2}$$

$$\omega_3 = (7.853)^2 \frac{c}{l^2}$$

$$\omega_4 = (10.996)^2 \frac{c}{l^2}$$

The function $\phi(x)$ obtained in this example is expressed in terms of the constants A_5 and A_6. The relationship between these two constants is

$$\frac{A_6}{A_5} = -\frac{\sinh \eta l - \sin \eta l}{\cosh \eta l - \cos \eta l} = -\frac{\cosh \eta l - \cos \eta l}{\sinh \eta l + \sin \eta l} = \bar{D}$$

Therefore, the space-dependent function ϕ can be expressed in terms of one constant as

$$\phi_j(x) = A_{5j}[\sinh \eta_j x + \sin \eta_j x + \bar{D}_j(\cosh \eta_j x + \cos \eta_j x)], \qquad j = 1, 2, 3, \ldots$$

The first three deformation mode shapes of the beam are shown in Fig. 10.

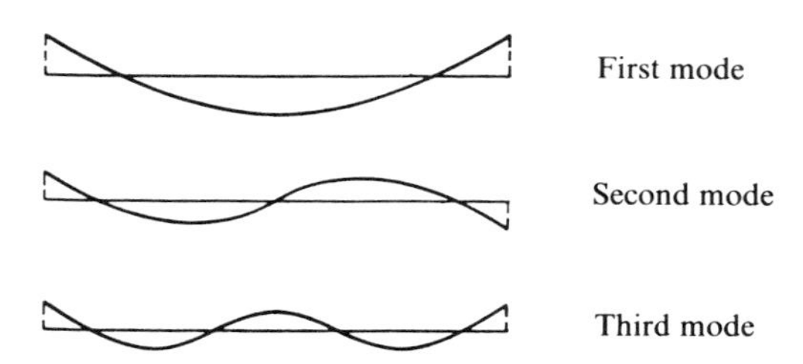

FIG. 4.10. First three deformation mode shapes of a beam with free ends.

Example 4.6

Neglecting the effect of the rotary inertia, find the first four natural frequencies of the transverse vibration of the beam in the preceding example if both ends of the beam are fixed.

Solution. For a beam with fixed ends the boundary conditions are

$$\phi(0) = 0, \qquad \phi'(0) = 0, \qquad \phi(l) = 0, \qquad \phi'(l) = 0$$

Substituting these boundary conditions into Eq. 104, and following the procedure described in the preceding example, one can show that the frequency equation, in this case, is given by

$$\cos \eta l \cosh \eta l = 1$$

The first four roots of this equation are

$$\eta_1 l = 1.875, \qquad \eta_2 l = 4.694, \qquad \eta_3 l = 7.855, \qquad \eta_4 l = 10.996$$

The space-dependent function $\phi_j(x)$ is given in this case by

$$\phi_j(x) = A_{6j}[\sin \eta_j x - \sinh \eta_j x + \bar{D}_j(\cos \eta_j x - \cosh \eta_j x)], \qquad j = 1, 2, 3, \ldots$$

where $\bar{D}_j$ is a constant defined as

$$\bar{D}_j = -\frac{\sin \eta_j l - \sinh \eta_j l}{\cos \eta_j l - \cosh \eta_j l} = -\frac{\cosh \eta_j l - \cos \eta_j l}{\sin \eta_j l + \sinh \eta_j l}$$

The first three mode shapes of vibration, in the case in which both ends are fixed, are shown in Fig. 11.

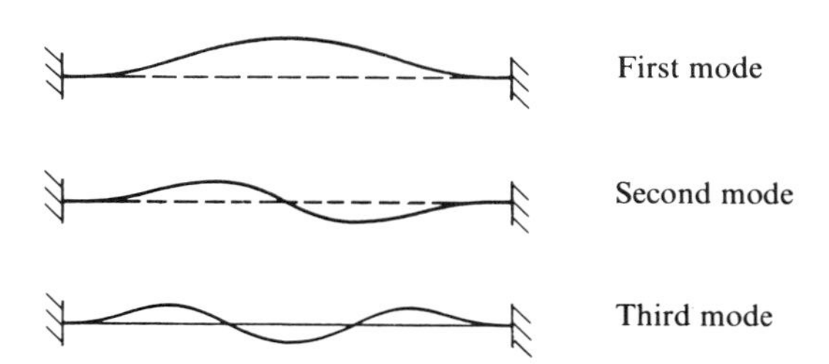

FIG. 4.11. First three mode shapes of a beam with fixed ends.

Table 5 shows the frequency equations and eigenfunctions of the transverse vibrations of beams with different end conditions.

Effect of the Rotary Inertia In the analysis presented in this section it was assumed that the cross sections of the beam during deformation remain perpendicular to the neutral axis, and accordingly the rotation of the cross section of the beam is not taken into account. This assumption leads to the simple relationship between the shear force and the moment given by Eq. 85.

TABLE 4.5. Frequency equations and mode shapes of the transverse vibrations $\left(\eta = \sqrt{\frac{\omega}{c}}, c = \sqrt{\frac{EI_z}{\rho A}}, A_j = \text{constant}\right)$

Boundary conditions	Frequency equation	Mode shapes
Simply Supported	$\sin \eta l = 0$ $\omega_j = \frac{j^2 \pi^2 c}{l^2}$	$\phi_j = A_j \sin \eta_j x$
Cantilever Beam	$\cos \eta l \cosh \eta l = -1$ $\eta_j l \approx (j - \frac{1}{2})\pi$ $\omega_j = \eta_j^2 c$	$\phi_j(x) = A_j[\sin \eta_j x - \sinh \eta_j x + D_j(\cos \eta_j x - \cosh \eta_j x)]$ $D_j = \frac{\cos \eta_j l + \cosh \eta_j l}{\sin \eta_j l - \sinh \eta_j l}$
Free-Free	$\cos \eta l \cosh \eta l = 1$	$\phi_j(x) = A_j[\sinh \eta_j x + \sin \eta_j x + D_j(\cosh \eta_j x + \cos \eta_j x)]$ $D_j = -\frac{\cosh \eta_j l - \cos \eta_j l}{\sinh \eta_j l + \sin \eta_j l}$
Clamped-Clamped	$\cos \eta l \cosh \eta l = 1$	$\phi_j(x) = A_j[\sin \eta_j x - \sinh \eta_j x + D_j(\cos \eta_j x - \cosh \eta_j x)]$ $D_j = -\frac{\cosh \eta_j l - \cos \eta_j l}{\sin \eta_j l + \sinh \eta_j l}$
Fixed-Simply Supported	$\tanh \eta l \cot \eta l = 1$	$\phi_j(x) = A_j[\sinh \eta_j x - \sin \eta_j x + D_j(\cosh \eta_j x - \cos \eta_j x)]$ $D_j = -\frac{\sinh \eta_j l - \sin \eta_j l}{\cosh \eta_j l - \cos \eta_j l}$
M m Fixed-Mass	$\eta l \frac{\cosh \eta l \sin \eta l - \sinh \eta l \cos \eta l}{\cosh \eta l \cos \eta l + 1} = -\frac{M}{m}$	$\phi_j(x) = A_j[\sinh \eta_j x - \sin \eta_j x + D_j(\cosh \eta_j x - \cos \eta_j x)]$ $D_j = -\frac{\sinh \eta_j l + \sin \eta_j l}{\cosh \eta_j l + \cos \eta_j l}$
Pin-Free	$\tanh \eta l \cot \eta l = 1$	$\phi_j(x) = A_j(\sin \eta_j x + D_j \sinh \eta_j x)$ $D_j = \frac{\sin \eta_j l}{\sinh \eta_j l}$

This simple beam theory in which the rotation of the cross section is assumed to be insignificant as compared to the translational displacement is valid only when the height of the beam is small compared to its length. If the height is denoted as h and the length as l, the accuracy of the simple beam theory is acceptable when $h/l \leq 1/10$.

Lord Rayleigh (1894) in his study of wave propagation obtained more satisfactory results when he took into account the effect of the rotary inertia of the cross section. If we make the assumption of small deformation, the rotation of the cross section about the z-axis can be defined as

$$\alpha = \frac{\partial v}{\partial x}$$

In this case, the sum of the moments given by Eq. 85a must be equal to the rotary inertia. This leads to

$$\frac{\partial M}{\partial x} - V = \rho I_z \frac{\partial^2 \alpha}{\partial t^2} = \rho I_z \frac{\partial^3 v}{\partial x \partial t^2}$$

where I_z is the second moment of area of the cross section. This equation, upon differentiation with respect to x, leads to

$$\frac{\partial V}{\partial x} = \frac{\partial^2 M}{\partial x^2} - \rho I_z \frac{\partial^4 v}{\partial x^2 \partial t^2}$$

Substituting this equation into Eq. 87, one obtains

$$\rho A \frac{\partial^2 v}{\partial t^2} = -\frac{\partial^2 M}{\partial x^2} + \rho I_z \frac{\partial^4 v}{\partial x^2 \partial t^2} + F(x, t)$$

which upon the use of Eq. 84 leads to

$$\rho A \frac{\partial^2 v}{\partial t^2} = -\frac{\partial^2}{\partial x^2}\left(EI_z \frac{\partial^2 v}{\partial x^2}\right) + \rho I_z \frac{\partial^4 v}{\partial x^2 \partial t^2} + F(x, t)$$

The second term on the right-hand side of this partial differential equation represents the effect of the rotary inertia. Observe that if the effect of rotary inertia is neglected, this equation reduces to Eq. 89 which was obtained earlier in this section.

It is clear from the results presented in the preceding example that the first mode shape has no nodes. That is, zero displacements occur only at the fixed ends. On the other hand, the second mode has one node and the third mode has two nodes. The analysis of higher modes shows that as the mode number increases, the number of nodes increases. Consequently the change in the slope of the curve that describes the shape of a high frequency mode is much faster than the change in the slope of the curve of the low-frequency modes. Therefore, for high-frequency modes, the rotation of the cross sections can be significant and the use of the simple beam theory that neglects the effect of the rotary inertia can lead to a significant error in the analysis of these modes in particular and in the analysis of the wave motion in general.

4.4 ORTHOGONALITY OF THE EIGENFUNCTIONS

In this section, we study in more detail the important property of the orthogonality of the eigenfunctions of the continuous systems. This property can be used to obtain an infinite number of decoupled second-order ordinary differential equations whose solution can be presented in a simple closed form. This development can be used to justify the use of approximate techniques in later sections to obtain a finite-dimensional model that represents, to a certain degree of accuracy, the vibration of the continuous systems. Furthermore, the use of the orthogonality of the eigenfunctions leads to the important definitions of the *modal mass*, *modal stiffness*, and *modal force coefficients* for the continuous systems. As will be seen in this section, there are an infinite number of such coefficients since a continuous system has an infinite number of degrees of freedom.

Longitudinal and Torsional Vibration of Rods The partial differential equations that govern the longitudinal and torsional vibration of rods have the same form, and consequently the resulting eigenfunctions are the same for similar end conditions. Therefore, in the following we discuss only the orthogonality of the eigenfunctions of the longitudinal vibration of rods.

It was shown, in Section 1, that the partial differential equation for the longitudinal vibration of rods can be written as (see Eq. 7)

$$\rho A \frac{\partial^2 u}{\partial t^2} = \frac{\partial}{\partial x}\left(EA \frac{\partial u}{\partial x}\right) \tag{4.114}$$

where $u = u(x, t)$ is the longitudinal displacement, ρ and A are, respectively, the mass density and cross-sectional area, and E is the modulus of elasticity. The solution of Eq. 114, which was obtained by using the separation of variables technique, can be expressed as

$$u(x, t) = \phi(x)q(t) \tag{4.115}$$

where $\phi(x)$ is a space-dependent function, and $q(t)$ depends only on time and can be expressed as

$$q(t) = B_1 \sin \omega t + B_2 \cos \omega t \tag{4.116}$$

By using Eqs. 115 and 116, the acceleration $\partial^2 u/\partial t^2$ can be written as

$$\frac{\partial^2 u}{\partial t^2} = -\omega^2 \phi(x)q(t) \tag{4.117}$$

Hence, Eq. 114 can be written as

$$-\rho A \omega^2 \phi(x)q(t) = (EA\ \phi'(x))'q(t)$$

which leads to

$$-\rho A \omega^2\ \phi(x) = (EA\phi'(x))' \tag{4.118}$$

For the jth eigenfunction ϕ_j, $j = 1, 2, 3, \ldots$, Eq. 118 yields

$$(EA\phi_j'(x))' = -\rho A\omega_j^2\phi_j(x) \tag{4.119}$$

Multiplying both sides of this equation by $\phi_k(x)$ and integrating over the length, one obtains

$$\int_0^l (EA\phi_j'(x))'\phi_k(x)\,dx = -\omega_j^2\int_0^l \rho A\phi_j(x)\phi_k(x)\,dx$$

Integrating the integral on the left-hand side of this equation by parts, one obtains

$$EA\phi_j'(x)\phi_k(x)\Big|_0^l - \int_0^l EA\phi_j'(x)\phi_k'(x)\,dx = -\omega_j^2\int_0^l \rho A\phi_j(x)\phi_k(x)\,dx \tag{4.120}$$

where l is the length of the rod.

For *simple end conditions* such as *free ends* ($\phi_j'(l) = 0$) or *fixed ends* ($\phi_j(l) = 0$), the first term in Eq. 120 is identically zero, and this equation reduces to

$$\int_0^l EA\phi_j'\phi_k'\,dx = \omega_j^2\int_0^l \rho A\phi_j\phi_k\,dx \tag{4.121}$$

Similarly, for the kth eigenfunction, we have

$$\int_0^l EA\phi_k'\phi_j'\,dx = \omega_k^2\int_0^l \rho A\phi_k\phi_j\,dx \tag{4.122}$$

Subtracting Eq. 122 from Eq. 121, one obtains

$$(\omega_j^2 - \omega_k^2)\int_0^l \rho A\phi_j\phi_k\,dx = 0 \tag{4.123a}$$

Assuming that ω_j and ω_k are distinct eigenvalues, that is, $\omega_j \neq \omega_k$, Eq. 123a yields for $j \neq k$

$$\left.\begin{aligned}\int_0^l \rho A\phi_j\phi_k\,dx &= 0\\ \int_0^l EA\phi_j'\phi_k'\,dx &= 0\end{aligned}\right\} \tag{4.123b}$$

and for $j = k$ we have

$$\left.\begin{aligned}\int_0^l \rho A\phi_j^2\,dx &= m_j\\ \int_0^l EA\phi_j'^2\,dx &= k_j\end{aligned}\right\} \tag{4.123c}$$

The coefficients m_j and k_j, $j = 1, 2, 3, \ldots$, are called, respectively, the *modal mass* and *modal stiffness* coefficients. They have, respectively, the units of mass and stiffness. It is also clear from Eq. 121 that

$$k_j = \omega_j^2 m_j, \qquad j = 1, 2, 3, \ldots \tag{4.124}$$

That is, the jth natural frequency ω_j is defined by

$$\omega_j^2 = \frac{k_j}{m_j} = \frac{\int_0^l EA\phi_j'^2\,dx}{\int_0^l \rho A\phi_j^2\,dx}, \qquad j = 1, 2, 3, \ldots \tag{4.125}$$

For *torsional systems* one can follow a similar procedure to show that the jth natural frequency of the torsional oscillations is given by

$$\omega_j^2 = \frac{k_j}{m_j} = \frac{\int_0^l GI_p\phi_j'^2\,dx}{\int_0^l \rho I_p\phi_j^2\,dx}, \qquad j = 1, 2, 3, \ldots \tag{4.126}$$

where G is the modulus of rigidity and I_p is the polar moment of inertia.

If the end conditions are not simple the definitions of the modal mass and stiffness coefficients given by Eq. 123c must be modified. To this end, the general relationship of Eq. 120 must be used to define the modal coefficients as demonstrated by the following example.

Example 4.7

Find the orthogonality relationships of the mode shapes of the longitudinal vibration of the rod shown in Fig. 12.

Solution. The boundary conditions for this system are

$$u(0, t) = 0$$

$$m\frac{\partial^2 u(l, t)}{\partial t^2} = -ku(l, t) - EA\frac{\partial u(l, t)}{\partial x}$$

These conditions yield

$$\phi(0) = 0$$

and

$$(k - \omega^2 m)\phi(l) = -EA\phi'(l)$$

The general orthogonality relationship of Eq. 120 can be written as

$$EA\phi_j'(l)\phi_k(l) - EA\phi_j'(0)\phi_k(0) - \int_0^l EA\phi_j'(x)\phi_k'(x)\,dx$$

$$= -\omega_j^2 \int_0^l \rho A\phi_j(x)\phi_k(x)\,dx$$

By using the boundary conditions of this example, this orthogonality relationship

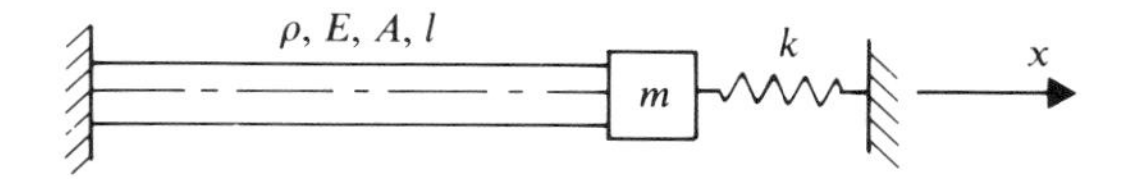

FIG. 4.12. Longitudinal vibration of bars.

can be written as

$$-(k - \omega_j^2 m)\phi_j(l)\phi_k(l) - \int_0^l EA\phi_j'(x)\phi_k'(x)\,dx = -\omega_j^2 \int_0^l \rho A\phi_j(x)\phi_k(x)\,dx$$

or

$$k\phi_j(l)\phi_k(l) + \int_0^l EA\phi_j'(x)\phi_k'(x)\,dx = \omega_j^2\left[m\phi_j(l)\phi_k(l) + \int_0^l \rho A\phi_j(x)\phi_k(x)\,dx\right]$$

Similar relationship can be obtained for mode k as

$$k\phi_j(l)\phi_k(l) + \int_0^l EA\phi_j'(x)\phi_k'(x)\,dx = \omega_k^2\left[m\phi_j(l)\phi_k(l) + \int_0^l \rho A\phi_j(x)\phi_k(x)\,dx\right]$$

Subtracting this equation from the one associated with mode j, we obtain the following orthogonality relationships for $j \neq k$:

$$m\phi_j(l)\phi_k(l) + \int_0^l \rho A\phi_j(x)\phi_k(x)\,dx = 0$$

$$k\phi_j(l)\phi_k(l) + \int_0^l EA\phi_j'(x)\phi_k'(x)\,dx = 0$$

and for $j = k$ we have

$$m\phi_j^2(l) + \int_0^l \rho A\phi_j^2(x)\,dx = m_j$$

$$k\phi_j^2(l) + \int_0^l EA\phi_j'^2(x)\,dx = k_j$$

where m_j and k_j are, respectively, the modal mass and stiffness coefficients. The jth natural frequency of the system can be defined as

$$\omega_j^2 = \frac{k_j}{m_j} = \frac{k\phi_j^2(l) + \int_0^l EA\phi_j'^2(x)\,dx}{m\phi_j^2(l) + \int_0^l \rho A\phi_j^2(x)\,dx}$$

If m and k approach zero, the natural frequency ω_j approaches the value obtained by Eq. 125 for the simple end conditions.

Transverse Vibration It was shown in Section 3 that the partial differential equation that governs the free transverse vibration of beams is given by

$$\rho A\frac{\partial^2 v}{\partial t^2} = -\frac{\partial^2}{\partial x^2}\left(EI_z\frac{\partial^2 v}{\partial x^2}\right) \tag{4.127a}$$

where $v = v(x, t)$ is the transverse displacement, ρ is the mass density, A is the cross-sectional area, E is the modulus of elasticity, and I_z is the moment of inertia of the cross section about the z-axis. The solution of this equation can be written using the separation of variables technique as

$$v = \phi(x)q(t) \tag{4.127b}$$

where $\phi(x)$ and $q(t)$ are, respectively, space- and time-dependent functions. The function $q(t)$ is defined as

$$q(t) = B_1 \sin \omega t + B_2 \cos \omega t$$

Substituting Eq. 127b into Eq. 127a yields

$$\omega^2 \rho A \phi(x) q(t) = (EI_z \phi''(x))'' \, q(t)$$

or

$$\omega^2 \rho A \phi(x) = (EI_z \phi''(x))''$$

Therefore, for the jth natural frequency ω_j, one has

$$(EI_z \phi_j'')'' = \omega_j^2 \rho A \phi_j$$

Multiplying this equation by ϕ_k and integrating over the length, we have

$$\int_0^l (EI_z \phi_j'')'' \phi_k \, dx = \omega_j^2 \int_0^l \rho A \phi_j \phi_k \, dx \qquad j = 1, 2, 3, \ldots \tag{4.128}$$

The integral on the left-hand side of this equation can be integrated by parts to yield

$$\int_0^l (EI_z \phi_j'')'' \phi_k \, dx = (EI_z \phi_j'')' \phi_k \Big|_0^l - EI_z \phi_j'' \phi_k' \Big|_0^l + \int_0^l EI_z \phi_j'' \phi_k'' \, dx$$

Substituting into Eq. 128, one obtains

$$(EI_z \phi_j'')' \phi_k \Big|_0^l - EI_z \phi_j'' \phi_k' \Big|_0^l + \int_0^l EI_z \phi_j'' \phi_k'' \, dx = \omega_j^2 \int_0^l \rho A \phi_j \phi_k \, dx, \qquad j = 1, 2, 3, \ldots \tag{4.129}$$

This is the general expression for the orthogonality condition of the eigenfunctions of the transverse vibration of beams.

One can show that if the beam has *simple end conditions* such as fixed ends, free ends, or simply supported ends, the orthogonality condition of Eq. 129 reduces to

$$\int_0^l EI_z \phi_j'' \phi_k'' \, dx = \omega_j^2 \int_0^l \rho A \phi_j \phi_k \, dx \tag{4.130}$$

Similarly, for the kth natural frequency ω_k, we have

$$\int_0^l EI_z\phi_j''\phi_k''\,dx = \omega_k^2\int_0^l \rho A\phi_j\phi_k\,dx \tag{4.131}$$

Subtracting Eq. 131 from Eq. 130, we obtain the following relatinships for the simple end conditions in the case $j \neq k$:

$$\left.\begin{aligned} \int_0^l \rho A\phi_j\phi_k\,dx = 0 \\ \int_0^l EI_z\phi_j''\phi_k''\,dx = 0 \end{aligned}\right\} \tag{4.132}$$

and for $j = k$

$$\left.\begin{aligned} \int_0^l \rho A\phi_j^2\,dx = m_j \\ \int_0^l EI_z\phi_j''^2\,dx = k_j \end{aligned}\right\} \tag{4.133}$$

where m_j and k_j are, respectively, the modal mass and modal stiffness coefficients which from Eq. 130 are related by

$$k_j = \omega_j^2 m_j \tag{4.134}$$

or

$$\omega_j^2 = \frac{k_j}{m_j} = \frac{\int_0^l EI_z\phi_j''^2\,dx}{\int_0^l \rho A\phi_j^2\,dx} \tag{4.135}$$

If the end conditions are not simple, Eq. 129 can still be used to define the modal mass and stiffness coefficients as demonstrated by the following example.

Example 4.8

Obtain the orthogonality relationships of the transverse vibration of the beam shown in Fig. 13.

Solution. The boundary conditions for this system are

$$v(0, t) = 0, \qquad \frac{\partial v(0, t)}{\partial x} = 0$$

$$m\frac{\partial^2 v(l, t)}{\partial t^2} = (EI_z v''(l, t))' - kv$$

$$EI_z v''(l, t) + k_t v'(l, t) = 0$$

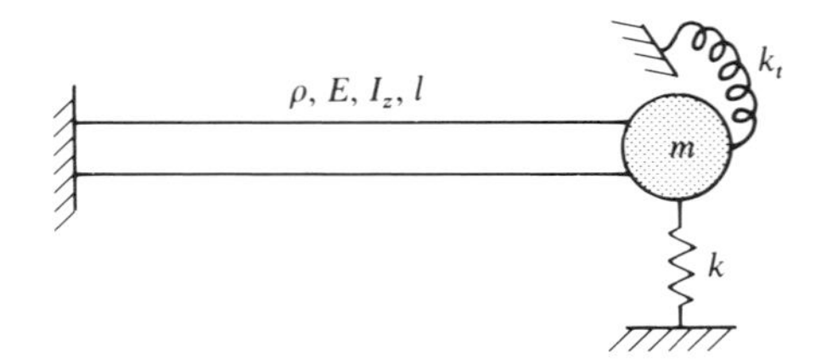

FIG. 4.13. Transverse vibration.

By substituting these boundary conditions in the general orthogonality relationship given by Eq. 129, one can verify that the orthogonality relationships for this example are given for $j \neq k$ by

$$\int_0^l \rho A \phi_j \phi_k \, dx + m\phi_j(l)\phi_k(l) = 0$$

$$\int_0^l EI_z \phi_j'' \phi_k'' \, dx + k\phi_j(l)\phi_k(l) + k_t \phi_j'(l)\phi_k'(l) = 0$$

and for $j = k$

$$\int_0^l \rho A \phi_j^2 \, dx + m\phi_j^2(l) = m_j$$

$$\int_0^l EI_z \phi_j''^2 \, dx + k\phi_j^2(l) + k_t \phi_j'^2(l) = k_j$$

where m_j and k_j, $j = 1, 2, 3, \ldots$, are, respectively, the modal mass and stiffness coefficients. Therefore, the jth natural frequency ω_j is defined as

$$\omega_j^2 = \frac{k_j}{m_j} = \frac{\int_0^l EI_z \phi_j''^2 \, dx + k\phi_j^2(l) + k_t \phi_j'^2(l)}{\int_0^l \rho A \phi_j^2 \, dx + m\phi_j^2(l)}$$

If m, k, and k_t are equal to zero, the natural frequency ω_j becomes the same as the natural frequency defined by Eq. 135 for the cases of simple boundary conditions.

Rigid-Body Modes Unrestrained continuous systems have rigid-body modes. An example is the beam with free ends discussed in Example 5. It was shown that the first natural frequency of the system is equal to zero and corresponds to the rigid-body mode

$$\phi_1(x) = c_1 + c_2 x$$

where c_1 and c_2 are constants. The expression of the rigid-body mode ϕ_1 is the sum of a translational and rotational motions. The rigid-body translation is represented by the constant c_1, while the contribution of the rigid-body rotation is represented by $c_2 = \phi_1'(x)$. In the case of pure translation $c_2 = 0$, while in the case of pure rotation $c_1 = 0$. The general displacement of the beam, however, is the combination of the rigid-body and deformation modes. A rigid-body mode does not contribute to the change in the system potential energy. In order to demonstrate this fact, we use the orthogonality condition of Eq. 133 which is applicable to the case of simple end conditions. The modal stiffness coefficient associated with the rigid-body mode of a beam with free ends is defined as

$$\begin{aligned} k_1 &= \int_0^l EI_z \phi_j''^2 \, dx \\ &= \int_0^l EI_z \frac{\partial^2}{\partial x^2}(c_1 + c_2 x) \, dx \end{aligned}$$

where E, I_z, and l are, respectively, the modulus of elasticity, the second moment of area of the cross section, and the length of the beam. Clearly, the modal stiffness coefficient associated with the rigid-body mode of the beam is equal to zero, that is, if the beam moves as a rigid body there is no change in the strain energy. One, however, can show using the orthogonality conditions, that the modal mass coefficient associated with a rigid-body mode is not equal to zero, since there can be no rigid-body motion of the beam with zero kinetic energy. Since the square of a natural frequency is defined to be the modal stiffness coefficient divided by the modal mass coefficient, the natural frequency associated with a rigid-body mode is identically equal to zero. In analogy with the definitions used in the analysis of multi-degree of freedom systems, a continuous system which has a rigid-body mode is said to be *positive semidefinite*, while a continuous system which does not have a rigid-body mode is said to be *positive definite.* Observe from the orthogonality conditions presented in this section for the cases of longitudinal, torsional, and transverse vibrations, that the modal mass and stiffness coefficients must be nonnegative numbers. In the case of a positive-definite system, both the kinetic and strain energies must be positive for nonzero displacements, and equal to zero only when the displacements are equal to zero. In the case of a positive-semidefinite system, the kinetic energy of the system must be greater than zero for nonzero velocities and equal to zero only when the velocities are zero. The strain energy of a positive-semidefinite system on the other hand is positive but it can be equal to zero for a nonzero rigid-body displacement.

4.5 FORCED VIBRATIONS

In this section, we use an analytical approach for developing the differential equations of the forced vibrations of continuous systems. As shown in the preceding sections, the vibrations of continuous systems are governed by partial differential equations expressed in terms of variables that are space and time dependent. In this section, the orthogonality of the eigenfunctions (mode shapes) are used to convert the partial differential equation to an infinite number of uncoupled second-order ordinary differential equations expressed in terms of the modal coordinates. These equations are similar to the equations that govern the vibration of single degree of freedom systems.

Longitudinal Vibration It was shown in Section 1 that the partial differential equation that governs the longitudinal forced vibration of rods is given by

$$\rho A \frac{\partial^2 u}{\partial t^2} = \frac{\partial}{\partial x}\left(EA \frac{\partial u}{\partial x}\right) + F(x, t) \tag{4.136}$$

where ρ, A, and E are, respectively, the mass density, cross-sectional area, and modulus of elasticity, $u = u(x, t)$ is the longitudinal displacement, and $F(x, t)$ is a space- and time-dependent axial forcing function.

Using the technique of the separation of variables, the displacement u can be written as

$$u(x, t) = \sum_{j=1}^{\infty} \phi_j(x) q_j(t) \tag{4.137}$$

where ϕ_j is the jth space-dependent eigenfunction (mode shape) and q_j is the time-dependent modal coordinate. A virtual change in the longitudinal displacement u is

$$\delta u = \sum_{j=1}^{\infty} \phi_j \delta q_j \tag{4.138}$$

Multiplying Eq. 136 by δu and integrating over the length of the beam leads to

$$\int_0^l \rho A \frac{\partial^2 u}{\partial t^2} \delta u \, dx = \int_0^l \frac{\partial}{\partial x}\left(EA \frac{\partial u}{\partial x}\right) \delta u \, dx + \int_0^l F(x, t) \delta u \, dx \tag{4.139}$$

Substituting Eqs. 137 and 138 into Eq. 139 yields

$$\sum_{j=1}^{\infty} \sum_{k=1}^{\infty} \int_0^l [\rho A \phi_k(x) \phi_j(x) \ddot{q}_k - (EA\phi_k'(x))' \phi_j(x) q_k - Q_j] \, dx \, \delta q_j = 0 \tag{4.140}$$

where

$$Q_j = F(x, t) \phi_j(x) \tag{4.141}$$

By using the integration by parts, one has

$$\int_0^l (EA\phi_k'(x))' \phi_j(x) \, dx = EA\phi_k'(x) \phi_j(x) \Big|_0^l - \int_0^l EA\phi_k'(x) \phi_j'(x) \, dx \tag{4.142}$$

Substituting this equation into Eq. 140 yields

$$\sum_{j=1}^{\infty} \left[\sum_{k=1}^{\infty} \int_0^l [\rho A \phi_k(x) \phi_j(x) \ddot{q}_k + EA\phi_k'(x) \phi_j'(x) q_k - Q_j] \, dx \right.$$

$$\left. - EA\phi_k'(x) \phi_j(x) \Big|_0^l q_k \right] \delta q_j = 0 \tag{4.143}$$

Using the boundary conditions and the orthogonality relationship of the eigenfunctions, one can show in general that

$$\sum_{k=1}^{\infty}\left[\int_0^l (\rho A\phi_k(x)\phi_j(x)\ddot{q}_k + EA\phi_k'(x)\phi_j'(x)q_k)\,dx - EA\phi_k'(x)\phi_j(x)\Big|_0^l q_k\right] = m_j\ddot{q}_j + k_jq_j \tag{4.144}$$

where m_j and k_j are, respectively, the modal mass and stiffness coefficients that depend on the boundary conditions and can be defined using the orthogonality relationships of the eigenfunctions. One may substitute Eq. 144 into Eq. 143. This yields

$$\sum_{j=1}^{\infty}[m_j\ddot{q}_j + k_jq_j - Q_j]\delta q_j = 0 \tag{4.145}$$

Since the virtual changes δq_j are linearly independent, Eq. 145 yields

$$m_j\ddot{q}_j + k_jq_j = Q_j, \qquad j = 1, 2, 3, \ldots \tag{4.146}$$

These are uncoupled second-order ordinary differential equations which are in the same form as the vibration equations of the single degree of freedom systems. Therefore, their solution can be obtained using Duhamel's integral as

$$q_j = q_{j0}\cos\omega_jt + \frac{\dot{q}_{j0}}{\omega_j}\sin\omega_jt + \frac{1}{m_j\omega_j}\int_0^\tau Q_j(\tau)\sin\omega_j(t-\tau)\,d\tau, \qquad j = 1, 2, 3, \ldots \tag{4.147}$$

where q_{j0} and $\dot{q}_{j0}$ are the initial modal displacements and velocities and ω_j is the jth natural frequency defined as

$$\omega_j = \sqrt{\frac{k_j}{m_j}} \qquad j = 1, 2, 3, \ldots$$

Having determined q_j using Eq. 147, the longitudinal displacement $u(x, t)$ can be determined by using Eq. 137.

In the case of free vibration, the modal force Q_j is equal to zero and Eq. 146 reduces to

$$m_j\ddot{q}_j + k_jq_j = 0, \qquad j = 1, 2, 3, \ldots$$

The solution of these equations is

$$q_j = q_{j0}\cos\omega_jt + \frac{\dot{q}_{j0}}{\omega_j}\sin\omega_jt, \qquad j = 1, 2, 3, \ldots$$

Example 4.9

If the rod in Example 7 is subjected to a distributed axial force of the form $F(x, t)$, determine the equations of forced longitudinal vibration of this system.

Solution. The boundary conditions in this example are

$$u(0, t) = 0$$

$$m\frac{\partial^2 u}{\partial t^2} = -ku(l, t) - EA\frac{\partial u(l, t)}{\partial x}$$

which yield

$$\phi(0) = 0$$

$$EA\phi'(l)q(t) = -k\phi(l)q(t) - m\phi(l)\ddot{q}(t)$$

That is,

$$\phi_k(0)\phi_j(0) = 0$$

$$EA\phi_k'(l)\phi_j(l)q_k(t) = -k\phi_k(l)\phi_j(l)q_k(t) - m\phi_k(l)\phi_j(l)\ddot{q}_k(t)$$

The last term on the left-hand side of Eq. 144 can be written as

$$EA\phi_k'(x)\phi_j(x)\Big|_0^l q_k = (EA\phi_k'(l)\phi_j(l) - EA\phi_k'(0)\phi_j(0))q_k$$

which upon using the boundary conditions yields

$$EA\phi_k'(x)\phi_j(x)\Big|_0^l q_k = -k\phi_k(l)\phi_j(l)q_k - m\phi_k(l)\phi_j(l)\ddot{q}_k$$

Therefore, Eq. 144 can be written as

$$\sum_{k=1}^{\infty}\int_0^l (\rho A\phi_k(x)\phi_j(x)\ddot{q}_k + EA\phi_k'(x)\phi_j'(x)q_k)\,dx + k\phi_k(l)\phi_j(l)q_k + m\phi_k(l)\phi_j(l)\ddot{q}_k$$

$$= \sum_{k=1}^{\infty}\left[m\phi_k(l)\phi_j(l) + \int_0^l \rho A\phi_k(x)\phi_j(x)\,dx\right]\ddot{q}_k$$

$$+ \sum_{k=1}^{\infty}\left[k\phi_k(l)\phi_j(l) + \int_0^l EA\phi_k'(x)\phi_j'(x)\,dx\right]q_k$$

Comparing this equation with the orthogonality relationships obtained in Example 7, it is clear that

$$\sum_{k=1}^{\infty}\left[m\phi_k(l)\phi_j(l) + \int_0^l \rho A\phi_k(x)\phi_j(x)\,dx\right]\ddot{q}_k$$

$$+ \sum_{k=1}^{\infty}\left[k\phi_k(l)\phi_j(l) + \int_0^l EA\phi_k'(x)\phi_j'(x)\,dx\right]q_k = m_j\ddot{q}_j + k_jq_j$$

where m_j and k_j are the modal mass and stiffness coefficients defined by

$$m_j = m\phi_j^2(l) + \int_0^l \rho A\phi_j^2(x)\,dx$$

$$k_j = k\phi_k^2(l) + \int_0^l EA\phi_j'^2(x)\,dx, \qquad j = 1, 2, 3, \ldots$$

Therefore, the equations of motion of the forced longitudinal vibration of the rod, expressed in the modal coordinates, are

$$m_j \ddot{q}_j + k_j q_j = Q_j, \qquad j = 1, 2, 3, \ldots$$

Concentrated Loads If the force F is a concentrated load that acts at a point p on the beam, that is, $F = F(t)$, there is no need to carry out integration in order to obtain the modal forces. In this case the virtual work of this force is given by

$$\begin{aligned} \delta W &= F(t)\delta u(x_p, t) \\ &= F(t) \sum_{j=1}^{\infty} \phi_j(x_p)\delta q_j(t) \\ &= \sum_{j=1}^{\infty} Q_j \delta q_j(t) \end{aligned}$$

where Q_j, $j = 1, 2, 3, \ldots$, is the modal force associated with the jth modal coordinate and defined as

$$Q_j = F(t)\phi_j(x_p)$$

in which $\phi_j(x_p)$ is the jth mode shape evaluated at point p on the beam.

Initial Conditions As in the case of the multi-degree of freedom systems, the orthogonality conditions can be utilized in determining the initial conditions q_{jo} and $\dot{q}_{jo}$, $j = 1, 2, 3, \ldots$. Let $f(x, 0)$ and $g(x, 0)$ be, respectively, the initial displacements and velocities of the rod, that is,

$$f(x, 0) = u(x, 0) = \sum_{j=1}^{\infty} \phi_j(x) q_{jo} \tag{4.148}$$

$$g(x, 0) = \dot{u}(x, 0) = \sum_{j=1}^{\infty} \phi_j(x) \dot{q}_{jo} \tag{4.149}$$

Multiplying these equations by $\rho A \phi_k$ and integrating over the length yields

$$\int_0^l \rho A \phi_k(x) f(x, 0)\, dx = \sum_{j=1}^{\infty} \left[\int_0^l \rho A \phi_k(x) \phi_j(x)\, dx \right] q_{jo} \tag{4.150}$$

$$\int_0^l \rho A \phi_k(x) g(x, 0)\, dx = \sum_{j=1}^{\infty} \left[\int_0^l \rho A \phi_k(x) \phi_j(x)\, dx \right] \dot{q}_{jo} \tag{4.151}$$

These two equations can be used to determine the initial modal displacements and velocities. To this end, the orthogonality conditions may be utilized. For example, in the case of simple boundary conditions, one has

$$\int_0^l \rho A \phi_k(x) \phi_j(x)\, dx = \begin{cases} 0 & \text{if} \quad j \neq k \\ m_j & \text{if} \quad j = k \end{cases}$$

Substituting this orthogonality relationship into Eqs. 150 and 151, we obtain

$$q_{jo} = \frac{1}{m_j} \int_0^l \rho A \phi_j(x) f(x, 0)\, dx$$

$$\dot{q}_{jo} = \frac{1}{m_j} \int_0^l \rho A \phi_j(x) g(x, 0)\, dx \qquad j = 1, 2, 3, \ldots$$

If the end conditions are not simple, Eqs. 150 and 151 can still be used to solve for the initial modal displacements and velocities. For instance, if we consider the problem of Example 7, one of the orthogonality conditions was given by

$$m\phi_j(l)\phi_k(l) + \int_0^l \rho A \phi_j(x)\phi_k(x)\, dx = \begin{cases} 0 & \text{if } j \neq k \\ m_j & \text{if } j = k \end{cases}$$

where m is the concentrated mass attached to the end of the rod. The preceding equation may be written as

$$\sum_{j=1}^{\infty} \left[m\phi_j(l)\phi_k(l) q_j + \left(\int_0^l \rho A \phi_j(x)\phi_k(x)\, dx \right) q_j \right] = m_k q_k$$

which can be rewritten as

$$m\phi_k(l) u(l, t) + \sum_{j=1}^{\infty} \left(\int_0^l \rho A \phi_j(x)\phi_k(x)\, dx \right) q_j = m_k q_k$$

which at $t = 0$ leads to

$$\sum_{j=1}^{\infty} \left[\int_0^l \rho A \phi_j(x)\phi_k(x)\, dx \right] q_{jo} = m_k q_{ko} - m\phi_k(l) f(l, 0)$$

$$\sum_{j=1}^{\infty} \left[\int_0^l \rho A \phi_j(x)\phi_k(x)\, dx \right] \dot{q}_{jo} = m_k \dot{q}_{ko} - m\phi_k(l) g(l, 0)$$

Substituting these two equations into Eqs. 150 and 151, one obtains

$$q_{ko} = \frac{1}{m_k} \left[\int_0^l \rho A \phi_k(x) f(x, 0)\, dx + m\phi_k(l) f(l, 0) \right]$$

$$\dot{q}_{ko} = \frac{1}{m_k} \left[\int_0^l \rho A \phi_k(x) g(x, 0)\, dx + m\phi_k(l) g(l, 0) \right], \qquad k = 1, 2, 3, \ldots$$

Torsional Vibration The equation for the forced torsional vibration takes a similar form to the equation of forced longitudinal vibration. This equation was given in Section 2 as

$$\rho I_p \frac{\partial^2 \theta}{\partial t^2} = \frac{\partial}{\partial x} \left(G I_p \frac{\partial \theta}{\partial x} \right) + T_e(x, t) \tag{4.152}$$

where ρ, I_p, and G are, respectively, the mass density, polar moment of inertia, and modulus of rigidity, $\theta = \theta(x, t)$ is the angle of torsional oscillation, and

$T_e(x, t)$ is the external torque which is time and space dependent. The solution of Eq. 152 can be expressed using the separation of variables technique as

$$\theta(x, t) = \sum_{j=1}^{\infty} \phi_j(x) q_j(t) \tag{4.153}$$

Following the same procedure as in the case of longitudinal vibration, one can show that the equations of the torsional vibration of the shaft in terms of the modal coordinates are given by

$$m_j \ddot{q}_j + k_j q_j = Q_j, \qquad j = 1, 2, 3, \ldots \tag{4.154}$$

where m_j and k_j are, respectively, the modal mass and stiffness coefficients and

$$Q_j = \int_0^l T_e(x, t) \phi_j(x)\, dx \tag{4.155}$$

The solution of Eq. 154 is defined by Eq. 147. Therefore, the mathematical treatment of the linear torsional oscillations of shafts is the same as the one used for the longitudinal vibration of rods.

Transverse Vibration The partial differential equation of the forced transverse vibration of the beams was given by Eq. 89 as

$$\rho A \frac{\partial^2 v}{\partial t^2} + \frac{\partial^2}{\partial x^2}\left(E I_z \frac{\partial^2 v}{\partial x^2} \right) = F(x, t) \tag{4.156}$$

where ρ, A, I_z, and E are, respectively, the mass density, cross-sectional area, second moment of area, and modulus of elasticity, $v(x, t)$ is the transverse displacement, and $F(x, t)$ is the forcing function which may depend on the spatial coordinate x and time t. By using the technique of separation of variables, we may write the transverse displacement v as

$$v = \sum_{j=1}^{\infty} \phi_j(x) q_j(t) \tag{4.157}$$

Multiplying both sides of Eq. 156 by the virtual displacement δv and integrating over the length of the beam we obtain

$$\int_0^l \left[\rho A \frac{\partial^2 v}{\partial t^2} \delta v + \frac{\partial^2}{\partial x^2}\left(E I_z \frac{\partial^2 v}{\partial x^2} \right) \delta v \right] dx = \int_0^l F(x, t) \delta v\, dx$$

which upon using Eq. 157 yields

$$\sum_{j=1}^{\infty} \sum_{k=1}^{\infty} \int_0^l [\rho A \phi_k(x) \phi_j(x) \ddot{q}_k + (E I_z \phi_k''(x))'' \phi_j(x) q_k] \delta q_j\, dx$$
$$= \sum_{j=1}^{\infty} \int_0^l F(x, t) \phi_j(x) \delta q_j\, dx \tag{4.158}$$

Integration by parts yields

$$\int_0^l (EI_z\phi_k''(x))''\phi_j(x)\,dx = (EI_z\phi_k'')'\phi_j\Big|_0^l - EI_z\phi_k''\phi_j'\Big|_0^l + \int_0^l EI_z\phi_k''\phi_j''\,dx \quad (4.159)$$

Substituting this equation into Eq. 158 and using the boundary conditions and the orthogonality relationships of the eigenfunctions, one gets

$$\sum_{j=1}^{\infty} [m_j\ddot{q}_j + k_jq_j - Q_j]\delta q_j = 0 \quad (4.160)$$

where m_j and k_j are the modal mass and modal stiffness coefficients that depend on the boundary conditions and can be defined using the orthogonality of the mode shapes, and Q_j is the modal forcing function defined as

$$Q_j = \int_0^l F(x, t)\phi_j(x)\,dx \quad (4.161)$$

Since the modal coordinates $q_j, j = 1, 2, 3, \ldots$, are independent, Eq. 160 yields the following uncoupled second-order ordinary differential equations:

$$m_j\ddot{q}_j + k_jq_j = Q_j, \qquad j = 1, 2, 3, \ldots \quad (4.162)$$

These equations are in a form similar to the one obtained in the preceding section for the two cases of longitudinal and torsional vibrations.

Example 4.10

For the system given in Example 8 obtain the uncoupled second-order differential equations of transverse vibration.

Solution. The boundary conditions for this system are

$$v(0, t) = 0, \qquad \frac{\partial v(0, t)}{\partial x} = 0$$

$$m\frac{\partial^2 v(l, t)}{\partial t^2} = (EI_z v''(l, t))' - kv(l, t)$$

and

$$EI_z v''(l, t) + k_t v'(l, t) = 0$$

which yield

$$\phi(0) = 0, \qquad \phi'(0) = 0$$

$$m\phi(l)\ddot{q}(t) = (EI_z\phi''(l))'q(t) - k\phi(l)q(t)$$

$$EI_z\phi''(l)q(t) + k_t\phi'(l)q(t) = 0$$

Substituting the boundary conditions at the fixed end into Eq. 159 yields

$$\int_0^l (EI_z\phi_k''(x))''\phi_j(x)\,dx = (EI_z\phi_k'')'\phi_j\Big|_0^l - EI_z\phi_k''\phi_j'\Big|_0^l + \int_0^l EI_z\phi_k''\phi_j''\,dx$$

$$= (EI_z\phi_k''(l))'\phi_j(l) - (EI_z\phi_k''(0))'\phi_j(0) - EI_z\phi_k''(l)\phi_j'(l) + EI_z\phi_k''(0)\phi_j'(0) + \int_0^l EI_z\phi_k''\phi_j''\,dx$$

$$= (EI_z\phi_k''(l))'\phi_j(l) - EI_z\phi_k''(l)\phi_j'(l) + \int_0^l EI_z\phi_k''\phi_j''\,dx$$

Multiplying this equation by $q_k(t)$ and use the remaining boundary conditions, we obtain

$$\int_0^l (EI_z\phi_k''(x))''\phi_j(x)\,dxq_k = (EI_z\phi_k''(l))'\phi_j(l)q_k - EI_z\phi_k''(l)\phi_j'(l)q_k + \int_0^l EI_z\phi_k''\phi_j''\,dxq_k$$

$$= m\phi_k(l)\phi_j(l)\ddot{q}_k(t) + k\phi_k(l)\phi_j(l)q_k(t) + k_t\phi_k'(l)\phi_j'(l)q_k(t) + \int_0^l EI_z\phi_k''\phi_j''\,dxq_k(t)$$

Substituting this equation into Eq. 158 yields

$$\sum_{j=1}^{\infty}\left\{\sum_{k=1}^{\infty}\int_0^l [\rho A\phi_k(x)\phi_j(x)\ddot{q}_k(t) + EI_z\phi_k''\phi_j''q_k(t)]\,dx + m\phi_k(l)\phi_j(l)\ddot{q}_k(t) + k\phi_k(l)\phi_j(l)q_k(t) + k_t\phi_k'(l)\phi_j'(l)q_k(t)\right\}\delta q_j = \sum_{j=1}^{\infty}\int_0^l F(x,t)\phi_j(x)\,dx\,\delta q_j$$

By using the orthogonality relationships obtained in Example 8, the preceding equation can be written as

$$\sum_{j=1}^{\infty}[m_j\ddot{q}_j + k_jq_j - Q_j]\delta q_j = 0$$

or

$$m_j\ddot{q}_j + k_jq_j = Q_j, \qquad j = 1, 2, 3, \dots$$

where

$$m_j = \int_0^l \rho A\phi_j^2(x)\,dx + m\phi_j^2(l)$$

$$k_j = \int_0^l EI_z\phi_j''^2(x)\,dx + k\phi_j^2(l) + k_t\phi_j'^2(l)$$

$$Q_j = \int_0^l F(x,t)\phi_j(x)\,dx$$

4.6 INHOMOGENEOUS BOUNDARY CONDITIONS

The solution of the vibration equations of continuous systems subject to homogeneous boundary conditions, which are not time dependent, was obtained and examined in the preceding sections. In this section, we discuss the case of inhomogeneous, time-dependent boundary conditions. To this end, we consider the partial differential equation that governs the longitudinal vibration of bars and is given by

$$\frac{\partial^2 u}{\partial t^2} - c^2 \frac{\partial^2 u}{\partial x^2} = F(x, t) \tag{4.163}$$

subject to the initial conditions

$$u(x, 0) = f(x), \qquad u_t(x, 0) = g(x) \tag{4.164}$$

and the inhomogeneous boundary conditions

$$\left.\begin{aligned} a_1 u(0, t) - b_1 u'(0, t) = h_1(t) \\ a_2 u(l, t) + b_1 u'(l, t) = h_2(t) \end{aligned}\right\} \tag{4.165}$$

where c, a_1, a_2, b_1, and b_2 are constants and h_1 and h_2 are given functions.

In order to solve Eq. 163 subject to the inhomogeneous boundary conditions of Eq. 165, the problem may be reduced to one with homogeneous boundary conditions. This can be achieved by introducing an auxiliary assumed function $W(x, t)$ which is twice differentiable and satisfies the boundary conditions of Eq. 165. It is not necessary, however, that the function W satisfies the partial differential equation. In terms of the new function W, the function $u(x, t)$ can be written as

$$u(x, t) = W(x, t) + Z(x, t) \tag{4.166}$$

where $Z(x, t)$ is an unknown function which will be determined. Substituting Eq. 166 into Eqs. 163–165 leads to

$$\ddot{Z} - c^2 Z'' = F(x, t) - \ddot{W} + c^2 W'' \tag{4.167}$$

subject to the initial conditions

$$\left.\begin{aligned} Z(x, 0) = f(x) - W(x, 0) \\ \dot{Z}(x, 0) = g(x) - \dot{W}(x, 0) \end{aligned}\right\} \tag{4.168}$$

and the boundary conditions

$$\left.\begin{aligned} a_1 Z(0, t) - b_1 Z'(0, t) = 0 \\ a_2 Z(l, t) + b_2 Z'(l, t) = 0 \end{aligned}\right\} \tag{4.169}$$

Observe that $Z(x, t)$ satisfies the new inhomogeneous partial differential equation given by Eq. 167 subject to the homogeneous boundary conditions given by Eq. 169.

The method described in this section can be used to solve a large class of vibration problems in which one point on the continuous media is subjected to a specified excitation. This is demonstrated by the following example.

Example 4.11

Consider the longitudinal vibration of a bar with length l. One end of the bar is assumed to be fixed, while the other end has a specified sinusoidal motion. If the initial conditions are assumed to be zero, find the series solution of the vibration equation of this system.

Solution. The partial differential equation of the system is

$$u_{tt} = c^2 u_{xx}, \qquad 0 < x < l, \quad t > 0$$

subject to the inhomogeneous boundary conditions

$$u(0, t) = 0$$
$$u(l, t) = \sin \omega t$$

and the initial conditions

$$u(x, 0) = 0$$
$$\dot{u}(x, 0) = 0$$

Observe that the function

$$W = W(x) \sin \omega t$$

is twice differentiable and satisfies the inhomogeneous boundary conditions. Let $\omega = kc$ for some constant k. This choice of the function W satisfies the differential equation as well as the boundary conditions. Clearly,

$$\frac{d^2 W}{dx^2} + k^2 W = 0, \qquad W(0) = 0, \qquad W(l) = 1$$

which implies that

$$W(x) = \frac{\sin kx}{\sin kl}$$

and accordingly

$$W(x, t) = W(x) \sin \omega t$$
$$= \frac{\sin kx}{\sin kl} \sin kct$$

The function $u(x, t)$ can then be written as

$$u(x, t) = Z(x, t) + W(x, t)$$

where $Z(x, t)$ is an unknown twice differentiable function. Substituting the preceding equation into the partial differential equation, one obtains the following partial differential equation:

$$\ddot{Z} - c^2 Z'' = 0$$

subject to the boundary conditions

$$Z(0, t) = 0, \qquad Z(l, t) = 0$$

and the initial conditions

$$Z(x, 0) = -W(x, 0) = 0$$
$$\dot{Z}(x, 0) = -\dot{W}(x, 0) = \frac{-kc \sin kx}{\sin kl}$$

Note that in this case the boundary conditions are homogeneous, and also the right-hand side of the resulting partial differential equation is zero because the selected function $W(x, t)$ satisfies the original partial differential equation. The function $W(x, t)$ does not have to satisfy this condition which leads in this example to a simplified equation. The new partial differential equation in Z can be solved by the methods described in the preceding sections. The solution u can then be obtained using Eq. 166, and it is left to the reader to verify that

$$u(x, t) = \frac{\sin kx}{\sin kl} \sin \omega t + \frac{2c\omega}{l} \sum_{n=1}^{\infty} \frac{(-1)^{n+1}}{\omega^2 - (n\pi c/l)^2} \sin \frac{n\pi x}{l} \sin \frac{n\pi ct}{l}$$

4.7 VISCOELASTIC MATERIALS

In the case of undamped free and forced vibration of continuous systems, the effect of discrete damping elements as well as structural damping is not considered. It is, however, clear at this point that the effect of external damping as the result of discrete damping elements such as dashpots can be introduced to the vibration equations by developing an expression for the virtual work of the damping forces. Internal damping, however, is present in most materials as the result of friction between the particles. It has been observed that the amplitude of free vibration of a solid specimen decreases with time even in the cases in which the specimen is isolated from any source of external damping. Such materials are called *viscoelastic*. There are various models for describing the behavior of viscoelastic materials. The simplest model is called the *Kelvin–Voigt* model in which the stress acting on the body is assumed to be proportional to the strain and its time derivative, that is,

$$\sigma = E\varepsilon + \gamma\dot{\varepsilon} \tag{4.170}$$

where σ and ε are, respectively, the stress and strain and E and γ are constants that depend on the properties of the material.

In order to demonstrate the use of the simple Kelvin–Voigt model in the vibration analysis of viscoelastic continuous systems, we consider the vibration of a uniform rod of length l and cross-sectional area A. The elastic force acting on the rod can be written as

$$P = \sigma A \tag{4.171}$$

The strain displacement relationship for this simple example is given by

$$\varepsilon = \frac{\partial u}{\partial x} \tag{4.172}$$

By using this equation, the stress of Eq. 170 can be expressed in terms of the displacement as

$$\sigma = E\frac{\partial u}{\partial x} + \gamma \frac{\partial^2 u}{\partial x \partial t} \tag{4.173}$$

Substituting this equation into Eq. 171 one obtains

$$P = EA\frac{\partial u}{\partial x} + \gamma A\frac{\partial^2 u}{\partial x \partial t} \tag{4.174}$$

Substituting this expression for P into Eq. 1 of Section 1, one obtains

$$\rho A\frac{\partial^2 u}{\partial t^2} = \frac{\partial}{\partial x}\left(EA\frac{\partial u}{\partial x} + \gamma A\frac{\partial^2 u}{\partial x \partial t}\right) + F(x, t) \tag{4.175}$$

where ρ is the mass density and $F(x, t)$ is the external force. If the cross-sectional area is assumed to be constant, Eq. 175 reduces to

$$\frac{\partial^2 u}{\partial t^2} = c^2\frac{\partial^2 u}{\partial x^2} + b^2\frac{\partial^3 u}{\partial x^2 \partial t} + F(x, t)/\rho A \tag{4.176}$$

where

$$c = \sqrt{\frac{E}{\rho}}, \qquad b = \sqrt{\frac{\gamma}{\rho}}$$

If $\gamma = 0$, Eq. 176 reduces to the same equation obtained for the undamped longitudinal vibrations of rods.

If the external forces are absent, Eq. 176 reduces to

$$\frac{\partial^2 u}{\partial t^2} = c^2\frac{\partial^2 u}{\partial x^2} + b^2\frac{\partial^3 u}{\partial x^2 \partial t} \tag{4.177}$$

A solution of this equation can be obtained using the separation of variables technique. To this end, we assume a solution of the form

$$u(x, t) = \phi(x)q(t) \tag{4.178}$$

Substituting this equation into Eq. 177 leads to

$$\begin{aligned}\phi\ddot{q} &= c^2\phi''q + b^2\phi''\dot{q}\\ &= (c^2 q + b^2\dot{q})\phi''\end{aligned}$$

This equation can be rewritten as

$$\frac{\ddot{q}}{c^2 q + b^2\dot{q}} = \frac{\phi''}{\phi}$$

Since the left-hand side of this equation depends on t only, and the right-hand side depends on x only, one must have

$$\frac{\ddot{q}}{c^2 q + b^2\dot{q}} = \frac{\phi''}{\phi} = -\beta^2 \tag{4.179}$$

where β^2 is a constant. One therefore has

$$\ddot{q} + 2\xi\omega\dot{q} + \omega^2 q = 0 \tag{4.180}$$

$$\phi'' + \beta^2\phi = 0 \tag{4.181}$$

where

$$\omega = c\beta \tag{4.182}$$

$$\xi = \frac{b^2\beta}{2c} \tag{4.183}$$

The solution of Eqs. 180 and 181 in the case of lightly damped system can be expressed as

$$q = X_1 e^{-\xi\omega t}\sin(\omega_d t + \psi_1) \tag{4.184}$$

$$\phi = X_2 \sin(\beta x + \psi_2) \tag{4.185}$$

where $\omega_d = \omega\sqrt{1-\xi^2}$ and X_1, ψ_1, X_2, and ψ_2 are constants which can be determined using the boundary and initial conditions as described in the preceding sections. The use of this procedure leads to the definition of the eigenfunctions ϕ_j. The general solution can then be written as the sum of the normal modes as

$$u(x, t) = \sum_{j=1}^{\infty} \phi_j(x) X_{1j} e^{-\xi_j\omega_j t}\sin(\omega_{dj}t + \psi_{1j}) \tag{4.186}$$

where subscript j refers to the jth mode of vibration.

Wave motion One may observe that each term in Eq. 186 is the product of two harmonic functions, one of them depends on x only, whilst the other depends on t only. Using the definition of $\phi(x)$ given by Eq. 185, each term in the series of Eq. 186 can be written as

$$u_j(x, t) = X_{1j}X_{2j}e^{-\xi_j\omega_j t}\sin(\beta_j x + \psi_{2j})\sin(\omega_{dj}t + \psi_{1j}) \tag{4.187}$$

By using the following trigonometric identity

$$\sin\alpha_1 \sin\alpha_2 = \tfrac{1}{2}\{\cos(\alpha_1 - \alpha_2) - \cos(\alpha_1 + \alpha_2)\},$$

Eq. 187 can be written as

$$u_j(x, t) = A_j e^{-\xi_j\omega_j t}\{\cos(\beta_j x - \omega_{dj}t + \psi_{mj}) - \cos(\beta_j x + \omega_{dj}t + \psi_{pj})\} \tag{4.188}$$

where

$$A_j = \tfrac{1}{2}X_{1j}X_{2j}$$
$$\psi_{mj} = \psi_{2j} - \psi_{1j} \tag{4.189}$$
$$\psi_{pj} = \psi_{2j} + \psi_{1j}$$

The terms between the brackets in Eq. 188 represent two harmonic waves, traveling in opposite directions with the same speed ω_{dj}/β_j. The amplitude of these two waves is multiplied by the exponential decay $e^{-\xi_j\omega_j t}$. Consequently, the amplitude of the resulting wave motion will decrease with time as the result of the damping effect. If the damping is equal to zero, the amplitude of the two harmonic waves remains constant. Recall from the analysis presented in Section 1, that in the case of undamped elastic rod, harmonic waves with different frequencies travel with the same *phase velocity* which is constant and equal to $c = \sqrt{E/\rho}$. In this case the elastic medium is said to be *nondispersive* and the *group velocity* is constant and equal to the phase velocity. This is not, however, the case when viscoelastic materials are considered. The phase velocity of harmonic waves that have different frequencies are not in general equal. Consequently, the group velocity is no longer equal to the phase velocity and it becomes dependent on the wave number.

4.8 ENERGY METHODS

Lagrange's equation, which utilizes scalar energy quantities, can be used as an alternative to the methods described in the preceding sections for the formulation of the vibration equations of continuous systems. We demonstrate the use of Lagrange's equation by considering the longitudinal vibration of rods.

It was shown, in the case of longitudinal vibration, that the solution can be assumed, using the technique of the separation of variables, in the form

$$u(x, t) = \sum_{j=1}^{\infty} \phi_j(x) q_j(t) \tag{4.190}$$

For simple boundary conditions, the kinetic and strain energies and virtual work are given, respectively, by

$$T = \tfrac{1}{2} \int_0^l \rho A \dot{u}^2 \, dx \tag{4.191}$$

$$U = \tfrac{1}{2} \int_0^l EAu'^2 \, dx \tag{4.192}$$

$$\delta W = \int_0^l F(x, t) \delta u \, dx \tag{4.193}$$

Substituting for $\dot{u}$, u', and δu from Eq. 190 and using the appropriate orthogonality conditions, the kinetic and strain energies and the virtual work can be written as

$$T = \tfrac{1}{2} \sum_{j=1}^{\infty} m_j \dot{q}_j^2 \tag{4.194}$$

$$U = \tfrac{1}{2} \sum_{j=1}^{\infty} k_j q_j^2 \tag{4.195}$$

$$\delta W = \sum_{j=1}^{\infty} Q_j \delta q_j \tag{4.196}$$

where m_j and k_j are, respectively, the modal mass and stiffness coefficients, and Q_j is the generalized force associated with the jth modal coordinate. By using the expressions for the kinetic and strain energies, the virtual work, and the following form of Lagrange's equation of motion,

$$\frac{d}{dt}\left(\frac{\partial T}{\partial \dot{q}_j}\right) - \frac{\partial T}{\partial q_j} + \frac{\partial U}{\partial q_j} = Q_j, \qquad j = 1, 2, 3, \ldots \tag{4.197}$$

one can show that the same system of differential equations of motion developed in the preceding sections can be obtained. This system of equations is written as

$$m_j\ddot{q}_j + k_j q_j = Q_j, \qquad j = 1, 2, 3, \ldots \tag{4.198}$$

Clearly, if the boundary conditions are not simple, the expressions for the kinetic and strain energies may take a different form. For example, if a mass m is attached to the end of the rod, the kinetic energy expression becomes

$$T = \tfrac{1}{2}\int_0^l \rho A\dot{u}^2(x, t)\, dx + \tfrac{1}{2}m\dot{u}^2(l, t) \tag{4.199}$$

Similarly, if a spring of stiffness k is attached to the end of the rod, the strain energy expression becomes

$$U = \tfrac{1}{2}\int_0^l EAu'^2(x, t)\, dx + \tfrac{1}{2}ku^2(l, t) \tag{4.200}$$

In the case of torsional oscillations, the expressions for the kinetic and strain energies in the case of simple boundary conditions are

$$T = \tfrac{1}{2}\int_0^l \rho I_p\dot{\theta}^2(x, t)\, dx \tag{4.201}$$

$$U = \tfrac{1}{2}\int_0^l GJ\theta'^2(x, t)\, dx \tag{4.202}$$

and in the case of transverse vibrations

$$T = \tfrac{1}{2}\int_0^l \rho A\dot{v}^2(x, t)\, dx \tag{4.203}$$

$$U = \tfrac{1}{2}\int_0^l EIv''^2(x, t)\, dx \tag{4.204}$$

Conservation of Energy From the analysis presented in this section it is clear that regardless of the type of vibration (longitudinal, torsional, or transverse), the kinetic energy and strain energy of the system can be written, respectively, as

$$T = \frac{1}{2}\sum_{j=1}^{\infty} m_j\dot{q}_j^2$$

$$U = \frac{1}{2}\sum_{j=1}^{\infty} k_j q_j^2$$

where m_j and k_j are, respectively, the modal mass and stiffness coefficients which can be defined using the orthogonality relationships, and q_j and $\dot{q}_j$ are, respectively, the modal coordinates and velocities.

It was shown in the case of *free undamped vibration* that the modal coordinate q_j can be expressed in the form

$$q_j = B_{1j} \sin \omega_j t + B_{2j} \cos \omega_j t$$

which can be rewritten in another alternate form as

$$q_j = C_j \sin(\omega_j t + \psi_j)$$

where B_{1j}, B_{2j}, C_j, and ψ_j are constants that can be determined by using the initial conditions.

The modal velocity $\dot{q}_j$ can also be written as

$$\dot{q}_j = \omega_j C_j \cos(\omega_j t + \psi_j)$$

Therefore, the kinetic and strain energies can be written in a more explicit form as

$$T = \frac{1}{2} \sum_{j=1}^{\infty} \omega_j^2 C_j^2 m_j \cos^2 (\omega_j t + \psi_j)$$

$$U = \frac{1}{2} \sum_{j=1}^{\infty} C_j^2 k_j \sin^2 (\omega_j t + \psi_j)$$

Recall that $k_j = \omega_j^2 m_j$, and consequently, the kinetic energy can be expressed in terms of the modal stiffness coefficients as

$$T = \frac{1}{2} \sum_{j=1}^{\infty} C_j^2 k_j \cos^2 (\omega_j t + \psi_j)$$

The sum of the kinetic and strain energies can then be written as

$$\begin{aligned} T + U &= \frac{1}{2} \sum_{j=1}^{\infty} C_j^2 k_j \cos^2 (\omega_j t + \psi_j) + \frac{1}{2} \sum_{j=1}^{\infty} C_j^2 k_j \sin^2 (\omega_j t + \psi_j) \\ &= \frac{1}{2} \sum_{j=1}^{\infty} C_j^2 k_j \left[\cos^2 (\omega_j t + \psi_j) + \sin^2 (\omega_j t + \psi_j) \right] \\ &= \frac{1}{2} \sum_{j=1}^{\infty} C_j^2 k_j \end{aligned}$$

That is, the sum of the kinetic and strain energies is a constant that can be written in the following form

$$T + U = \frac{1}{2} \sum_{j=1}^{\infty} C_j^2 \omega_j^2 m_j$$

Since both the kinetic and strain energies are nonnegative numbers, the strain energy becomes maximum when the kinetic energy is minimum. Similarly, the kinetic energy becomes maximum when the strain energy is minimum. The contribution of each mode to the total energy of the system depends on the initial conditions as represented by the constant C_j, the natural frequency of this mode, and the modal mass and stiffness coefficients.

As discussed in Chapter 2, the fact that the energy of the system, during the undamped free vibration, is conserved can be utilized to derive the equations of motion. To this end, we write

$$\frac{d}{dt}(T + U) = 0$$

which can be written more explicitly as

$$\frac{d}{dt}\left[\frac{1}{2}\sum_{j=1}^{\infty}(m_j\dot{q}_j^2 + k_jq_j^2)\right] = 0$$

Since m_j and k_j are constant coefficients, taking the derivatives in the preceding equation, one obtains

$$\sum_{j=1}^{\infty}(m_j\ddot{q}_j + k_jq_j)\dot{q}_j = 0$$

If we assume $\dot{q}_j, j = 1, 2, 3, \ldots$, to be independent velocities, their coefficients in the preceding equation must be equal to zero. This yields

$$m_j\ddot{q}_j + k_jq_j = 0, \qquad j = 1, 2, 3, \ldots$$

which is the same system of equations that can be obtained using Lagrange's equation in the case of undamped free vibration. These are uncoupled differential equations expressed in terms of the modal coordinates. Since there is no coupling between the modes, one may expect that there is no exchange of energy between different modes, and as a consequence, the total energy associated with each mode is conserved. In order to see this, the total energy associated with an arbitrary mode j can be written as

$$T_j + U_j = \tfrac{1}{2}m_j\dot{q}_j^2 + \tfrac{1}{2}k_jq_j^2$$

where T_j and U_j are, respectively, the kinetic and strain energies associated with the jth modal coordinate. Since in the undamped free vibration, the modal coordinates and modal velocities are harmonic functions, one has

$$T_j = \tfrac{1}{2}\omega_j^2C_j^2m_j\cos^2(\omega_jt + \psi_j)$$
$$U_j = \tfrac{1}{2}C_j^2k_j\sin^2(\omega_jt + \psi_j)$$

which yields

$$T_j + U_j = \tfrac{1}{2}C_j^2k_j$$
$$= \tfrac{1}{2}C_j^2\omega_j^2m_j$$

That is, the total energy associated with each mode is indeed constant.

Owing to the fact that the modal coordinates and velocities are simple harmonics, the total energy associated with a particular mode j can be written as

$$T_j + U_j = \tfrac{1}{2}C_j^2 k_j = U_j^* \tag{4.205}$$

or alternatively as

$$T_j + U_j = \tfrac{1}{2}C_j^2 \omega_j^2 m_j = T_j^* \tag{4.206}$$

where T_j^* and U_j^* are the maximum kinetic and strain energies of the mode j. Therefore, the total kinetic energy associated with a mode is constant and is equal to the maximum kinetic energy or the maximum strain energy.

4.9 APPROXIMATION METHODS

It is clear from the analysis presented thus far that the exact solution of the free and forced vibration of continuous systems is represented by an infinite series expressed in terms of the principal modes of vibration. In many applications, high-frequency modes of vibration may not have a significant effect on the solution of the vibration equations such that the contribution of these high-frequency modes can be neglected and the solution may be represented in terms of a finite number of modes or in terms of assumed polynomials that describe the shape of deformation of the continuous systems.

In this section and the following sections we discuss several approximate methods that can be used to determine the fundamental natural frequencies of the continuous systems. In some of these methods a finite-dimensional model, which is in a form similar to the mathematical model that governs the free vibration of the multi-degree of freedom system, is first developed. The resulting reduced system of equations can be used to determine a finite set of natural frequencies and mode shapes.

We shall present first an approximate method, called *Rayleigh's method*, that can be used to determine the fundamental natural frequency of the continuous systems. Recall from the analysis presented in the preceding section that in the case of free undamped vibration, the energy associated with any mode of vibration is conserved. The sum of the kinetic energy and strain energy of a mode remains constant and equal to the maximum kinetic energy or equal to the maximum strain energy. Consequently, for the jth mode of vibration, the conservation of energy in the case of undamped free vibration leads to

$$T_j^* = U_j^*, \qquad j = 1, 2, 3, \ldots \tag{4.207}$$

where T_j^* and U_j^* are, respectively, the maximum kinetic and strain energies associated with mode j. By using Eqs. 205 and 206, it is clear that the natural frequency of the jth mode can be defined using the following equation

$$\omega_j^2 = \frac{U_j^*}{\bar{T}^*} = \frac{k_j}{m_j}, \qquad j = 1, 2, 3, \ldots \tag{4.208}$$

where $T_j^* = \omega_j^2 \bar{T}_j^*$. Note that in the case of *longitudinal vibration*, the preceding equation and Eq. 120 yield

$$\omega_j^2 = \frac{\int_0^l EA\phi_j'^2(x)\,dx - EA\phi_j'^2(x)|_0^l}{\int_0^l \rho A\phi_j^2(x)\,dx} \tag{4.209a}$$

where E, ρ, A, and l are, respectively, the modulus of elasticity, mass density, cross-sectional area, and length of the rod. In the case of simple end conditions, Eq. 209a reduces to

$$\omega_j^2 = \frac{\int_0^l EA\phi_j'^2(x)\,dx}{\int_0^l \rho A\phi_j^2(x)\,dx} \tag{4.209b}$$

In the case of transverse vibration, the conservation of energy of the mode j and Eq. 208 lead to

$$\omega_j^2 = \frac{\int_0^l EI_z\phi_j''^2\,dx + [(EI_z\phi_j'')'\phi_j - EI_z\phi_j''\phi_j']|_0^l}{\int_0^l \rho A\phi_j^2\,dx} \tag{4.210a}$$

where E, ρ, I_z, and l are, respectively, the modulus of elasticity, mass density, moment of inertia of the cross section, and length of the beam. In the case of simple end conditions, Eq. 210a reduces to

$$\omega_j^2 = \frac{\int_0^l EI_z\phi_j''^2\,dx}{\int_0^l \rho A\phi_j^2\,dx} \tag{4.210b}$$

Note that the numerators in Eqs. 209a–210b are proportional to the strain energy of mode j while the denominator is proportional to the kinetic energy of that particular mode. The expressions on the right-hand side of these equations are called *Rayleigh quotients.*

In many practical applications, it is difficult to determine the exact mode shapes. Nonetheless, due to the fact that in many of these applications the shape of the first mode is simple, a guess, with reasonable degree of accuracy, can be made for a shape that resembles the first mode. In this case the Rayleigh quotient can be used to determine an approximate value for the fundamental natural frequency. In order to demonstrate this procedure, let us consider the case of a cantilever beam and choose the following function to approximate the fundamental mode shape

$$\bar{\phi}_1 = b\left(1 - \cos\frac{\pi x}{2l}\right) \tag{4.211}$$

where b is a constant. Since we have simple end conditions in the case of a cantilever beam, the Rayleigh quotient of Eq. 210b can be used to obtain an estimate for the fundamental natural frequency ω_1. Substituting Eq. 211 into Eq. 210b, one obtains

$$\omega_1^2 \approx \bar{\omega}_1^2 = \frac{\int_0^l EI_z \bar{\phi}_1''^2(x)\, dx}{\int_0^l \rho A \bar{\phi}_1^2(x)\, dx} = \frac{\int_0^l EI_z \left(\frac{\pi^2}{4l^2} \cos \frac{\pi x}{2l}\right)^2 dx}{\int_0^l \rho A \left(1 - \cos \frac{\pi x}{2l}\right)^2 dx} \tag{4.212}$$

where $\bar{\omega}_1$ is the approximate fundamental natural frequency. If we assume the case of a uniform beam which is made of homogeneous material, one can show that Eq. 212 leads to

$$\omega_1 = \bar{\omega}_1 \approx \frac{3.667}{l^2} \sqrt{\frac{EI_z}{\rho A}}$$

The exact fundamental natural frequency of a cantilever beam is

$$\omega_1 = \frac{3.5156}{l^2} \sqrt{\frac{EI_z}{\rho A}} \tag{4.213}$$

This equation shows that the error in the solution obtained by using the approximate Rayleigh method is less than 5%.

The assumed mode of Eq. 211 satisfies all the geometric boundary conditions at the clamped end, but satisfies only one of the natural boundary conditions at the free end ($\phi''(l, t) = 0$). In Rayleigh's method, the assumed functions must satisfy only the geometric boundary conditions, it is not necessary that these functions satisfy the natural boundary conditions. The error in the fundamental natural frequency obtained using the Rayleigh quotient becomes smaller if the integrals

$$\int_0^l EI_z(\bar{\phi}_1''^2 - \phi_1''^2)\, dx$$

and

$$\int_0^l \rho A(\bar{\phi}_1^2 - \phi_1^2)\, dx$$

are small.

The use of the Rayleigh quotient to predict the fundamental natural frequency always leads to an approximate eigenvalue which is higher than the exact one, that is, $(\bar{\omega}_1 - \omega_1)$ is always greater than or equal to zero. In order to prove this result, let the assumed eigenfunction of the beam have an arbitrary shape that can be represented as a linear combination of the exact eigenfunctions, that is

$$\begin{aligned} \bar{\phi}_1 &= \alpha_1 \phi_1 + \alpha_2 \phi_2 + \cdots \\ &= \sum_{j=1}^{\infty} \alpha_j \phi_j \end{aligned}$$

where $\alpha_j, j = 1, 2, \ldots$, are constants. In this case, the Rayleigh quotient for the beam can be written as

$$R = \frac{\int_0^l EI_z \bar{\phi}_1''^2(x)\,dx}{\int_0^l \rho A \bar{\phi}^2\,dx} = \frac{\int_0^l EI_z \left(\sum_{j=1}^{\infty} \alpha_j \phi_j''\right)^2 dx}{\int_0^l \rho A \left(\sum_{j=1}^{\infty} \alpha_j \phi_j\right)^2 dx}$$

which upon using the orthogonality conditions leads to

$$R = \frac{\sum_{j=1}^{\infty} \alpha_j^2 k_j}{\sum_{j=1}^{\infty} \alpha_j^2 m_j} = \frac{\sum_{j=1}^{\infty} \alpha_j^2 \omega_j^2 m_j}{\sum_{j=1}^{\infty} \alpha_j^2 m_j}$$

Since $\omega_1 < \omega_2 < \cdots$, and the modal mass coefficients are always positive, one has

$$R = \frac{\sum_{j=1}^{\infty} \alpha_j^2 \omega_j^2 m_j}{\sum_{j=1}^{\infty} \alpha_j^2 m_j} \geq \frac{\omega_1^2 \sum_{j=1}^{\infty} \alpha_j^2 m_j}{\sum_{j=1}^{\infty} \alpha_j^2 m_j} = \omega_1^2$$

This equation indicates that the exact fundamental natural frequency is indeed the minimum of the Rayleigh quotient, and any assumed shape for the fundamental mode will lead to an approximate value for the natural frequency that is higher than the exact one. This fact can be demonstrated by using another example. We consider again the case of a cantilever beam and assume the fundamental mode to be the static deflection of the beam. In this case the function $\bar{\phi}_1$ can be chosen as

$$\bar{\phi}_1 = b[x^4 - 4lx^3 + 6l^2x^2]$$

where b is a constant defined as

$$b = \frac{\rho A}{24EI_z}$$

It follows that

$$\bar{\phi}_1' = b[4x^3 - 12lx^2 + 12l^2x]$$

$$\bar{\phi}_1'' = 12b[x^2 - 2lx + l^2]$$

Substituting $\bar{\phi}_1$ and $\bar{\phi}_1''$ into the Rayleigh quotient of Eq. 210b, one obtains

$$\omega_1 \approx \bar{\omega}_1 = \frac{3.53}{l^2}\sqrt{\frac{EI_z}{\rho A}}$$

Comparing this estimated value with the exact solution of Eq. 213, it is clear that the error in the estimated fundamental natural frequency, using the static deflection as the assumed mode, does not exceed 0.5%.

Effect of the Rotary Inertia The Rayleigh quotient can be used to examine the effect of the rotary inertia on the fundamental frequency as well as higher natural frequencies. To this end, we consider the same example as the one discussed in Dym and Shames (1974). We have previously shown that the exact eigenfunctions of a simply supported beam are given by

$$\phi_j = \sin\frac{j\pi x}{l}, \qquad j = 1, 2, \ldots \tag{4.214}$$

In this case, one has

$$\left.\begin{aligned} \int_0^l \phi_j^2(x)\,dx &= \int_0^l \sin^2\frac{j\pi x}{l}\,dx = \frac{l}{2} \\ \int_0^l \phi_j'^2(x)\,dx &= \left(\frac{j\pi}{l}\right)^2 \int_0^l \cos^2\frac{j\pi x}{l}\,dx = \frac{(j\pi)^2}{2l} \\ \int_0^l \phi_j''^2(x)\,dx &= \left(\frac{j\pi}{l}\right)^4 \int_0^l \sin^2\frac{j\pi x}{l}\,dx = \frac{(j\pi)^4}{2l^3} \end{aligned}\right\} \tag{4.215}$$

The Rayleigh quotient that accounts for the effect of the rotary inertia can be written for mode j as

$$\omega_j^2 = \frac{\displaystyle\int_0^l EI_z\phi_j''^2\,dx}{\displaystyle\int_0^l \rho A\phi_j^2\,dx + \int_0^l \rho I_z\phi_j'^2\,dx} \tag{4.216}$$

Assuming the case of a uniform beam with constant modulus of elasticity and using the identities of Eq. 215, one obtains

$$\omega_j^2 = \frac{EI_z\dfrac{(j\pi)^4}{2l^3}}{\rho A\dfrac{l}{2} + \rho I_z\dfrac{(j\pi)^2}{2l}}$$

which can be written as

$$\omega_j^2 = \frac{EI_z(j\pi)^4/\rho A l^4}{\left[1 + \left(\dfrac{j^2\pi^2}{12}\right)\left(\dfrac{h}{l}\right)^2\right]} \tag{4.217}$$

where h is the height of the cross section.

The exact natural frequency ω_j, neglecting the effect of the rotary inertia, is

$$\omega_j = \left(\frac{j\pi}{l}\right)^2\sqrt{\frac{EI_z}{\rho A}} \tag{4.218}$$

The fundamental natural frequency obtained by including the effect of the rotary inertia is given by Eq. 217 with $j = 1$ as

$$\omega_1^2 = \frac{EI_z\pi^4/\rho A l^4}{\left[1 + \left(\frac{\pi^2}{12}\right)\left(\frac{h}{l}\right)^2\right]} \tag{4.219}$$

Clearly, in the case of a long thin beam ($(h/l) \ll 1$), the effect of the rotary inertia on the fundamental mode of vibration is not significant as compared to the frequencies of higher modes of vibration.

Rayleigh–Ritz Method In the Rayleigh method discussed in this section, the fundamental natural frequency of the system was predicted by assuming an approximate shape for the first mode. The Rayleigh–Ritz method can be considered as an extension of the Rayleigh method. It allows us, not only to obtain a more accurate value of the fundamental natural frequency, but also to determine the higher frequencies and the associated mode shapes.

In the Rayleigh–Ritz method, the shape of deformation of the continuous system is approximated using a series of *trial shape functions* that must satisfy the geometric boundary conditions of the problem. The shape of deformation of the continuous system can be written as

$$\begin{aligned} w(x) &= c_1\bar{\phi}_1(x) + c_2\bar{\phi}_2(x) + \cdots + c_n\bar{\phi}_n(x) \\ &= \sum_{j=1}^{n} c_j\bar{\phi}_j(x) \end{aligned} \tag{4.220}$$

where $\bar{\phi}_1, \bar{\phi}_2, \ldots,$ and $\bar{\phi}_n$ are the trial shape functions which can be the eigenfunctions, a set of assumed mode shapes, or a set of polynomials, and $c_1, c_2, \ldots, c_n$ are constant coefficients called the *Ritz coefficients.* In Eq. 220, the functions $\bar{\phi}_1, \bar{\phi}_2, \ldots,$ and $\bar{\phi}_n$ are assumed to be known, while the coefficients $c_1, c_2, \ldots,$ and c_n are adjusted by minimizing the Rayleigh quotient with respect to each of these coefficients. This procedure leads to a homogeneous system of n algebraic equations. For this system of equations to have a nontrivial solution, the determinant of the coefficient matrix must be equal to zero. This defines an equation of order n in ω^2, and the roots of this equation define the approximate natural frequencies, $\omega_1, \omega_2, \ldots,$ and ω_n. These approximate natural frequencies can be used to determine a set of approximate mode shapes for the system. By using this procedure, the continuous system which has an infinite number of degrees of freedom is represented by a model which has n degrees of freedom. The accuracy of the finite-dimensional model in representing the actual infinite-dimensional model depends on the choice of the trial shape functions.

We have shown previously that the Rayleigh quotient can be written as

$$\omega^2 = R = \frac{U^*}{T^*} \tag{4.221}$$

where, by using the expression of Eq. 220, $\bar{T}^*$ and U^* can be written as

$$\bar{T}^* = \frac{1}{2} \sum_{i=1}^{n} \sum_{j=1}^{n} m_{ij} c_i c_j \tag{4.222}$$

$$U^* = \frac{1}{2} \sum_{i=1}^{n} \sum_{j=1}^{n} k_{ij} c_i c_j \tag{4.223}$$

where m_{ij} and k_{ij} are mass and stiffness coefficients that depend on the shape functions. For example, in the case of longitudinal vibration of a rod with simple boundary conditions, the mass and stiffness coefficients are

$$m_{ij} = \int_0^l \rho A \bar{\phi}_i \bar{\phi}_j \, dx \qquad \text{and} \qquad k_{ij} = \int_0^l EA \bar{\phi}_i' \bar{\phi}_j' \, dx$$

whereas in the case of transverse vibration of a beam with simple boundary conditions, the mass and stiffness coefficients are

$$m_{ij} = \int_0^l \rho A \bar{\phi}_i \bar{\phi}_j \, dx \qquad \text{and} \qquad k_{ij} = \int_0^l EI_z \bar{\phi}_i'' \bar{\phi}_j'' \, dx$$

Note that if the shape functions are chosen such that they satisfy some orthogonality relationships, the mass coefficients m_{ij} or the stiffness coefficients k_{ij} can be equal to zero if $i \neq j$.

By using matrix notation, Eqs. 222 and 223 can be written as

$$\bar{T}^* = \tfrac{1}{2} \mathbf{c}^{\mathrm{T}} \mathbf{M} \mathbf{c} \tag{4.224}$$

$$U^* = \tfrac{1}{2} \mathbf{c}^{\mathrm{T}} \mathbf{K} \mathbf{c} \tag{4.225}$$

where $\mathbf{M}$ and $\mathbf{K}$ are, respectively, the system mass and stiffness matrices, and $\mathbf{c}$ is the vector of Ritz coefficients defined as

$$\mathbf{c} = [c_1 \quad c_2 \quad \cdots \quad c_n]^{\mathrm{T}}$$

Substituting Eqs. 224 and 225 into the Rayleigh quotient of Eq. 221, one obtains

$$R = \frac{U^*}{\bar{T}^*} = \frac{\tfrac{1}{2} \mathbf{c}^{\mathrm{T}} \mathbf{K} \mathbf{c}}{\tfrac{1}{2} \mathbf{c}^{\mathrm{T}} \mathbf{M} \mathbf{c}} \tag{4.226}$$

In order to minimize the Rayleigh quotient with respect to the Ritz coefficients, we differentiate R with respect to these coefficients and set the result of the differentiation equal to zero. This leads to

$$\frac{\partial R}{\partial \mathbf{c}} = \frac{\bar{T}^* \dfrac{\partial U^*}{\partial \mathbf{c}} - U^* \dfrac{\partial \bar{T}^*}{\partial \mathbf{c}}}{[\bar{T}^*]^2} = \mathbf{0}$$

That is

$$\bar{T}^* \frac{\partial U^*}{\partial \mathbf{c}} - U^* \frac{\partial \bar{T}^*}{\partial \mathbf{c}} = \mathbf{0}$$

which upon dividing by $\bar{T}^*$ and using Eq. 221, one obtains

$$\frac{\partial U^*}{\partial \mathbf{c}} - \omega^2 \frac{\partial \bar{T}^*}{\partial \mathbf{c}} = \mathbf{0} \tag{4.227}$$

Observe that

$$\frac{\partial U^*}{\partial \mathbf{c}} = \mathbf{c}^{\mathrm{T}}\mathbf{K}, \qquad \frac{\partial \bar{T}^*}{\partial \mathbf{c}} = \mathbf{c}^{\mathrm{T}}\mathbf{M}$$

and, consequently, Eq. 227 leads to

$$[\mathbf{K} - \omega^2\mathbf{M}]\mathbf{c} = \mathbf{0} \tag{4.228}$$

This is a system of n algebraic homogeneous equations that has a nontrivial solution if the determinant of the coefficient matrix is equal to zero. That is

$$|\mathbf{K} - \omega^2\mathbf{M}| = 0 \tag{4.229}$$

This leads to a nonlinear equation of order n in ω^2. The roots of this equation define the system natural frequencies $\omega_1^2, \omega_2^2, \ldots, \omega_n^2$. Associated with each natural frequency, ω_j, there is a vector of Ritz coefficients $\mathbf{c}_j$ which can be determined to within an arbitrary constant by solving the equation

$$[\mathbf{K} - \omega_j^2\mathbf{M}]\mathbf{c}_j = \mathbf{0} \tag{4.230a}$$

which can be written more explicitly as

$$\begin{bmatrix} (k_{11} - \omega_j^2 m_{11}) & (k_{12} - \omega_j^2 m_{12}) & \cdots & (k_{1n} - \omega_j^2 m_{1n}) \\ (k_{21} - \omega_j^2 m_{21}) & (k_{22} - \omega_j^2 m_{22}) & \cdots & (k_{2n} - \omega_j^2 m_{2n}) \\ \vdots & \vdots & \ddots & \vdots \\ (k_{n1} - \omega_j^2 m_{n1}) & (k_{n2} - \omega_j^2 m_{n2}) & \cdots & (k_{nn} - \omega_j^2 m_{nn}) \end{bmatrix} \begin{bmatrix} c_1 \\ c_2 \\ \vdots \\ c_n \end{bmatrix}_j = \begin{bmatrix} 0 \\ 0 \\ \vdots \\ 0 \end{bmatrix},$$

$$j = 1, 2, \ldots, n \tag{4.230b}$$

The approximate mode shape associated with the frequency ω_j can be obtained by using Eq. 220 as

$$\begin{aligned} w_j(x) &= c_{1j}\bar{\phi}_1(x) + c_{2j}\bar{\phi}_2(x) + \cdots + c_{nj}\bar{\phi}_n(x) \\ &= \sum_{k=1}^{n} c_{kj}\bar{\phi}_k(x), \qquad j = 1, 2, \ldots, n \end{aligned} \tag{4.231}$$

Once the natural frequencies and mode shapes are determined, the continuous system which has an infinite number of degrees of freedom can be represented by an equivalent multi-degree of freedom system. In this case, the vibration of the continuous system can be described using the separation of variables as

$$w(x, t) = \sum_{j=1}^{n} w_j(x)q_j(t) \tag{4.232}$$

where $q_j(t)$, $j = 1, 2, \ldots, n$, are the time-dependent coefficients.

It can be shown that if the assumed shape functions happen to be the exact eigenfunctions, the Rayleigh–Ritz method yields the exact natural frequencies and the exact mode shapes. In order to demonstrate this, let us consider the case of a simply supported beam with a uniform cross section and constant modulus of elasticity. We choose the shape functions to be the exact eigenfunctions and consider the case in which the number of Ritz coefficients is equal to three, that is

$$\bar{\phi}_j(x) = \sin\frac{j\pi x}{l}, \qquad j = 1, 2, 3$$

The mass coefficients m_{ij} and the stiffness coefficients k_{ij} are

$$m_{ij} = \int_0^l \rho A \bar{\phi}_i \bar{\phi}_j \, dx = \begin{cases} m/2 & \text{if } i = j \\ 0 & \text{if } i \neq j \end{cases}$$

$$k_{ij} = \int_0^l EI_z \bar{\phi}_i'' \bar{\phi}_j'' \, dx = \begin{cases} (i\pi)^4 EI_z/2l^3 & \text{if } i = j \\ 0 & \text{if } i \neq j \end{cases}$$

where $m = \rho A l$ is the total mass of the beam. The matrices $\mathbf{M}$ and $\mathbf{K}$ are given by

$$\mathbf{M} = \frac{m}{2}\begin{bmatrix} 1 & 0 & 0 \\ 0 & 1 & 0 \\ 0 & 0 & 1 \end{bmatrix}, \qquad \mathbf{K} = \frac{EI_z\pi^4}{2l^3}\begin{bmatrix} 1 & 0 & 0 \\ 0 & 16 & 0 \\ 0 & 0 & 81 \end{bmatrix}$$

Substituting the matrices $\mathbf{M}$ and $\mathbf{K}$ into Eq. 229, one obtains

$$\begin{vmatrix} 1-\beta & 0 & 0 \\ 0 & 16-\beta & 0 \\ 0 & 0 & 81-\beta \end{vmatrix} = 0$$

where $\beta = \omega^2(ml^3/EI_z\pi^4)$. The preceding equation leads to

$$(1-\beta)(16-\beta)(81-\beta) = 0$$

It is clear that the roots of this equation are

$$\beta_j = j^4$$

which define the three fundamental natural frequencies of the beam as

$$\omega_j = (j\pi)^2\sqrt{\frac{EI_z}{ml^3}}, \qquad j = 1, 2, 3$$

These are the same as the exact natural frequencies of the simply supported beam. The Ritz coefficients associated with each natural frequency can be obtained by using Eq. 230 which can be written for this system as

$$\begin{bmatrix} 1-\beta_j & 0 & 0 \\ 0 & 16-\beta_j & 0 \\ 0 & 0 & 81-\beta_j \end{bmatrix}\begin{bmatrix} c_{1j} \\ c_{2j} \\ c_{3j} \end{bmatrix} = \begin{bmatrix} 0 \\ 0 \\ 0 \end{bmatrix}$$

The solution of this equation for $j = 1, 2, 3$ is

$$\mathbf{c}_1 = \begin{bmatrix} c_{11} \\ 0 \\ 0 \end{bmatrix}, \qquad \mathbf{c}_2 = \begin{bmatrix} 0 \\ c_{22} \\ 0 \end{bmatrix}, \qquad \mathbf{c}_3 = \begin{bmatrix} 0 \\ 0 \\ c_{33} \end{bmatrix}$$

Therefore, the first mode shape is

$$\begin{aligned} w_1 &= c_{11}\bar{\phi}_1(x) + c_{21}\bar{\phi}_2(x) + c_{31}\bar{\phi}_3(x) \\ &= c_{11}\bar{\phi}_1(x) = c_{11} \sin \frac{\pi x}{l} \end{aligned}$$

Similarly, the second and third mode shapes are

$$w_2(x) = c_{22} \sin \frac{2\pi x}{l}$$

$$w_3(x) = c_{33} \sin \frac{3\pi x}{l}$$

Which are the exact eigenfunctions.

Example 4.12

Consider the case of a cantilever beam which has a uniform cross-sectional area and a constant modulus of elasticity. Assume that the shape of deformation of the beam is in the form

$$w(x) = c_1\bar{\phi}_1(x) + c_2\bar{\phi}_2(x)$$

where

$$\bar{\phi}_1(x) = \xi^2 \qquad \text{and} \qquad \bar{\phi}_2(x) = \xi^3$$

where $\xi = x/l$ and l is the length of the beam. The assumed shape functions $\bar{\phi}_1(x)$ and $\bar{\phi}_2(x)$ satisfy the geometric boundary conditions, that is, the deflection and slope are equal to zero at the fixed end ($x = l$).

Using the assumed shape functions $\bar{\phi}_1(x)$ and $\bar{\phi}_2(x)$, the mass coefficients are

$$\begin{aligned} m_{11} &= \int_0^l \rho A \bar{\phi}_1^2(x)\, dx = \int_0^l \rho A \xi^4\, dx \\ &= \int_0^1 m\xi^4\, d\xi = \frac{m}{5} \end{aligned}$$

$$\begin{aligned} m_{12} = m_{21} &= \int_0^l \rho A \bar{\phi}_1(x)\bar{\phi}_2(x)\, dx = \int_0^1 m\xi^5\, d\xi \\ &= \frac{m}{6} \end{aligned}$$

$$m_{22} = \int_0^l \rho A \bar{\phi}_2^2(x)\, dx = \int_0^1 m\xi^6\, d\xi = \frac{m}{7}$$

where m is the total mass of the beam.

Similarly, the stiffness coefficients are

$$k_{11} = \int_0^l EI_z \bar{\phi}_1''^2(x)\, dx = \int_0^l EI_z \left[\frac{1}{l^2}\frac{\partial^2 \bar{\phi}_1(x)}{\partial \xi^2}\right]^2 dx$$

$$= \int_0^1 EI_z l \left[\frac{2}{l^2}\right]^2 d\xi = \frac{4EI_z}{l^3}$$

$$k_{12} = k_{21} = \int_0^l EI_z \bar{\phi}_1''(x)\bar{\phi}_2''(x)\, dx = \int_0^l EI_z \left[\frac{1}{l^4}\frac{\partial^2 \bar{\phi}_1(x)}{\partial \xi^2}\frac{\partial^2 \bar{\phi}_2(x)}{\partial \xi^2}\right] dx$$

$$= \int_0^l EI_z l \left[\frac{12\xi}{l^4}\right] d\xi = \frac{6EI_z}{l^3}$$

$$k_{22} = \int_0^l EI_z \bar{\phi}_2''^2(x)\, dx = \int_0^l EI_z \left[\frac{1}{l^2}\frac{\partial^2 \bar{\phi}_2(x)}{\partial \xi^2}\right] dx$$

$$= \int_0^l EI_z l \left[\frac{6\xi^2}{l^2}\right]^2 d\xi = \frac{12EI_z}{l^3}$$

Therefore, the matrices $\mathbf{M}$ and $\mathbf{K}$ can be defined as

$$\mathbf{M} = m\begin{bmatrix} \frac{1}{5} & \frac{1}{6} \\ \frac{1}{6} & \frac{1}{7} \end{bmatrix}, \qquad \mathbf{K} = \frac{EI_z}{l^3}\begin{bmatrix} 4 & 6 \\ 6 & 12 \end{bmatrix}$$

Substituting these two matrices into Eq. 229, one obtains

$$\begin{vmatrix} 4 - \beta/5 & 6 - \beta/6 \\ 6 - \beta/6 & 12 - \beta/7 \end{vmatrix} = 0$$

where

$$\beta = \omega^2 \frac{ml^3}{EI_z}$$

One, therefore, has the frequency equation

$$\left(4 - \frac{\beta}{5}\right)\left(12 - \frac{\beta}{7}\right) - \left(6 - \frac{\beta}{6}\right)^2 = 0$$

which yields the quadratic equation

$$\beta^2 - 1224\beta + 15120 = 0$$

This equation has the roots

$$\beta_1 = 12.4802, \qquad \beta_2 = 1211.51981$$

From which the first two fundamental frequencies can be determined as

$$\omega_1 = 3.53273\sqrt{\frac{EI_z}{ml^3}}, \qquad \omega_2 = 34.8069\sqrt{\frac{EI_z}{ml^3}}$$

The exact fundamental natural frequency is 3.5156 $\sqrt{EI_z/ml^3}$. The error in the frequency estimated using the Rayleigh–Ritz method is less than 0.5%. The exact value of the second natural frequency of the cantilever beam is 22.0336 $\sqrt{EI_z/ml^3}$. Therefore, the error in the second natural frequency is about 58%. A significant improvement in the second natural frequency can be obtained by increasing the number of shape functions.

4.10 GALERKIN'S METHOD

Galerkin's method is a technique used for obtaining an approximate solution for both linear and nonlinear partial differential equations. In order to apply Galerkin's method an approximate solution for the displacement field is first assumed. Since this approximate solution is not, in general, the same as the exact solution, substitution of the assumed solution into the partial differential equations and the boundary conditions leads to some error called a *residual*. In *Galerkin's method*, the criterion for selecting the assumed solution is to make the error small. In order to demonstrate the use of Galerkin's method, we write the partial differential equations of the systems discussed in this chapter in the following general form

$$D(w(x, t)) - F(x, t) = 0 \tag{4.233}$$

where $F(x, t)$ is an arbitrary forcing function and $D(w(x, t))$ is a function that depends on the type of the problem considered. For example, in the case of *longitudinal vibration* of rods, the function $D(w(x, t))$ is given by

$$D(w(x, t)) = \frac{\partial^2 u}{\partial t^2} - c^2 \frac{\partial^2 u}{\partial x^2}$$

where in this case $w(x, t) = u(x, t)$ is the longitudinal displacement of the rod, and D is the differential operator defined as

$$D = \frac{\partial^2}{\partial t^2} - c^2 \frac{\partial^2}{\partial x^2}$$

and $c = \sqrt{E/\rho}$. Similar definitions can be made in the case of the torsional oscillation of shafts. In the case of transverse vibrations of beams, the function $D(w(x, t))$ can be recognized as

$$D(w(x, t)) = \frac{\partial^2 v}{\partial t^2} + c^2 \frac{\partial^4 v}{\partial x^4}$$

where $w(x, t) = v(x, t)$ is the transverse displacement of the beam, $c = \sqrt{EI_z/\rho A}$, and D is a differential operator that can be defined as

$$D = \frac{\partial^2}{\partial t^2} + c^2 \frac{\partial^4}{\partial x^4}$$

In Galerkin's method the unknown exact solution $w(x, t)$ is approximated by the function $\bar{w}(x, t)$ which can be written as

$$\bar{w}(x, t) = \sum_{j=1}^{n} \bar{\phi}_j(x) q_j(t) \tag{4.234}$$

where $\bar{\phi}_j(x)$ are assumed shape functions and $q_j(t)$ are unknown coordinates that depend on time. The assumed functions $\bar{\phi}_j(x)$ are chosen to satisfy the boundary conditions of the problem. Due to the fact that the assumed approximate solution is not the same as the unknown exact solution, Eq. 234 will not satisfy the partial differential equation of Eq. 233, that is

$$D(\bar{w}(x, t)) - F(x, t) \neq 0$$

Therefore, the preceding equation can be written as

$$D(\bar{w}(x, t)) - F(x, t) = R_e \tag{4.235}$$

where R_e is the residual or the error that results from the use of the approximate solution. By taking a virtual change in the assumed approximate solution of Eq. 234, one obtains

$$\delta\bar{w}(x, t) = \sum_{j=1}^{n} \bar{\phi}_j(x)\, \delta q_j(t) \tag{4.236}$$

Multiplying both sides of Eq. 235 and integrating over the solution domain yields

$$\sum_{j=1}^{n} \left\{ \int_0^l [D(\bar{w}(x, t)) - F(x, t) - R_e]\bar{\phi}_j(x)\, dx \right\} \delta q_j(t) = 0$$

Since the coordinates $q_j(t)$ are assumed to be linearly independent, the preceding equation yields

$$\int_0^l [D(\bar{w}(x, t)) - F(x, t)]\bar{\phi}_j(x)\, dx = \int_0^l R_e\bar{\phi}_j(x)\, dx, \qquad j = 1, 2, \ldots, n \tag{4.237}$$

In Galerkin's method, the assumed shape functions $\bar{\phi}_j(x)$ are chosen such that

$$\int_0^l R_e\bar{\phi}_j(x)\, dx \approx 0, \qquad j = 1, 2, \ldots, n \tag{4.238}$$

Consequently, Eq. 237 yields n differential equations which can be written as

$$\int_0^l [D(\bar{w}(x, t)) - F(x, t)]\bar{\phi}_j(x)\, dx = 0, \qquad j = 1, 2, \ldots, n \tag{4.239}$$

It is clear that in Galerkin's method the assumed shape functions $\bar{\phi}_j(x)$ can be considered as weighting functions selected such that the error R_e over the solution domain is small. Consequently, Galerkin's method can be considered as a special case of the *weighted residual techniques.* In the weighted residual techniques there are broad choices of the weighting functions or error distribution principles, whereas in Galerkin's method the weighting functions are selected to satisfy the criterion of Eq. 238.

Example 4.13

In order to demonstrate the use of Galerkin's method, we consider the longitudinal forced vibration of a prismatic rod. The governing partial differential equation of the system is

$$\rho A \frac{\partial^2 u}{\partial t^2} - EA \frac{\partial^2 u}{\partial x^2} = F(x, t)$$

We assume a solution in the form

$$\bar{u}(x, t) = \bar{\phi}_1(x)q_1(t) + \bar{\phi}_2(x)q_2(t)$$

where

$$\bar{\phi}_1(x) = 1 - \xi, \qquad \bar{\phi}_2(x) = \xi$$

in which $\xi = x/l$. It follows that

$$\delta\bar{u}(x, t) = \bar{\phi}_1\,\delta q_1 + \bar{\phi}_2\,\delta q_2$$

Multiplying both sides of the partial differential equation by $\delta\bar{u}$ and integrating over the volume, we obtain

$$\int_0^l \left[\rho A \frac{\partial^2 \bar{u}}{\partial t^2} - EA \frac{\partial^2 \bar{u}}{\partial x^2} - F(x, t)\right] \delta\bar{u}\,dx = 0 \tag{4.240a}$$

The first term in Eq. 240a represents the virtual work of the inertia forces and can be written as

$$\int_0^l \rho A \frac{\partial^2 \bar{u}}{\partial t^2} \delta\bar{u}\,dx = \int_0^l \rho A[\bar{\phi}_1\ddot{q}_1 + \bar{\phi}_2\ddot{q}_2][\bar{\phi}_1\,\delta q_1 + \bar{\phi}_2\,\delta q_2]\,dx$$

which can be written using matrix notation as

$$\int_0^l \rho A \frac{\partial^2 \bar{u}}{\partial t^2}\,\delta\bar{u}\,dx = \ddot{\mathbf{q}}^{\mathrm{T}}\mathbf{M}\,\delta\mathbf{q} \tag{4.240b}$$

where $\mathbf{q} = [q_1 \quad q_2]^{\mathrm{T}}$, and $\mathbf{M}$ is the mass matrix defined as

$$\mathbf{M} = \int_0^l \rho A \begin{bmatrix} \bar{\phi}_1^2 & \bar{\phi}_1\bar{\phi}_2 \\ \bar{\phi}_1\bar{\phi}_2 & \bar{\phi}_2^2 \end{bmatrix} dx = m \begin{bmatrix} \frac{1}{3} & \frac{1}{6} \\ \frac{1}{6} & \frac{1}{3} \end{bmatrix}$$

where $m = \rho A l$ is the mass of the rod.

The second term in Eq. 240a represents the virtual work of the elastic forces. This term can be integrated by parts to yield

$$-\int_0^l EA \frac{\partial^2 \bar{u}}{\partial x^2} \delta\bar{u}\,dx = -EA \frac{\partial \bar{u}}{\partial x} \delta\bar{u}\bigg|_0^l + \int_0^l EA \frac{\partial \bar{u}}{\partial x} \frac{\partial}{\partial x}(\delta\bar{u})\,dx$$

While this equation automatically accounts for any set of boundary conditions, in this example, for simplicity, we consider simple boundary conditions. Consequently

$$EA \frac{\partial \bar{u}}{\partial x} \delta\bar{u}\bigg|_0^l = 0$$

One, therefore, has

$$\begin{aligned} -\int_0^l EA \frac{\partial^2 \bar{u}}{\partial x^2} \delta\bar{u}\,dx &= \int_0^l EA \frac{\partial \bar{u}}{\partial x} \frac{\partial}{\partial x}(\delta\bar{u})\,dx \\ &= \int_0^l EA[\bar{\phi}_1' q_1 + \bar{\phi}_2' q_2][\bar{\phi}_1'\delta q_1 + \bar{\phi}_2'\,\delta q_2]\,dx \\ &= \mathbf{q}^{\mathrm{T}}\mathbf{K}\,\delta\mathbf{q} \end{aligned} \tag{4.240c}$$

where the stiffness matrix $\mathbf{K}$ is defined as

$$\mathbf{K} = \int_0^l EA\begin{bmatrix} \bar{\phi}_1'^2 & \bar{\phi}_1'\bar{\phi}_2' \\ \bar{\phi}_1'\bar{\phi}_2' & \bar{\phi}_2' \end{bmatrix} dx = \frac{EA}{l}\begin{bmatrix} 1 & -1 \\ -1 & 1 \end{bmatrix}$$

The third term in Eq. 240a is the virtual work of the external forces and can be written as

$$\begin{aligned} -\int_0^l F(x, t)\, \delta\bar{u}\, dx &= -\int_0^l F(x, t)[\bar{\phi}_1\, \delta q_1 + \bar{\phi}_2\, \delta q_2]\, dx \\ &= Q_1\, \delta q_1 + Q_2\, \delta q_2 \end{aligned}$$

in which

$$Q_1 = -\int_0^l F(x, t)\bar{\phi}_1(x)\, dx$$

$$Q_2 = -\int_0^l F(x, t)\bar{\phi}_2(x)\, dx$$

One can then write the virtual work of the external forces as

$$-\int_0^l F(x, t)\, \delta\bar{u}\, dx = \mathbf{Q}^{\mathrm{T}}\, \delta\mathbf{q} \tag{4.240d}$$

where

$$\mathbf{Q} = [Q_1 \quad Q_2]^{\mathrm{T}}$$

Substituting Eqs. 240b–d into Eq. 240a, one obtains the differential equations of the rod written in a matrix form as

$$[\ddot{\mathbf{q}}^{\mathrm{T}}\mathbf{M} + \mathbf{q}^{\mathrm{T}}\mathbf{K} - \mathbf{Q}^{\mathrm{T}}]\, \delta\mathbf{q} = 0$$

Since the components of the vector $\mathbf{q}$ are assumed to be linearly independent, the preceding equation leads to

$$\mathbf{M}\ddot{\mathbf{q}} + \mathbf{K}\mathbf{q} = \mathbf{Q}$$

which can be written more explicitly as

$$m\begin{bmatrix} \frac{1}{3} & \frac{1}{6} \\ \frac{1}{6} & \frac{1}{3} \end{bmatrix}\begin{bmatrix} \ddot{q}_1 \\ \ddot{q}_2 \end{bmatrix} + \frac{EA}{l}\begin{bmatrix} 1 & -1 \\ -1 & 1 \end{bmatrix}\begin{bmatrix} q_1 \\ q_2 \end{bmatrix} = \begin{bmatrix} Q_1 \\ Q_2 \end{bmatrix}$$

4.11 ASSUMED-MODES METHOD

The assumed modes method is closely related to the Rayleigh–Ritz and Galerkin methods discussed in the preceding sections. In fact, it leads to the same dynamic formulation as Galerkin's method if the same assumed displacement field is used. In the assumed modes method, the shape of deformation of the continuous system is approximated using a set of assumed shape functions. This approach can be used in the dynamic analysis of structures with complex geometrical shapes and complex boundary conditions. For such systems, it is difficult to determine the exact eigenfunctions. If, however, the deflected shape of the structures resembles the shape of deformation of some of the simple systems as the ones discussed in this chapter, the eigenfunctions of the simple systems can be assumed as the shape functions of the more complex structure. Another alternative is to use experimental testing to determine the shape of deformation by measuring the displacements at selected nodal points. The shape of deformation between these selected nodal points can be defined using polynomial functions. As was shown in this chapter, the vibration of continuous systems can be represented in terms of an infinite number of vibration modes. Each of these modes can be excited independently by the use of appropriate initial conditions or forcing functions. Therefore, the determination of the shape of the fundamental modes of vibration by experimental measurements is possible. In fact, the use of modal testing to determine the mode shapes and natural frequencies of continuous systems is an integral part in the design of many mechanical systems. In addition to the mode shapes and natural frequencies, modal testing also provides the modal stiffness, mass, and damping coefficients associated with these modes. By using these coefficients, a finite-dimensional model can be developed for the continuous systems.

Let us assume that the shape functions that define the deflected shape of the continuous system can be obtained experimentally or by inspection. In terms of these shape functions, an approximation for the displacement field can be assumed as

$$w(x, t) \approx \sum_{j=1}^{n} \bar{\phi}_j(x) q_j(t) \tag{4.241}$$

where $w(x, t) = u(x, t)$ in the case of longitudinal vibration, $w(x, t) = \theta(x, t)$ in the case of torsional vibration, and $w(x, t) = v(x, t)$ in the case of transverse vibration, $\bar{\phi}_j(x)$ may be the jth eigenfunction or an assumed polynomial, q_j is a time-dependent coordinate, and n is the number of elastic degrees of freedom used in the assumed solution. By using a finite number of terms in the solution, the kinetic energy and strain energy can be written as

$$T = \frac{1}{2} \sum_{j=1}^{n} \sum_{k=1}^{n} m_{jk} \dot{q}_j \dot{q}_k \tag{4.242}$$

$$U = \frac{1}{2} \sum_{j=1}^{n} \sum_{k=1}^{n} k_{jk} q_j q_k \tag{4.243}$$

The equations of vibration can be developed using these energy expressions and the following form of Lagrange's equation:

$$\frac{d}{dt}\left(\frac{\partial T}{\partial \dot{q}_j}\right) - \frac{\partial T}{\partial q_j} + \frac{\partial U}{\partial q_j} = Q_j, \qquad j = 1, 2, 3, \ldots, n \tag{4.244}$$

This procedure leads to the following n ordinary differential equations of motion:

$$\sum_{k=1}^{n} (m_{jk} \ddot{q}_k + k_{jk} q_k) = Q_j, \qquad j = 1, 2, 3, \ldots, n \tag{4.245}$$

These equations can be written compactly in a matrix form as

$$\mathbf{M}\ddot{\mathbf{q}} + \mathbf{K}\mathbf{q} = \mathbf{Q} \tag{4.246}$$

where $\mathbf{M}$ and $\mathbf{K}$ are, respectively, the $n \times n$ mass and stiffness matrices, $\mathbf{q} = [q_1 \; q_2 \; \dots \; q_n]^{\mathrm{T}}$ is the vector of generalized coordinates, and $\mathbf{Q} = [Q_1 \; Q_2 \; \dots \; Q_n]^{\mathrm{T}}$ is the vector of generalized forces. The preceding matrix equation is in a form similar to the matrix equation of the multi-degree of freedom systems, and therefore, the techniques discussed in the preceding chapter can be applied to obtain a solution for this matrix equation. Clearly, Eq. 246 represents an equivalent finite-dimensional model for the infinite degrees of freedom continuous system. An equivalent single degree of freedom model can be obtained if n is equal to 1. The use of this coordinate reduction procedure is demonstrated by the following example.

Example 4.14

If the transverse vibration of a beam with one end fixed and the other free is approximated by

$$v(x, t) = \bar{\phi}_1(x)q_1(t) + \bar{\phi}_2(x)q_2(t)$$

where

$$\bar{\phi}_1(x) = 3\xi^2 - 2\xi^3$$

$$\bar{\phi}_2(x) = l(\xi^3 - \xi^2)$$

in which l is the length of the beam and $\xi = x/l$, obtain the matrix equation of motion of this system.

Solution. The displacement and velocity of the beam can be expressed as

$$v(x, t) = [\bar{\phi}_1(x) \quad \bar{\phi}_2(x)] \begin{bmatrix} q_1(t) \\ q_2(t) \end{bmatrix}$$

$$\dot{v}(x, t) = [\bar{\phi}_1(x) \quad \bar{\phi}_2(x)] \begin{bmatrix} \dot{q}_1(t) \\ \dot{q}_2(t) \end{bmatrix}$$

Also

$$v''(x, t) = [\bar{\phi}_1''(x) \quad \bar{\phi}_2''(x)] \begin{bmatrix} q_1(t) \\ q_2(t) \end{bmatrix}$$

$$= \begin{bmatrix} \frac{6}{l^2}(1 - 2\xi) & \frac{2}{l}(3\xi - 1) \end{bmatrix} \begin{bmatrix} q_1(t) \\ q_2(t) \end{bmatrix}$$

Therefore, the kinetic energy and strain energy are given by

$$T = \tfrac{1}{2}\int_0^l \rho A \dot{v}^2 dx = \tfrac{1}{2}\dot{\mathbf{q}}^{\mathrm{T}}\mathbf{M}\dot{\mathbf{q}}$$

$$U = \tfrac{1}{2}\int_0^l EI v''^2 dx = \tfrac{1}{2}\mathbf{q}^{\mathrm{T}}\mathbf{K}\mathbf{q}$$

where $\mathbf{q} = [q_1 \ \ q_2]^{\mathrm{T}}$ and **M** and **K** are the mass and stiffness matrices defined as

$$\mathbf{M} = \int_0^l \rho A \begin{bmatrix} \bar{\phi}_1 \\ \bar{\phi}_2 \end{bmatrix} [\bar{\phi}_1 \ \ \bar{\phi}_2]\, dx = \int_0^l \rho A \begin{bmatrix} \bar{\phi}_1^2 & \bar{\phi}_1 \bar{\phi}_2 \\ \bar{\phi}_2 \bar{\phi}_1 & \bar{\phi}_2^2 \end{bmatrix} dx$$

$$= \rho A l \begin{bmatrix} 13/35 & -11l/210 \\ -11l/210 & l^2/105 \end{bmatrix}$$

$$\mathbf{K} = \int_0^l EI \begin{bmatrix} \bar{\phi}_1'' \\ \bar{\phi}_2'' \end{bmatrix} [\bar{\phi}_1'' \ \ \bar{\phi}_2'']\, dx = \int_0^l EI \begin{bmatrix} \bar{\phi}_1''^2 & \bar{\phi}_1'' \bar{\phi}_2'' \\ \bar{\phi}_2'' \bar{\phi}_1'' & \bar{\phi}_2''^2 \end{bmatrix} dx$$

$$= \frac{EI}{l} \begin{bmatrix} 12/l^2 & -6/l \\ -6/l & 4 \end{bmatrix}$$

The virtual work of the external force is

$$\delta W = \int_0^l F(x, t) \delta v \, dx = \int_0^l F(x, t) [\bar{\phi}_1 \ \ \bar{\phi}_2] \begin{bmatrix} \delta q_1 \\ \delta q_2 \end{bmatrix} dx = [Q_1 \ \ Q_2] \begin{bmatrix} \delta q_1 \\ \delta q_2 \end{bmatrix}$$

where

$$Q_1 = \int_0^l F(x, t) \bar{\phi}_1 \, dx$$

$$Q_2 = \int_0^l F(x, t) \bar{\phi}_2 \, dx$$

Therefore, the matrix equation for the equivalent two degree of freedom system can be written as

$$\rho A l \begin{bmatrix} 13/35 & -11l/210 \\ -11l/210 & l^2/105 \end{bmatrix} \begin{bmatrix} \ddot{q}_1 \\ \ddot{q}_2 \end{bmatrix} + \frac{EI}{l} \begin{bmatrix} 12/l^2 & -6/l \\ -6/l & 4 \end{bmatrix} \begin{bmatrix} q_1 \\ q_2 \end{bmatrix} = \begin{bmatrix} Q_1 \\ Q_2 \end{bmatrix}$$

Problems

4.1. Determine the equation of motion, boundary conditions, and frequency equation of the longitudinal vibration of the system shown in Fig. P1, assuming that the rod has a uniform cross-sectional area.

4.2. The system shown in Fig. P2 consists of a uniform rod with a spring attached to its end. Derive the partial differential equation of motion and determine the boundary conditions and frequency equation of the longitudinal vibration.

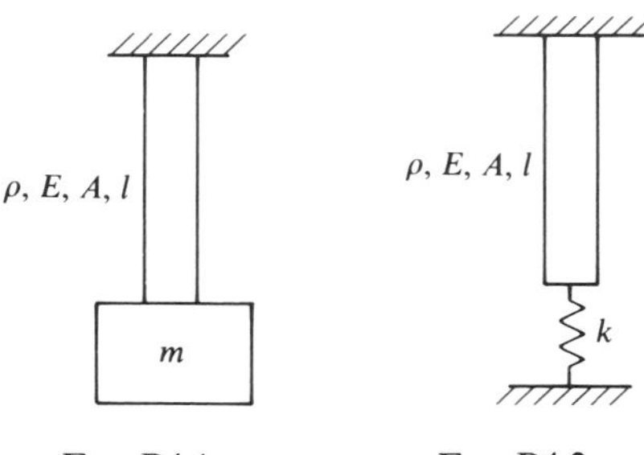

Fig. P4.1 Fig. P4.2

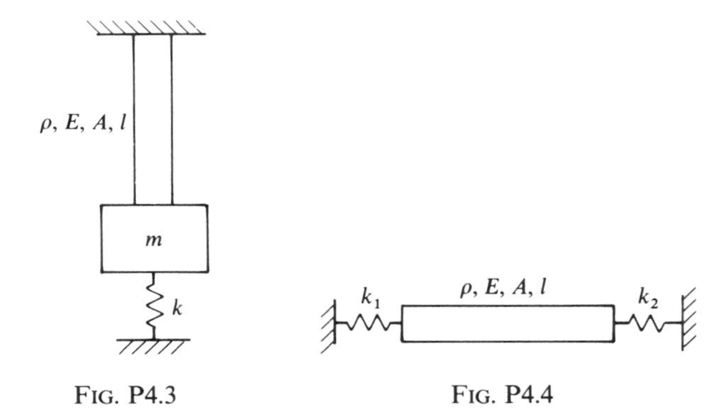

Fig. P4.3

Fig. P4.4

4.3. Obtain the equation of motion, boundary conditions, and frequency equation of the system shown in Fig. P3.

4.4. Determine the equation of motion, boundary conditions, and frequency equation of longitudinal vibration of a uniform rod with a mass m attached to each end. Check the fundamental frequency by reducing the uniform rod to a spring with end masses.

4.5. Derive the equation of motion of longitudinal vibration of the system shown in Fig. P4. Obtain the boundary conditions and the frequency equation for the case in which $k_1 = k_2 = k$.

4.6. Determine the equation of motion, boundary conditions, and natural frequencies of a torsional system which consists of a uniform shaft of mass moment of inertia I and a disk having mass moment of inertia I_1 attached to each end of the shaft.

4.7. Determine the equation of torsional oscillation of a uniform shaft with one end fixed and the other end attached to a disk with inertia I_1. Obtain the boundary conditions and the frequency equations.

4.8. Obtain the equation of torsional oscillation of a uniform shaft with a torsional spring of stiffness k attached to each end.

4.9. Derive the partial differential equation of the torsional oscillations of a uniform shaft clamped at the middle and free at the two ends. The shaft has length l, modulus of rigidity G, mass density ρ, and cross-sectional area A.

4.10. For the system shown in Fig. P5, obtain the partial differential equation of the longitudinal vibration, the boundary conditions, and the frequency equation.

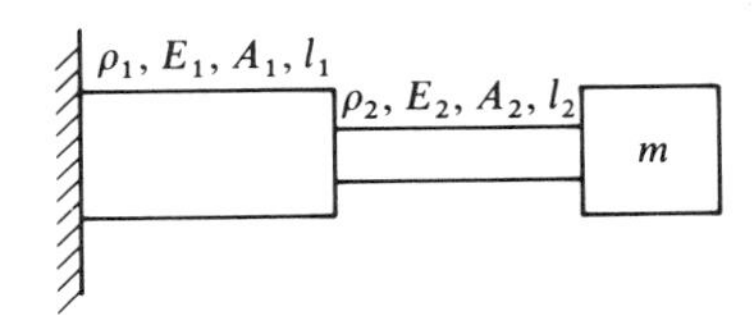

Fig. P4.5

Check the results by letting the two rods have the same dimensions and by comparing with the results obtained by solving problem 1.

4.11. Determine the equation of transverse vibration, boundary conditions, and frequency equation of a uniform beam of length l clamped at one end and pinned at the other end.

4.12. Obtain the frequency equation of the transverse vibration of the beam shown in Fig. P6.

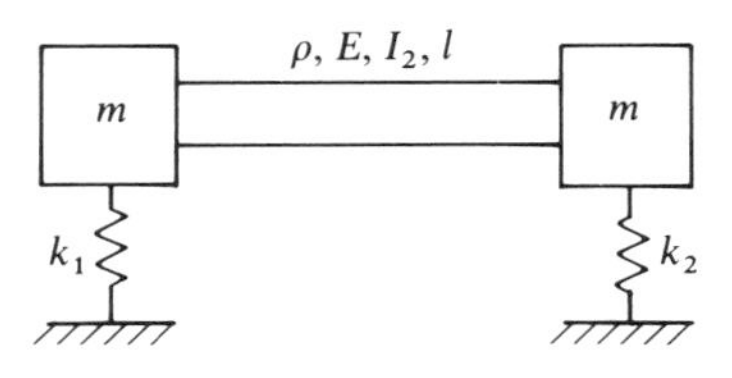

FIG. P4.6

4.13. A uniform rod which is fixed at one end and free at the other end has the following initial conditions:

$$u(x, 0) = A_0 \sin \frac{\pi x}{2l}$$
$$\dot{u}(x, 0) = 0$$

Obtain the general solution of the longitudinal vibration.

4.14. Repeat Problem 13 for the following initial conditions:

$$u(x, 0) = 0$$
$$\dot{u}(x, 0) = V_0 \sin \frac{\pi x}{2l}$$

4.15. For the system of Problem 13, examine the validity of using the initial conditions

$$u(x, 0) = A_0 \cos \frac{\pi x}{2l}$$
$$\dot{u}(x, 0) = 0$$

What is the expected solution? Comment on the results.

4.16. A uniform shaft which is fixed at one end and free at the other end has the following initial conditions:

$$\theta(x, 0) = A_0 \sin \frac{n\pi x}{2l}$$
$$\dot{\theta}(x, 0) = 0$$

where n is a fixed odd number. Obtain the solution of the free torsional vibration of this system.

4.17. Repeat Problem 13 in the case where the initial conditions are given by

$$u(x, 0) = A_0 x$$
$$\dot{u}(x, 0) = 0$$

4.18. Repeat Problem 16 by considering the following initial conditions:

$$\theta(x, 0) = A_0(x + x^2)$$
$$\dot{\theta}(x, 0) = 0$$

4.19. For a uniform rod with both ends free, obtain the solution for the longitudinal vibration if the initial conditions are given by

$$u(x, 0) = A_0 \cos \frac{n\pi x}{l}$$
$$\dot{u}(x, 0) = 0$$

4.20. Repeat Problem 19 using the following initial conditions:

$$u(x, 0) = A_0 x^2$$
$$\dot{u}(x, 0) = 0$$

Comment on the validity of using these initial conditions.

4.21. A uniform rod with one end fixed and the other end free is subjected to a distributed axial load in the form

$$F(x, t) = x \sin 5t$$

Determine the response of the system as the result of application of this forcing function.

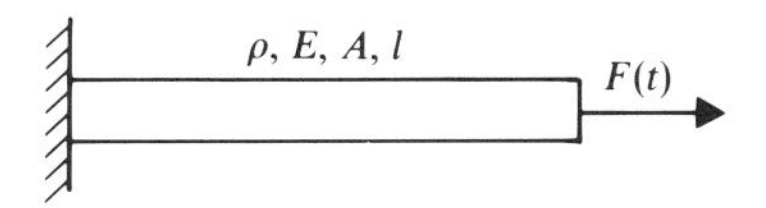

FIG. P4.7

4.22. Determine the response of the system shown in Fig. P7 to the concentrated harmonic forcing function $F(t)$.

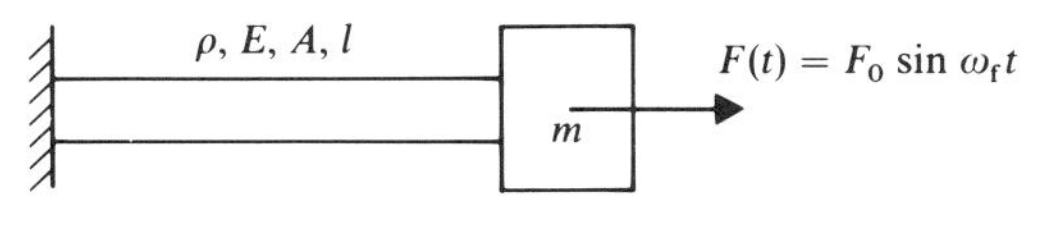

FIG. P4.8

4.23 The system shown in Fig. P8 is subject to a concentrated axial force at the mass attached to the end of the rod. Determine the response of this system to this end excitation.

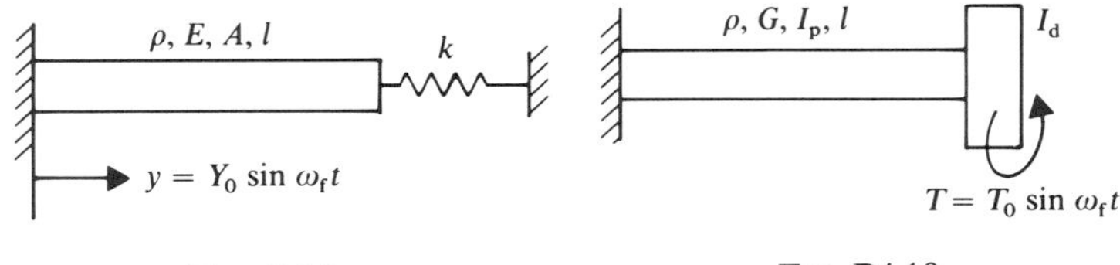

FIG. P4.9

FIG. P4.10

4.24. Determine the response of the system shown in Fig. P9 to the harmonic motion of the support.

4.25. Determine the response of the torsional system shown in Fig. P10 to the harmonic torque excitation.

4.26. Repeat Problem 22 if the forcing function $F(t)$ is the periodic function shown in Fig. P11.

4.27. Determine the response of the beam shown in Fig. P12 to the forcing function

$$F(t) = F_0 \sin \omega_f t$$

4.28. Repeat Problem 27 if $F(t)$ is given by the periodic function defined in Problem 26.

4.29. The system shown in Fig. P13 consists of a uniform beam with one end pinned and the other end free. Determine the response of this system to the harmonic support excitation.

4.30. Repeat Problem 22 if the force $F(t)$ is given by the rectangular pulse shown in Fig. P14.

4.31. If the longitudinal displacement of the system of Problem 22 is represented by

$$u(x, t) = \phi_1(x)q_1 + \phi_2(x)q_2$$

where

$$\phi_1(x) = \sin\frac{\pi x}{2l}, \qquad \phi_2(x) = \sin\frac{3\pi x}{2l}$$

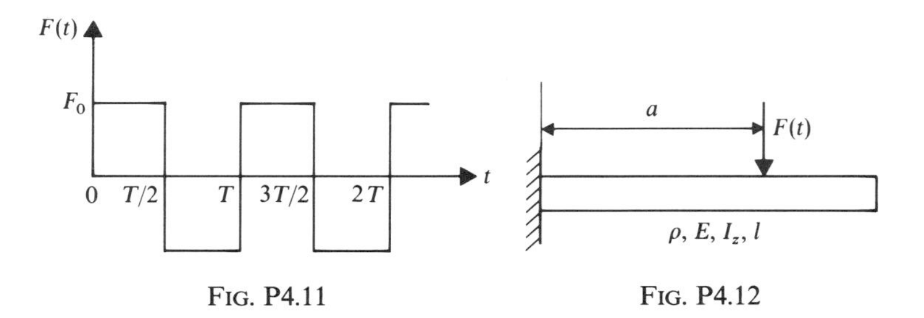

FIG. P4.11

FIG. P4.12

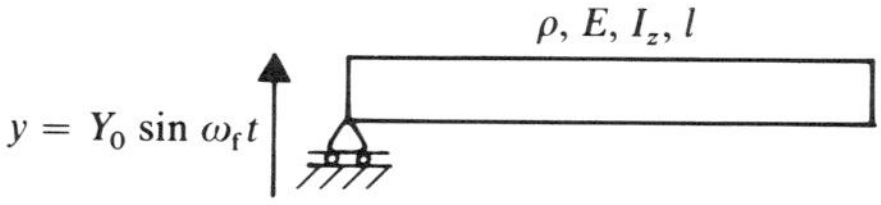

FIG. P4.13

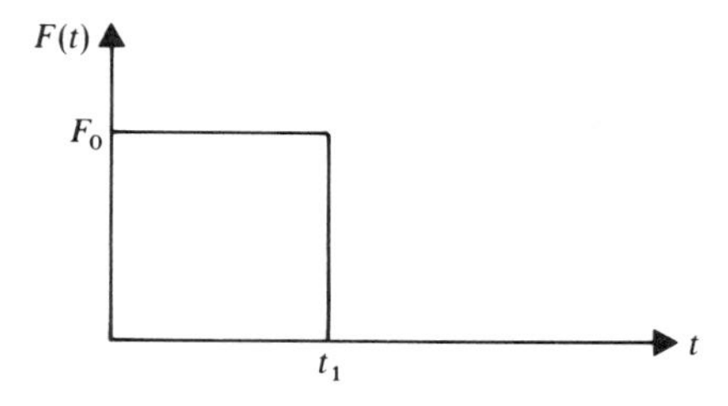

FIG. P4.14

and q_1 and q_2 are time-dependent coordinates, obtain the matrix equation that governs the forced vibration of this system.

4.32. Repeat Problem 31 if

$$\phi_1(x) = 0, \qquad \phi_2(x) = x$$

4.33. Obtain the solution of the longitudinal vibration of a uniform rod with one end fixed and the other end condition is

$$u(l, t) = d \cos \omega t$$

where d is a constant. Assume zero initial conditions.

4.34. By using the Rayleigh–Ritz method determine the natural frequencies and mode shapes of a beam with free ends. Assume that the modulus of elasticity and the cross-sectional area of the beam are constant. Use the following assumed function

$$\bar{v}(x) = c_1 + c_2 x + c_3 x^2 + c_4 x^3$$

Compare the results obtained using the Rayleigh–Ritz method and the exact solution.

4.35. Determine the natural frequencies and mode shapes of the system shown in Fig. P2 using the Rayleigh–Ritz method. Assume that the trial functions are the exact eigenfunctions of a rod with one end fixed and the other end free. Consider the two cases in which $n = 2, 3$. Plot the approximate mode shapes and compare the results with the exact ones.

4.36. Repeat Problem 35 assuming that the trial function is:

$$\bar{u}(x) = \sum_{j=1}^{n} c_j x^j$$

Comment on the results.

4.37. Using Galerkin's method obtain a three degree of freedom mathematical model that represents the system of Problem 34. Use the mode shapes obtained in Problem 34 as the assumed shape functions.

4.38. Use Galerkin's method to obtain a two and three degree of freedom mathematical model for the system of Problem 35. Use the mode shapes obtained in Problem 35 as the assumed shape functions.

5
The Finite-Element Method

The approximate methods presented at the end of the preceding chapter for the solution of the vibration problems of continuous systems are based on the assumption that the shape of the deformation of the continuous system can be described by a set of assumed functions. By using this approach, the vibration of the continuous system which has an infinite number of degrees of freedom is described by a finite number of ordinary differential equations. This approach, however, can be used in the case of structural elements with simple geometrical shapes such as rods, beams, and plates. In large-scale systems with complex geometrical shapes, difficulties may be encountered in defining the assumed shape functions. In order to overcome these problems the finite-element method has been widely used in the dynamic analysis of large-scale structural systems. The finite-element method is a numerical approach that can be used to obtain approximate solutions to a large class of engineering problems. In particular, the finite-element method is well suited for problems with complex geometries.

In the finite-element method, the stucture is discretized to relatively small regions called *elements* which are rigidly interconnected at selected *nodal points*. The deformation within each element can then be described by interpolating polynomials. The coefficients of these polynomials are defined in terms of physical coordinates called the *element nodal coordinates* that describe the displacements and slopes of selected nodal points on the element. Therefore, the displacement of the element can be expressed using the separation of variables as the product of space-dependent functions and time-dependent nodal coordinates. By using the connectivity between elements, the assumed displacement field can be written in terms of the element shape function and the nodal corrdinates of the structure. Using the assumed displacement field, the kinetic and strain energy of each element can be developed, thus defining the finite-element mass and stiffness matrices. The energy expressions of the structure can be obtained by summing the energy expressions of its elements. This leads to the definition of the structure mass and stiffness matrices.

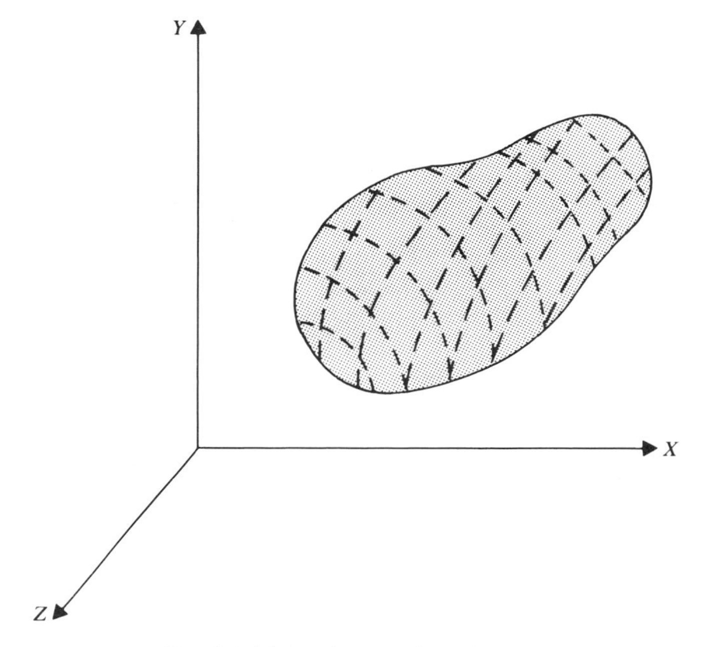

FIG. 5.1. Finite element discretization.

Figure 1 shows a continuous system or a structure which is divided to a finite number of elements which collectively approximate the physical description of the system. These elements are assumed to be rigidly connected at selected nodal points whose displacements and slopes are the elastic coordinates of the system. Figure 2 shows some typical finite elements which are commonly used in the dynamic analysis of the structural systems. Observe that different number of nodes can be used for different types of elements. The nodal coordinates of these elements are also shown in Fig. 2.

In this chapter, we discuss the formulation of the mass and stiffness matrices of the continuous systems using the finite-element method. In Sections 1–3, the general procedure for defining the assumed displacement field in terms of the element shape function and the nodal coordinates of the continuous system is described. In Section 4, the formulation of the mass matrix is developed using the kinetic energy of the elements. In Section 5, the element strain energy is used to define the stiffness matrix of the structure. The formulation of the structure equations of motion is presented in Section 6, while the convergence of the finite-element solution is examined in Section 7. Sections 8 and 9 are devoted to the analysis of higher-order and spatial elements, respectively. Examples are presented throughout the chapter in order to demonstrate the use of the formulation presented for different types of elements.

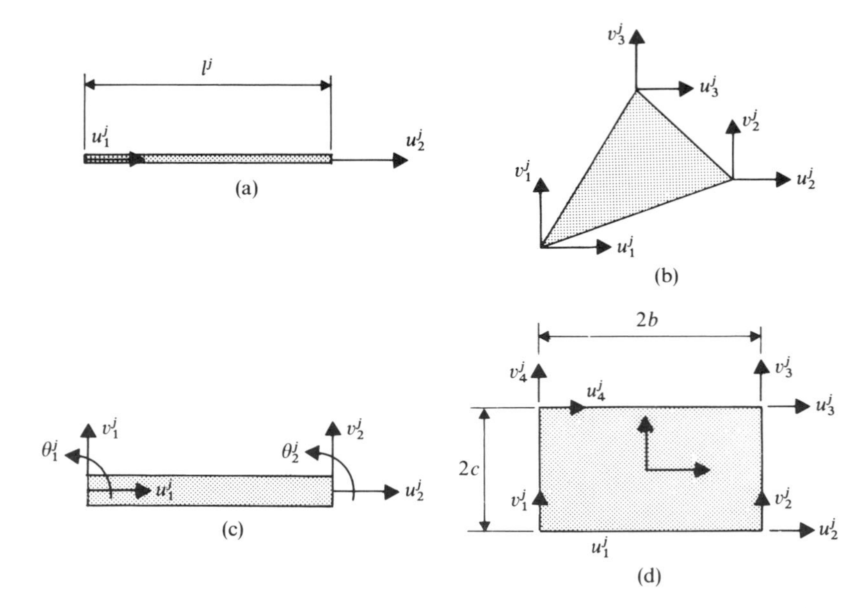

FIG. 5.2. Finite elements: (a) truss element; (b) triangular element; (c) beam element; (d) rectangular element.

5.1 ASSUMED DISPLACEMENT FIELD

Each finite element represents a continuous system which has an infinite number of degrees of freedom. However, by choosing the size of the element to be small, the deformation of the finite element can be approximated by using relatively low-order polynomials. The coefficients of these polynomials in the dynamic case are time dependent. Therefore, in a selected coordinate system of element j, the displacement field is defined as

$$\mathbf{u}^j = \mathbf{S}_1^j \mathbf{a}^j, \qquad j = 1, 2, \ldots, n_e \tag{5.1}$$

where $\mathbf{S}_1^j$ is a space-dependent matrix, $\mathbf{a}^j$ is the vector of the time-dependent coefficients of the polynomials, and n_e is the total number of elements used to discretize the continuous system. The displacement $\mathbf{u}^j$ can be a scalar, two-dimensional or three-dimensional vector depending on the type of element used.

The time-dependent coefficients in Eq. 1 lack an obvious physical meaning. In the finite-element method, these coefficients are expressed in terms of the time-dependent element nodal coordinates. This relationship can be written as

$$\mathbf{a}^j = \mathbf{S}_2^j \mathbf{q}^j, \qquad j = 1, 2, \ldots, n_e \tag{5.2}$$

where $\mathbf{S}_2^j$ is a constant matrix which can be obtained by defining the displacements and slopes at the nodal points and $\mathbf{q}^j$ is the vector of the element nodal coordinates.

By substituting Eq. 2 into Eq. 1, the displacement field of the element can be expressed in terms of the element nodal coordinates as

$$\mathbf{u}^j = \mathbf{S}_1^j \mathbf{S}_2^j \mathbf{q}^j, \qquad j = 1, 2, \ldots, n_e \tag{5.3}$$

which can be written as

$$\mathbf{u}^j = \mathbf{S}^j \mathbf{q}^j, \qquad j = 1, 2, \ldots, n_e \tag{5.4}$$

where $\mathbf{S}^j$ is the space-dependent element *shape function* defined as

$$\mathbf{S}^j = \mathbf{S}_1^j \mathbf{S}_2^j, \qquad j = 1, 2, \ldots, n_e \tag{5.5}$$

The procedure described in this section for writing the displacement field in terms of the space-dependent element shape function and the time-dependent nodal coordinates is a general procedure which can be used for many elements. The use of this procedure is demonstrated by the following examples.

Truss Element The truss element shown in Fig. 2(a) is assumed to carry only tension or compression loads. Therefore, only axial displacement is allowed. The displacement field of this element is expressed using a first-order polynomial as

$$u^j = a_1^j + a_2^j x \tag{5.6}$$

where a_1^j and a_2^j are time-dependent coefficients. Equation 6 can be written in the form of Eq. 1 as

$$u^j = [1 \quad x] \begin{bmatrix} a_1^j \\ a_2^j \end{bmatrix} \tag{5.7}$$

where the space-dependent matrix $\mathbf{S}_1^j$ and the time-dependent vector $\mathbf{a}^j$ are recognized as

$$\mathbf{S}_1^j = [1 \quad x] \tag{5.8a}$$

$$\mathbf{a}^j = [a_1^j \quad a_2^j]^{\mathrm{T}} \tag{5.8b}$$

As shown in Fig. 2(a), this element has two nodal points. Each nodal point has one degree of freedom which represents the axial displacement of this node. One can, therefore, write the following conditions:

$$u^j(x = 0) = u_1^j \tag{5.9a}$$

$$u^j(x = l^j) = u_2^j \tag{5.9b}$$

where u_1^j and u_2^j are the element nodal coordinates and l^j is the length of the element. Substituting the conditions of Eq. 9 into Eq. 7, yields

$$u^j(x = 0) = [1 \quad 0] \begin{bmatrix} a_1^j \\ a_2^j \end{bmatrix} = u_1^j$$

and

$$u^j(x = l^j) = [1 \quad l^j]\begin{bmatrix} a_1^j \\ a_2^j \end{bmatrix} = u_2^j$$

These two equations can be written in one matrix equation as

$$\begin{bmatrix} 1 & 0 \\ 1 & l^j \end{bmatrix}\begin{bmatrix} a_1^j \\ a_2^j \end{bmatrix} = \begin{bmatrix} u_1^j \\ u_2^j \end{bmatrix}$$

The solution of this equation defines the coefficients a_1^j and a_2^j in terms of the nodal coordinates u_1^j and u_2^j as

$$\begin{bmatrix} a_1^j \\ a_2^j \end{bmatrix} = \frac{1}{l^j}\begin{bmatrix} l^j & 0 \\ -1 & 1 \end{bmatrix}\begin{bmatrix} u_1^j \\ u_2^j \end{bmatrix}$$

where $\mathbf{S}_2^j$ of Eq. 4 can be recognized as

$$\mathbf{S}_2^j = \frac{1}{l^j}\begin{bmatrix} l^j & 0 \\ -1 & 1 \end{bmatrix} \tag{5.10}$$

The element shape function $\mathbf{S}^j$ of Eq. 5 can then be defined using Eqs. 8 and 10 as

$$\mathbf{S}^j = \mathbf{S}_1^j\mathbf{S}_2^j = [1 \quad x]\begin{bmatrix} 1 & 0 \\ -\dfrac{1}{l^j} & \dfrac{1}{l^j} \end{bmatrix} = \left[\left(1 - \frac{x}{l^j}\right) \quad \frac{x}{l^j}\right] \tag{5.11a}$$

which can also be written as

$$\mathbf{S}^j = [(1 - \xi) \quad \xi] \tag{5.11b}$$

where ξ is the dimensionless parameter

$$\xi = \frac{x}{l^j}$$

One can then write the displacement of the element in terms of the element nodal coordinates as

$$\begin{aligned} \mathbf{u}^j &= \mathbf{S}^j\mathbf{q}^j \\ &= [(1 - \xi) \quad \xi]\begin{bmatrix} u_1^j \\ u_2^j \end{bmatrix} \end{aligned} \tag{5.11c}$$

This equation can also be written as

$$\mathbf{u}^j = \phi_1^j(\xi)u_1^j + \phi_2^j(\xi)u_2^j$$

where

$$\phi_1^j(\xi) = 1 - \xi \qquad \text{and} \qquad \phi_2^j(\xi) = \xi$$

Note that when $x = 0$, that is, $\xi = 0$, $\phi_1^j(0) = 1$, and $\phi_2^j(0) = 0$, while when $x = l^j$, that is, $\xi = 1$, $\phi_1^j(1) = 0$ and $\phi_2^j(1) = 1$. That is, the element of the shape function ϕ_i^j takes the value one at node i and the value zero at the other node.

Triangular Element Several finite-element formulations for the two-dimensional triangular elements can be found in the literature. In this section, we consider the simplest one which is called the *constant-strain triangular element*. The element is shown in Fig. 2(b) and has three nodal coordinates. Each node has two degrees of freedom which represent the horizontal and vertical displacements of the node as shown in Fig. 2(b). Therefore, the element has six nodal coordinates given by

$$\mathbf{q}^j = [u_1^j \quad v_1^j \quad u_2^j \quad v_2^j \quad u_3^j \quad v_3^j]^{\mathrm{T}} \tag{5.12}$$

The displacement field is approximated by the following polynomials:

$$u^j = a_1^j + a_2^j x + a_3^j y \tag{5.13a}$$

$$v^j = a_4^j + a_5^j x + a_6^j y \tag{5.13b}$$

This element in which the displacement field is defined by the preceding equation is called *constant strain* since

$$\varepsilon_x^j = \frac{\partial u^j}{\partial x} = a_2^j$$

$$\varepsilon_y^j = \frac{\partial v^j}{\partial y} = a_6^j$$

$$\gamma_{xy}^j = \frac{\partial u^j}{\partial y} + \frac{\partial v^j}{\partial x} = a_3^j + a_5^j$$

That is, the strain is the same at every point on the element.

Let (x_1, y_1), (x_2, y_2), and (x_3, y_3) be the coordinates of the nodal points of the element. It is clear that

$$\left.\begin{aligned} u_k^j &= a_1^j + a_2^j x_k + a_3^j y_k \\ v_k^j &= a_4^j + a_5^j x_k + a_6^j y_k, \qquad k = 1, 2, 3 \end{aligned}\right\} \tag{5.14}$$

where u_k^j and v_k^j are the nodal coordinates of the jth node. By using a similar procedure to the one described in the case of the truss element, Eq. 14 can be used to write the coefficients $a_1^j, a_2^j, \ldots, a_6^j$ in terms of the nodal coordinates $u_1^j, v_1^j, u_2^j, v_2^j, u_3^j, v_3^j$. This leads to the following displacement field:

$$\mathbf{u}^j = \begin{bmatrix} u^j \\ v^j \end{bmatrix} = \begin{bmatrix} N_1 & 0 & N_2 & 0 & N_3 & 0 \\ 0 & N_1 & 0 & N_2 & 0 & N_3 \end{bmatrix} \begin{bmatrix} u_1^j \\ v_1^j \\ u_2^j \\ v_2^j \\ u_3^j \\ v_3^j \end{bmatrix} \tag{5.15a}$$

where the scalars N_1, N_2, and N_3 are defined by

$$\left.\begin{aligned} N_1 &= \frac{1}{2A^j}[x_2 y_3 - x_3 y_2 + x(y_2 - y_3) + y(x_3 - x_2)] \\ N_2 &= \frac{1}{2A^j}[x_3 y_1 - x_1 y_3 + x(y_3 - y_1) + y(x_1 - x_3)] \\ N_3 &= \frac{1}{2A^j}[x_1 y_2 - x_2 y_1 + x(y_1 - y_2) + y(x_2 - x_1)] \end{aligned}\right\} \tag{5.15b}$$

in which A^j is the area of the element defined by the determinant.

$$A^j = \frac{1}{2}\begin{vmatrix} 1 & x_1 & y_1 \\ 1 & x_2 & y_2 \\ 1 & x_3 & y_3 \end{vmatrix}$$

Consequently, the matrix $\mathbf{S}^j$ of Eq. 4 can be recognized as

$$\mathbf{S}^j = \begin{bmatrix} N_1 & 0 & N_2 & 0 & N_3 & 0 \\ 0 & N_1 & 0 & N_2 & 0 & N_3 \end{bmatrix} \tag{5.16}$$

Observe that

$$N_k(x_i, y_i) = \begin{cases} 0 & \text{if } k \neq i \\ 1 & \text{if } k = i \end{cases}$$

Beam Element The beam element shown in Fig. 2(c) has two nodal points. Each node has three nodal coordinates that represent the longitudinal and transverse displacements and the slope. These nodal coordinates are

$$\mathbf{q}^j = [u_1^j \quad v_1^j \quad \theta_1^j \quad u_2^j \quad v_2^j \quad \theta_2^j]^{\mathrm{T}} \tag{5.17}$$

Since we selected six nodal coordinates, our displacement field must contain six coefficients. The longitudinal displacement is approximated by a first-degree polynomial, while the transverse displacement is approximated by a cubic polynomial. The displacement field is, therefore, given for element j by

$$u^j = a_1^j + a_2^j x \tag{5.18a}$$

$$v^j = a_3^j + a_4^j x + a_5^j x^2 + a_6^j x^3 \tag{5.18b}$$

Note that the axial displacement is described by a first-order polynomial. That is, the axial strain is constant within the element and $\partial^2 u^j/\partial x^2 = 0$. On the other hand, the transverse displacement is described using a cubic polynomial and consequently the shear force is assumed to be constant within the element.

At the nodal points, one has

$$
\begin{aligned}
u_1^j &= a_1^j \\
u_2^j &= a_1^j + a_2^j l^j \\
v_1^j &= a_3^j \\
v_2^j &= a_3^j + a_4^j l^j + a_5^j (l^j)^2 + a_6^j (l^j)^3 \\
\theta_1^j &= a_4^j \\
\theta_2^j &= a_4^j + 2a_5^j l^j + 3a_6^j (l^j)^2
\end{aligned}
$$

where l^j is the length of the element. The preceding equations can be used to define the displacement field in terms of the nodal coordinates. This leads to

$$
\mathbf{u}^j = \begin{bmatrix} u^j \\ v^j \end{bmatrix} = \mathbf{S}^j \mathbf{q}^j \tag{5.19}
$$

where $\mathbf{q}^j$ is the vector of the element nodal coordinates defined by Eq. 17, and $\mathbf{S}^j$ is the space-dependent element shape function defined by

$$
\mathbf{S}^j = \begin{bmatrix} \left(1 - \frac{x}{l^j}\right) & 0 & 0 & \frac{x}{l^j} & 0 & 0 \\ 0 & 1 - 3\left(\frac{x}{l^j}\right)^2 + 2\left(\frac{x}{l^j}\right)^3 & l^j\left[\frac{x}{l^j} - 2\left(\frac{x}{l^j}\right)^2 + \left(\frac{x}{l^j}\right)^3\right] & 0 & 3\left(\frac{x}{l^j}\right)^2 - 2\left(\frac{x}{l^j}\right)^3 & l^j\left[\left(\frac{x}{l^j}\right)^3 - \left(\frac{x}{l^j}\right)^2\right] \end{bmatrix} \tag{5.20}
$$

Note that the shape function of the truss element can be obtained from the shape function of the beam element as a special case in which the transverse deformation is neglected.

Rectangular Element The rectangular element shown in Fig. 2(d) has four nodes. Each node has two degrees of freedom that represent the horizontal and vertical displacements. The nodal coordinates of the element are defined by the vector $\mathbf{q}^j$ as

$$
\mathbf{q}^j = [u_1^j \quad v_1^j \quad u_2^j \quad v_2^j \quad u_3^j \quad v_3^j \quad u_4^j \quad v_4^j]^{\mathrm{T}}
$$

The assumed displacement field of this rectangular element can be written as

$$
\begin{aligned}
u^j &= a_1^j + a_2^j x + a_3^j y + a_4^j xy \\
v^j &= a_5^j + a_6^j x + a_7^j y + a_8^j xy
\end{aligned}
$$

Following the procedure described in the case of the truss, triangular, and beam elements, one can show that the shape function of the four-node rectangular element is given by

$$\mathbf{S}^j = \begin{bmatrix} N_1 & 0 & N_2 & 0 & N_3 & 0 & N_4 & 0 \\ 0 & N_1 & 0 & N_2 & 0 & N_3 & 0 & N_4 \end{bmatrix}$$

where

$$N_1 = \frac{1}{4bc}(b - x)(c - y)$$

$$N_2 = \frac{1}{4bc}(b + x)(c - y)$$

$$N_3 = \frac{1}{4bc}(b + x)(c + y)$$

$$N_4 = \frac{1}{4bc}(b - x)(c + y)$$

Observe again that

$$N_k(x_i, y_i) = \begin{cases} 0 & \text{if } k \neq i \\ 1 & \text{if } k = i \end{cases}$$

5.2 COMMENTS ON THE ELEMENT SHAPE FUNCTIONS

The assumed displacement field of the element is represented by simple functions expressed in the form of polynomials. The use of these simple functions guarantees that the displacement field within the element is continuous. Furthermore, the assumed displacement field must include low-order terms if the approximate solution is to be close enough to the exact solution. For example, consider the assumed displacement field of the beam element given by Eq. 18. If the beam is subjected to a static axial end load F, the exact solution is

$$u^j = \left(\frac{F}{E^j A^j}\right)x, \qquad v^j = 0$$

where E^j and A^j are, respectively, the modulus of elasticity and the cross-sectional area of the element j. Note that if we omit the term $a_2^j x$ in the assumed axial displacement of Eq. 18a, we delete the term that contains the exact answer. It is, therefore, important that the order of the polynomial is selected in such a manner that the approximate solution converges to the exact solution. This property is called the *completeness*. The completeness requirement guarantees that the approximate solution converges to the exact solution as the number of elements increases. The order of the polynomials in the assumed displacement field can be increased by increasing the number of nodes of the element or/and by increasing the number of coordinates of the nodes by considering higher derivatives of the displacements.

It is clear that as a part of the completeness requirements, the assumed displacement field of the element must be able to represent the state of constant strain. The rigid-body motion without strain is a special case of the state of constant strain. Therefore, rigid-body modes must be present in the assumed displacement field. In this section, we examine this property for the elements discussed in the preceding section. The procedure for doing this is to assume that the element undergoes an arbitrary rigid-body motion. The values of the nodal coordinates as the result of this motion can be determined and substituted into the displacement field of the element in order to check whether this assumed displacement field can represent this rigid-body motion.

Truss Element If the truss element whose assumed displacement field is defined by Eq. 11 undergoes a rigid-body translation d_x in the direction of the x-coordinate, the values of the nodal coordinates as the result of this rigid-body motion become

$$[u_1^j \quad u_2^j]^{\mathrm{T}} = [d_x \quad l^j + d_x]^{\mathrm{T}}$$

Substituting this vector into Eq. 11c, the displacement of an arbitrary point on the element as the result of this rigid-body motion is given by

$$\begin{aligned} u^j &= [(1-\xi) \quad \xi]\begin{bmatrix} u_1^j \\ u_2^j \end{bmatrix} \\ &= [(1-\xi) \quad \xi]\begin{bmatrix} d_x \\ l^j + d_x \end{bmatrix} = l^j + d_x \end{aligned}$$

which indicates that each point on the element indeed translates by the amount d_x. Consequently, the truss-element shape function as defined by Eq. 11b can be used to describe an arbitrary rigid-body translation along the x-coordinate of the element.

Rectangular Element While the truss element is used to describe only axial displacements, the linear rectangular element is used to describe an arbitrary planar displacement. An arbitrary rigid-body motion is described by a translation and a rotation. If the linear rectangular element undergoes an arbitrary rigid-body translation d_x in the horizontal direction, the element nodal coordinates as the result of this rigid-body displacement become

$$\mathbf{q}^j = [-b + d_x \quad -c \quad b + d_x \quad -c \quad b + d_x \quad c \quad -b + d_x \quad c]^{\mathrm{T}}$$

Substituting this vector into the displacement field presented in the preceding section, one obtains the displacement of an arbitrary point on the element as the result of this rigid-body motion as

$$\mathbf{u}^j = \begin{bmatrix} u^j \\ v^j \end{bmatrix} = \begin{bmatrix} N_1 & 0 & N_2 & 0 & N_3 & 0 & N_4 & 0 \\ 0 & N_1 & 0 & N_2 & 0 & N_3 & 0 & N_4 \end{bmatrix} \begin{bmatrix} -b + d_x \\ -c \\ b + d_x \\ -c \\ b + d_x \\ c \\ -b + d_x \\ c \end{bmatrix}$$

$$= \begin{bmatrix} (-N_1 + N_2 + N_3 - N_4)b + (N_1 + N_2 + N_3 + N_4)\, d_x \\ (-N_1 - N_2 + N_3 + N_4)c \end{bmatrix}$$

Observe that

$$N_1 + N_2 + N_3 + N_4 = 1$$

$$-N_1 - N_2 + N_3 + N_4 = \frac{y}{c}$$

$$-N_1 + N_2 + N_3 - N_4 = \frac{x}{b}$$

and accordingly

$$\mathbf{u}^j = [x + d_x \quad y]^{\mathrm{T}} = \begin{bmatrix} x \\ y \end{bmatrix} + \begin{bmatrix} d_x \\ 0 \end{bmatrix}$$

That is, the displacement of an arbitrary point on the element as the result of this rigid-body translation is d_x. It is, therefore, clear that the element shape function as defined in the preceding section can be used to describe an arbitrary rigid-body translation in the x-direction. By using a similar procedure one can also show that this element shape function can also be used to describe an arbitrary rigid-body translation in the y-direction.

If the element undergoes a rigid-body rotation α, the value of the nodal coordinates as the result of this rigid-body rotation is

$$\begin{aligned} \mathbf{q}^j = [&(-b \cos\alpha + c \sin\alpha) \quad (-b \sin\alpha - c \cos\alpha) \quad (b \cos\alpha + c \sin\alpha) \\ &(b \sin\alpha - c \cos\alpha) \quad (b \cos\alpha - c \sin\alpha) \quad (b \sin\alpha + c \cos\alpha) \\ &(-b \cos\alpha - c \sin\alpha) \quad (-b \sin\alpha + c \cos\alpha)]^{\mathrm{T}} \end{aligned}$$

Substituting this vector in the assumed displacement field of the bilinear rectangular element defined in the preceding section, one obtains

$$\mathbf{u}^j = \begin{bmatrix} u^j \\ v^j \end{bmatrix} = \begin{bmatrix} (-N_1 + N_2 + N_3 - N_4)b \cos\alpha + (N_1 + N_2 - N_3 - N_4)c \sin\alpha \\ (-N_1 + N_2 + N_3 - N_4)b \sin\alpha + (-N_1 - N_2 + N_3 + N_4)c \cos\alpha \end{bmatrix}$$

$$= \begin{bmatrix} x\cos\alpha - y\sin\alpha \\ x\sin\alpha + y\cos\alpha \end{bmatrix} = \begin{bmatrix} \cos\alpha & -\sin\alpha \\ \sin\alpha & \cos\alpha \end{bmatrix} \begin{bmatrix} x \\ y \end{bmatrix}$$

which indicates that the element shape function of the bilinear rectangular element can be used to describe an arbitrary rigid-body rotation. This rotation can be finite or infinitesimal.

By using similar procedures, one can also show that the shape function of the *constant strain triangular element* can be used to describe an arbitrary rigid-body motion.

Beam Element Following the procedure described in this section, it is an easy matter to show that the shape function of the beam element presented in the preceding section can be used to describe an arbitrary rigid-body translation. In order to examine the ability of this shape function in describing an arbitrary rigid-body rotation, we consider the case in which the element rotates a rigid-body rotation defined by the angle α. In this case, one must have

$$\begin{bmatrix} u^j \\ v^j \end{bmatrix} = \begin{bmatrix} \cos\alpha & -\sin\alpha \\ \sin\alpha & \cos\alpha \end{bmatrix} \begin{bmatrix} x \\ 0 \end{bmatrix} = \begin{bmatrix} x\cos\alpha \\ x\sin\alpha \end{bmatrix}$$

where x is the coordinate of an arbitrary point on the element. In this case

$$\theta^j = \frac{\partial v^j}{\partial x} = \sin\alpha$$

and accordingly the vector of nodal coordinates as the result of this rigid-body rotation is given by

$$\mathbf{q}^j = [0 \quad 0 \quad \sin\alpha \quad l^j\cos\alpha \quad l^j\sin\alpha \quad \sin\alpha]^{\mathrm{T}}$$

Substituting this vector into the displacement field of the beam element as defined by Eqs. 19 and 20, one obtains

$$\begin{bmatrix} u^j \\ v^j \end{bmatrix} = \begin{bmatrix} x\cos\alpha \\ x\sin\alpha \end{bmatrix} = \begin{bmatrix} \cos\alpha & -\sin\alpha \\ \sin\alpha & \cos\alpha \end{bmatrix} \begin{bmatrix} x \\ 0 \end{bmatrix}$$

Note that, in order to arrive at this result, we assumed that the slope $\theta^j = \partial v^j/\partial x = \sin\alpha$ instead of the angle α or $\tan\alpha$. Clearly,

$$\sin\alpha \approx \tan\alpha \approx \alpha$$

if the rigid-body rotation α is infinitesimal. In the classical finite element literature, the slopes at the nodes are assumed to represent infinitesimal rotations in order to allow treating the element coordinates using vector algebra. Transforming the nodal coordinates between different frames is then possible. The use of infinitesimal rotations as nodal coordinates, however, leads to difficulties in the large rotation and deformation analysis of structural systems as discussed in Section 10 of this chapter.

5.3 CONNECTIVITY BETWEEN ELEMENTS

In the preceding sections, the displacement field of the element was defined in the element coordinate system. In the finite-element approach, a structural component is discretized using a number of elements, each of which has its own coordinate system. Because of geometrical considerations imposed by the shape of the structural components, the axes of the element coordinate system may not be parallel to the axes of the global coordinate system. The connectivity conditions between elements, however, require that when two elements are rigidly connected at a nodal point, the coordinates of this point as defined by the nodal coordinates of one element must be the same as defined by the coordinates of the other element. Therefore, before applying the connectivity conditions, the nodal coordinates of the elements must be defined in one global system.

Coordinate Transformation As shown in Fig. 3, let β^j be the angle which the X^j-axis of the element j coordinate system makes with the X-axis of the global coordinate system. Let $\mathbf{C}^j$ be the transformation matrix from the element coordinate system to the global coordinate system. The transformation matrix $\mathbf{C}^j$ is given by

$$\mathbf{C}^j = \begin{bmatrix} \cos\beta^j & -\sin\beta^j \\ \sin\beta^j & \cos\beta^j \end{bmatrix} \tag{5.21}$$

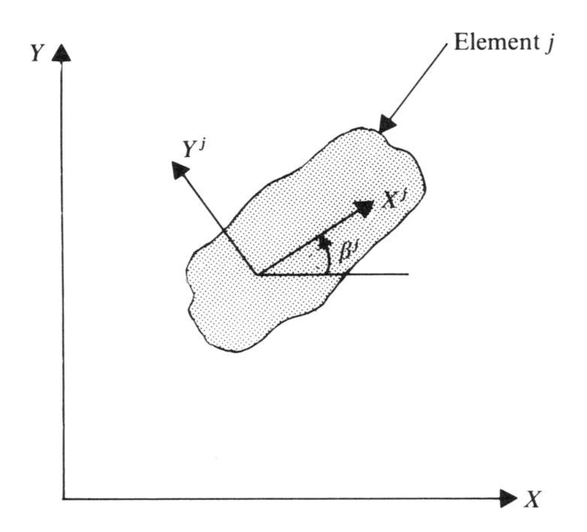

FIG. 5.3. Connectivity conditions.

Therefore, the displacement field $\mathbf{u}^j$ of Eq. 4 can be written in the global coordinate system as

$$\mathbf{u}_g^j = \mathbf{C}^j\mathbf{u}^j = \mathbf{C}^j\mathbf{S}^j\mathbf{q}^j \tag{5.22}$$

In this equation, the vector of nodal coordinates $\mathbf{q}^j$ is defined in the jth element coordinate system. This vector can be expressed in terms of the global coordinates as

$$\mathbf{q}^j = \bar{\mathbf{C}}^j\mathbf{q}_g^j \tag{5.23}$$

where $\mathbf{q}_g^j$ is the vector of nodal coordinates defined in the global coordinate system and $\bar{\mathbf{C}}^j$ is an orthogonal transformation matrix. In the case of the triangular element the matrix $\bar{\mathbf{C}}^j$ is given by

$$\bar{\mathbf{C}}^j = \begin{bmatrix} \mathbf{C}^{j\mathrm{T}} & \mathbf{0} & \mathbf{0} \\ \mathbf{0} & \mathbf{C}^{j\mathrm{T}} & \mathbf{0} \\ \mathbf{0} & \mathbf{0} & \mathbf{C}^{j\mathrm{T}} \end{bmatrix} \tag{5.24}$$

where $\mathbf{C}^j$ is defined by Eq. 21.

In the case of the beam element the matrix $\bar{\mathbf{C}}^j$ is given by

$$\bar{\mathbf{C}}^j = \begin{bmatrix} \mathbf{C}_1^{j\mathrm{T}} & \mathbf{0} \\ \mathbf{0} & \mathbf{C}_1^{j\mathrm{T}} \end{bmatrix} \tag{5.25}$$

where $\mathbf{C}_1^j$ is the matrix

$$\mathbf{C}_1^j = \begin{bmatrix} \cos\beta^j & -\sin\beta^j & 0 \\ \sin\beta^j & \cos\beta^j & 0 \\ 0 & 0 & 1 \end{bmatrix} = \begin{bmatrix} \mathbf{C}^j & \mathbf{0} \\ \mathbf{0}^\mathrm{T} & 1 \end{bmatrix} \tag{5.26}$$

Substituting Eq. 23 into Eq. 22, the global displacement vector can be expressed in terms of the global element nodal coordinates as

$$\mathbf{u}_g^j = \mathbf{C}^j\mathbf{S}^j\bar{\mathbf{C}}^j\mathbf{q}_g^j \tag{5.27}$$

which can be written as

$$\mathbf{u}_g^j = \mathbf{S}_g^j\mathbf{q}_g^j \tag{5.28}$$

where

$$\mathbf{S}_g^j = \mathbf{C}^j\mathbf{S}^j\bar{\mathbf{C}}^j \tag{5.29}$$

Connectivity Conditions Having defined the element nodal coordinates in the global system, one can proceed a step further to write the displacement vector in terms of the total vector of nodal coordinates of the structure. To this end, we define the total vector of nodal coordinates of the structure as

$$\mathbf{q}_g = [q_1 \quad q_2 \quad \cdots \quad q_n]^\mathrm{T} \tag{5.30}$$

where n is the total number of nodal coordinates of the structure. One can then write the vector of nodal coordinates of the element in terms of the structure nodal coordinates as

$$\mathbf{q}_g^j = \mathbf{B}^j\mathbf{q}_g \tag{5.31}$$

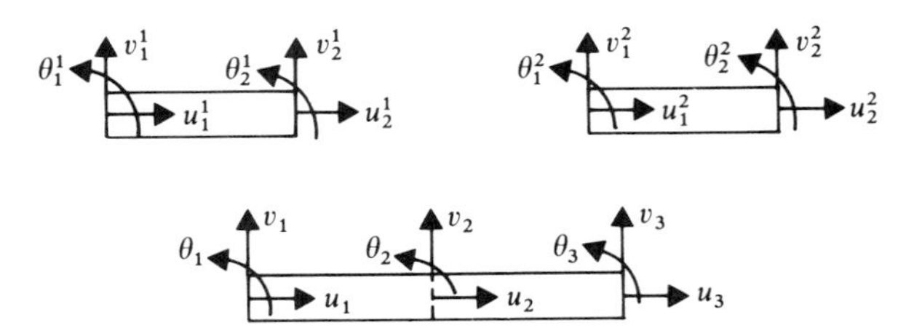

FIG. 5.4. Connectivity between two beam elements.

where $\mathbf{B}^j$ is a *Boolean matrix* whose elements are either zeros or ones. For example, we consider the beam shown in Fig. 4. The beam is divided into two finite beam elements. For these two elements the nodal coordinates are

$$\mathbf{q}_g^1 = [u_1^1 \quad v_1^1 \quad \theta_1^1 \quad u_2^1 \quad v_2^1 \quad \theta_2^1]^{\mathrm{T}}$$
$$\mathbf{q}_g^2 = [u_1^2 \quad v_1^2 \quad \theta_1^2 \quad u_2^2 \quad v_2^2 \quad \theta_2^2]^{\mathrm{T}}$$

Since the structure has three nodes, the total vector of the structure nodal coordinates is given by

$$\mathbf{q}_g = [u_1 \quad v_1 \quad \theta_1 \quad u_2 \quad v_2 \quad \theta_2 \quad u_3 \quad v_3 \quad \theta_3]^{\mathrm{T}}$$

The vector $\mathbf{q}_g^1$ of the nodal coordinates of the first element can be written as

$$\mathbf{q}_g^1 = \mathbf{B}^1 \mathbf{q}_g$$

where $\mathbf{B}^1$ is the matrix

$$\mathbf{B}^1 = \begin{bmatrix} 1 & 0 & 0 & 0 & 0 & 0 & 0 & 0 & 0 \\ 0 & 1 & 0 & 0 & 0 & 0 & 0 & 0 & 0 \\ 0 & 0 & 1 & 0 & 0 & 0 & 0 & 0 & 0 \\ 0 & 0 & 0 & 1 & 0 & 0 & 0 & 0 & 0 \\ 0 & 0 & 0 & 0 & 1 & 0 & 0 & 0 & 0 \\ 0 & 0 & 0 & 0 & 0 & 1 & 0 & 0 & 0 \end{bmatrix}$$

Similarly, for the second element

$$\mathbf{q}_g^2 = \mathbf{B}^2 \mathbf{q}_g$$

where

$$\mathbf{B}^2 = \begin{bmatrix} 0 & 0 & 0 & 1 & 0 & 0 & 0 & 0 & 0 \\ 0 & 0 & 0 & 0 & 1 & 0 & 0 & 0 & 0 \\ 0 & 0 & 0 & 0 & 0 & 1 & 0 & 0 & 0 \\ 0 & 0 & 0 & 0 & 0 & 0 & 1 & 0 & 0 \\ 0 & 0 & 0 & 0 & 0 & 0 & 0 & 1 & 0 \\ 0 & 0 & 0 & 0 & 0 & 0 & 0 & 0 & 1 \end{bmatrix}$$

That is, the structure of the matrices $\mathbf{B}^1$ and $\mathbf{B}^2$ defines the second node as a common node for the two elements in this example. Note that, in general, the matrix $\mathbf{B}^j$ has the number of rows equal to the number of element coordinates, and the number of columns equal to the number of nodal coordinates of the structure.

The global displacement vector $\mathbf{u}_g^j$ of element j can now be expressed in terms of the total vector of the nodal coordinates of the structure. To this end, we substitute Eq. 31 into Eq. 28 to obtain

$$\mathbf{u}_g^j = \mathbf{S}_g^j \mathbf{B}^j \mathbf{q}_g, \qquad j = 1, 2, \ldots, n_e \tag{5.32a}$$

This equation and Eq. 29 will be used in the following sections to develop the element mass and stiffness matrices. The structure mass and stiffness matrices can then be obtained by assembling the element matrices.

Kinematic Constraints In many structural applications some of the nodal coordinates may be specified. For example, one end of a beam or a truss in a structural system may be fixed. Such conditions can be used to reduce the number of the elastic degrees of freedom of the structure. In these cases the vector of nodal coordinates $\mathbf{q}_g$ can be written as

$$\mathbf{q}_g = [\mathbf{q}_f^{\mathrm{T}} \quad \mathbf{q}_s^{\mathrm{T}}]^{\mathrm{T}}$$

where $\mathbf{q}_s$ is the vector of *specified nodal coordinates* and $\mathbf{q}_f$ is the vector of *free nodal coordinates* or the *system degrees of freedom.* If the specified nodes are fixed, one can write the vector $\mathbf{q}_g$ in terms of the system degrees of freedom as

$$\mathbf{q}_g = \mathbf{B}_c \mathbf{q}_f \tag{5.32b}$$

where $\mathbf{B}_c$ is an appropriate matrix whose entries are either zeros or ones. If no nodal coordinates are fixed, $\mathbf{B}_c$ is the identity matrix. Substituting Eq. 32b into Eq. 32a yields

$$\mathbf{u}_g^j = \mathbf{S}_g^j \mathbf{B}^j \mathbf{B}_c \mathbf{q}_f \tag{5.32c}$$

The use of the development presented in this section is demonstrated by the following examples.

Example 5.1

For the structural system shown in Fig. 5, obtain the connectivity matrices. Assume that the system is divided into two beam elements.

Solution. As shown in the figure, the structure has three nodal points. The nodal coordinates of the structure defined in the global coordinate system are

$$\mathbf{q}_g = [u_1 \quad v_1 \quad \theta_1 \quad u_2 \quad v_2 \quad \theta_2 \quad u_3 \quad v_3 \quad \theta_3]^{\mathrm{T}}$$

The coordinate systems of the elements are also shown in the figure. For the first element it is clear that $\beta^1 = 90°$ and, accordingly, the matrix $\mathbf{C}^j$ of Eq. 21 is given by

$$\mathbf{C}^1 = \begin{bmatrix} \cos\beta^1 & -\sin\beta^1 \\ \sin\beta^1 & \cos\beta^1 \end{bmatrix} = \begin{bmatrix} \cos 90° & -\sin 90° \\ \sin 90° & \cos 90° \end{bmatrix}$$
$$= \begin{bmatrix} 0 & -1 \\ 1 & 0 \end{bmatrix}$$

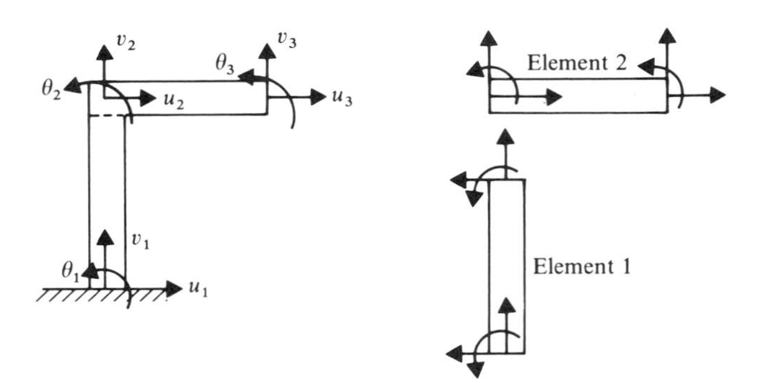

FIG. 5.5. Structural system.

For the second element $\beta^2 = 0$ and, accordingly, the matrix $\mathbf{C}^2$ is the identity matrix, that is,

$$\mathbf{C}^2 = \begin{bmatrix} 1 & 0 \\ 0 & 1 \end{bmatrix}$$

Therefore, the matrices $\mathbf{C}_1^j, j = 1, 2$, of Eq. 26 are given by

$$\mathbf{C}_1^1 = \begin{bmatrix} 0 & -1 & 0 \\ 1 & 0 & 0 \\ 0 & 0 & 1 \end{bmatrix}$$

$$\mathbf{C}_1^2 = \begin{bmatrix} 1 & 0 & 0 \\ 0 & 1 & 0 \\ 0 & 0 & 1 \end{bmatrix}$$

We can then write the nodal coordinates of the two elements as

$$\mathbf{q}^j = \bar{\mathbf{C}}^j \mathbf{q}_g^j, \qquad j = 1, 2$$

where $\bar{\mathbf{C}}^j$ is the matrix

$$\bar{\mathbf{C}}^j = \begin{bmatrix} \mathbf{C}_1^{j\mathrm{T}} & \mathbf{0} \\ \mathbf{0} & \mathbf{C}_1^{j\mathrm{T}} \end{bmatrix}$$

That is, the nodal coordinates of the element defined in the global system are

$$\mathbf{q}_g^1 = [-v_1^1 \quad u_1^1 \quad \theta_1^1 \quad -v_2^1 \quad u_2^1 \quad \theta_2^1]^\mathrm{T}$$
$$\mathbf{q}_g^2 = [u_1^2 \quad v_1^2 \quad \theta_1^2 \quad u_2^2 \quad v_2^2 \quad \theta_2^2]^\mathrm{T}$$

Since these two vectors are defined in the global coordinate system, we also have

$$\mathbf{q}_g^1 = [u_1 \quad v_1 \quad \theta_1 \quad u_2 \quad v_2 \quad \theta_2]^\mathrm{T}$$
$$\mathbf{q}_g^2 = [u_2 \quad v_2 \quad \theta_2 \quad u_3 \quad v_3 \quad \theta_3]^\mathrm{T}$$

That is, the connectivity matrices $\mathbf{B}^1$ and $\mathbf{B}^2$ are defined for this system as

$$\mathbf{B}^1 = \begin{bmatrix} 1 & 0 & 0 & 0 & 0 & 0 & 0 & 0 & 0 \\ 0 & 1 & 0 & 0 & 0 & 0 & 0 & 0 & 0 \\ 0 & 0 & 1 & 0 & 0 & 0 & 0 & 0 & 0 \\ 0 & 0 & 0 & 1 & 0 & 0 & 0 & 0 & 0 \\ 0 & 0 & 0 & 0 & 1 & 0 & 0 & 0 & 0 \\ 0 & 0 & 0 & 0 & 0 & 1 & 0 & 0 & 0 \end{bmatrix}$$

$$\mathbf{B}^2 = \begin{bmatrix} 0 & 0 & 0 & 1 & 0 & 0 & 0 & 0 & 0 \\ 0 & 0 & 0 & 0 & 1 & 0 & 0 & 0 & 0 \\ 0 & 0 & 0 & 0 & 0 & 1 & 0 & 0 & 0 \\ 0 & 0 & 0 & 0 & 0 & 0 & 1 & 0 & 0 \\ 0 & 0 & 0 & 0 & 0 & 0 & 0 & 1 & 0 \\ 0 & 0 & 0 & 0 & 0 & 0 & 0 & 0 & 1 \end{bmatrix}$$

Furthermore, since in this example node 1 is fixed, one has

$$u_1 = v_1 = \theta_1 = 0$$

and the degrees of freedom of the system are given by

$$\mathbf{q}_f = [u_2 \quad v_2 \quad \theta_2 \quad u_3 \quad v_3 \quad \theta_3]^T$$

The vector of nodal coordinates of the structure can be expressed in terms of these degrees of freedom as

$$\begin{bmatrix} u_1 \\ v_1 \\ \theta_1 \\ u_2 \\ v_2 \\ \theta_2 \\ u_3 \\ v_3 \\ \theta_3 \end{bmatrix} = \begin{bmatrix} 0 & 0 & 0 & 0 & 0 & 0 \\ 0 & 0 & 0 & 0 & 0 & 0 \\ 0 & 0 & 0 & 0 & 0 & 0 \\ 1 & 0 & 0 & 0 & 0 & 0 \\ 0 & 1 & 0 & 0 & 0 & 0 \\ 0 & 0 & 1 & 0 & 0 & 0 \\ 0 & 0 & 0 & 1 & 0 & 0 \\ 0 & 0 & 0 & 0 & 1 & 0 \\ 0 & 0 & 0 & 0 & 0 & 1 \end{bmatrix} \begin{bmatrix} u_2 \\ v_2 \\ \theta_2 \\ u_3 \\ v_3 \\ \theta_3 \end{bmatrix}$$

where the matrix $\mathbf{B}_c$ in this example can be recognized as

$$\mathbf{B}_c = \begin{bmatrix} 0 & 0 & 0 & 0 & 0 & 0 \\ 0 & 0 & 0 & 0 & 0 & 0 \\ 0 & 0 & 0 & 0 & 0 & 0 \\ 1 & 0 & 0 & 0 & 0 & 0 \\ 0 & 1 & 0 & 0 & 0 & 0 \\ 0 & 0 & 1 & 0 & 0 & 0 \\ 0 & 0 & 0 & 1 & 0 & 0 \\ 0 & 0 & 0 & 0 & 1 & 0 \\ 0 & 0 & 0 & 0 & 0 & 1 \end{bmatrix}$$

By using the matrices $\mathbf{C}^1$, $\bar{\mathbf{C}}^1$, $\mathbf{B}^1$, $\mathbf{B}_c$, and the vector $\mathbf{q}_f$ defined in this example, the displacement field of the first element can be defined by using Eqs. 29 and 32c as

$$\mathbf{u}^1 = \mathbf{C}^1\mathbf{S}^1\bar{\mathbf{C}}^1\mathbf{B}^1\mathbf{B}_c\mathbf{q}_f$$

Similarly, the displacement field of the second element is

$$\mathbf{u}^2 = \mathbf{C}^2\mathbf{S}^2\bar{\mathbf{C}}^2\mathbf{B}^2\mathbf{B}_c\mathbf{q}_f$$

5.4 FORMULATION OF THE MASS MATRIX

In this section, the element mass matrix is formulated using the kinetic energy expression. Two approaches are commonly used in the finite-element formulation of the mass matrix. The first approach is the *lumped-mass approach* in which the element mass matrix is obtained by lumping the element mass at the nodal points. One way of using the lumped-mass approach is to describe the inertia of the element by concentrated masses which have zero moment of inertia about their centers. This approach leads to a singular mass matrix in problems involving beams since there is no inertia assigned to the rotational degrees of freedom. Clearly, this approach has the drawback that the kinetic energy becomes a positive-semidefinite quadratic form instead of positive-definite. Another way of using the lumped mass approach is to describe the inertia of the element using small rigid bodies attached to the nodal points of the element. The process of doing this, however, is arbitrary.

The second approach which is used in the finite-element formulation of the mass matrix is the *consistent-mass formulation.* In this approach, the assumed displacement field of the element is used to define the element kinetic energy. Consequently, the resulting mass matrix is often nondiagonal and positive definite. This is the approach which we will discuss in this section for the formulation of the element mass matrix.

Element Mass Matrix The kinetic energy of the element j is defined as

$$T^j = \frac{1}{2}\int_{V^j} \rho^j \dot{\mathbf{u}}_g^{jT}\dot{\mathbf{u}}_g^j \, dV^j, \qquad j = 1, 2, \ldots, n_e \tag{5.33}$$

where V^j and ρ^j are, respectively, the volume and mass density of the element j and $\dot{\mathbf{u}}_g^j$ is the global velocity vector which is defined using Eq. 32c as

$$\dot{\mathbf{u}}_g^j = \mathbf{S}_g^j\mathbf{B}^j\mathbf{B}_c\dot{\mathbf{q}}_f, \qquad j = 1, 2, \ldots, n_e \tag{5.34}$$

where $\mathbf{B}^j$ is the Boolean matrix that defines the connectivity of the element, $\mathbf{B}_c$ is the matrix of the kinematic constraints, and $\mathbf{S}_g^j$ is the shape matrix defined by Eq. 29.

Substituting Eq. 34 into Eq. 33, one obtains

$$T^j = \frac{1}{2}\dot{\mathbf{q}}_f^{\mathrm{T}}\mathbf{B}_c^{\mathrm{T}}\mathbf{B}^{j\mathrm{T}}\left[\int_{V^j} \rho^j \mathbf{S}_g^{j\mathrm{T}}\mathbf{S}_g^j \, dV^j\right]\mathbf{B}^j\mathbf{B}_c\dot{\mathbf{q}}_f$$

which can also be written as

$$T^j = \frac{1}{2}\dot{\mathbf{q}}_f^T \mathbf{M}_f^j \dot{\mathbf{q}}_f, \qquad j = 1, 2, \ldots, n_e \tag{5.35}$$

where $\mathbf{M}_f^j$ is recognized as the jth element mass matrix and it is defined as

$$\mathbf{M}_f^j = \mathbf{B}_c^T \mathbf{B}^{jT} \left[\int_{V^j} \rho^j \mathbf{S}_g^{jT} \mathbf{S}_g^j \, dV^j \right] \mathbf{B}^j \mathbf{B}_c, \qquad j = 1, 2, \ldots, n_e \tag{5.36}$$

Observe that this matrix has dimension equal to the number of degrees of freedom of the structure. Using the definition of the matrix $\mathbf{S}_g^j$ of Eq. 29 and the orthogonality of the transformation matrix $\mathbf{C}^j$, the mass matrix $\mathbf{M}_f^j$ of Eq. 36 can be written as

$$\mathbf{M}_f^j = \mathbf{B}_c^T \mathbf{B}^{jT} \bar{\mathbf{C}}^{jT} \mathbf{M}^j \bar{\mathbf{C}}^j \mathbf{B}^j \mathbf{B}_c, \qquad j = 1, 2, \ldots, n_e \tag{5.37}$$

where $\mathbf{M}^j$ is the element mass matrix defined in the element coordinate system. This element mass matrix is given by

$$\mathbf{M}^j = \int_{V^j} \rho^j \mathbf{S}^{jT} \mathbf{S}^j \, dV^j, \qquad j = 1, 2, \ldots, n_e \tag{5.38}$$

The mass matrix $\mathbf{M}^j$ is symmetric and has dimension equal to the number of the element nodal coordinates. The global element mass matrix $\mathbf{M}_f^j$, on the other hand, has a dimension equal to the number of free coordinates of the structure.

Structure Mass Matrix The mass matrix of the structure can be defined by summing the kinetic energies of its elements. To this end, we write

$$T = \sum_{j=1}^{n_e} T^j \tag{5.39}$$

where T is the kinetic energy of the structure.

Substituting Eq. 35 into Eq. 39 leads to

$$T = \frac{1}{2}\dot{\mathbf{q}}_f^T \left[\sum_{j=1}^{n_e} \mathbf{M}_f^j \right] \dot{\mathbf{q}}_f$$

which can be rewritten as

$$T = \frac{1}{2}\dot{\mathbf{q}}_f^T \mathbf{M}_f \dot{\mathbf{q}}_f \tag{5.40}$$

where $\mathbf{M}_f$ is recognized as the structure mass matrix and is defined as

$$\mathbf{M}_f = \sum_{j=1}^{n_e} \mathbf{M}_f^j \tag{5.41}$$

By using Eq. 37, the structure mass matrix $\mathbf{M}_f$ can be written in a more explicit form as

$$\mathbf{M}_f = \sum_{j=1}^{n_e} \mathbf{B}_c^T \mathbf{B}^{jT} \bar{\mathbf{C}}^{jT} \mathbf{M}^j \bar{\mathbf{C}}^j \mathbf{B}^j \mathbf{B}_c \tag{5.42}$$

where the element mass matrix $\mathbf{M}^j$ is defined by Eq. 38.

Example 5.2

For the truss element with the shape function defined by Eq. 11, the element mass matrix $\mathbf{M}^j$ defined in the element coordinate system is given by

$$\mathbf{M}^j = \int_{V^j} \rho^j \mathbf{S}^{j\mathrm{T}} \mathbf{S}^j \, dV^j = \frac{m^j}{6}\begin{bmatrix} 2 & 1 \\ 1 & 2 \end{bmatrix}$$

where m^j is the mass of the element j. The system shown in Fig. 6 consists of two truss elements which have equal masses. The transformation matrices $\mathbf{C}^j$ and $\bar{\mathbf{C}}^j$ for these elements are the identity matrices. The connectivity matrices $\mathbf{B}^1$ and $\mathbf{B}^2$ are

$$\mathbf{B}^1 = \begin{bmatrix} 1 & 0 & 0 \\ 0 & 1 & 0 \end{bmatrix}$$

$$\mathbf{B}^2 = \begin{bmatrix} 0 & 1 & 0 \\ 0 & 0 & 1 \end{bmatrix}$$

Therefore, the matrices $\mathbf{M}_f^1$ and $\mathbf{M}_f^2$ are

$$\mathbf{M}_f^1 = \mathbf{B}^{1\mathrm{T}}\bar{\mathbf{C}}^{1\mathrm{T}}\mathbf{M}^1\bar{\mathbf{C}}^1\mathbf{B}^1 = \mathbf{B}^{1\mathrm{T}}\mathbf{M}^1\mathbf{B}^1$$

$$= \frac{m^1}{6}\begin{bmatrix} 1 & 0 \\ 0 & 1 \\ 0 & 0 \end{bmatrix}\begin{bmatrix} 2 & 1 \\ 1 & 2 \end{bmatrix}\begin{bmatrix} 1 & 0 & 0 \\ 0 & 1 & 0 \end{bmatrix} = \frac{m^1}{6}\begin{bmatrix} 2 & 1 & 0 \\ 1 & 2 & 0 \\ 0 & 0 & 0 \end{bmatrix}$$

$$\mathbf{M}_f^2 = \mathbf{B}^{2\mathrm{T}}\bar{\mathbf{C}}^{2\mathrm{T}}\mathbf{M}^2\bar{\mathbf{C}}^2\mathbf{B}^2 = \mathbf{B}^{2\mathrm{T}}\mathbf{M}^2\mathbf{B}^2$$

$$= \frac{m^2}{6}\begin{bmatrix} 0 & 0 \\ 1 & 0 \\ 0 & 1 \end{bmatrix}\begin{bmatrix} 2 & 1 \\ 1 & 2 \end{bmatrix}\begin{bmatrix} 0 & 1 & 0 \\ 0 & 0 & 1 \end{bmatrix} = \frac{m^2}{6}\begin{bmatrix} 0 & 0 & 0 \\ 0 & 2 & 1 \\ 0 & 1 & 2 \end{bmatrix}$$

If we assume that $m^1 = m^2 = m$, the structure mass matrix $\mathbf{M}_f$ can then be defined as

$$\mathbf{M}_f = \sum_{j=1}^{2} \mathbf{M}_f^j = \mathbf{M}_f^1 + \mathbf{M}_f^2$$

$$= \frac{m}{6}\begin{bmatrix} 2 & 1 & 0 \\ 1 & 2 & 0 \\ 0 & 0 & 0 \end{bmatrix} + \frac{m}{6}\begin{bmatrix} 0 & 0 & 0 \\ 0 & 2 & 1 \\ 0 & 1 & 2 \end{bmatrix}$$

$$= \frac{m}{6}\begin{bmatrix} 2 & 1 & 0 \\ 1 & 4 & 1 \\ 0 & 1 & 2 \end{bmatrix}$$

FIG. 5.6. Two-truss member.

Example 5.3

For the beam element with the shape function defined by Eq. 20, the element mass matrix $\mathbf{M}^j$ is given by

$$\mathbf{M}^j = \int_{V^j} \rho^j \mathbf{S}^{jT} \mathbf{S}^j \, dV^j$$

$$= m^j \begin{bmatrix} 1/3 & & & & & \text{Symmetric} \\ 0 & 13/35 & & & & \\ 0 & 11l/210 & l^2/105 & & & \\ 1/6 & 0 & 0 & 1/3 & & \\ 0 & 9/70 & 13l/420 & 0 & 13/35 & \\ 0 & -13/420 & -l^2/140 & 0 & -11l/210 & l^2/105 \end{bmatrix}^j$$

where m^j and l^j are, respectively, the mass and length of the element and superscript j on the major bracket implies that all quantities inside the bracket are superscripted with j. If we consider the structural system of Example 1, for the first element, one can show by direct matrix multiplication that

$$\bar{\mathbf{C}}^1 \mathbf{B}^1 \mathbf{B}_c = \begin{bmatrix} 0 & 0 & 0 & 0 & 0 & 0 \\ 0 & 0 & 0 & 0 & 0 & 0 \\ 0 & 0 & 0 & 0 & 0 & 0 \\ 0 & -1 & 0 & 0 & 0 & 0 \\ 1 & 0 & 0 & 0 & 0 & 0 \\ 0 & 0 & 1 & 0 & 0 & 0 \end{bmatrix}$$

where the matrices $\bar{\mathbf{C}}^1$, $\mathbf{B}^1$, and $\mathbf{B}_c$ are evaluated in Example 1. If we assume that the two elements have the same mass m and equal length l, one can then write the mass matrix of the first element using Eq. 37 as

$$\mathbf{M}_f^1 = \mathbf{B}_c^T \mathbf{B}^{1T} \bar{\mathbf{C}}^{1T} \mathbf{M}^1 \bar{\mathbf{C}}^1 \mathbf{B}^1 \mathbf{B}_c$$

$$= m \begin{bmatrix} 13/35 & 0 & -11l/210 & 0 & 0 & 0 \\ 0 & 1/3 & 0 & 0 & 0 & 0 \\ -11l/210 & 0 & l^2/105 & 0 & 0 & 0 \\ 0 & 0 & 0 & 0 & 0 & 0 \\ 0 & 0 & 0 & 0 & 0 & 0 \\ 0 & 0 & 0 & 0 & 0 & 0 \end{bmatrix}$$

For the second element one has

$$\bar{\mathbf{C}}^2 \mathbf{B}^2 \mathbf{B}_c = \begin{bmatrix} 1 & 0 & 0 & 0 & 0 & 0 \\ 0 & 1 & 0 & 0 & 0 & 0 \\ 0 & 0 & 1 & 0 & 0 & 0 \\ 0 & 0 & 0 & 1 & 0 & 0 \\ 0 & 0 & 0 & 0 & 1 & 0 \\ 0 & 0 & 0 & 0 & 0 & 1 \end{bmatrix}$$

Therefore, the mass matrix $\mathbf{M}_f^2$ of the second element is

$$\mathbf{M}_f^2 = \mathbf{B}_c^T \mathbf{B}^{2T} \bar{\mathbf{C}}^{2T} \mathbf{M}^2 \bar{\mathbf{C}}^2 \mathbf{B}^2 \mathbf{B}_c$$

$$= m \begin{bmatrix} 1/3 & 0 & 0 & 1/6 & 0 & 0 \\ 0 & 13/35 & 11l/210 & 0 & 9/70 & -13/420 \\ 0 & 11l/210 & l^2/105 & 0 & 13l/420 & -l^2/140 \\ 1/6 & 0 & 0 & 1/3 & 0 & 0 \\ 0 & 9/70 & 13l/420 & 0 & 13/35 & -11l/210 \\ 0 & -13/420 & -l^2/140 & 0 & -11l/210 & l^2/105 \end{bmatrix}$$

The composite mass matrix of the structure can then be obtained as

$$\mathbf{M}_f = \sum_{j=1}^{2} \mathbf{M}_f^j = \mathbf{M}_f^1 + \mathbf{M}_f^2$$

$$= m \begin{bmatrix} 74/105 & 0 & -11l/210 & 1/6 & 0 & 0 \\ 0 & 74/35 & 11l/210 & 0 & 9/70 & -13/420 \\ -11l/210 & 11l/210 & 2l^2/105 & 0 & 13l/420 & -l^2/140 \\ 1/6 & 0 & 0 & 1/3 & 0 & 0 \\ 0 & 9/70 & 13l/420 & 0 & 13/35 & -11l/210 \\ 0 & -13/420 & -l^2/140 & 0 & -11l/210 & l^2/105 \end{bmatrix}$$

5.5 FORMULATION OF THE STIFFNESS MATRIX

There are several techniques for developing the stiffness matrix of the finite element. In the analysis presented in this section, the strain energy is used to formulate the stiffness matrices. To this end, some basic concepts and definitions which are important in the analysis of continuous systems are briefly discussed.

Stress-Strain Relationships For an element j, the strain energy is given by

$$U^j = \frac{1}{2} \int_{V^j} \boldsymbol{\sigma}^{jT} \boldsymbol{\varepsilon}^j \, dV^j \tag{5.43}$$

where V^j is the element volume and $\boldsymbol{\sigma}^j$ and $\boldsymbol{\varepsilon}^j$ are, respectively, the vectors of stress and strain. The stress and strain vectors are in general six-dimensional vectors which can be written as

$$\boldsymbol{\sigma}^j = [\sigma_1 \quad \sigma_2 \quad \sigma_3 \quad \sigma_4 \quad \sigma_5 \quad \sigma_6]^T$$
$$= [\sigma_x \quad \sigma_y \quad \sigma_z \quad \sigma_{xy} \quad \sigma_{yz} \quad \sigma_{zx}]^T$$

and

$$\boldsymbol{\varepsilon}^j = [\varepsilon_1 \quad \varepsilon_2 \quad \varepsilon_3 \quad \varepsilon_4 \quad \varepsilon_5 \quad \varepsilon_6]^T$$
$$= [\varepsilon_x \quad \varepsilon_y \quad \varepsilon_z \quad \varepsilon_{xy} \quad \varepsilon_{yz} \quad \varepsilon_{zx}]^T$$

For an *isotropic material*, the stress components can be expressed in terms of the strain components as

$$\sigma_x = \frac{E}{(1+v)(1-2v)}[(1-2v)\varepsilon_x + v\varepsilon_t]$$

$$\sigma_y = \frac{R}{(1+v)(1-2v)}[(1-2v)\varepsilon_y + v\varepsilon_t]$$

$$\sigma_z = \frac{E}{(1+v)(1-2v)}[(1-2v)\varepsilon_z + v\varepsilon_t]$$

$$\sigma_{xy} = G\varepsilon_{xy}, \qquad \sigma_{yz} = G\varepsilon_{yz}, \qquad \sigma_{zx} = G\varepsilon_{zx}$$

where E, G, and v are, respectively, the modulus of elasticity, modulus of rigidity, and Poisson's ratio, and $\varepsilon_t = \varepsilon_x + \varepsilon_y + \varepsilon_z$. The stress vector $\boldsymbol{\sigma}^j$ can be written compactly using the *constitutive equations* as

$$\boldsymbol{\sigma}^j = \mathbf{E}^j\boldsymbol{\varepsilon}^j \tag{5.44}$$

where $\mathbf{E}^j$ is the matrix of elastic coefficients. The preceding equation is the *generalized Hooke's law*. The matrix $\mathbf{E}^j$ of the elastic coefficients depends on the material properties of the structure. This matrix can be written explicitly in terms of the modulus of elasticity, modulus of rigidity, and Poisson's ratio as

$$\mathbf{E}^j = \frac{E}{(1+v)(1-2v)}\begin{bmatrix} 1-v & v & v & 0 & 0 & 0 \\ v & 1-v & v & 0 & 0 & 0 \\ v & v & 1-v & 0 & 0 & 0 \\ 0 & 0 & 0 & (1-2v)/2 & 0 & 0 \\ 0 & 0 & 0 & 0 & (1-2v)/2 & 0 \\ 0 & 0 & 0 & 0 & 0 & (1-2v)/2 \end{bmatrix}$$

The inverse of this matrix is

$$(\mathbf{E}^j)^{-1} = \frac{1}{E}\begin{bmatrix} 1 & -v & -v & 0 & 0 & 0 \\ -v & 1 & -v & 0 & 0 & 0 \\ -v & -v & 1 & 0 & 0 & 0 \\ 0 & 0 & 0 & 2(1+v) & 0 & 0 \\ 0 & 0 & 0 & 0 & 2(1+v) & 0 \\ 0 & 0 & 0 & 0 & 0 & 2(1+v) \end{bmatrix}$$

Accordingly, the strain components can be expressed in terms of the stress components as

$$\varepsilon_x = \frac{1}{E}[\sigma_x - v(\sigma_y + \sigma_z)]$$

$$\varepsilon_y = \frac{1}{E}[\sigma_y - v(\sigma_x + \sigma_z)]$$

$$\varepsilon_z = \frac{1}{E}[\sigma_z - \nu(\sigma_x + \sigma_y)]$$

$$\varepsilon_{xy} = \frac{\sigma_{xy}}{G}, \qquad \varepsilon_{yz} = \frac{\sigma_{yz}}{G}, \qquad \varepsilon_{zx} = \frac{\sigma_{zx}}{G}$$

where the modulus of rigidity G can be expressed in terms of the modulus of elasticity E and Poisson's ratio ν as

$$G = \frac{E}{2(1 + \nu)}$$

For *plane strain*, $\varepsilon_z = \varepsilon_{xz} = \varepsilon_{yz} = 0$, the stress–strain relationships reduce in this special case to

$$\sigma_x = \frac{E}{(1 + \nu)(1 - 2\nu)}[(1 - \nu)\varepsilon_x + \nu\varepsilon_y]$$

$$\sigma_y = \frac{E}{(1 + \nu)(1 - 2\nu)}[(1 - \nu)\varepsilon_y + \nu\varepsilon_x]$$

$$\sigma_z = \frac{\nu E}{(1 + \nu)(1 - 2\nu)}(\varepsilon_x + \varepsilon_y), \qquad \sigma_{xy} = G\varepsilon_{xy}$$

which can be written in a matrix form as

$$\begin{bmatrix} \sigma_x \\ \sigma_y \\ \sigma_z \\ \sigma_{xy} \end{bmatrix} = \frac{E}{(1 + \nu)(1 - 2\nu)} \begin{bmatrix} 1 - \nu & \nu & 0 \\ \nu & 1 - \nu & 0 \\ \nu & \nu & 0 \\ 0 & 0 & (1 - 2\nu)/2 \end{bmatrix} \begin{bmatrix} \varepsilon_x \\ \varepsilon_y \\ \varepsilon_{xy} \end{bmatrix}$$

Observe that in the case of plane strain, one has

$$\sigma_z = \nu(\sigma_x + \sigma_y)$$

The inverse relationships are given by

$$\varepsilon_x = \frac{1 + \nu}{E}[\sigma_x - \nu(\sigma_x + \sigma_y)]$$

$$\varepsilon_y = \frac{1 + \nu}{E}[\sigma_y - \nu(\sigma_x + \sigma_y)]$$

$$\varepsilon_{xy} = \frac{2(1 + \nu)}{E}\sigma_{xy}$$

which can be written in a matrix form as

$$\begin{bmatrix} \varepsilon_x \\ \varepsilon_y \\ \varepsilon_{xy} \end{bmatrix} = \frac{(1 + \nu)}{E} \begin{bmatrix} 1 - \nu & -\nu & 0 \\ -\nu & 1 - \nu & 0 \\ 0 & 0 & 2 \end{bmatrix} \begin{bmatrix} \sigma_x \\ \sigma_y \\ \sigma_{xy} \end{bmatrix}$$

Another special case is the case of *plane stress*. In this case, $\sigma_z = \sigma_{yz} = \sigma_{zx} = 0$. By using a similar procedure to the case of plane strain, the constitutive relationships in the case of plane stress can be obtained.

Strain-Displacement Relationships For a linear elastic material, the strain-displacement relationships are

$$\varepsilon_x = \frac{\partial u}{\partial x}, \qquad \varepsilon_y = \frac{\partial v}{\partial y}, \qquad \varepsilon_z = \frac{\partial w}{\partial z}$$

$$\varepsilon_{xy} = \frac{\partial u}{\partial y} + \frac{\partial v}{\partial x}$$

$$\varepsilon_{yz} = \frac{\partial v}{\partial z} + \frac{\partial w}{\partial y}$$

$$\varepsilon_{xz} = \frac{\partial w}{\partial x} + \frac{\partial u}{\partial z}$$

where u, v, and w are the translational displacements of an arbitrary point on the elastic medium in the x, y, and z directions, respectively. That is

$$\mathbf{u}^j = [u \quad v \quad w]^{\mathrm{T}}$$

The strain-displacement relationships can be written compactly as

$$\boldsymbol{\varepsilon}^j = \mathbf{D}^j \mathbf{u}^j \tag{5.45}$$

where $\mathbf{D}^j$ is a differential operator defined as

$$\mathbf{D}^j = \begin{bmatrix} \partial/\partial x & 0 & 0 \\ 0 & \partial/\partial y & 0 \\ 0 & 0 & \partial/\partial z \\ \partial/\partial y & \partial/\partial x & 0 \\ 0 & \partial/\partial z & \partial/\partial y \\ \partial/\partial z & 0 & \partial/\partial x \end{bmatrix}$$

Element Stiffness Matrix The constitutive equations and the strain-displacement relationships can be used to express the strain vector in terms of the derivatives of the displacement.

Substituting Eq. 45 into Eq. 44 yields

$$\boldsymbol{\sigma}^j = \mathbf{E}^j \mathbf{D}^j \mathbf{u}^j \tag{5.46}$$

which can be written in terms of the nodal coordinates of the structure as

$$\boldsymbol{\sigma}^j = \mathbf{E}^j \mathbf{D}^j \mathbf{S}^j \overline{\mathbf{C}}^j \mathbf{B}^j \mathbf{B}_{\mathrm{c}} \mathbf{q}_{\mathrm{f}} \tag{5.47}$$

Similarly, the strain $\boldsymbol{\varepsilon}^j$ can be written as

$$\boldsymbol{\varepsilon}^j = \mathbf{D}^j\mathbf{S}^j\bar{\mathbf{C}}^j\mathbf{B}^j\mathbf{B}_c\mathbf{q}_f \tag{5.48}$$

Substituting Eqs. 47 and 48 into the strain energy of Eq. 43, one obtains

$$U^j = \frac{1}{2}\int_{V^j} \mathbf{q}_f^T\mathbf{B}_c^T\mathbf{B}^{jT}\bar{\mathbf{C}}^{jT}(\mathbf{D}^j\mathbf{S}^j)^T\mathbf{E}^j\mathbf{D}^j\mathbf{S}^j\bar{\mathbf{C}}^j\mathbf{B}^j\mathbf{B}_c\mathbf{q}_f\, dV^j \tag{5.49}$$

where the symmetry of the matrix of elastic coefficients has been utilized.

Equation 49 can be written in a simple form as

$$U^j = \frac{1}{2}\mathbf{q}_f^T\mathbf{K}_f^j\mathbf{q}_f \tag{5.50}$$

where $\mathbf{K}_f^j$ is the global element stiffness matrix defined as

$$\mathbf{K}_f^j = \mathbf{B}_c^T\mathbf{B}^{jT}\bar{\mathbf{C}}^{jT}\left[\int_{V^j}(\mathbf{D}^j\mathbf{S}^j)^T\mathbf{E}^j\mathbf{D}^j\mathbf{S}^j\, dV^j\right]\bar{\mathbf{C}}^j\mathbf{B}^j\mathbf{B}_c \tag{5.51}$$

The global stiffness matrix $\mathbf{K}_f^j$ can be written as

$$\mathbf{K}_f^j = \mathbf{B}_c^T\mathbf{B}^{jT}\bar{\mathbf{C}}^{jT}\mathbf{K}^j\bar{\mathbf{C}}^j\mathbf{B}^j\mathbf{B}_c, \qquad j = 1, 2, \ldots, n_e \tag{5.52}$$

where $\mathbf{K}^j$ is the element stiffness matrix defined in the element coordinate system and is given by

$$\mathbf{K}^j = \int_{V^j}(\mathbf{D}^j\mathbf{S}^j)^T\mathbf{E}^j\mathbf{D}^j\mathbf{S}^j\, dV^j, \qquad j = 1, 2, \ldots, n_e \tag{5.53}$$

In order to illustrate the procedure discussed in this section for formulating the element stiffness matrix, we consider the four node, eight degree of freedom planar rectangular element. The shape function of this element is defined in Section 1 as

$$\mathbf{S}^j = \begin{bmatrix} N_1 & 0 & N_2 & 0 & N_3 & 0 & N_4 & 0 \\ 0 & N_1 & 0 & N_2 & 0 & N_3 & 0 & N_4 \end{bmatrix}$$

where

$$N_1 = \frac{1}{4bc}(b - x)(c - y)$$

$$N_2 = \frac{1}{4bc}(b + x)(c - y)$$

$$N_3 = \frac{1}{4bc}(b + x)(c + y)$$

$$N_4 = \frac{1}{4bc}(b - x)(c + y)$$

where $2b$ and $2c$ are the dimensions of the element as shown in Fig. 2(d). In the case of two-dimensional displacement, the differential operator $\mathbf{D}^j$ is given by

$$\mathbf{D}^j = \begin{bmatrix} \dfrac{\partial}{\partial x} & 0 \\ 0 & \dfrac{\partial}{\partial y} \\ \dfrac{\partial}{\partial y} & \dfrac{\partial}{\partial x} \end{bmatrix}$$

Therefore,

$$\mathbf{D}^j\mathbf{S}^j = \begin{bmatrix} \dfrac{\partial N_1}{\partial x} & 0 & \dfrac{\partial N_2}{\partial x} & 0 & \dfrac{\partial N_3}{\partial x} & 0 & \dfrac{\partial N_4}{\partial x} & 0 \\ 0 & \dfrac{\partial N_1}{\partial y} & 0 & \dfrac{\partial N_2}{\partial y} & 0 & \dfrac{\partial N_3}{\partial y} & 0 & \dfrac{\partial N_4}{\partial y} \\ \dfrac{\partial N_1}{\partial y} & \dfrac{\partial N_1}{\partial x} & \dfrac{\partial N_2}{\partial y} & \dfrac{\partial N_2}{\partial x} & \dfrac{\partial N_3}{\partial y} & \dfrac{\partial N_3}{\partial x} & \dfrac{\partial N_4}{\partial y} & \dfrac{\partial N_4}{\partial x} \end{bmatrix}$$

in which

$$\frac{\partial N_1}{\partial x} = -\frac{(c-y)}{4bc}, \qquad \frac{\partial N_1}{\partial y} = -\frac{(b-x)}{4bc}$$

$$\frac{\partial N_2}{\partial x} = \frac{(c-y)}{4bc}, \qquad \frac{\partial N_2}{\partial y} = -\frac{(b+x)}{4bc}$$

$$\frac{\partial N_3}{\partial x} = \frac{(c+y)}{4bc}, \qquad \frac{\partial N_3}{\partial y} = \frac{(b+x)}{4bc}$$

$$\frac{\partial N_4}{\partial x} = -\frac{(c+y)}{4bc}, \qquad \frac{\partial N_4}{\partial y} = \frac{(b-x)}{4bc}$$

Therefore, the matrix $\mathbf{D}^j\mathbf{S}^j$ is given by

$$\mathbf{D}^j\mathbf{S}^j = \frac{1}{4bc}\begin{bmatrix} -(c-y) & 0 & (c-y) & 0 & (c+y) & 0 & -(c+y) & 0 \\ 0 & -(b-x) & 0 & -(b+x) & 0 & (b+x) & 0 & (b-x) \\ -(b-x) & -(c-y) & -(b+x) & (c-y) & (b+x) & (c+y) & (b-x) & -(c+y) \end{bmatrix} \tag{5.54}$$

If we consider the case of plane stress, the matrix of elastic coefficients $\mathbf{E}^j$ is given by

$$\mathbf{E}^j = \frac{E}{(1-\nu^2)}\begin{bmatrix} 1 & \nu & 0 \\ \nu & 1 & 0 \\ 0 & 0 & (1-\nu)/2 \end{bmatrix}$$

where E and ν are, respectively, the modulus of elasticity and Poisson's ratio of the element. Assuming that the element has a constant thickness and constant modulus of elasticity and Poisson's ratio, the use of Eq. 53 leads to

$$\mathbf{K}^j = \int_{-c}^{c}\int_{-b}^{b} (\mathbf{D}^j\mathbf{S}^j)^{\mathrm{T}}\mathbf{E}^j\mathbf{D}^j\mathbf{S}^j t \, dx \, dy$$

where t is the thickness of the element. Using the matrices $\mathbf{D}^j\mathbf{S}^j$ and $\mathbf{E}^j$, the stiffness matrix of the four node rectangular element can be written explicitly as

$$\mathbf{K}^j = \frac{Et}{12(1-\nu^2)}\begin{bmatrix} k_{11} & & & & & & & \\ k_{21} & k_{22} & & & & & & \\ k_{31} & k_{32} & k_{33} & & & \text{symmetric} & & \\ k_{41} & k_{42} & k_{43} & k_{44} & & & & \\ k_{51} & k_{52} & k_{53} & k_{54} & k_{55} & & & \\ k_{61} & k_{62} & k_{63} & k_{64} & k_{65} & k_{66} & & \\ k_{71} & k_{72} & k_{73} & k_{74} & k_{75} & k_{76} & k_{77} & \\ k_{81} & k_{82} & k_{83} & k_{84} & k_{85} & k_{86} & k_{87} & k_{88} \end{bmatrix}$$

in which the stiffness coefficients k_{ij} are

$$k_{11} = k_{33} = k_{55} = k_{77} = 4r + 2(1-\nu)r^{-1}$$

$$k_{22} = k_{44} = k_{66} = k_{88} = 4r^{-1} + 2(1-\nu)r$$

$$k_{21} = -k_{61} = -k_{52} = -k_{43} = k_{83} = k_{74} = k_{65} = -k_{87} = \tfrac{3}{2}(1+\nu)$$

$$k_{31} = k_{75} = -4r + (1-\nu)r^{-1}$$

$$k_{41} = -k_{81} = -k_{32} = k_{72} = k_{63} = -k_{54} = k_{85} = -k_{76} = -\tfrac{3}{2}(1-3\nu)$$

$$k_{51} = k_{73} = -2r - (1-\nu)r^{-1}$$

$$k_{71} = k_{53} = 2r - 2(1-\nu)r^{-1}$$

$$k_{42} = k_{86} = 2r^{-1} - 2(1-\nu)r$$

$$k_{62} = k_{84} = -2r^{-1} - (1-\nu)r$$

$$k_{82} = k_{64} = -4r^{-1} + (1-\nu)r$$

where $r = c/b$.

Structure Stiffness Matrix The structure stiffness matrix can be defined by summing the strain energy expressions of its elements, that is,

$$U = \sum_{j=1}^{n_e} U^j \tag{5.55a}$$

where U is the strain energy of the structure. Substituting Eq. 50 into Eq. 55a yields

$$U = \tfrac{1}{2}\mathbf{q}_f^{\mathrm{T}}\left[\sum_{j=1}^{n_e} \mathbf{K}_f^j\right]\mathbf{q}_f \tag{5.55b}$$

which can be written as

$$U = \tfrac{1}{2}\mathbf{q}_f^{\mathrm{T}}\mathbf{K}_f\mathbf{q}_f \tag{5.56}$$

where $\mathbf{K}_f$ is the stiffness matrix of the structure which is defined as

$$\mathbf{K}_f = \sum_{j=1}^{n_e} \mathbf{K}_f^j \tag{5.57}$$

or in a more explicit form as

$$\mathbf{K}_f = \sum_{j=1}^{n_e} \mathbf{B}_c^{\mathrm{T}}\mathbf{B}^{j\mathrm{T}}\bar{\mathbf{C}}^{j\mathrm{T}}\mathbf{K}^j\bar{\mathbf{C}}^j\mathbf{B}^j\mathbf{B}_c \tag{5.58}$$

where $\mathbf{K}^j$ is the jth element stiffness matrix defined in the element coordinate system.

Example 5.4

The strain energy for the truss element of Example 2 is

$$U^j = \tfrac{1}{2}\int_{V^j} \sigma_x^j \varepsilon_x^j \, dV^j$$

where σ_x^j and ε_x^j are, respectively, the stress and strain components. The stress–strain relationship in this simple case is given by

$$\sigma_x^j = E^j\varepsilon_x^j$$

where E^j is Young's modulus of the element material. The strain displacement relationship is

$$\varepsilon_x^j = \frac{\partial u^j}{\partial x} = D^j u^j$$

where the differential operator of Eq. 45 reduces in this case to

$$D^j = \frac{\partial}{\partial x}$$

The displacement u^j can be written in terms of the nodal coordinates as

$$u^j = \mathbf{S}^j\mathbf{q}^j$$

where $\mathbf{S}^j$ is the element shape function defined as

$$\mathbf{S}^j = \left[\left(1 - \frac{x}{l^j}\right) \quad \frac{x}{l^j}\right]$$

That is,

$$D^j\mathbf{S}^j = \frac{1}{l^j}[-1 \quad 1]$$

Therefore, the stiffness matrix of Eq. 53 can be written as

$$\mathbf{K}^j = \int_{V^j} (\mathbf{D}^j\mathbf{S}^j)^{\mathrm{T}}\mathbf{E}^j\mathbf{D}^j\mathbf{S}^j\, dV^j$$

$$= A^j \int_0^{l^j} \frac{1}{l^{j2}} \begin{bmatrix} -1 \\ 1 \end{bmatrix} E^j [-1 \quad 1]\, dx$$

$$= \frac{E^jA^j}{l^j} \begin{bmatrix} 1 & -1 \\ -1 & 1 \end{bmatrix}$$

where A^j is the cross-sectional area of the truss element j. The stiffness matrix $\mathbf{K}^j$ is defined in the element coordinate system. The global matrix can be obtained using a procedure similar to the one described for the mass matrix of the system discussed in Example 2.

Example 5.5

Using the assumptions of the elementary beam theory and neglecting rotary inertia and shear deformation, one can show that the strain energy for a beam element j can be written as

$$U^j = \frac{1}{2} \int_0^{l^j} \left[E^jA^j \left(\frac{\partial u^j}{\partial x} \right)^2 + E^jI^j \left(\frac{\partial^2 v^j}{\partial x^2} \right)^2 \right] dx$$

where E^j is the modulus of elasticity, A^j is the cross-sectional area, I^j is the second moment of area, and u^j and v^j are, respectively, the axial and transverse displacements. The strain energy can be written in the following form:

$$U^j = \frac{1}{2} \int_0^{l^j} [(u^j)' \quad (v^j)''] \begin{bmatrix} E^jA^j & 0 \\ 0 & E^jI^j \end{bmatrix} \begin{bmatrix} (u^j)' \\ (v^j)'' \end{bmatrix} dx$$

where

$$\begin{bmatrix} (u^j)' \\ (v^j)'' \end{bmatrix} = \begin{bmatrix} \mathbf{S}_x^{j\prime} \\ \mathbf{S}_y^{j\prime\prime} \end{bmatrix} \mathbf{q}^j$$

where $\mathbf{S}_x^j$ and $\mathbf{S}_y^j$ are the rows of the element shape function defined by Eq. 20, that is,

$$\mathbf{S}^j = \begin{bmatrix} \mathbf{S}_x^j \\ \mathbf{S}_y^j \end{bmatrix}$$

One can then write the strain energy as

$$U^j = \tfrac{1}{2}\mathbf{q}^{j\mathrm{T}}\mathbf{K}^j\mathbf{q}^j$$

where the stiffness matrix $\mathbf{K}^j$ is defined as

$$\mathbf{K}^j = \int_0^{l^j} [E^jA^j\mathbf{S}_x^{j\prime\mathrm{T}}\mathbf{S}_x^{j\prime} + E^jI^j\mathbf{S}_y^{j\prime\prime\mathrm{T}}\mathbf{S}_y^{j\prime\prime}]\, dx$$

Using the shape function of Eq. 20, it can be shown that the stiffness matrix of the beam element is given explicitly by

$$\mathbf{K}^j = \frac{E^j I^j}{l^j}\begin{bmatrix} A/I & & & & & \\ 0 & 12/l^2 & & & \text{Symmetric} & \\ 0 & 6/l & 4 & & & \\ -A/I & 0 & 0 & A/I & & \\ 0 & -12/l^2 & -6/l & 0 & 12/l^2 & \\ 0 & 6/l & 2 & 0 & -6/l & 4 \end{bmatrix}^j$$

where superscript j on the major bracket implies that all the variables inside the bracket are superscripted with j. Note that $\mathbf{K}^j$ is the element stiffness matrix defined in the element coordinate system.

For the structural system of Example 1, if we assume that the two elements have the same dimensions and material properties and follow a procedure similar to the one used for the mass matrix in Example 3, one can show that the stiffness matrices $\mathbf{K}_f^1$ and $\mathbf{K}_f^2$ of the two elements are given, respectively, by

$$K_f^1 = \frac{EI}{l}\begin{bmatrix} 12/l^2 & 0 & 6/l & 0 & 0 & 0 \\ 0 & A/I & 0 & 0 & 0 & 0 \\ 6/l & 0 & 4 & 0 & 0 & 0 \\ 0 & 0 & 0 & 0 & 0 & 0 \\ 0 & 0 & 0 & 0 & 0 & 0 \\ 0 & 0 & 0 & 0 & 0 & 0 \end{bmatrix}$$

and

$$\mathbf{K}_f^2 = \frac{EI}{l}\begin{bmatrix} A/I & 0 & 0 & -A/I & 0 & 0 \\ 0 & 12/l^2 & 6/l & 0 & -12/l^2 & 6/l \\ 0 & 6/l & 4 & 0 & -6/l & 2 \\ -A/I & 0 & 0 & A/I & 0 & 0 \\ 0 & -12/l^2 & -6/l & 0 & 12/l^2 & -6/l \\ 0 & 6/l & 2 & 0 & -6/l & 4 \end{bmatrix}$$

where E, I, l, and A, are, respectively, the modulus of elasticity, second moment of area, length, and area of the elements.

The composite stiffness matrix of the structure can then be obtained using Eq. 57 as

$$\mathbf{K}_f = \sum_{j=1}^{2} \mathbf{K}_f^j = \mathbf{K}_f^1 + \mathbf{K}_f^2$$

That is,

$$\mathbf{K}_f = \frac{EI}{l}\begin{bmatrix} (A/I + 12/l^2) & 0 & 6/l & -A/I & 0 & 0 \\ 0 & (A/I + 12/l^2) & 6/l & 0 & -12/l^2 & 6/l \\ 6/l & 6/l & 8 & 0 & -6/l & 2 \\ -A/I & 0 & 0 & A/I & 0 & 0 \\ 0 & -12/l^2 & -6/l & 0 & 12/l^2 & -6/l \\ 0 & 6/l & 2 & 0 & -6/l & 4 \end{bmatrix}$$

5.6 EQUATIONS OF MOTION

In the preceding sections, expressions for the element kinetic and strain energies were obtained and used to identify, respectively, the element mass and stiffness matrices. The structure kinetic and strain energies can be obtained by adding the energy expressions of its elements. This leads to the definition of the structure mass and stiffness matrices. In this section, the energy expressions are used with the virtual work of the external forces to define the equations of motion of the structure.

Let $\mathbf{F}^j$ be the vector of the external forces that acts at a point p^j of element j. If $\mathbf{F}^j$ is defined in the global coordinate system, the virtual work of this force vector is given by

$$\delta W^j = \mathbf{F}^{j\mathrm{T}} \delta \mathbf{u}_{\mathrm{g}}^j \tag{5.59}$$

Using Eq. 32c, the virtual change $\delta \mathbf{u}_{\mathrm{g}}^j$ can be written as

$$\begin{aligned} \delta \mathbf{u}_{\mathrm{g}}^j &= \mathbf{S}_{\mathrm{g}}^j \mathbf{B}^j \mathbf{B}_{\mathrm{c}} \delta \mathbf{q}_{\mathrm{f}} \\ &= \mathbf{C}^j \mathbf{S}^j \bar{\mathbf{C}}^j \mathbf{B}^j \mathbf{B}_{\mathrm{c}} \delta \mathbf{q}_{\mathrm{f}} \end{aligned} \tag{5.60}$$

where $\mathbf{S}^j$ is evaluated at the point of application of the force. Substituting Eq. 60 into Eq. 59 yields

$$\delta W^j = \mathbf{F}^{j\mathrm{T}} \mathbf{C}^j \mathbf{S}^j \bar{\mathbf{C}}^j \mathbf{B}^j \mathbf{B}_{\mathrm{c}} \delta \mathbf{q}_{\mathrm{f}}$$

which can be written as

$$\delta W^j = \mathbf{Q}_{\mathrm{f}}^{j\mathrm{T}} \delta \mathbf{q}_{\mathrm{f}} \tag{5.61}$$

in which $\mathbf{Q}_{\mathrm{f}}^j$ is the generalized force vector given by

$$\mathbf{Q}_{\mathrm{f}}^j = \mathbf{B}_{\mathrm{c}}^{\mathrm{T}} \mathbf{B}_j^{\mathrm{T}} \bar{\mathbf{C}}^{j\mathrm{T}} \mathbf{S}^{j\mathrm{T}} \mathbf{C}^{j\mathrm{T}} \mathbf{F}^j, \qquad j = 1, 2, \ldots, n_{\mathrm{e}} \tag{5.62}$$

The virtual work of the forces acting on the structure is given by

$$\begin{aligned} \delta W &= \sum_{j=1}^{n_{\mathrm{e}}} \delta W^j \\ &= \sum_{j=1}^{n_{\mathrm{e}}} \mathbf{Q}_{\mathrm{f}}^{j\mathrm{T}} \delta \mathbf{q}_{\mathrm{f}} = \mathbf{Q}_{\mathrm{f}}^{\mathrm{T}} \delta \mathbf{q}_{\mathrm{f}} \end{aligned} \tag{5.63}$$

where $\mathbf{Q}_{\mathrm{f}}$ is the vector of generalized forces associated with the generalized nodal coordinates of the structure and is defined as

$$\mathbf{Q}_{\mathrm{f}} = \sum_{j=1}^{n_{\mathrm{e}}} \mathbf{Q}_{\mathrm{f}}^j \tag{5.64}$$

Having defined in the preceding sections the kinetic and strain energies of the structure and having obtained the generalized forces associated with the structure nodal coordinates, one can use Lagrange's equation to develop the equations of vibration of the structure as

$$\mathbf{M}_{\mathrm{f}} \ddot{\mathbf{q}}_{\mathrm{f}} + \mathbf{K}_{\mathrm{f}} \mathbf{q}_{\mathrm{f}} = \mathbf{Q}_{\mathrm{f}} \tag{5.65}$$

where $\mathbf{M}_f$ and $\mathbf{K}_f$ are, respectively, the symmetric mass and stiffness matrices developed in the preceding sections, $\mathbf{q}_f$ is the vector of nodal coordinates defined in the global coordinate system, and $\mathbf{Q}_f$ is the vector of generalized forces defined by Eq. 64. If the system is viscously damped, Eq. 65 can be modified in order to include the damping effect. In this case, the matrix equation of motion takes the form

$$\mathbf{M}_f\ddot{\mathbf{q}}_f + \mathbf{C}_f\dot{\mathbf{q}}_f + \mathbf{K}_f\mathbf{q}_f = \mathbf{Q}_f \tag{5.66}$$

where $\mathbf{C}_f$ is the damping matrix.

Equations 65 and 66 are in a form similar to the matrix equation obtained for the multi-degree of freedom systems. Therefore, the solution techniques discussed in Chapter 3 can be used to obtain a solution for the matrix equation of Eqs. 65 and 66. If the number of nodal coordinates of the structure is very large, modal truncation methods can be used in order to reduce the number of degrees of freedom.

The use of the procedure described in this section for formulating the equations of motion of structural systems is demonstrated by the following example.

Example 5.6

Figure 7 shows the structural system discussed in Examples 1, 3, and 5. The force vector $\mathbf{F}(t) = [F\cos\alpha \quad F\sin\alpha]^{\mathrm{T}}$ acts at point B as shown in the figure. The virtual work of this force is given by

$$\delta W^2 = \mathbf{F}^{\mathrm{T}}\delta\mathbf{u}_{gB}^2$$

where $\mathbf{u}_{gB}^2$ is the position vector of point B defined in the global system. The virtual change $\delta\mathbf{u}_{gB}^2$ can be obtained using Eq. 60 as

$$\delta\mathbf{u}_{gB}^2 = \mathbf{C}^2\mathbf{S}^2\bar{\mathbf{C}}^2\mathbf{B}^2\mathbf{B}_c\delta\mathbf{q}_f$$

Observe that in this example $\mathbf{C}^2$ and $\bar{\mathbf{C}}^2$ are identity matrices and, accordingly, $\delta\mathbf{u}_{gB}^2$ reduces to

$$\delta\mathbf{u}_{gB}^2 = \mathbf{S}^2\mathbf{B}^2\mathbf{B}_c\delta\mathbf{q}_f$$

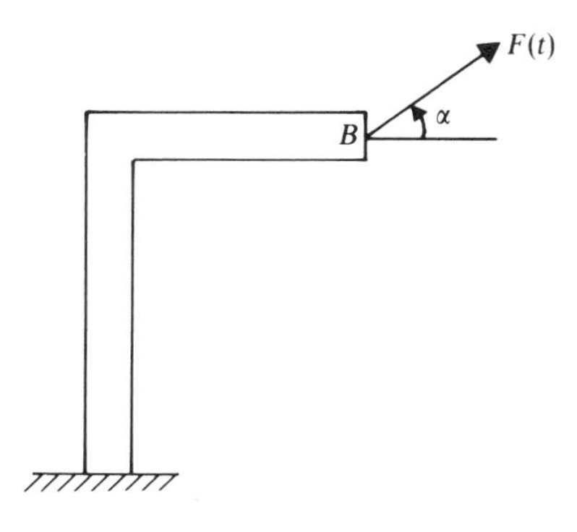

FIG. 5.7. Forced vibration.

in which the shape function $\mathbf{S}^2$ must be evaluated at point B at which $x = l$. By using the shape function of Eq. 20 and substituting $x = l$, one obtains

$$\mathbf{S}^2 = \begin{bmatrix} 0 & 0 & 0 & 1 & 0 & 0 \\ 0 & 0 & 0 & 0 & 1 & 0 \end{bmatrix}$$

Furthermore, from the results of Example 5, $\mathbf{B}^2\mathbf{B}_c$ is the identity matrix. Therefore, the virtual work reduces to

$$\begin{aligned} \delta W^2 &= \mathbf{F}^{\mathrm{T}}\mathbf{S}^2\delta\mathbf{q}_f \\ &= [F\cos\alpha \quad F\sin\alpha]\begin{bmatrix} 0 & 0 & 0 & 1 & 0 & 0 \\ 0 & 0 & 0 & 0 & 1 & 0 \end{bmatrix}\begin{bmatrix} \delta u_2 \\ \delta v_2 \\ \delta\theta_2 \\ \delta u_3 \\ \delta v_3 \\ \delta\theta_3 \end{bmatrix} \\ &= [0 \quad 0 \quad 0 \quad F\cos\alpha \quad F\sin\alpha \quad 0]\begin{bmatrix} \delta u_2 \\ \delta v_2 \\ \delta\theta_2 \\ \delta u_3 \\ \delta v_3 \\ \delta\theta_3 \end{bmatrix} \end{aligned}$$

which can be written compactly as

$$\delta W^2 = \mathbf{Q}_f^{2\mathrm{T}}\delta\mathbf{q}_f$$

where $\mathbf{Q}_f^2$ is the vector of generalized forces resulting from the application of the force $\mathbf{F}$ and is given by

$$\mathbf{Q}_f^2 = [0 \quad 0 \quad 0 \quad F\cos\alpha \quad F\sin\alpha \quad 0]^{\mathrm{T}}$$

Since there are no forces acting on the first element one has

$$\mathbf{Q}_f^1 = \mathbf{0}$$

By using Eq. 64, the generalized forces of the structure can then be obtained as

$$\mathbf{Q}_f = \mathbf{Q}_f^1 + \mathbf{Q}_f^2 = \mathbf{Q}_f^2$$

The mass and stiffness matrices of this structural system were obtained, respectively, in Examples 3 and 5. Using the results presented in these two examples and Eq. 65, the vibration equations of this structural system can be written as

$$\mathbf{M}_f\ddot{\mathbf{q}}_f + \mathbf{K}_f\mathbf{q}_f = \mathbf{Q}_f$$

or in a more explicit form as

$$m\begin{bmatrix} 74/105 & 0 & -11l/210 & 1/6 & 0 & 0 \\ 0 & 74/35 & 11l/210 & 0 & 9/70 & -13/420 \\ -11l/210 & 11l/210 & 2l^2/105 & 0 & 13l/420 & -l^2/140 \\ 1/6 & 0 & 0 & 1/3 & 0 & 0 \\ 0 & 9/70 & 13l/420 & 0 & 13/35 & -11l/210 \\ 0 & -13/420 & -l^2/140 & 0 & -11l/210 & l^2/105 \end{bmatrix}\begin{bmatrix} \ddot{u}_2 \\ \ddot{v}_2 \\ \ddot{\theta}_2 \\ \ddot{u}_3 \\ \ddot{v}_3 \\ \ddot{\theta}_3 \end{bmatrix}$$

$$
+\frac{EI}{l}\begin{bmatrix}
(A/I + 12/l^2) & 0 & 6/l & -A/I & 0 & 0 \\
0 & (A/I + 12/l^2) & 6/l & 0 & -12/l^2 & 6/l \\
6/l & 6/l & 8 & 0 & -6/l & 2 \\
-A/I & 0 & 0 & A/I & 0 & 0 \\
0 & -12/l^2 & -6/l & 0 & 12/l^2 & -6/l \\
0 & 6/l & 2 & 0 & -6/l & 4
\end{bmatrix}\begin{bmatrix} u_2 \\ v_2 \\ \theta_2 \\ u_3 \\ v_3 \\ \theta_3 \end{bmatrix}
$$

$$
= \begin{bmatrix} 0 \\ 0 \\ 0 \\ F\cos\alpha \\ F\sin\alpha \\ 0 \end{bmatrix}
$$

Remarks The close relationship between the finite-element method and the classical approximation techniques, discussed at the end of the preceding chapter, should be apparent by now. In fact, the finite-element method can be regarded as a special case of the Rayleigh–Ritz method. While in both techniques the separation of variables is used to write the displacement field in terms of a set of assumed shape functions and a set of time-dependent coordinates, there are some important differences between the two techniques. For example, in the finite-element method, the time-dependent coordinates represent physical variables such as the displacements and slopes at selected nodal points. By defining these variables in a global coordinate system, the assembly of the finite elements is possible. In the Rayleigh–Ritz method, the coefficients in the assumed displacement field may lack any physical meaning. Furthermore, in the Rayleigh–Ritz method, the assumed displacement field is defined for the whole structure. In many practical applications, especially in structures with complex geometry, difficulties are encountered in defining a suitable displacement field. This basic problem is avoided by using the finite-element method, since the assumed displacement field is defined over the domain of a small element. Consequently, the global deflected shape of the structure is defined by simply selecting the appropriate boundary conditions imposed on the element nodal coordinates. By assuming the displacement field for relatively small elements, low-order interpolating functions can be used, leading to a simple formulation for the element mass and stiffness matrices. The element matrices, which are often computer generated, are assembled in order to obtain the mass and stiffness matrices of the structure.

5.7 CONVERGENCE OF THE FINITE-ELEMENT SOLUTION

The finite-element theory guarantees that the approximate finite-element solution converges to the exact solution by increasing the number of elements.

The use of more elements, however, leads to an increase in the number of degrees of freedom and, accordingly, an increase in the number of differential equations which must be solved. In this section we examine the accuracy of the finite-element solution. In particular, we consider the finite-element solution of the eigenvalue problem and compare the obtained results with the exact solution.

In the case of free undamped vibration, Eq. 65 or Eq. 66 leads to

$$\mathbf{M}_{\mathrm{f}}\ddot{\mathbf{q}}_{\mathrm{f}} + \mathbf{K}_{\mathrm{f}}\mathbf{q}_{\mathrm{f}} = \mathbf{0} \tag{5.67}$$

were $\mathbf{M}_{\mathrm{f}}$ and $\mathbf{K}_{\mathrm{f}}$ are, respectively, the mass and stiffness matrices of the structure formulated using the finite-element method, and $\mathbf{q}_{\mathrm{f}}$ is the vector of free nodal coordinates. As in the case of multi-degree of freedom systems we assume a solution to Eq. 67 in the form

$$\mathbf{q}_{\mathrm{f}} = \mathbf{A}\sin(\omega t + \phi) \tag{5.68}$$

where $\mathbf{A}$ is the vector of amplitudes, ω is the frequency, t is time, and ϕ is the phase angle.

Substituting, Eq. 68 into Eq. 67, one obtains the standard eigenvalue problem

$$[\mathbf{K}_{\mathrm{f}} - \omega^2\mathbf{M}_{\mathrm{f}}]\mathbf{A} = \mathbf{0} \tag{5.69}$$

For this system of equations to have a nontrivial solution, the determinant of the coefficient matrix must be equal to zero, that is

$$|\mathbf{K}_{\mathrm{f}} - \omega^2\mathbf{M}_{\mathrm{f}}| = 0 \tag{5.70}$$

This is the characteristic equation which is of order n in ω^2, where n is the number of degrees of freedom of the system. The roots of the characteristic equation define the eigenvalues $\omega_1^2, \omega_2^2, \ldots, \omega_n^2$. The eigenvector or mode shape associated with the natural frequency ω_i is obtained using Eq. 69 as

$$[\mathbf{K}_{\mathrm{f}} - \omega_i^2\mathbf{M}_{\mathrm{f}}]\mathbf{A}_i = \mathbf{0} \tag{5.71}$$

In order to examine the convergence of the finite-element solution, we consider the simple example of the longitudinal vibration of a prismatic rod with one end fixed. It was shown in the preceding chapter that the exact natural frequencies and mode shapes of the rod are

$$\omega_k = \frac{(2k-1)\pi}{2l}\sqrt{\frac{E}{\rho}}$$

$$\phi_k = A_{1k}\sin\frac{\omega_k}{c}x, \qquad k = 1, 2, 3, \ldots$$

where E, ρ, and l are, respectively, the modulus of elasticity, mass density, and length of the rod; A_{1k} is an arbitrary constant and c is the wave velocity without dispersion defined as

$$c = \sqrt{\frac{E}{\rho}}$$

Let us consider the case in which the rod is represented by one truss element. In this case, since one end of the rod ($x = 0$) is fixed, the assumed displacement field is

$$u(x, t) = \xi q(t)$$

where $\xi = x/l$ and $q(t)$ is the axial displacement of the free end. The kinetic energy of the rod in this case is

$$T = \frac{1}{2}\int_0^l \rho A \dot{u}^2 \, dx = \tfrac{1}{2}\rho A l \dot{q}^2 \int_0^1 \xi^2 \, d\xi$$

$$= \frac{1}{2}\frac{\rho A l}{3}\dot{q}^2$$

The strain energy is

$$U = \frac{1}{2}\int_0^l EA(u')^2 \, dx = \frac{1}{2}\frac{EA}{l}q^2$$

Clearly, in this case, the mass and stiffness matrices reduce to scalars since the system has one degree of freedom. The mass and stiffness coefficients are

$$m_{11} = \frac{\rho A l}{3}, \qquad k_{11} = \frac{EA}{l}$$

Therefore, the approximate solution for the first natural frequency is

$$\omega_1 = \sqrt{\frac{k_{11}}{m_{11}}} = \frac{1.73205}{l}\sqrt{\frac{E}{\rho}}$$

The exact solution is

$$\omega_1 = \frac{1.570796}{l}\sqrt{\frac{E}{\rho}},$$

that is, the error in the finite-element solution using one element is about 10%. Observe that the estimated solution is higher than the exact solution since the continuous system is represented by one degree of freedom.

If the same rod is divided into two equal elements, each has length $l/2$, the system has two degrees of freedom, and the resulting mass and stiffness matrices of the rod are

$$\mathbf{M}_f = \frac{\rho A l}{12}\begin{bmatrix} 4 & 1 \\ 1 & 2 \end{bmatrix}, \qquad \mathbf{K}_f = \frac{2EA}{l}\begin{bmatrix} 2 & -1 \\ -1 & 1 \end{bmatrix}$$

Equation 70 then defines the characteristic equation as

$$\begin{vmatrix} 2 - 4\beta & -1 - \beta \\ -1 - \beta & 1 - 2\beta \end{vmatrix} = 0$$

where $\beta = (\rho l^2/24E)\omega^2$. The characteristic equation can be expressed in terms of β as

$$(2 - 4\beta)(1 - 2\beta) - (1 + \beta)^2 = 0$$

which can be written as

$$7\beta^2 - 10\beta + 1 = 0$$

The roots of this equation are $\beta_1 = 0.108194$ and $\beta_2 = 1.320377$. Since

$$\omega_i^2 = \frac{24E\beta_i}{\rho l^2}, \qquad i = 1, 2,$$

the approximate natural frequencies are

$$\omega_1 = \frac{1.6114}{l}\sqrt{\frac{E}{\rho}}$$

$$\omega_2 = \frac{5.6292}{l}\sqrt{\frac{E}{\rho}}$$

Observe that the use of two finite elements instead of one element leads to an improvement in the fundamental natural frequency. The error in the case of the two elements is less than 2.6%. The exact value of the second natural frequency is

$$\omega_2 = \frac{4.71239}{l}\sqrt{\frac{E}{\rho}}.$$

The error in the second natural frequency using two finite elements is about 19.5%. By increasing the number of elements, a better approximation can be obtained for the second natural frequency. Roughly speaking, a reasonable approximation for a set of low-frequency modes can be achieved by using a number of degrees of freedom equal to twice the number of the desired natural frequencies. Table 1 and Fig. 8 show the computer results obtained using the finite-element method. The number of degrees of freedom used in this model is 150 degrees of freedom. In Table 1, the obtained finite-element results for the modal mass and stiffness coefficients as well as the natural frequencies are compared with the exact solution. These results are presented for a rod with length 3.6 m, modulus of elasticity $E = 2 \times 10^{11}$ N/m^2, and mass density $\rho =$ 7870 kg/m^3. The rod has a circular cross-sectional area with diameter 0.0185 m. Observe that by using a large number of nodal coordinates, the finite-element method can be used to obtain very good approximations for the fundamental natural frequencies. Note also, from the results presented in Fig. 8, the good agreement between the approximate mode shapes obtained using the finite-element method and the exact eigenfunctions obtained by solving the partial differential equation.

Semidefinite Systems Systems which admit rigid-body motion are called *semidefinite systems*. The strain energy of such systems is a positive-semidefinite quadratic form in the displacement coordinates. That is, the strain energy can be zero for nonzero values of the system coordinates. It was shown previously that the element shape functions can describe rigid-body displacements. The shape function of the truss element has one rigid-body mode that describes the rigid-body translation along the axis of the element. The rect-

TABLE 5.1. Modal coefficients.

	Modal mass coefficients		Modal stiffness coefficients		Natural frequencies	
Mode number	Exact solution	Finite element solution	Exact solution	Finite element solution	Exact solution	Finite element solution
1	3.808	3.807	1.84234×10^{7}	1.84231×10^{7}	2199.56	2199.83
2	3.808	3.807	1.65811×10^{8}	1.65796×10^{8}	6598.70	6599.27
3	3.808	3.806	4.60585×10^{8}	4.60476×10^{8}	10997.82	10999.40
4	3.808	3.804	9.02747×10^{8}	9.02334×10^{8}	15396.95	15401.51
5	3.808	3.802	1.49230×10^{9}	1.49118×10^{9}	19796.10	19804.28
6	3.808	3.799	2.22923×10^{9}	2.22675×10^{9}	24195.19	24210.35
7	3.808	3.796	3.11355×10^{9}	3.10873×10^{9}	28594.30	28617.28
8	3.808	3.792	4.14527×10^{9}	4.13672×10^{9}	32993.47	33028.85
9	3.808	3.788	5.32436×10^{9}	5.31032×10^{9}	37392.56	37441.68
10	3.808	3.783	6.65085×10^{9}	6.62888×10^{9}	41791.71	41860.26
11	3.808	3.777	8.12472×10^{9}	8.09199×10^{9}	46190.83	46286.48
12	3.808	3.771	9.74598×10^{9}	9.69896×10^{9}	50589.95	50714.75
13	3.808	3.765	1.15146×10^{10}	1.14489×10^{10}	54989.02	55144.14
14	3.808	3.757	1.34307×10^{10}	1.33414×10^{10}	59388.29	59590.92
15	3.808	3.750	1.54941×10^{10}	1.53752×10^{10}	63787.37	64031.66
16	3.808	3.742	1.77049×10^{10}	1.75498×10^{10}	68186.48	68483.22
17	3.808	3.733	2.00631×10^{10}	1.98641×10^{10}	72585.61	72946.67
18	3.808	3.724	2.25687×10^{10}	2.23169×10^{10}	76984.77	77412.68

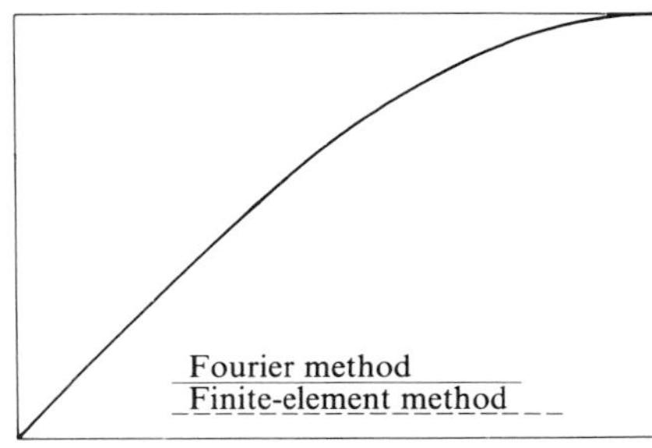

Mode 1

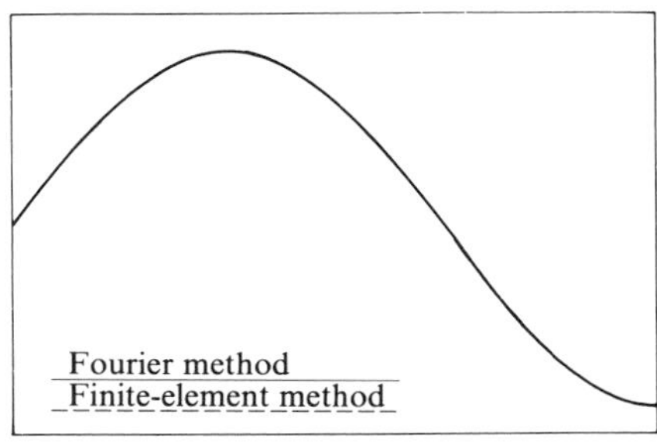

Mode 2

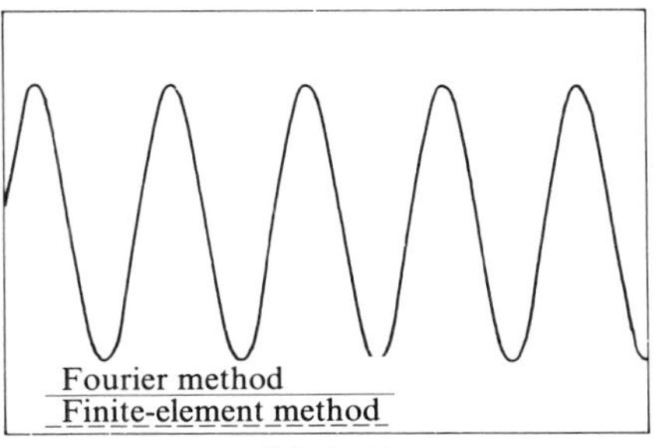

Mode 10

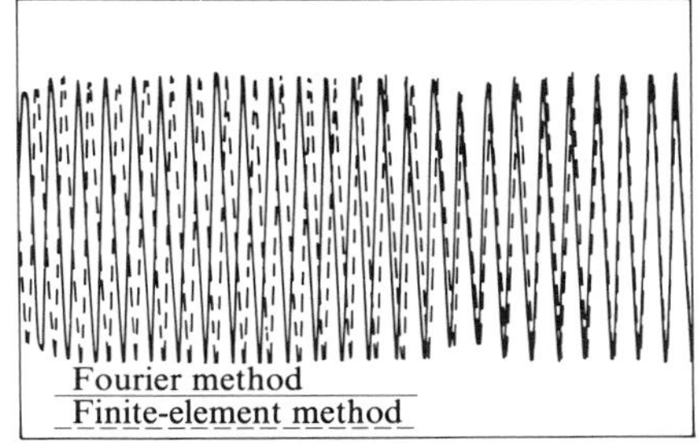

Mode 50

FIG. 5.8. Mode shapes.

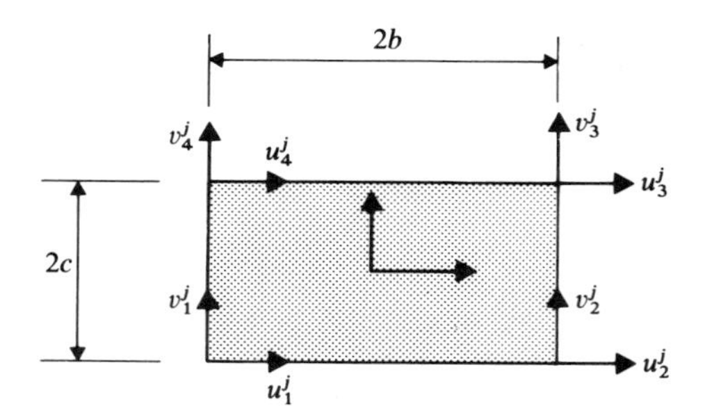

FIG. 5.9. Rectangular element.

angular and triangular element shape functions have three rigid-body modes that describe an arbitrary rigid-body planar motion. While the beam element shape function has three rigid-body modes, it can be used to describe only infinitesimal rigid-body rotations and arbitrary rigid-body translations in two directions. Clearly, if the number of the fixed coordinates of a structure discretized using the finite-element method is less than the number of the rigid-body modes of the element shape functions, the structure will have rigid-body degrees of freedom. Consequently, the strain energy of the system is a positive-semidefinite quadratic form.

Let us consider the case of the four node, eight degree of freedom bilinear rectangular element shown in Fig. 9. The shape function of this element was developed in the first section of this chapter and its stiffness matrix was developed in the preceding section. The vector of nodal coordinates of this element is

$$\mathbf{q}^j = [u_1^j \quad v_1^j \quad u_2^j \quad v_2^j \quad u_3^j \quad v_3^j \quad u_4^j \quad v_4^j]^{\mathrm{T}}$$

As previously mentioned, the shape function of the bilinear rectangular element has three rigid-body modes. It can be shown that possible three rigid-body modes are

$$\begin{aligned}\mathbf{A}_1^j &= [1 \quad 0 \quad 1 \quad 0 \quad 1 \quad 0 \quad 1 \quad 0]^{\mathrm{T}}\\ \mathbf{A}_2^j &= [0 \quad 1 \quad 0 \quad 1 \quad 0 \quad 1 \quad 0 \quad 1]^{\mathrm{T}}\\ \mathbf{A}_3^j &= [c \quad -b \quad c \quad b \quad -c \quad b \quad -c \quad -b]^{\mathrm{T}}\end{aligned}$$

where $2b$ and $2c$ are the dimensions of the element as shown in Fig. 9. Note that the vectors $\mathbf{A}_1^j$ and $\mathbf{A}_2^j$ describe rigid-body translations in two orthogonal directions, while the vector $\mathbf{A}_3^j$ describes a rigid-body rotation. Note also that the vectors $\mathbf{A}_1^j$, $\mathbf{A}_2^j$, and $\mathbf{A}_3^j$ are orthogonal, that is

$$\mathbf{A}_i^{j\mathrm{T}}\mathbf{A}_k^j = 0 \qquad \text{for} \quad i \neq k$$

Clearly, a rigid-body motion of the bilinear rectangular element can be expressed as a linear combination of the vectors $\mathbf{A}_1^j$, $\mathbf{A}_2^j$, and $\mathbf{A}_3^j$.

By using the stiffness matrix of the bilinear rectangular element developed in the preceding section, one can show that the rigid-body modes $\mathbf{A}_1^j$, $\mathbf{A}_2^j$, and $\mathbf{A}_3^j$ have no contribution to the elastic forces. In fact, the use of direct matrix multiplications shows that

$$\mathbf{K}^j\mathbf{A}_k^j = \mathbf{0}, \qquad k = 1, 2, 3$$

Consequently, for a more general rigid-body displacement that can be expressed as a linear combination of $\mathbf{A}_1^j$, $\mathbf{A}_2^j$, and $\mathbf{A}_3^j$, the vector of elastic forces is equal to zero. Furthermore, since $\mathbf{K}^j\mathbf{A}_k^j$ is equal to zero, for $k = 1, 2, 3$, one has

$$\mathbf{A}_k^{j\mathrm{T}}\mathbf{K}^j\mathbf{A}_k^j = 0, \qquad k = 1, 2, 3$$

That is, the stiffness coefficients associated with the rigid-body modes $\mathbf{A}_1^j$, $\mathbf{A}_2^j$, and $\mathbf{A}_3^j$ are equal to zero. Therefore, the strain energy can be zero for nonzero values of the coordinates and it is indeed a positive-semidefinite quadratic form when the element has one or more of its rigid-body modes.

The mass coefficients associated with the rigid-body modes $\mathbf{A}_1^j$, $\mathbf{A}_2^j$, and $\mathbf{A}_3^j$ are not equal to zero. By using the shape function of the bilinear rectangular element developed in the first section of this chapter, it can be shown that the element mass matrix is given by

$$\mathbf{M}^j = \frac{m^j}{36}\begin{bmatrix} 4 & 0 & 2 & 0 & 1 & 0 & 2 & 0 \\ 0 & 4 & 0 & 2 & 0 & 1 & 0 & 2 \\ 2 & 0 & 4 & 0 & 2 & 0 & 1 & 0 \\ 0 & 2 & 0 & 4 & 0 & 2 & 0 & 1 \\ 1 & 0 & 2 & 0 & 4 & 0 & 2 & 0 \\ 0 & 1 & 0 & 2 & 0 & 4 & 0 & 2 \\ 2 & 0 & 1 & 0 & 2 & 0 & 4 & 0 \\ 0 & 2 & 0 & 1 & 0 & 2 & 0 & 4 \end{bmatrix}$$

where m^j is the mass of the element. Observe that

$$\mathbf{A}_1^{j\mathrm{T}}\mathbf{M}^j\mathbf{A}_1^j = m^j$$

$$\mathbf{A}_2^{j\mathrm{T}}\mathbf{M}^j\mathbf{A}_2^j = m^j$$

$$\mathbf{A}_3^{j\mathrm{T}}\mathbf{M}^j\mathbf{A}_3^j = \frac{m^j}{3}(b^2 + c^2)$$

That is, the mass coefficients associated with the rigid-body modes of the element are not equal to zero. Consequently, the kinetic energy of the element is a positive-definite quadratic form in the velocities. In fact, the mass coefficients associated with the rigid-body modes $\mathbf{A}_1^j$, $\mathbf{A}_2^j$, and $\mathbf{A}_3^j$ are the exact mass coefficients that are used in the dynamics of rigid bodies. Moreover, the vectors $\mathbf{A}_1^j$, $\mathbf{A}_2^j$, and $\mathbf{A}_3^j$ are orthogonal with respect to the mass matrix, that is

$$\mathbf{A}_i^{j\mathrm{T}}\mathbf{M}^j\mathbf{A}_k^j = 0 \qquad \text{if} \quad i \neq k$$

5.8 HIGHER-ORDER ELEMENTS

Thus far, we have considered simple elements in order to demonstrate the use of the powerful finite-element technique in the dynamic analysis of structural systems. It was shown that the degree of the polynomial used to describe the element displacement field depends on the number of nodal coordinates of this element. In fact, the number of coefficients of the interpolating polynomial must be equal to the number of the degrees of freedom of the element. Therefore, the order of these polynomials can be increased by increasing the number of nodal coordinates. Two approaches can be used to achieve this goal. In the first approach, the number of nodal points of the element is increased using the same type of nodal coordinates. In the second approach, the same number of nodal points is used, and the number of nodal coordinates is increased by using higher derivatives of the displacement variables. This leads to an increase in the number of degrees of freedom of each nodal point.

While the use of higher-order elements leads to an increase in the dimensions of the element matrices and consequently larger number of calculations are required, it has the advantage that fewer higher-order elements are needed in order to obtain the same degree of accuracy. Furthermore, the use of higher-order elements produces more accurate results in applications where the gradient of the displacements cannot be approximated by low-order polynomials.

Illustrative Example To demonstrate the development of the kinematic and dynamic equations of higher-order elements, we consider the truss element shown in Fig. 10. This element has three nodal points, each node has one degree of freedom that represents the axial displacement at this node. The second node is assumed to be centrally located between the end points of the element. Since this element has three degrees of freedom, a quadratic interpolating polynomial with three coefficients can be used. This polynomial can be written as

$$u^j = a_1^j + a_2^j x + a_3^j x^2 \tag{5.72}$$

which can also be written as

$$u^j = [1 \quad x \quad x^2] \begin{bmatrix} a_1^j \\ a_2^j \\ a_3^j \end{bmatrix} \tag{5.73}$$

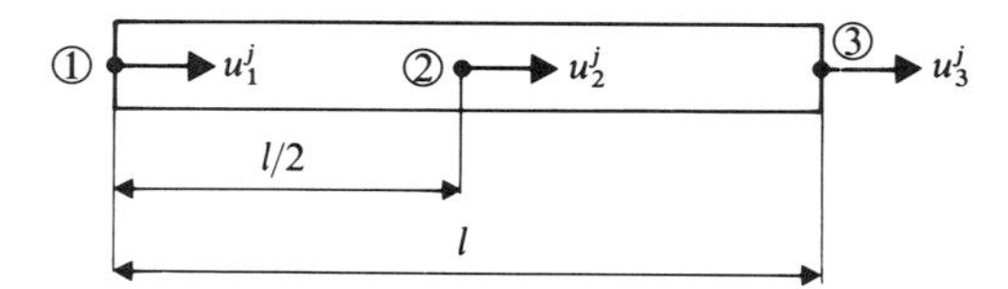

FIG. 5.10. Three-node truss element.

where a_1^j, a_2^j, and a_3^j are three coefficients which can be determined using the conditions

$$\left.\begin{aligned} u^j(x=0) &= u_1^j \\ u^j(x=l/2) &= u_2^j \\ u^j(x=l) &= u_3^j \end{aligned}\right\} \tag{5.74}$$

where l is the length of the truss element.

Substituting Eq. 74 into Eq. 73, one obtains

$$\begin{bmatrix} u_1^j \\ u_2^j \\ u_3^j \end{bmatrix} = \begin{bmatrix} 1 & 0 & 0 \\ 1 & l/2 & l^2/4 \\ 1 & l & l^2 \end{bmatrix} \begin{bmatrix} a_1^j \\ a_2^j \\ a_3^j \end{bmatrix}$$

This equation can be solved for the coefficients a_1^j, a_2^j, a_3^j in terms of the nodal coordinates u_1^j, u_2^j, u_3^j as

$$\begin{bmatrix} a_1^j \\ a_2^j \\ a_3^j \end{bmatrix} = \begin{bmatrix} 1 & 0 & 0 \\ -3/l & 4/l & -1/l \\ -2/l^2 & -4/l^2 & 2/l^2 \end{bmatrix} \begin{bmatrix} u_1^j \\ u_2^j \\ u_3^j \end{bmatrix}$$

Substituting this equation into Eq. 73, the displacement field of the element can be written in terms of the nodal coordinates as

$$u^j = [1 \quad x \quad x^2] \begin{bmatrix} 1 & 0 & 0 \\ -3/l & 4/l & -1/l \\ -2/l^2 & -4/l^2 & 2/l^2 \end{bmatrix} \begin{bmatrix} u_1^j \\ u_2^j \\ u_3^j \end{bmatrix}$$

This equation leads to

$$u^j = N_1 u_1^j + N_2 u_2^j + N_3 u_3^j \tag{5.75}$$

where

$$\left.\begin{aligned} N_1 &= (1-2\xi)(1-\xi) \\ N_2 &= 4\xi(1-\xi) \\ N_3 &= -\xi(1-2\xi) \end{aligned}\right\} \tag{5.76}$$

in which $\xi = x/l$. Observe that

$$N_1 + N_2 + N_3 = \sum_{i=1}^{3} N_i = 1 \tag{5.77}$$

and

$$N_i = \begin{cases} 1 & \text{at node } i \\ 0 & \text{at all nodes other than } i \end{cases} \tag{5.78}$$

The kinetic energy of the three-node truss element is

$$T^j = \frac{1}{2} \int_0^l \rho A (\dot{u}^j)^2 \, dx \tag{5.79}$$

where ρ and A are, respectively, the mass density and cross-sectional area of the element, and

$$\dot{u}^j = [N_1 \quad N_2 \quad N_3] \begin{bmatrix} \dot{u}_1^j \\ \dot{u}_2^j \\ \dot{u}_3^j \end{bmatrix} \tag{5.80}$$

Substituting Eq. 80 into Eq. 79 yields

$$T^j = \tfrac{1}{2}\dot{\mathbf{q}}^{j^{\mathrm{T}}}\mathbf{M}^j\dot{\mathbf{q}}^j \tag{5.81}$$

where $\mathbf{q}^j = [u_1^j \ u_2^j \ u_3^j]^{\mathrm{T}}$, and $\mathbf{M}^j$ is the 3×3 element mass matrix defined as

$$\mathbf{M}^j = \int_0^l \rho A \begin{bmatrix} N_1^2 & N_1 N_2 & N_1 N_3 \\ N_2 N_1 & N_2^2 & N_2 N_3 \\ N_3 N_1 & N_3 N_2 & N_3^2 \end{bmatrix} dx$$

which can be written as

$$\mathbf{M}^j = m \int_0^1 \begin{bmatrix} N_1^2 & N_1 N_2 & N_1 N_3 \\ N_2 N_1 & N_2^2 & N_2 N_3 \\ N_3 N_1 & N_3 N_2 & N_3^2 \end{bmatrix} d\xi \tag{5.82}$$

where m is the total mass of the element. Upon carrying out the integrations in Eq. 82, it can be shown that the element mass matrix $\mathbf{M}^j$ is given by

$$\mathbf{M}^j = \frac{m}{30} \begin{bmatrix} 4 & 2 & -1 \\ 2 & 16 & 2 \\ -1 & 2 & 4 \end{bmatrix} \tag{5.83}$$

The strain energy of the three-node truss element is given by

$$U^j = \frac{1}{2} \int_0^l EA(u^{j'})^2 \, dx \tag{5.84}$$

in which (′) denotes differentiation with respect to x and

$$u^{j'} = N_1' u_1^j + N_2' u_2^j + N_3' u_3^j \tag{5.85}$$

Substituting Eq. 85 into Eq. 84, the strain energy of the three-node truss element can be written as

$$U^j = \tfrac{1}{2}\mathbf{q}^{j^{\mathrm{T}}}\mathbf{K}^j\mathbf{q}^j \tag{5.86}$$

where $\mathbf{K}^j$ is the 3×3 element stiffness matrix defined as

$$\mathbf{K}^j = \int_0^l EA \begin{bmatrix} N_1'^2 & N_1' N_2' & N_1' N_3' \\ N_2' N_1' & N_2'^2 & N_2' N_3' \\ N_3' N_1' & N_3' N_2' & N_3'^2 \end{bmatrix} dx$$

$$= \int_0^1 EAl \begin{bmatrix} N_1'^2 & N_1' N_2' & N_1' N_3' \\ N_2' N_1' & N_2'^2 & N_2' N_3' \\ N_3' N_1' & N_3' N_2' & N_3'^2 \end{bmatrix} d\xi$$

which upon integration yields

$$\mathbf{K}^j = \frac{EA}{3l}\begin{bmatrix} 7 & -8 & 1 \\ -8 & 16 & -8 \\ 1 & -8 & 7 \end{bmatrix}$$

Let us consider the case of a rod with one end fixed and the other end free. If we assume that only one three-node truss element is used, the system has two degrees of freedom and the eigenvalue problem is given by

$$\mathbf{KA} = \omega^2\mathbf{MA}$$

where $\mathbf{A}$ is the vector of amplitudes, ω is the frequency, and $\mathbf{M}$ and $\mathbf{K}$ are defined in this case as

$$\mathbf{M} = \frac{m}{15}\begin{bmatrix} 8 & 1 \\ 1 & 2 \end{bmatrix}, \qquad \mathbf{K} = \frac{EA}{3l}\begin{bmatrix} 16 & -8 \\ -8 & 7 \end{bmatrix}$$

Therefore, the characteristic equation of the system can be written as

$$\begin{vmatrix} 16 - 8\beta & -8 - \beta \\ -8 - \beta & 7 - 2\beta \end{vmatrix} = 0$$

in which $\beta = (ml/5EA)\omega^2$. It follows that

$$(16 - 8\beta)(7 - 2\beta) - (8 + \beta)^2 = 0$$

which can be written as

$$15\beta^2 - 104\beta + 48 = 0$$

This quadratic equation has two roots

$$\beta_1 = 0.497193 \qquad \text{and} \qquad \beta_2 = 6.43614$$

which define the first two natural frequencies of the system as

$$\omega_i = \sqrt{\frac{5EA\beta_i}{ml}}, \qquad i = 1, 2$$

That is,

$$\omega_1 = 1.57669\sqrt{\frac{EA}{ml}} = \frac{1.57669}{l}\sqrt{\frac{E}{\rho}}$$

$$\omega_2 = 5.67280\sqrt{\frac{EA}{ml}} = \frac{5.67280}{l}\sqrt{\frac{E}{\rho}}$$

Comparing these results with the results obtained in the preceding section using the two-node truss element, it is clear that the three-node truss element gives better accuracy for the fundamental natural frequency. This is an expected result since higher-order interpolating polynomials and more degrees of freedom are used in the case of the three-node truss element as compared to the two-node truss element.

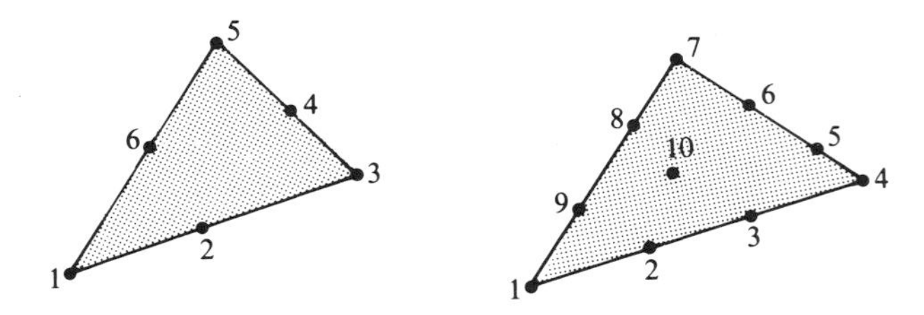

FIG. 5.11. Higher order triangular element.

While only the three-node truss element is discussed in this section as an example, other higher-order elements can be developed using a similar procedure. For example, in the case of a four-node truss element a cubic interpolating polynomial with four coefficients must be used. The element in this case has four degrees of freedom and, consequently, the resulting mass and stiffness matrices are 4×4. Other examples of higher-order elements are the six-node and the ten-node triangular elements shown in Fig. 11. The six-node triangular element has 12 degrees of freedom. The displacement field of this element is approximated using a quadratic polynomial, that is

$$u^j = a_1^j + a_2^j x + a_3^j y + a_4^j x^2 + a_5^j xy + a_6^j y^2$$

$$v^j = a_7^j + a_8^j x + a_9^j y + a_{10}^j x^2 + a_{11}^j xy + a_{12}^j y^2$$

The ten-node triangular element has 18 degrees of freedom. The displacement field within this element is approximated using cubic polynomials. In this case the displacements u^j and v^j are defined by

$$u^j = a_1^j + a_2^j x + a_3^j y + a_4^j x^2 + a_5^j xy + a_6^j y^2 + a_7^j x^3 + a_8^j x^2 y + a_9^j xy^2 + a_{10}^j y^3$$

$$v^j = a_{11}^j + a_{12}^j x + a_{13}^j y + a_{14}^j x^2 + a_{15}^j xy + a_{16}^j y^2 + a_{17}^j x^3 + a_{18}^j x^2 y + a_{19}^j xy^2 + a_{20}^j y^3$$

Observe that an internal node (node 10) is introduced for this element in order to have complete cubic polynomials.

5.9 SPATIAL ELEMENTS

The procedure for developing the kinematic and dynamic equations for spatial elements is very similar to that of the planar elements. There are slight differences in the analytical development of the spatial finite elements as compared to the planar elements. While in planar elements, the displacement of a nodal point is at most two-dimensional; axial and transverse displacements, in spatial elements the displacement of a nodal point can be three-

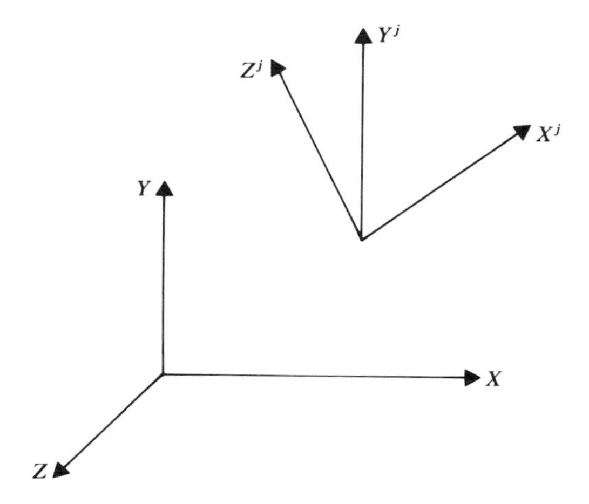

FIG. 5.12. Spatial transformation.

dimensional. Furthermore, bending and extension of the three-dimensional element in more than one plane (as well as torsion) are possible. Therefore, in general, using the same order of approximation, three-dimensional elements have a larger number of nodal coordinates as compared to planar elements.

Another difference between the spatial and planar elements lies in the transformation of coordinates from the element coordinate system to the global coordinate system. In planar elements, the transformation depends only on one independent angle. This transformation is given by Eq. 21 as

$$\mathbf{C}^j = \begin{bmatrix} \cos\beta^j & -\sin\beta^j \\ \sin\beta^j & \cos\beta^j \end{bmatrix}$$

where β^j is the angle the X^j-axis of the element makes with the X-axis of the global system. In the spatial analysis, however, three independent orientational coordinates must be used. Figure 12 shows the coordinate system $X^jY^jZ^j$ of the element j. The orientation of this coordinate system in the XYZ global coordinate system can be described using the *direction cosines*. Let $\mathbf{i}^j$, $\mathbf{j}^j$, and $\mathbf{k}^j$ be unit vectors along the axes X^j, Y^j, and Z^j of the element, and let $\mathbf{i}$, $\mathbf{j}$, and $\mathbf{k}$ be unit vectors along the global axes X, Y, and Z, respectively. Let β_1^j be the angle between X^j and X, let β_2^j be the angle between X^j and Y, and let β_3^j be the angle between X^j and Z. The components of the unit vector $\mathbf{i}^j$ along the X-, Y-, and Z-axes are given, respectively, by

$$\left.\begin{aligned} \alpha_{11} &= \cos\beta_1^j = \mathbf{i}^j\cdot\mathbf{i} \\ \alpha_{12} &= \cos\beta_2^j = \mathbf{i}^j\cdot\mathbf{j} \\ \alpha_{13} &= \cos\beta_3^j = \mathbf{i}^j\cdot\mathbf{k} \end{aligned}\right\} \tag{5.87}$$

where α_{11}, α_{12}, and α_{13} are the direction cosines of the X^j-axis with respect to the X, Y, and Z-axes, respectively. In a similar manner the direction cosines α_{21}, α_{22}, and α_{23} of the axis Y^j and the direction cosines α_{31}, α_{32}, and α_{33} of the axis Z^j can be defined as

$$\left.\begin{aligned} \alpha_{21} &= \mathbf{j}^j \cdot \mathbf{i} \\ \alpha_{22} &= \mathbf{j}^j \cdot \mathbf{j} \\ \alpha_{23} &= \mathbf{j}^j \cdot \mathbf{k} \end{aligned}\right\} \tag{5.88}$$

and

$$\left.\begin{aligned} \alpha_{31} &= \mathbf{k}^j \cdot \mathbf{i} \\ \alpha_{32} &= \mathbf{k}^j \cdot \mathbf{j} \\ \alpha_{33} &= \mathbf{k}^j \cdot \mathbf{k} \end{aligned}\right\} \tag{5.89}$$

Since the direction cosines α_{ij} represent the components of the unit vectors $\mathbf{i}^j$, $\mathbf{j}^j$, and $\mathbf{k}^j$ along the axes X, Y, and Z, one has

$$\left.\begin{aligned} \mathbf{i}^j &= \alpha_{11}\mathbf{i} + \alpha_{12}\mathbf{j} + \alpha_{13}\mathbf{k} \\ \mathbf{j}^j &= \alpha_{21}\mathbf{i} + \alpha_{22}\mathbf{j} + \alpha_{23}\mathbf{k} \\ \mathbf{k}^j &= \alpha_{31}\mathbf{i} + \alpha_{32}\mathbf{j} + \alpha_{33}\mathbf{k} \end{aligned}\right\} \tag{5.90}$$

Let us now consider the vector $\mathbf{u}$ whose components in the element coordinate system are denoted as $\bar{u}$, $\bar{v}$, and $\bar{w}$. In the global coordinate system the components of the vector $\mathbf{u}$ are denoted as u, v, and w. The vector $\mathbf{u}$ can, therefore, have the following two different representations

$$\mathbf{u} = \bar{u}\mathbf{i}^j + \bar{v}\mathbf{j}^j + \bar{w}\mathbf{k}^j \tag{5.91}$$

or

$$\mathbf{u} = u\mathbf{i} + v\mathbf{j} + w\mathbf{k} \tag{5.92}$$

Substituting Eq. 90 into Eq. 91, one obtains

$$\begin{aligned} \mathbf{u} = {} & \bar{u}(\alpha_{11}\mathbf{i} + \alpha_{12}\mathbf{j} + \alpha_{13}\mathbf{k}) \\ & + \bar{v}(\alpha_{21}\mathbf{i} + \alpha_{22}\mathbf{j} + \alpha_{23}\mathbf{k}) \\ & + \bar{w}(\alpha_{31}\mathbf{i} + \alpha_{32}\mathbf{j} + \alpha_{33}\mathbf{k}) \end{aligned}$$

which leads to

$$\begin{aligned} \mathbf{u} = {} & (\alpha_{11}\bar{u} + \alpha_{21}\bar{v} + \alpha_{31}\bar{w})\mathbf{i} \\ & + (\alpha_{12}\bar{u} + \alpha_{22}\bar{v} + \alpha_{32}\bar{w})\mathbf{j} \\ & + (\alpha_{13}\bar{u} + \alpha_{23}\bar{v} + \alpha_{33}\bar{w})\mathbf{k} \end{aligned} \tag{5.93}$$

By comparing Eqs. 92 and 93, one concludes

$$\begin{aligned} u &= \alpha_{11}\bar{u} + \alpha_{21}\bar{v} + \alpha_{31}\bar{w} \\ v &= \alpha_{12}\bar{u} + \alpha_{22}\bar{v} + \alpha_{32}\bar{w} \\ w &= \alpha_{13}\bar{u} + \alpha_{23}\bar{v} + \alpha_{33}\bar{w} \end{aligned}$$

That is, the relationship between the global and local components can be written in the following matrix form

$$\begin{bmatrix} u \\ v \\ w \end{bmatrix} = \begin{bmatrix} \alpha_{11} & \alpha_{21} & \alpha_{31} \\ \alpha_{12} & \alpha_{22} & \alpha_{32} \\ \alpha_{13} & \alpha_{23} & \alpha_{33} \end{bmatrix} \begin{bmatrix} \bar{u} \\ \bar{v} \\ \bar{w} \end{bmatrix} \tag{5.94}$$

from which the transformation matrix $\mathbf{C}^j$ can be recognized as

$$\mathbf{C}^j = \begin{bmatrix} \alpha_{11} & \alpha_{21} & \alpha_{31} \\ \alpha_{12} & \alpha_{22} & \alpha_{32} \\ \alpha_{13} & \alpha_{23} & \alpha_{33} \end{bmatrix} \tag{5.95}$$

This transformation matrix is expressed in terms of nine direction cosines α_{ij}, $i, j = 1, 2, 3$. These nine components, however, are not independent. They are related by the six algebraic equations

$$\alpha_{k1}\alpha_{l1} + \alpha_{k2}\alpha_{l2} + \alpha_{k3}\alpha_{l3} = \delta_{kl}, \qquad k, l = 1, 2, 3 \tag{5.96}$$

where δ_{kl} is the *Kronecker delta*, that is

$$\delta_{kl} = \begin{cases} 1 & \text{if } k = l \\ 0 & \text{if } k \neq l \end{cases}$$

Therefore, there are only three independent direction cosines in the transformation matrix of Eq. 95. This transformation matrix can be used to define the nodal coordinates of the three-dimensional element in the global coordinate system.

In the remainder of this section, we discuss the assumed displacement field of some of the three-dimensional elements. The development of the mass and stiffness matrices of these elements is left as an exercise, since a similar procedure to the one previously discussed in this chapter can be used. The mass matrix can be defined using the kinetic energy expression, while the stiffness matrix can be defined using the strain energy expression and the general elasticity relationships presented in this chapter. In the case of beam and plate elements, the classical beam and plate theories may be used in order to simplify the elasticity equations.

Beam Element Figure 13 shows an example of a three-dimensional beam element. This element has two nodal points, one located at each end. Each node has six degrees of freedom u_i^j, v_i^j, w_i^j, $\theta_{x_i}^j$, $\theta_{y_i}^j$, $\theta_{z_i}^j$, $i = 1, 2$, where u_i^j, v_i^j, and w_i^j are the translational displacements of the node i and $\theta_{x_i}^j$, $\theta_{y_i}^j$, and $\theta_{z_i}^j$ are its rotations about the three perpendicular axes of the element. Therefore, the element has twelve degrees of freedom which are defined as

$$\mathbf{q}^j = [u_1^j \quad v_1^j \quad w_1^j \quad \theta_{x_1}^j \quad \theta_{y_1}^j \quad \theta_{z_1}^j \quad u_2^j \quad v_2^j \quad w_2^j \quad \theta_{x_2}^j \quad \theta_{y_2}^j \quad \theta_{z_2}^j]^{\mathrm{T}} \tag{5.97}$$

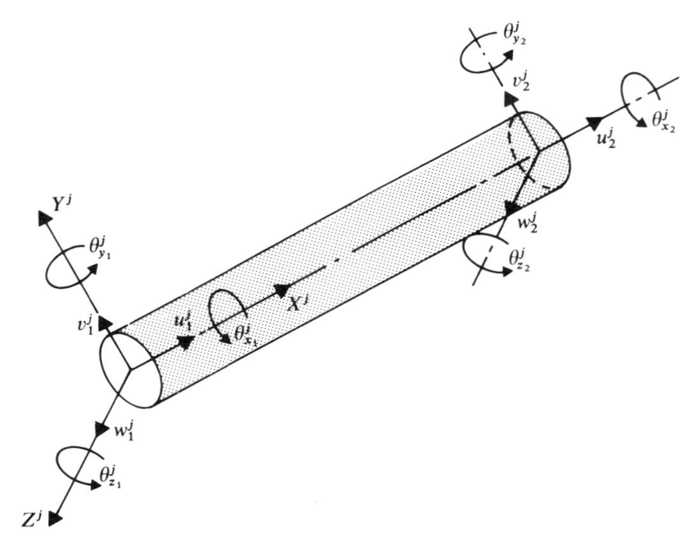

FIG. 5.13. Beam element.

The shape function $\mathbf{S}^j$ of this element is

$$\mathbf{S}^{j\mathrm{T}} = \begin{bmatrix} 1-\xi & 0 & 0 \\ 6(\xi-\xi^2)\eta & 1-3\xi^2+2\xi^3 & 0 \\ 6(\xi-\xi^2) & 0 & 1-3\xi^2+2\xi^3 \\ 0 & -(1-\xi)l\zeta & (1-\xi)l\eta \\ (1-4\xi+3\xi)l\zeta & 0 & (-\xi+2\xi^2-\xi^3)l \\ (-1+4\xi-3\xi^2)l\eta & (\xi-2\xi^2+\xi^3)l & 0 \\ \xi & 0 & 0 \\ 6(-\xi+\xi^2)\eta & 3\xi^2-2\xi^3 & 0 \\ 6(-\xi+\xi^2)\zeta & 0 & 3\xi^2-2\xi^3 \\ 0 & -l\xi\zeta & l\xi\eta \\ (-2\xi+3\xi^2)l\zeta & 0 & \xi^2-\xi^3 \\ (2\xi-3\xi^2)l\eta & (-\xi^2+\xi^3)l & 0 \end{bmatrix} \tag{5.98}$$

where $\xi = x/l$, $\eta = y/l$, and $\zeta = z/l$, and l is the length of the beam element.

Solid Element Figure 14 shows the eight-node *brick element*. This element has 24 degrees of freedom which represent the displacements of the nodes along three perpendicular axes. The shape function of the brick element may

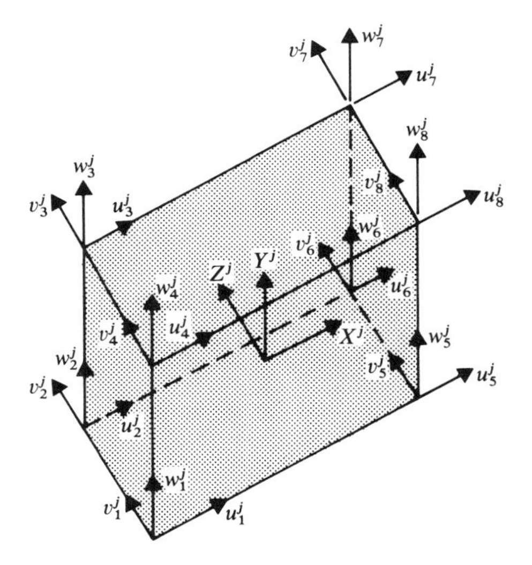

FIG. 5.14. Solid element.

be derived using the interpolating polynomials

$$\left.\begin{aligned} u^j &= a_1^j + a_2^j x + a_3^j y + a_4^j z + a_5^j xy + a_6^j yz + a_7^j zx + a_8^j xyz \\ v^j &= a_9^j + a_{10}^j x + a_{11}^j y + a_{12}^j z + a_{13}^j xy + a_{14}^j yz + a_{15}^j zx + a_{16}^j xyz \\ w^j &= a_{17}^j + a_{18}^j x + a_{19}^j y + a_{20}^j z + a_{21}^j xy + a_{22}^j yz + a_{23}^j zx + a_{24}^j xyz \end{aligned}\right\} \tag{5.99}$$

The 24 nodal coordinates are

$$\mathbf{q}^j = [u_1^j \quad v_1^j \quad w_1^j \quad u_2^j \quad v_2^j \quad w_2^j \quad u_3^j \quad v_3^j \quad w_3^j \quad u_4^j \quad v_4^j \quad w_4^j \quad u_5^j \quad v_5^j \quad w_5^j \quad u_6^j \quad v_6^j \quad w_6^j \quad u_7^j \quad v_7^j \quad w_7^j \quad u_8^j \quad v_8^j \quad w_8^j]^{\mathrm{T}} \tag{5.100}$$

By using Eqs. 99 and 100, one can show that the shape function of the solid element is given by

$$\mathbf{S}^j = \begin{bmatrix} N_1 & 0 & 0 & N_2 & 0 & 0 & \dots & N_8 & 0 & 0 \\ 0 & N_1 & 0 & 0 & N_2 & 0 & \dots & 0 & N_8 & 0 \\ 0 & 0 & N_1 & 0 & 0 & N_2 & \dots & 0 & 0 & N_8 \end{bmatrix} \tag{5.101}$$

where

$$\begin{aligned} N_1 &= \tfrac{1}{8}(1 - \xi)(1 - \eta)(1 - \zeta), & N_5 &= \tfrac{1}{8}(1 + \xi)(1 - \eta)(1 - \zeta) \\ N_2 &= \tfrac{1}{8}(1 - \xi)(1 + \eta)(1 - \zeta), & N_6 &= \tfrac{1}{8}(1 + \xi)(1 + \eta)(1 - \zeta) \\ N_3 &= \tfrac{1}{8}(1 - \xi)(1 + \eta)(1 + \zeta), & N_7 &= \tfrac{1}{8}(1 + \xi)(1 + \eta)(1 + \zeta) \\ N_4 &= \tfrac{1}{8}(1 - \xi)(1 - \eta)(1 + \zeta), & N_8 &= \tfrac{1}{8}(1 + \xi)(1 - \eta)(1 + \zeta) \end{aligned}$$

in which $\xi = x/a$, $\eta = y/b$, and $\zeta = z/c$, where $2a$, $2b$, and $2c$ are, respectively, the dimensions of the solid element in the X-, Y-, and Z-directions.

Tetrahedral Element A typical *constant strain tetrahedral element* is shown in Fig. 15. The element has straight sides with four nodes, one at each corner. Each node has three nodal coordinates that represent the translational displacements of the nodes along three perpendicular axes. Therefore, the element has 12 degrees of freedom. The displacement field of the element may be described using the following interpolating polynomials

$$\left.\begin{aligned} u^j &= a_1^j + a_2^j x + a_3^j y + a_4^j z \\ v^j &= a_5^j + a_6^j x + a_7^j y + a_8^j z \\ w^j &= a_9^j + a_{10}^j x + a_{11}^j y + a_{12}^j z \end{aligned}\right\} \tag{5.102}$$

The vector of nodal coordinates of the tetrahedral element is

$$\mathbf{q}^j = [u_1^j \quad v_1^j \quad w_1^j \quad u_2^j \quad v_2^j \quad w_2^j \quad u_3^j \quad v_3^j \quad w_3^j \quad u_4^j \quad v_4^j \quad w_4^j]^{\mathrm{T}} \tag{5.103}$$

By using these nodal coordinates and the assumed displacement field of Eq. 102, it can be shown that the element shape function is

$$\mathbf{S}^j = \begin{bmatrix} N_1 & 0 & 0 & N_2 & 0 & 0 & N_3 & 0 & 0 & N_4 & 0 & 0 \\ 0 & N_1 & 0 & 0 & N_2 & 0 & 0 & N_3 & 0 & 0 & N_4 & 0 \\ 0 & 0 & N_1 & 0 & 0 & N_2 & 0 & 0 & N_3 & 0 & 0 & N_4 \end{bmatrix} \tag{5.104}$$

where the scalars N_i, $i = 1, 2, 3, 4$, are

$$\begin{aligned} N_1(x, y, z) &= C_{11} + C_{21}x + C_{31}y + C_{41}z \\ N_2(x, y, z) &= C_{12} + C_{22}x + C_{32}y + C_{42}z \\ N_3(x, y, z) &= C_{13} + C_{23}x + C_{33}y + C_{43}z \\ N_4(x, y, z) &= C_{14} + C_{24}x + C_{34}y + C_{44}z \end{aligned}$$

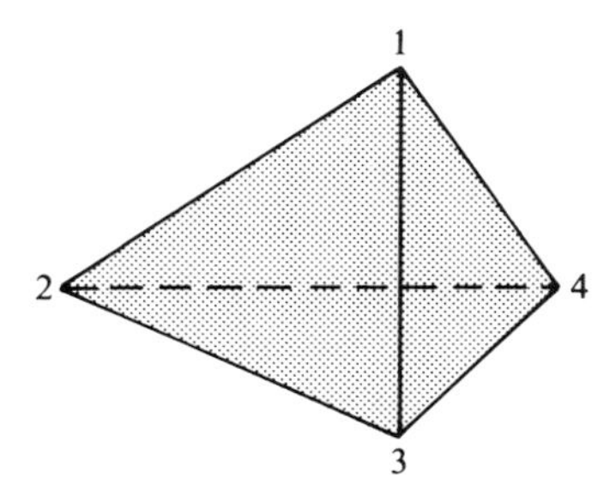

FIG. 5.15. Four-node tetrahedral element.

in which

$$C_{11} = \frac{1}{6V^j}\begin{vmatrix} x_2 & y_2 & z_2 \\ x_3 & y_3 & z_3 \\ x_4 & y_4 & z_4 \end{vmatrix}, \qquad C_{12} = -\frac{1}{6V^j}\begin{vmatrix} x_1 & y_1 & z_1 \\ x_3 & y_3 & z_3 \\ x_4 & y_4 & z_4 \end{vmatrix}$$

$$C_{13} = \frac{1}{6V^j}\begin{vmatrix} x_1 & y_1 & z_1 \\ x_2 & y_2 & z_2 \\ x_4 & y_4 & z_4 \end{vmatrix}, \qquad C_{14} = -\frac{1}{6V^j}\begin{vmatrix} x_1 & y_1 & z_1 \\ x_2 & y_2 & z_2 \\ x_3 & y_3 & z_3 \end{vmatrix}$$

$$C_{21} = -\frac{1}{6V^j}\begin{vmatrix} 1 & y_2 & z_2 \\ 1 & y_3 & z_3 \\ 1 & y_4 & z_4 \end{vmatrix}, \qquad C_{22} = \frac{1}{6V^j}\begin{vmatrix} 1 & y_1 & z_1 \\ 1 & y_3 & z_3 \\ 1 & y_4 & z_4 \end{vmatrix}$$

$$C_{23} = -\frac{1}{6V^j}\begin{vmatrix} 1 & y_1 & z_1 \\ 1 & y_2 & z_2 \\ 1 & y_4 & z_4 \end{vmatrix}, \qquad C_{24} = \frac{1}{6V^j}\begin{vmatrix} 1 & y_1 & z_1 \\ 1 & y_2 & z_2 \\ 1 & y_3 & z_3 \end{vmatrix}$$

$$C_{31} = \frac{1}{6V^j}\begin{vmatrix} 1 & x_2 & z_2 \\ 1 & x_3 & z_3 \\ 1 & x_4 & z_4 \end{vmatrix}, \qquad C_{32} = -\frac{1}{6V^j}\begin{vmatrix} 1 & x_1 & z_1 \\ 1 & x_3 & z_3 \\ 1 & x_4 & z_4 \end{vmatrix}$$

$$C_{33} = \frac{1}{6V^j}\begin{vmatrix} 1 & x_1 & z_1 \\ 1 & x_2 & z_2 \\ 1 & x_4 & z_4 \end{vmatrix}, \qquad C_{34} = -\frac{1}{6V^j}\begin{vmatrix} 1 & x_1 & z_1 \\ 1 & x_2 & z_2 \\ 1 & x_3 & z_3 \end{vmatrix}$$

$$C_{41} = -\frac{1}{6V^j}\begin{vmatrix} 1 & x_2 & y_2 \\ 1 & x_3 & y_3 \\ 1 & x_4 & y_4 \end{vmatrix}, \qquad C_{42} = \frac{1}{6V^j}\begin{vmatrix} 1 & x_1 & y_1 \\ 1 & x_3 & y_3 \\ 1 & x_4 & y_4 \end{vmatrix}$$

$$C_{43} = -\frac{1}{6V^j}\begin{vmatrix} 1 & x_1 & y_1 \\ 1 & x_2 & y_2 \\ 1 & x_4 & y_4 \end{vmatrix}, \qquad C_{44} = \frac{1}{6V^j}\begin{vmatrix} 1 & x_1 & y_1 \\ 1 & x_2 & y_2 \\ 1 & x_3 & y_3 \end{vmatrix}$$

and V^j is the volume of the tetrahedral element defined as

$$V^j = \frac{1}{6}\begin{vmatrix} 1 & x_1 & y_1 & z_1 \\ 1 & x_2 & y_2 & z_2 \\ 1 & x_3 & y_3 & z_3 \\ 1 & x_4 & y_4 & z_4 \end{vmatrix}$$

Plate Element Figure 16 shows a plate element that has four nodal points, one at each corner. Each nodal point has five degrees of freedom, three displacements along three perpendicular directions and two rotations about the X^j- and Y^j-axes. This plate element, therefore, has 20 nodal coordinates which can be written in vector form as

$$\mathbf{q}^j = [u_1^j \quad v_1^j \quad w_1^j \quad \theta_{x_1}^j \quad \theta_{y_1}^j \quad u_2^j \quad v_2^j \quad w_2^j \quad \theta_{x_2}^j \quad \theta_{y_2}^j$$
$$u_3^j \quad v_3^j \quad w_3^j \quad \theta_{x_3}^j \quad \theta_{y_3}^j \quad u_4^j \quad v_4^j \quad w_4^j \quad \theta_{x_4}^j \quad \theta_{y_4}^j]^{\mathrm{T}} \tag{5.105}$$

The extension of the plate (membrane effect) may be described using the interpolating polynomials of the planar bilinear rectangular element, that is

$$\left.\begin{aligned} u^j &= a_1^j + a_2^j x + a_3^j y + a_4^j xy \\ v^j &= a_5^j + a_6^j x + a_7^j y + a_8^j xy \end{aligned}\right\} \tag{5.106}$$

By using these interpolating polynomials and the shape functions of plate bending presented by Prezemieniecki (1968), a plate element shape function that accounts for the membrane and bending effect can be defined as

$$\mathbf{S}^{j^{\mathrm{T}}} = \begin{bmatrix} (1-\xi)(1-\eta) & 0 & 0 \\ 0 & (1-\xi)(1-\eta) & 0 \\ 0 & 0 & (1+2\xi)(1-\xi)^2(1+2\eta)(-\eta)^2 \\ 0 & 0 & (1+2\xi)(1-\xi)^2(1-\eta^2)\eta b \\ 0 & 0 & -\xi a(1-\xi)^2(1+2\eta)(1-\eta)^2 \\ (1-\xi)\eta & 0 & 0 \\ 0 & (1-\xi)\eta & 0 \\ 0 & 0 & (1+2\xi)(1-\xi)^2(3-2\eta)\eta^2 \\ 0 & 0 & -(1+2\xi)(1-\xi)^2(1-\eta)\eta^2 b \\ 0 & 0 & -\xi(1-\xi)^2(3-2\eta)\eta^2 a \\ \xi\eta & 0 & 0 \\ 0 & \xi\eta & 0 \\ 0 & 0 & (3-2\xi)\xi^2(3-2\eta)\eta^2 \\ 0 & 0 & -(3-2\xi)\xi^2(1-\eta)\eta^2 b \\ 0 & 0 & (1-\xi)\xi^2(3-2\eta)\eta^2 a \\ \xi(1-\eta) & 0 & 0 \\ 0 & \xi(1-\eta) & 0 \\ 0 & 0 & (3-2\xi)\xi^2(1+2\eta)(1-\eta)^2 \\ 0 & 0 & (3-2\xi)\xi^2(1-\eta)^2\eta b \\ 0 & 0 & (1-\xi)\xi^2(1+2\eta)(1-\eta)^2 a \end{bmatrix} \tag{5.107}$$

where $\xi = x/a$ and $\eta = y/b$, and a and b are, respectively, the dimensions of the plate element along the X^j and Y^j element axes as shown in Fig. 16.

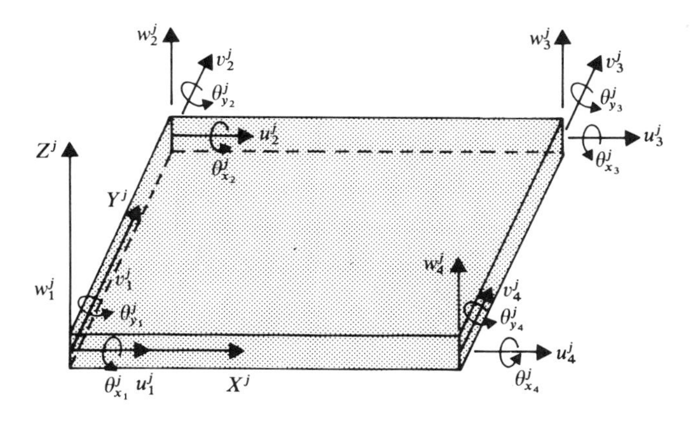

FIG. 5.16. Plate element.

5.10 LARGE ROTATIONS AND DEFORMATIONS

Rectangular, triangular, solid, and tetrahedral elements are referred to as *isoparametric elements*, since both the position and the deformation of these elements can be described using the same shape function. This is mainly due to the fact that the shape functions and the nodal coordinates of these isoparametric elements can be used to describe an arbitrary rigid body displacement as demonstrated in Section 2 of this chapter. In the classical finite element literature, beams and plates are not considered as isoparametric elements because the displacement gradients at the nodes are used to describe infinitesimal rotations. Such a linearization of the slopes leads to incorrect rigid body equations of motion when the beams and plates rotate as rigid bodies (Shabana, 1996a). Such a difficulty can be circumvented if no infinitesimal or finite rotations are used as nodal coordinates. In this section, we introduce a simple and efficient procedure referred to as the *absolute nodal coordinate formulation.* This formulation, in which no infinitesimal or finite rotations are used as nodal coordinates, can be efficiently used in many large deformation applications. In this formulation, the nodal coordinates are defined in the inertial frame in terms of absolute nodal displacements and displacement gradients, and as a consequence, no coordinate transformations are required to determine the element inertia properties. The absolute nodal coordinate formulation leads to a constant mass matrix, and as a consequence, an efficient procedure can be used for solving for the nodal accelerations. In the absolute nodal coordinate formulation, the beam and plate elements can be treated as isoparametric elements since their shape functions and nodal coordinates can be used to describe an arbitrary rigid body displacement. The displacement gradients at the nodes can be determined in the undeformed reference configuration using simple rigid body kinematics. In this section, we briefly describe the absolute nodal coordinate formulation, which can be used in the large rotation and deformation analysis of beam and plate structures,

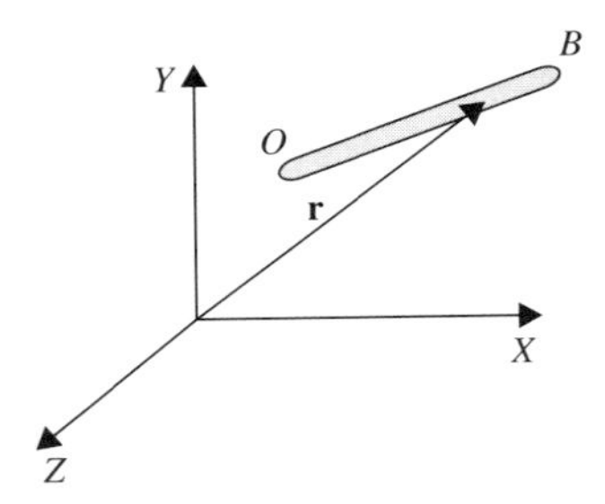

FIG. 5.17. Absolute nodal coordinate formulation

such as the vibrations of cables and flexible space antennas. For simplicity, in the development presented in this section, we drop the superscript j, which defines the element number.

Absolute Nodal Coordinate Formulation In the absolute coordinate formulation (Shabana, 1996b), the coordinates of the element are defined in an inertial coordinate system, and as such no coordinate transformation is required for the elements. Furthermore, no infinitesimal or finite rotations are used as nodal coordinates, and instead slopes that are obtained as the derivatives of the displacements are used. In this case, the matrix equation of motion of the finite element can be written in the simple form

$$\mathbf{M}\ddot{\mathbf{e}} + \mathbf{K}\mathbf{e} = \mathbf{Q} \tag{5.108}$$

where $\mathbf{M}$ is the element mass matrix, $\mathbf{K}$ is the element stiffness matrix, $\mathbf{e}$ is the vector of the element nodal coordinates, and $\mathbf{Q}$ is the vector of generalized nodal forces. In Eq. 108, the element mass matrix is constant, and it is the same matrix that appears in linear structural dynamics. The element stiffness matrix, on the other hand, is a nonlinear function of the nodal coordinates even in the case of small deformations.

In order to demonstrate the use of the absolute nodal coordinate formulation in the analysis of rigid body motion, we consider the uniform slender beam shown in Fig. 17. The beam has length l, cross-sectional area A, mass density ρ, volume V, and mass m. The coordinate system of this beam element is assumed to be initially attached to its left end, which is defined by point O as shown in the figure. Even though, in the absolute nodal coordinate formulation, the displacements in the orthogonal directions can be interpolated using the same polynomials, we use in this section, for the purpose of demonstration, the shape function of a planar beam element that was defined in Section 1 of this chapter as

$$\mathbf{S} = \begin{bmatrix} 1-\xi & 0 & 0 & \xi & 0 & 0 \\ 0 & 1-3\xi^2+2\xi^3 & l(\xi-2\xi^2+\xi^3) & 0 & 3\xi^2-2\xi^3 & l(\xi^3-\xi^2) \end{bmatrix} \tag{5.109}$$

where $\xi = x/l$. The vector of nodal coordinates associated with the shape function of Eq. 109 is

$$\mathbf{e} = [e_1 \quad e_2 \quad e_3 \quad e_4 \quad e_5 \quad e_6]^{\mathrm{T}} \tag{5.110}$$

where e_1 and e_2 are the translational coordinates at the node at O, e_4 and e_5 are the translational coordinates of the node at B, and e_3 and e_6 are the slopes at the two nodes. An arbitrary rigid body displacement of the beam is defined by the translation $\mathbf{R} = [R_x \quad R_y]^{\mathrm{T}}$ of the reference point O, and a rigid body rotation θ. As the result of this arbitrary rigid body displacement, the vector of nodal coordinates $\mathbf{e}$ can be defined in the global coordinate system as

$$\mathbf{e} = [R_x \quad R_y \quad \sin\theta \quad R_x + l\cos\theta \quad R_y + l\sin\theta \quad \sin\theta]^{\mathrm{T}} \tag{5.111}$$

Note that the slopes in the preceding equation are defined using simple rigid body kinematics, since in the case of a rigid body motion, the location of an arbitrary point on the element as the result of the translation $\mathbf{R}$ and the rotation θ can be defined as

$$\mathbf{r} = \begin{bmatrix} r_x \\ r_y \end{bmatrix} = \begin{bmatrix} R_x + x\cos\theta \\ R_y + x\sin\theta \end{bmatrix}$$

It follows that

$$\frac{\partial r_y}{\partial x} = \sin\theta$$

It can be demonstrated that the element shape function of Eq. 109 and the vector of nodal coordinates of Eq. 111 can be used to describe the exact rigid body motion of the element if the slopes are used instead of the infinitesimal rotations. This is clear from the following equation:

$$\mathbf{Se} = \begin{bmatrix} R_x + x\cos\theta \\ R_y + x\sin\theta \end{bmatrix} \tag{5.112}$$

which is the same as the vector $\mathbf{r}$ previously determined using simple rigid body kinematics. The preceding equations clearly demonstrate that the element shape function and the nodal coordinates can describe an arbitrary rigid body displacement provided that the coordinates are defined in the inertial frame and the slopes are defined in terms of trigonometric functions. Therefore, the conventional finite element shape function can be used to describe an exact rigid body displacement, and as a consequence, beam elements can be treated in the absolute nodal coordinate formulation as isoparametric elements.

Using the slopes as nodal coordinates, the element kinetic energy takes the following simple form:

$$T = \tfrac{1}{2}\dot{\mathbf{e}}^{\mathrm{T}}\mathbf{M}\dot{\mathbf{e}} \tag{5.113}$$

where $\mathbf{M}$ is the constant mass matrix that was previously defined in this chapter as

$$\mathbf{M} = \int_V \rho \mathbf{S}^{\mathrm{T}} \mathbf{S} dV$$

Strain Energy While in this section we consider only the case of small deformations for simplicity, the absolute nodal coordinate formulation can be used efficiently to solve large deformation problems. One needs only to change the form of the stiffness matrix, which is nonlinear even in the case of small deformations. If we select point O on the beam element as the reference point, the components of the relative displacement of an arbitrary point with respect to point O can be defined in the inertial coordinate system as

$$\mathbf{u} = \begin{bmatrix} u_x \\ u_y \end{bmatrix} = \begin{bmatrix} (\mathbf{S}_1 - \mathbf{S}_{1O})\mathbf{e} \\ (\mathbf{S}_2 - \mathbf{S}_{2O})\mathbf{e} \end{bmatrix} \tag{5.114}$$

where $\mathbf{S}_1$ and $\mathbf{S}_2$ are the rows of the element shape function matrix, and $\mathbf{S}_{1O}$ and $\mathbf{S}_{2O}$ are the rows of the shape function matrix defined at the reference point O. In order to define the longitudinal and transverse displacements of the beam, we first define the unit vector $\mathbf{i}$ along a selected beam axis as

$$\mathbf{i} = [i_x \quad i_y]^{\mathrm{T}} = \frac{\mathbf{r}_B - \mathbf{r}_O}{|\mathbf{r}_B - \mathbf{r}_O|} \tag{5.115}$$

where $\mathbf{r}_O$ and $\mathbf{r}_B$ are the global position vectors of the two end points of the beam. A unit vector $\mathbf{j}$ perpendicular to $\mathbf{i}$ can be obtained as

$$\mathbf{j} = [j_x \quad j_y]^{\mathrm{T}} = \mathbf{k} \times \mathbf{i} \tag{5.116}$$

where $\mathbf{k}$ is a unit vector along the Z axis. Then, the longitudinal and transverse deformations of the beam can be defined as

$$\mathbf{u}_d = \begin{bmatrix} u_l \\ u_t \end{bmatrix} = \begin{bmatrix} \mathbf{u}^{\mathrm{T}}\mathbf{i} - x \\ \mathbf{u}^{\mathrm{T}}\mathbf{j} \end{bmatrix} = \begin{bmatrix} u_x i_x + u_y i_y - x \\ u_x j_x + u_y j_y \end{bmatrix} \tag{5.117}$$

If we assume a linear elastic model, a simple expression for the strain energy U can be written as

$$U = \frac{1}{2} \int_0^l \left(EA \left(\frac{\partial u_l}{\partial x} \right)^2 + EI \left(\frac{\partial^2 u_t}{\partial x^2} \right)^2 \right) dx \tag{5.118}$$

where E is the modulus of elasticity and I is the second moment of area. It can be shown that the use of the absolute nodal coordinate formulation produces zero deformation in the case of an arbitrary rigid body motion. Using the simple definition of the strain energy presented in the preceding equation, one can show that the stiffness matrix in the case of a linear elastic model is nonlinear. In the case of large deformation analysis, another expression for the strain energy that is based on the nonlinear strain-displacement relationships must be used. Since the governing equations in the case of small and large deformation analysis are nonlinear and must be solved numerically, there is little to be gained by using the small strain assumptions.

Problems

5.1. The beam element shown in Fig. P1 has two nodal points. The coordinates of each node are given by v_i and θ_i, $i = 1, 2$. The element has length l. If the displacement is described using the polynomial

$$v = a_1 + a_2 x + a_3 x^2 + a_4 x^3,$$

obtain the element shape function.

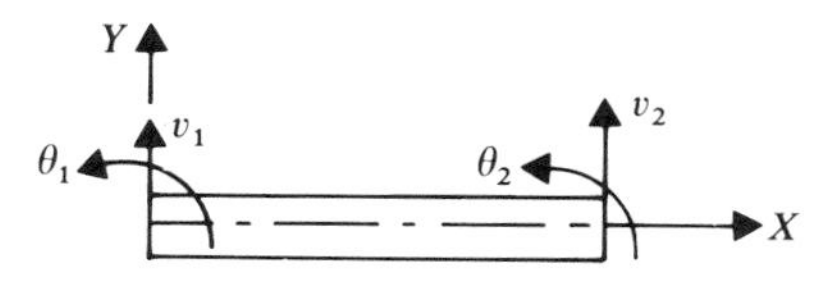

FIG. P5.1

5.2. For the structure shown in Fig. P2, obtain the displacement field of each element in terms of the structure nodal coordinates. Assume that the structure is discretized using the truss elements shown in the figure.

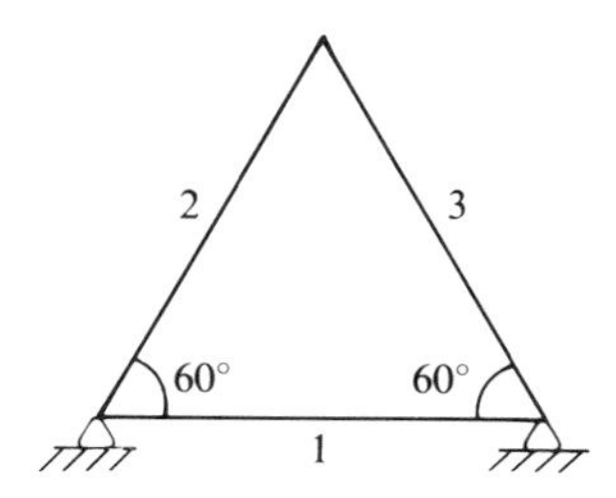

FIG. P5.2

5.3. Obtain the displacement field of each truss element of the structure shown in Fig. P3 in terms of the nodal coordinates of the structure. Identify the element transformation and connectivity matrices.

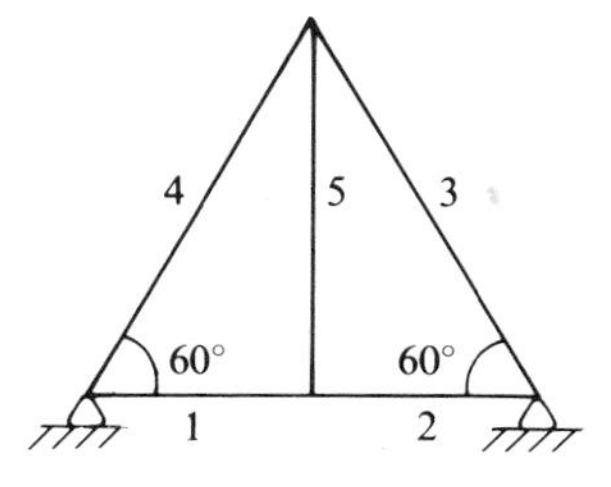

FIG. P5.3

5.4. Using the shape function obtained in Problem 1 for the beam elements, obtain the displacement field of each element of the structure shown in Fig. P4 in terms of the structure nodal coordinates. Identify the element transformation and connectivity matrices. Assume that the length of element j is l^j.

5.5. Repeat Problem 4 using the shape function of the beam element given in Section 1 of this chapter.

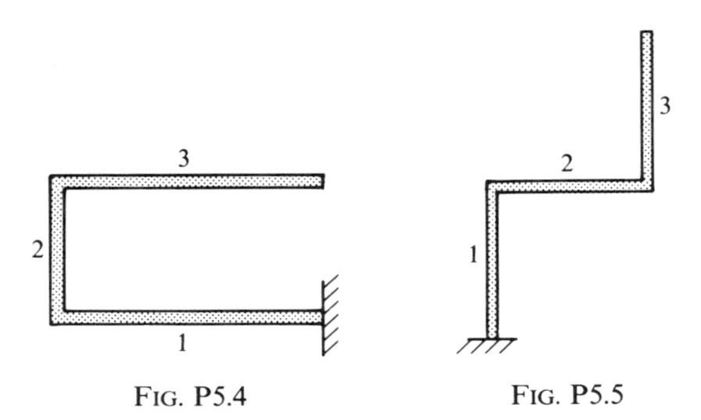

FIG. P5.4 FIG. P5.5

5.6. Using the beam element of Problem 1, obtain the displacement field of each element expressed in terms of the nodal coordinates of the structure shown in Fig. P5. Identify the element transformation and connectivity matrices.

5.7. Repeat Problem 6 using the shape function of the beam element given in Section 1 of this chapter.

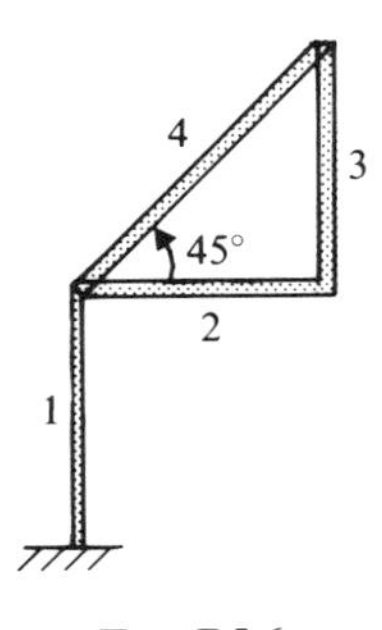

FIG. P5.6

5.8. Obtain the displacement field of the elements of the structure shown in Fig. P6 by using the element defined in Problem 1. Assume that the elements have equal lengths. Identify the element transformation matrices as well as the connectivity matrices.

5.9. Repeat Problem 8 using the finite beam element given in Section 1 of this chapter.

5.10. By using the finite-element method obtain the mass and stiffness matrices of the structure of Problem 2.

5.11. Derive the mass and stiffness matrices of the structure of Problem 3.

5.12. Obtain the mass and stiffness matrices of the structure of Problem 4.

5.13. Develop the mass and stiffness matrices of the structure of Problem 5.

5.14. For the structure given in Problem 6, define the structure mass and stiffness matrices.

5.15. For the structure of Problem 7, obtain the structure mass and stiffness matrices.

5.16. Using beam elements, obtain the mass and stiffness matrices of the structure of Problem 8.

5.17. Define the mass and stiffness matrices of the structure of Problem 9.

5.18. Obtain the generalized forces associated with the structure generalized coordinates as the result of the application of the force F shown in Fig. P7. Assume that the structure is discretized using two equal beam elements and the element shape functions are the same as obtained in Problem 1.

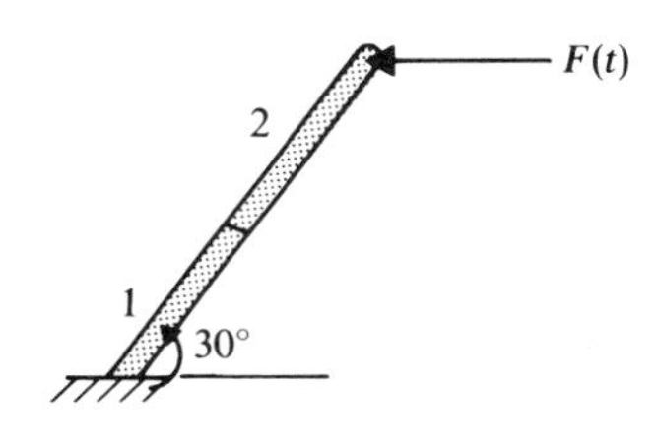

FIG. P5.7

5.19. Repeat Problem 18 using the shape function of the beam element given in Section 1 of this chapter.

5.20. The structure shown in Fig. P8 is discretized into three equal truss elements. Obtain the generalized forces associated with the structure generalized coordinates as the result of the application of the force F.

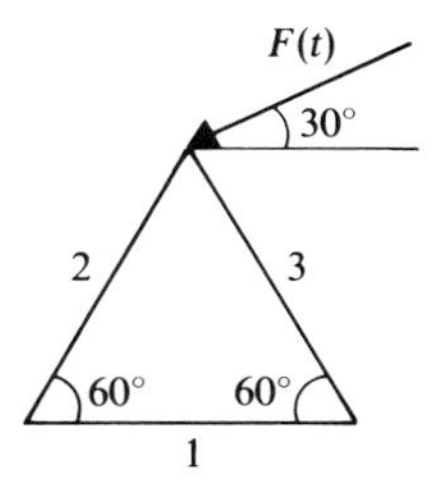

FIG. P5.8

5.21. The structure shown in Fig. P9 is discretized using two-dimensional beam elements. The length, mass, and moment of area of element j are denoted, respectively, by l^j, m^j, and I^j. Obtain the generalized forces associated with the structure generalized coordinates as the result of application of the forces F_1 and F_2. Use the beam element defined in Problem 1.

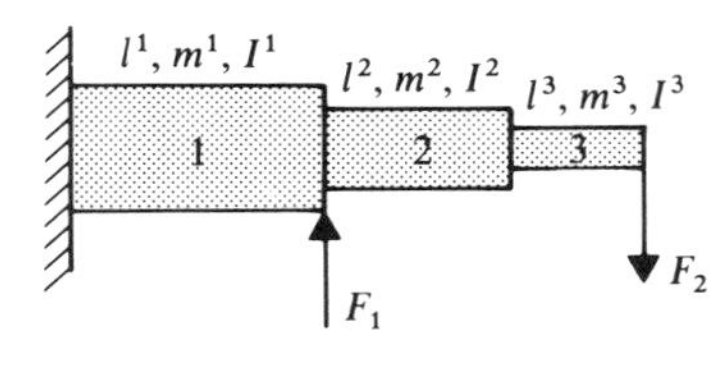

FIG. P5.9

5.22. Repeat Problem 21 using the two-dimensional beam element defined in Section 1 of this chapter.

5.23. Using the shape function of the beam element defined in Section 1 of this chapter, obtain the generalized forces associated with the structure generalized coordinates as the result of applying the forces F_1 and F_2 on the structure shown in Fig. P10.

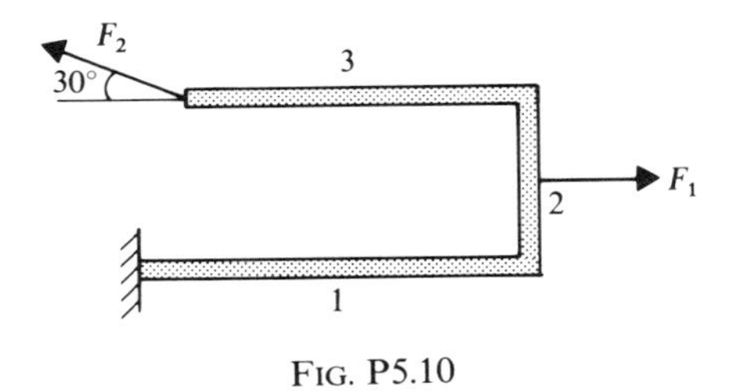

FIG. P5.10

5.24. Repeat Problem 23 using the beam element defined in Problem 1.

5.25. Derive the matrix differential equations of free vibration of the structure of Problem 2.

5.26. Obtain the differential equations of free vibration of the structure of Problem 6.

5.27. Obtain the equations of free vibration of the system of Problem 7.

5.28. Obtain the equations of motion of the system of Problem 20.

5.29. Derive the differential equations of motion of the system of Problem 21.

5.30. Derive the differential equations of motion of the system of Problem 22.

5.31. Identify a set of rigid-body modes for the two-dimensional beam element and discuss the orthogonality of these modes with respect to the mass and stiffness matrices.

5.32. Identify a set of rigid-body modes for the two-dimensional triangular element and discuss the orthogonality of these modes with respect to the element mass and stiffness matrices.

5.33. Determine the element shape function of the six-node triangular element.

5.34. Determine the element shape function of the ten-node triangular element.

5.35. Develop the mass and stiffness matrices of the three-dimensional beam element.

5.36. Obtain the mass and stiffness matrices of the solid element.

5.37. Obtain the mass and stiffness matrices of the tetrahedral element.

5.38. Identify a set of rigid-body modes for the three-dimensional beam element and discuss the orthogonality of these modes with respect to the element mass and stiffness matrices.

5.39. Identify a set of rigid-body modes for the solid element and discuss the orthogonality of these modes with respect to the element mass and stiffness matrices.

6
Methods for the Eigenvalue Analysis

In addition to the matrix-iteration method discussed in Chapter 3, there are several other computer methods that are widely used for solving the eigenvalue problem of vibration systems. Among these methods are the *Jacobi method* and the *QR* method. In these methods, which are based on the *similarity transformation*, a series of transformations that convert a given matrix to a diagonal matrix which has the same eigenvalues as the original matrix are used. Not every matrix, however, is similar to a diagonal matrix. and therefore we find it appropriate to devote several sections of this chapter to discuss the similarity transformation before we briefly discuss the computer methods used for solving the eigenvalue problem of vibration systems. Several definitions will be used repeatedly throughout the development presented in this chapter. Some of these definitions are summarized below.

Monic Polynomials A *monic polynomial* is a nonzero polynomial with a leading coefficient equal to one. An example of such a polynomial in λ is

$$f(\lambda) = \lambda^8 - 5\lambda^3 + \lambda^2 + 8$$

The coefficient of the term that has the highest power in λ in this nonzero polynomial is one, and therefore the preceding polynomial is a monic polynomial. The definition of the monic polynomial will be used repeatedly in this chapter, and for this reason we will slightly change the way we write our characteristic matrix for a matrix $\mathbf{B}$ to $(\lambda\mathbf{I} - \mathbf{B})$ instead of $(\mathbf{B} - \lambda\mathbf{I})$.

Quotient and Remainder If $f(\lambda)$ and $g(\lambda)$ are two polynomials such that $g(\lambda) \neq 0$, then there exist unique polynomials $q(\lambda)$ and $r(\lambda)$ such that

$$f(\lambda) = q(\lambda)g(\lambda) + r(\lambda)$$

with $r(\lambda) = 0$ or the degree of $r(\lambda)$ is less than the degree of $g(\lambda)$. The polynomial $q(\lambda)$ is called the *quotient* and $r(\lambda)$ is called the *remainder*. If $r(\lambda) = 0$, $g(\lambda)$ is said to be a *divisor* or a *factor* of $f(\lambda)$. We also say that $f(\lambda)$ is divisible by $g(\lambda)$. For example, the polynomial

$$f(\lambda) = \lambda^2 + 6\lambda + 5$$

can be written as

$$f(\lambda) = (\lambda + 5)(\lambda + 1)$$

It follows that the polynomials $(\lambda + 5)$ and $(\lambda + 1)$ are divisors or factors of $f(\lambda)$. If, on the other hand, we consider the polynomial $g(\lambda) = (\lambda + 2)$, then $f(\lambda)$ can be written as

$$f(\lambda) = (\lambda + 4)g(\lambda) - 3$$

where, in this case, the remainder $r(\lambda) = -3$, which is a polynomial of a degree less than the degree of $g(\lambda)$.

Greatest Common Divisor A monic polynomial $g(\lambda)$ is said to be the greatest common divisor of the nonzero polynomials $f_1(\lambda)$ and $f_2(\lambda)$ if the following two conditions are satisfied:

(1) $g(\lambda)$ divides both $f_1(\lambda)$ and $f_2(\lambda)$.
(2) If $h(\lambda)$ divides both $f_1(\lambda)$ and $f_2(\lambda)$, then $h(\lambda)$ divides $g(\lambda)$.

Least Common Multiple The *least common multiple* of two polynomials $f_1(\lambda)$ and $f_2(\lambda)$ is a polynomial $g(\lambda)$ such that:

(1) $f_1(\lambda)$ and $f_2(\lambda)$ are factors (divisors) of $g(\lambda)$.
(2) If $h(\lambda)$ is a polynomial such that $f_1(\lambda)$ and $f_2(\lambda)$ are divisors of $h(\lambda)$, then $g(\lambda)$ is a divisor of $h(\lambda)$.

6.1 SIMILARITY TRANSFORMATION

Two square matrices **B** and **D** are said to be similar if there exists a nonsingular matrix $\mathbf{\Phi}$ such that

$$\mathbf{B} = \mathbf{\Phi}^{-1}\mathbf{D}\mathbf{\Phi} \tag{6.1}$$

It is clear that if **B** is similar to **D**, then **D** is similar to **B** since

$$\mathbf{D} = \mathbf{\Phi}\mathbf{B}\mathbf{\Phi}^{-1}$$

Furthermore, if **B** is similar to **D** and **D** is similar to **E**, then **B** is similar to **E**, for

$$\mathbf{B} = \mathbf{\Phi}_1^{-1}\mathbf{D}\mathbf{\Phi}_1, \qquad \mathbf{D} = \mathbf{\Phi}_2^{-1}\mathbf{E}\mathbf{\Phi}_2$$

implies that

$$\mathbf{B} = (\mathbf{\Phi}_2\mathbf{\Phi}_1)^{-1}\mathbf{E}(\mathbf{\Phi}_2\mathbf{\Phi}_1)$$

It also can be verified that

$$\mathbf{\Phi}^{-1}\mathbf{B}^m\mathbf{\Phi} = (\mathbf{\Phi}^{-1}\mathbf{B}\mathbf{\Phi})^m \tag{6.2}$$

Example 6.1

Let

$$\mathbf{B} = \begin{bmatrix} 1 & 1 \\ 0 & 1 \end{bmatrix}, \qquad \mathbf{\Phi} = \begin{bmatrix} 1 & 2 \\ 0 & 1 \end{bmatrix}$$

It is clear that

$$\mathbf{B}^n = \begin{bmatrix} 1 & n \\ 0 & 1 \end{bmatrix}, \qquad \mathbf{\Phi}^{-1} = \begin{bmatrix} 1 & -2 \\ 0 & 1 \end{bmatrix}$$

Since

$$\mathbf{\Phi}^{-1}\mathbf{B}\mathbf{\Phi} = \begin{bmatrix} 1 & -2 \\ 0 & 1 \end{bmatrix}\begin{bmatrix} 1 & 1 \\ 0 & 1 \end{bmatrix}\begin{bmatrix} 1 & 2 \\ 0 & 1 \end{bmatrix} = \begin{bmatrix} 1 & 1 \\ 0 & 1 \end{bmatrix}$$

one has

$$(\mathbf{\Phi}^{-1}\mathbf{B}\mathbf{\Phi})^n = \begin{bmatrix} 1 & 1 \\ 0 & 1 \end{bmatrix}^n = \begin{bmatrix} 1 & n \\ 0 & 1 \end{bmatrix}$$

Characteristic Polynomial Similar matrices have the same characteristic polynomials. In order to demonstrate this, let **B** and **D** be two similar matrices such that

$$\mathbf{B} = \mathbf{\Phi}^{-1}\mathbf{D}\mathbf{\Phi}$$

The characteristic polynomial of **B** is

$$|\lambda\mathbf{I} - \mathbf{B}| = 0$$

where λ is the eigenvalue of the matrix **B**. It follows from the similarity transformation that

$$\begin{aligned} |\lambda\mathbf{I} - \mathbf{B}| &= |\lambda\mathbf{I} - \mathbf{\Phi}^{-1}\mathbf{D}\mathbf{\Phi}| = |\mathbf{\Phi}^{-1}(\lambda\mathbf{I} - \mathbf{D})\mathbf{\Phi}| \\ &= |\mathbf{\Phi}^{-1}||\lambda\mathbf{I} - \mathbf{D}||\mathbf{\Phi}| \end{aligned}$$

Since

$$|\mathbf{\Phi}^{-1}||\mathbf{\Phi}| = 1$$

one has

$$|\lambda\mathbf{I} - \mathbf{B}| = |\lambda\mathbf{I} - \mathbf{D}| \tag{6.3}$$

which implies that similar matrices have the same characteristic polynomial and, consequently, have the same eigenvalues. It also follows that similar matrices have the same trace and the same determinant.

The equality of the characteristic polynomial is a necessary, but not a sufficient, condition for the similarity of two matrices. For example, the two matrices

$$\mathbf{B} = \begin{bmatrix} 1 & 0 \\ 0 & 1 \end{bmatrix}, \qquad \mathbf{D} = \begin{bmatrix} 1 & 1 \\ 0 & 1 \end{bmatrix}$$

have the same characteristic polynomial $(1-\lambda)^2$, but **B** and **D** are not similar, since for any nonsingular matrix $\mathbf{\Phi}$ one has

$$\mathbf{\Phi}^{-1}\mathbf{B}\mathbf{\Phi} = \mathbf{\Phi}^{-1}\mathbf{\Phi} = \mathbf{I}$$

Note that because **B** and **D** have the same characteristic polynomial, they have the same eigenvalues $\lambda_1 = \lambda_2 = \lambda = 1$. Associated with these repeated eigenvalues, the matrix **B** has the two independent eigenvectors

$$\mathbf{A}_1 = \begin{bmatrix} 1 \\ 0 \end{bmatrix}, \qquad \mathbf{A}_2 = \begin{bmatrix} 0 \\ 1 \end{bmatrix}$$

This is because the characteristic matrix $(\lambda\mathbf{I} - \mathbf{B})$ has rank zero, and, as a consequence, there are two independent solutions for the homogeneous system of algebraic equations $(\lambda\mathbf{I} - \mathbf{B})\mathbf{A} = \mathbf{0}$. On the other hand, the matrix **D** does not have two independent eigenvectors associated with the repeated roots λ_1 and λ_2 since the characteristic matrix $(\lambda\mathbf{I} - \mathbf{D})$ has rank one, and, therefore, there is only one independent solution for the system of homogeneous algebraic equations $(\lambda\mathbf{I} - \mathbf{D})\mathbf{A} = \mathbf{0}$. The independent eigenvector of the matrix **D** takes the form

$$\mathbf{A} = \begin{bmatrix} 1 \\ 0 \end{bmatrix}$$

Recall that the number of dependent variables in a system of homogeneous algebraic equations is equal to the number of independent equations, which is the same as the rank of the coefficient matrix in this system of equations. It follows that the number of independent solutions is the same as the degree of singularity of the coefficient matrix, and is equal to the dimension n minus the rank r of the matrix.

In general, a matrix that has s repeated eigenvalues has s linearly independent eigenvectors associated with these repeated eigenvalues if the matrix is similar to a diagonal matrix. If a matrix is not similar to a diagonal matrix, the s repeated eigenvalues do not correspond to s linearly independent eigenvectors, but they do correspond to a set of *generalized eigenvectors*, which will be introduced in later sections.

As demonstrated by the example of the two matrices **B** and **D** previously presented in this section, having the same characteristic polynomial is not sufficient for the similarity of two matrices. In fact, one should not be able to find a nonsingular matrix $\mathbf{\Phi}$ such that $\mathbf{\Phi}^{-1}\mathbf{D}\mathbf{\Phi}$ is equal to a diagonal matrix.

6.2 POLYNOMIAL MATRICES

A polynomial in an arbitrary square matrix **B** can be written as

$$f(\mathbf{B}) = \alpha_0\mathbf{I} + \alpha_1\mathbf{B} + \alpha_2\mathbf{B}^2 + \cdots + \alpha_n\mathbf{B}^n \tag{6.4}$$

where $\alpha_0, \alpha_1, \ldots,$ and α_n are scalar coefficients. It can be shown that if $f_1(\mathbf{B})$

and $f_2(\mathbf{B})$ are two polynomials in the arbitrary matrix **B** such that

$$f_1(\mathbf{B}) = \alpha_0\mathbf{I} + \alpha_1\mathbf{B} + \alpha_2\mathbf{B}^2 + \cdots + \alpha_n\mathbf{B}^n$$
$$f_2(\mathbf{B}) = \beta_0\mathbf{I} + \beta_1\mathbf{B} + \beta_2\mathbf{B}^2 + \cdots + \beta_m\mathbf{B}^m$$

then

$$f_1(\mathbf{B}) + f_2(\mathbf{B}) = f_2(\mathbf{B}) + f_1(\mathbf{B}) \tag{6.5}$$
$$f_1(\mathbf{B})f_2(\mathbf{B}) = f_2(\mathbf{B})f_1(\mathbf{B}) \tag{6.6}$$

which imply that operations with polynomial matrices follow the same rules as scalar polynomials. This fact can be used to obtain matrix identities, as demonstrated by the following example.

Example 6.2

It is known from scalar polynomials that

$$\lambda^2 - 1 = (\lambda - 1)(\lambda + 1)$$
$$\lambda^3 + 1 = (\lambda + 1)(\lambda^2 - \lambda + 1)$$

It follows that for any arbitrary square matrix **B**, one has

$$\mathbf{B}^2 - \mathbf{I} = (\mathbf{B} - \mathbf{I})(\mathbf{B} + \mathbf{I})$$
$$\mathbf{B}^3 + \mathbf{I} = (\mathbf{B} + \mathbf{I})(\mathbf{B}^2 - \mathbf{B} + \mathbf{I})$$

Cayley–Hamilton Theorem The *Cayley–Hamilton theorem*, which we present here without proof, states that every symmetric matrix satisfies its characteristic polynomial.

Example 6.3

The matrix

$$\mathbf{B} = \begin{bmatrix} 4 & 1 & 2 \\ 1 & 0 & 0 \\ 2 & 0 & 0 \end{bmatrix}$$

has the characteristic polynomial

$$|\lambda\mathbf{I} - \mathbf{B}| = \lambda^3 - 4\lambda^2 - 5\lambda = 0$$

The matrices $\mathbf{B}^2$ and $\mathbf{B}^3$ are

$$\mathbf{B}^2 = \begin{bmatrix} 21 & 4 & 8 \\ 4 & 1 & 2 \\ 8 & 2 & 4 \end{bmatrix}, \quad \mathbf{B}^3 = \begin{bmatrix} 104 & 21 & 42 \\ 21 & 4 & 8 \\ 42 & 8 & 16 \end{bmatrix}$$

Using the characteristic polynomial, it can be shown that

$$\mathbf{B}^3 - 4\mathbf{B}^2 - 5\mathbf{B} = \mathbf{0}$$

The Cayley–Hamilton theorem also can be used to define the inverse of nonsingular square matrices. As an example, we consider the matrix

$$\mathbf{B} = \begin{bmatrix} 1 & 5 & 0 \\ -1 & 1 & 2 \\ -2 & 2 & -1 \end{bmatrix}$$

which has the characteristic polynomial

$$\lambda^3 - \lambda^2 + 30 = 0$$

It follows from the Cayley–Hamilton theorem that

$$\mathbf{B}^3 - \mathbf{B}^2 + 30\mathbf{I} = \mathbf{0}$$

or

$$\mathbf{I} = \frac{-1}{30}\{\mathbf{B}^3 - \mathbf{B}^2\}$$

Premultiplying this equation by $\mathbf{B}^{-1}$, one obtains

$$\mathbf{B}^{-1} = \frac{-1}{30}\{\mathbf{B}^2 - \mathbf{B}\} = \begin{bmatrix} \frac{1}{6} & \frac{-1}{6} & \frac{-1}{3} \\ \frac{1}{6} & \frac{1}{30} & \frac{1}{15} \\ 0 & \frac{2}{5} & \frac{-1}{5} \end{bmatrix}$$

Minimal Polynomial The *minimal polynomial* of an arbitrary square matrix $\mathbf{B}$ is the nonzero monic polynomial of least degree that has $\mathbf{B}$ as a zero. Every matrix has a unique minimal polynomial, for if $f_1(\mathbf{B})$ and $f_2(\mathbf{B})$ are two different minimal polynomials of $\mathbf{B}$, then $f_1(\mathbf{B}) - f_2(\mathbf{B})$ would be a nonzero polynomial of lower degree that has $\mathbf{B}$ as a zero. If this polynomial is divided by the leading coefficient, one obtains a monic minimal polynomial with lower degree. This contradicts the fact that $f_1(\mathbf{B})$ and $f_2(\mathbf{B})$ are minimal polynomials of $\mathbf{B}$. Therefore, $\mathbf{B}$ has a unique minimal polynomial.

Every polynomial that has the matrix $\mathbf{B}$ as a zero is divisible without remainder by the minimal polynomial of the matrix $\mathbf{B}$. In order to prove this fact, suppose, on the contrary, that $g(\mathbf{B}) = \mathbf{0}$ is a polynomial in $\mathbf{B}$ that is not divisible by the minimal polynomial $f(\mathbf{B})$ of the matrix $\mathbf{B}$. It follows that

$$g(\mathbf{B}) = f(\mathbf{B})h(\mathbf{B}) + r(\mathbf{B}) \tag{6.7}$$

where $h(\mathbf{B})$ is the quotient and $r(\mathbf{B})$ is the remainder that has a degree less than the degree of $f(\mathbf{B})$. Since $f(\mathbf{B}) = \mathbf{0}$ and $g(\mathbf{B}) = \mathbf{0}$, one also has $r(\mathbf{B}) = \mathbf{0}$. This contradicts the assumption that $f(\mathbf{B})$ is the minimal polynomial of the matrix $\mathbf{B}$. Consequently, the minimal polynomial of a matrix $\mathbf{B}$ is a divisor of any polynomial that has $\mathbf{B}$ as its zero. In particular, the minimal polynomial is a divisor of the characteristic polynomial of the matrix.

It was previously pointed out that if $\mathbf{B}$ is an arbitrary square matrix, and $\mathbf{\Phi}$ is any nonsingular matrix, then

$$\mathbf{\Phi}^{-1}\mathbf{B}^m\mathbf{\Phi} = (\mathbf{\Phi}^{-1}\mathbf{B}\mathbf{\Phi})^m$$

for any integer m. It follows from this identity that

$$\mathbf{\Phi}^{-1}f(\mathbf{B})\mathbf{\Phi} = f(\mathbf{\Phi}^{-1}\mathbf{B}\mathbf{\Phi}) \tag{6.8}$$

This identity can be used to show that similar matrices have the same minimal polynomial. In order to prove this assertion, suppose that $\mathbf{B}$ and $\mathbf{D}$ are two similar matrices such that $\mathbf{B} = \mathbf{\Phi}^{-1}\mathbf{D}\mathbf{\Phi}$. If $f(\mathbf{D}) = \mathbf{0}$, then

$$f(\mathbf{B}) = f(\mathbf{\Phi}^{-1}\mathbf{D}\mathbf{\Phi}) = \mathbf{\Phi}^{-1}f(\mathbf{D})\mathbf{\Phi} = \mathbf{0}$$

which implies that the set of polynomials having $\mathbf{B}$ as a zero is the same as the set of polynomials having $\mathbf{D}$ as a zero, and hence the minimal polynomials of two similar matrices are identical.

As in the case of the characteristic polynomial, the equality of the minimal polynomials is a necessary, but not sufficient, condition for the similarity of two matrices, as demonstrated by the following example.

Example 6.4

Consider the two matrices

$$\mathbf{B} = \begin{bmatrix} 1 & 0 & 0 \\ 0 & 2 & 0 \\ 0 & 0 & 2 \end{bmatrix}, \qquad \mathbf{D} = \begin{bmatrix} 1 & 0 & 0 \\ 0 & 1 & 2 \\ 0 & 0 & 2 \end{bmatrix}$$

The characteristic polynomials of these two matrices are

$$|\lambda\mathbf{I} - \mathbf{B}| = (\lambda - 1)(\lambda - 2)^2$$

$$|\lambda\mathbf{I} - \mathbf{D}| = (\lambda - 1)^2(\lambda - 2)$$

Since these two characteristic polynomials are not the same, $\mathbf{B}$ is not similar to $\mathbf{D}$. It can be shown, however, that $\mathbf{B}$ and $\mathbf{D}$ have the same minimal polynomials:

$$g(\mathbf{B}) = (\mathbf{B} - \mathbf{I})(\mathbf{B} - 2\mathbf{I}) = \mathbf{0}$$

$$g(\mathbf{D}) = (\mathbf{D} - \mathbf{I})(\mathbf{D} - 2\mathbf{I}) = \mathbf{0}$$

These are the monic polynomials of least degree that have $\mathbf{B}$ and $\mathbf{D}$ as zeros.

If $\mathbf{B}$ is the block diagonal matrix

$$\mathbf{B} = \begin{bmatrix} \mathbf{B}_1 & & & \\ & \mathbf{B}_2 & & \mathbf{0} \\ & & \ddots & \\ & \mathbf{0} & & \mathbf{B}_s \end{bmatrix} \tag{6.9}$$

then the characteristic polynomial of **B** is the product of the characteristic polynomials of its diagonal blocks, that is,

$$|\lambda\mathbf{I} - \mathbf{B}| = |\lambda\mathbf{I} - \mathbf{B}_1||\lambda\mathbf{I} - \mathbf{B}_2|\cdots|\lambda\mathbf{I} - \mathbf{B}_s| \tag{6.10}$$

The minimal polynomial of a block diagonal matrix is the *least common multiple* of the minimal polynomials of its diagonal blocks.

6.3 EQUIVALENCE OF THE CHARACTERISTIC MATRICES

It can be shown that the eigenvectors associated with distinct eigenvalues of a matrix are linearly independent, and hence a matrix that has distinct eigenvalues is similar to a diagonal matrix. For example, if **B** is an $n \times n$ matrix that has the distinct eigenvalues $\lambda_1, \lambda_2, \ldots, \lambda_n$, then the eigenvectors $\mathbf{A}_1$, $\mathbf{A}_2$, $\ldots$, $\mathbf{A}_n$ associated with these eigenvalues are linearly independent. Furthermore, since

$$\mathbf{BA}_i = \lambda_i \mathbf{A}_i \tag{6.11}$$

one has

$$\mathbf{B\Phi} = \mathbf{\Phi\Lambda} \tag{6.12}$$

where $\mathbf{\Phi}$ is the modal matrix, and $\mathbf{\Lambda}$ is a diagonal matrix that has the eigenvalues as the diagonal elements. The matrices $\mathbf{\Phi}$ and $\mathbf{\Lambda}$ are

$$\mathbf{\Phi} = [\mathbf{A}_1 \quad \mathbf{A}_2 \quad \ldots \quad \mathbf{A}_n] \tag{6.13a}$$

$$\mathbf{\Lambda} = \begin{bmatrix} \lambda_1 & & & \\ & \lambda_2 & & \mathbf{0} \\ & & \ddots & \\ & \mathbf{0} & & \lambda_n \end{bmatrix} \tag{6.13b}$$

Premultiplying both sides of Eq. 12 by $\mathbf{\Phi}^{-1}$, one gets

$$\mathbf{\Phi}^{-1}\mathbf{B\Phi} = \mathbf{\Lambda} \tag{6.14}$$

which implies that if **B** has linearly independent eigenvectors, then **B** is similar to a diagonal matrix whose diagonal elements are the eigenvalues of **B**. Furthermore, the matrix of the similarity transformation $\mathbf{\Phi}$ is the modal matrix whose columns are the linearly independent eigenvectors of **B**.

Condition for Similarity In the analysis presented in Chapter 3, it was pointed out that the number of independent eigenvectors associated with repeated roots depends on the rank of the characteristic matrix. In fact, two matrices **B** and **D** are similar if their characteristic matrices $(\lambda\mathbf{I} - \mathbf{B})$ and $(\lambda\mathbf{I} - \mathbf{D})$ are equivalent, that is, if $(\lambda\mathbf{I} - \mathbf{B})$ can be obtained from $(\lambda\mathbf{I} - \mathbf{D})$ by a series of elementary operations. Similarity implies the equivalence of the characteristic matrices since, if **B** is similar to **D**, then there exists a nonsingular matrix $\mathbf{\Phi}$ such that

$$\mathbf{D} = \mathbf{\Phi}^{-1}\mathbf{B}\mathbf{\Phi}$$

It follows that

$$(\lambda\mathbf{I} - \mathbf{D}) = \mathbf{\Phi}^{-1}(\lambda\mathbf{I} - \mathbf{B})\mathbf{\Phi}$$

which shows that the characteristic matrices of the two similar matrices **B** and **D** are equivalent. Conversely, if the characteristic matrices $(\lambda\mathbf{I} - \mathbf{B})$ and $(\lambda\mathbf{I} - \mathbf{D})$ are equivalent, then there exist two nonsingular matrices **V** and **W** such that

$$(\lambda\mathbf{I} - \mathbf{D}) = \mathbf{V}(\lambda\mathbf{I} - \mathbf{B})\mathbf{W}$$

By comparing the coefficients of λ and the constant terms, we conclude that

$$\mathbf{V}\mathbf{W} = \mathbf{I}, \qquad \mathbf{D} = \mathbf{V}\mathbf{B}\mathbf{W}$$

which imply that

$$\mathbf{D} = \mathbf{W}^{-1}\mathbf{B}\mathbf{W}$$

proving that **B** and **D** are similar if their characteristic matrices are equivalent.

Invariant Factors A characteristic matrix can be reduced by a series of elementary operations to the diagonal form

$$\begin{bmatrix} f_1(\lambda) & & & \\ & f_2(\lambda) & & \mathbf{0} \\ & & \ddots & \\ & \mathbf{0} & & f_n(\lambda) \end{bmatrix} \tag{6.15}$$

This diagonal matrix is called a *canonical diagonal λ-matrix* if each diagonal element $f_i(\lambda)$ is a divisor of the next element $f_{i+1}(\lambda)$ and if all nonzero diagonal elements are monic polynomials. Therefore, an example of a canonical diagonal matrix takes the form

$$\begin{bmatrix} 1 & & & & & & & & \\ & \ddots & & & \mathbf{0} & & & & \\ & & 1 & & & & & & \\ & & & f_1(\lambda) & & & & & \\ & & & & \ddots & & & & \\ & & & & & f_k(\lambda) & & & \\ & & & & & & 0 & & \\ & & & & \mathbf{0} & & & \ddots & \\ & & & & & & & & 0 \end{bmatrix} \tag{6.16}$$

where $f_1(\lambda), \ldots, f_k(\lambda)$ are monic polynomials such that $f_i(\lambda)$ is a divisor of $f_{i+1}(\lambda)$.

In addition to having the same characteristic and minimal polynomial, similar matrices have the same canonical diagonal matrix; that is, their characteristic matrices can be reduced by a series of elementary operations to the same diagonal form. Therefore, the diagonal elements of the canonical diagonal λ matrix, which are called the *invariant factors*, are unique. It follows that similar matrices have the same invariant factors.

Example 6.5

The matrices

$$\mathbf{B} = \begin{bmatrix} 4 & 1 & 2 \\ 1 & 0 & 0 \\ 2 & 0 & 0 \end{bmatrix}, \qquad \mathbf{D} = \begin{bmatrix} 3 & 4 & 6 \\ -1 & -1 & -1 \\ 2 & 2 & 2 \end{bmatrix}$$

are similar since there exists a nonsingular matrix

$$\mathbf{V} = \begin{bmatrix} 1 & -1 & 0 \\ 0 & 1 & -1 \\ 0 & 0 & 1 \end{bmatrix}$$

such that

$$\mathbf{D} = \mathbf{V}\mathbf{B}\mathbf{V}^{-1}$$

where

$$\mathbf{V}^{-1} = \begin{bmatrix} 1 & 1 & 1 \\ 0 & 1 & 1 \\ 0 & 0 & 1 \end{bmatrix}$$

The matrices **B** and **D** have the same eigenvalues:

$$\lambda_1 = -1, \qquad \lambda_2 = 5, \qquad \lambda_3 = 0$$

The eigenvectors of **B** associated with these eigenvalues are

$$\mathbf{A}_{B1} = \begin{bmatrix} 1 \\ -1 \\ -2 \end{bmatrix}, \qquad \mathbf{A}_{B2} = \begin{bmatrix} 5 \\ 1 \\ 2 \end{bmatrix}, \qquad \mathbf{A}_{B3} = \begin{bmatrix} 0 \\ 2 \\ -1 \end{bmatrix}$$

The eigenvectors of **D** are

$$\mathbf{A}_{D1} = \begin{bmatrix} 2 \\ 1 \\ -2 \end{bmatrix}, \qquad \mathbf{A}_{D2} = \begin{bmatrix} 4 \\ -1 \\ 2 \end{bmatrix}, \qquad \mathbf{A}_{D3} = \begin{bmatrix} -2 \\ 3 \\ -1 \end{bmatrix}$$

Note that because of the similarity of **B** and **D**,

$$\mathbf{A}_{Di} = \mathbf{V}\mathbf{A}_{Bi}$$

The characteristic matrices of **B** and **D** are

$$(\lambda\mathbf{I} - \mathbf{B}) = \begin{bmatrix} \lambda - 4 & -1 & -2 \\ -1 & \lambda & 0 \\ -2 & 0 & \lambda \end{bmatrix}$$

$$(\lambda\mathbf{I} - \mathbf{D}) = \begin{bmatrix} \lambda - 3 & -4 & -6 \\ 1 & \lambda + 1 & 1 \\ -2 & -2 & \lambda - 2 \end{bmatrix}$$

Define the two matrices

$$\mathbf{\Phi}_B = [\mathbf{A}_{B1} \quad \mathbf{A}_{B2} \quad \mathbf{A}_{B3}] = \begin{bmatrix} 1 & 5 & 0 \\ -1 & 1 & 2 \\ -2 & 2 & -1 \end{bmatrix}$$

$$\mathbf{\Phi}_D = [\mathbf{A}_{D1} \quad \mathbf{A}_{D2} \quad \mathbf{A}_{D3}] = \begin{bmatrix} 2 & 4 & -2 \\ 1 & -1 & 3 \\ -2 & 2 & -1 \end{bmatrix}$$

It can be shown that

$$\mathbf{\Phi}_B^{-1}(\lambda\mathbf{I} - \mathbf{B})\mathbf{\Phi}_B = \begin{bmatrix} \lambda + 1 & 0 & 0 \\ 0 & \lambda - 5 & 0 \\ 0 & 0 & \lambda \end{bmatrix}$$

$$\mathbf{\Phi}_D^{-1}(\lambda\mathbf{I} - \mathbf{D})\mathbf{\Phi}_D = \begin{bmatrix} \lambda + 1 & 0 & 0 \\ 0 & \lambda - 5 & 0 \\ 0 & 0 & \lambda \end{bmatrix}$$

In this case, the canonical diagonal λ-matrix, which can be obtained by elementary operations, is

$$\begin{bmatrix} 1 & 0 & 0 \\ 0 & 1 & 0 \\ 0 & 0 & \lambda(\lambda + 1)(\lambda - 5) \end{bmatrix}$$

Note that the matrix $\mathbf{\Phi}_B$ and its inverse, which reduce the characteristic matrix to a diagonal form, are the products of elementary matrices. These matrices can be written as

$$\mathbf{\Phi}_B = (\mathbf{E}_4\mathbf{E}_3\mathbf{E}_2\mathbf{E}_1)^{-1}$$

$$\mathbf{\Phi}_B^{-1} = (\mathbf{E}_4\mathbf{E}_3\mathbf{E}_2\mathbf{E}_1) = \begin{bmatrix} \frac{1}{6} & -\frac{1}{6} & -\frac{1}{3} \\ \frac{1}{6} & \frac{1}{30} & \frac{1}{15} \\ 0 & \frac{2}{5} & -\frac{1}{5} \end{bmatrix}$$

where $\mathbf{E}_1$, $\mathbf{E}_2$, $\mathbf{E}_3$, and $\mathbf{E}_4$ are the elementary matrices

$$\mathbf{E}_1 = \begin{bmatrix} 1 & 0 & 0 \\ 1 & 1 & 0 \\ 2 & 0 & 1 \end{bmatrix}, \qquad \mathbf{E}_2 = \begin{bmatrix} 1 & 0 & 0 \\ 0 & 1 & 2 \\ 0 & 0 & 1 \end{bmatrix}$$

$$\mathbf{E}_3 = \begin{bmatrix} 1 & -\frac{1}{6} & 0 \\ 0 & 1 & 0 \\ 0 & -\frac{2}{5} & 1 \end{bmatrix}, \qquad \mathbf{E}_4 = \begin{bmatrix} 1 & 0 & 0 \\ 0 & \frac{1}{30} & 0 \\ 0 & 0 & -1 \end{bmatrix}$$

The inverses of the elementary matrices are

$$\mathbf{E}_1^{-1} = \begin{bmatrix} 1 & 0 & 0 \\ -1 & 1 & 0 \\ -2 & 0 & 1 \end{bmatrix}, \qquad \mathbf{E}_2^{-1} = \begin{bmatrix} 1 & 0 & 0 \\ 0 & 1 & -2 \\ 0 & 0 & 1 \end{bmatrix}$$

$$\mathbf{E}_3^{-1} = \begin{bmatrix} 1 & \frac{1}{6} & 0 \\ 0 & 1 & 0 \\ 0 & \frac{2}{5} & 0 \end{bmatrix}, \qquad \mathbf{E}_4^{-1} = \begin{bmatrix} 1 & 0 & 0 \\ 0 & 30 & 0 \\ 0 & 0 & -1 \end{bmatrix}$$

Similar comments apply to the matrix $\mathbf{\Phi}_D$.

Each characteristic matrix can be reduced to a canonical diagonal λ-matrix whose diagonal elements are the invariant factors. It can be shown that the kth diagonal element (invariant factor) of the canonical diagonal matrix, which is a monic polynomial, is the greatest common divisor of all minors of order k divided by the greatest common divisors of all minors of order $k - 1$. For example, using the matrix $\mathbf{B}$ of the preceding example, one can show that the minors of order 2 of the characteristic matrix of $\mathbf{B}$ are

$$\begin{aligned} &M_{11} = \lambda^2, \qquad &&M_{12} = -\lambda, \qquad &&M_{13} = 2\lambda \\ &M_{21} = -\lambda, &&M_{22} = \lambda^2 - 4\lambda - 4, &&M_{23} = -2 \\ &M_{31} = 2\lambda, &&M_{32} = -2, &&M_{33} = \lambda^2 - 4\lambda - 1 \end{aligned}$$

The polynomial with a leading coefficient one, which represents the greatest common divisor of all the minors of the matrix $(\lambda\mathbf{I} - \mathbf{B})$, is 1. This is also the greatest common divisor of the minors of order 2 of the matrix $(\lambda\mathbf{I} - \mathbf{D})$. It can be shown that equivalent matrices have the same greatest common divisors of the kth-order minors.

6.4 JORDAN MATRICES

While similar matrices have equivalent characteristic matrices that can be reduced by a series of elementary operations to a unique diagonal form, not every matrix is similar to a diagonal matrix. A matrix $\mathbf{B}$ is similar to a diagonal matrix $\mathbf{\Lambda}$ if and only if $\mathbf{B}$ has a complete set of linearly independent eigenvectors. For, if $\mathbf{B}$ has linearly independent eigenvectors, the matrix $\mathbf{\Phi}$ whose columns are these independent eigenvectors is nonsingular. In this case, the eigenvalue problem leads to

$$\mathbf{B\Phi} = \mathbf{\Phi\Lambda}$$

where $\mathbf{\Lambda}$ is a diagonal matrix whose diagonal elements are the eigenvalues of $\mathbf{B}$. The preceding equation demonstrates that $\mathbf{B}$ and $\mathbf{\Lambda}$ are similar. Conversely, if **B** is similar to a diagonal matrix, then **B** has a complete set of linearly independent eigenvectors since

$$\mathbf{V}^{-1}\mathbf{B}\mathbf{V} = \mathbf{\Lambda}$$

implies that

$$\mathbf{B}\mathbf{V} = \mathbf{V}\mathbf{\Lambda}$$

which proves that **V** is the matrix of the linearly independent eigenvectors which can be determined to within an arbitrary constant. The question of the linear independence of the eigenvectors arises when a matrix has repeated eigenvalues, as is demonstrated by the following example.

Example 6.6

The matrix

$$\mathbf{B} = \begin{bmatrix} 1 & 1 & -2 & 0 \\ -5 & 7 & -10 & 0 \\ -2 & 2 & -2 & 0 \\ 0 & 0 & 0 & 1 \end{bmatrix}$$

has the characteristic equation

$$|\lambda \mathbf{I} - \mathbf{B}| = (\lambda - 2)^3(\lambda - 1) = 0$$

which defines the four eigenvalues

$$\lambda_1 = \lambda_2 = \lambda_3 = 2, \qquad \lambda_4 = 1$$

The matrix **B** has three repeated eigenvalues, λ_1, λ_2, and λ_3, and one distinct eigenvalue, λ_4. The matrix $(\lambda_4\mathbf{I} - \mathbf{B})$ has rank 3, and as such the system of homogeneous algebraic equations

$$(\lambda_4\mathbf{I} - \mathbf{B})\mathbf{A}_4 = \mathbf{0}$$

has one independent nontrivial solution:

$$\mathbf{A}_4 = \begin{bmatrix} 0 \\ 0 \\ 0 \\ 1 \end{bmatrix}$$

For the repeated roots, one can show that the matrix $(\lambda_1\mathbf{I} - \mathbf{B})$ has rank 2, and therefore the system of homogeneous algebraic equations

$$(\lambda_i\mathbf{I} - \mathbf{B})\mathbf{A}_i = 0, \qquad i = 1, 2, 3$$

defines the following two independent eigenvectors:

$$\mathbf{A}_1 = \begin{bmatrix} 1 \\ 5 \\ 2 \\ 0 \end{bmatrix}, \qquad \mathbf{A}_2 = \begin{bmatrix} 0 \\ 2 \\ 1 \\ 0 \end{bmatrix}$$

In this case, the matrix **B** is not similar to a diagonal matrix. One, however, can show that the matrix **B** is similar to the matrix **D** defined as

$$\mathbf{D} = \begin{bmatrix} 2 & 1 & 0 & 0 \\ 0 & 2 & 0 & 0 \\ 0 & 0 & 2 & 0 \\ 0 & 0 & 0 & 1 \end{bmatrix}$$

and

$$\mathbf{B} = \mathbf{V}\mathbf{D}\mathbf{V}^{-1}$$

where

$$\mathbf{V} = \begin{bmatrix} 1 & 0 & 0 & 0 \\ 5 & 1 & 2 & 0 \\ 2 & 0 & 1 & 0 \\ 0 & 0 & 0 & 1 \end{bmatrix}, \qquad \mathbf{V}^{-1} = \begin{bmatrix} 1 & 0 & 0 & 0 \\ -1 & 1 & -2 & 0 \\ -2 & 0 & 1 & 0 \\ 0 & 0 & 0 & 1 \end{bmatrix}$$

The preceding example demonstrates that not every matrix is similar to a diagonal matrix. In particular, matrices that do not have a complete set of independent eigenvectors are not similar to diagonal matrices, but instead are similar to matrices that belong to a class called *Jordan matrices.*

A matrix of order m in the form

$$\mathbf{G}_b = \begin{bmatrix} \gamma & 1 & & & & & \\ & \gamma & 1 & & & \mathbf{0} & \\ & & \gamma & 1 & & & \\ & & & \ddots & \ddots & & \\ & & \mathbf{0} & & \ddots & \ddots & \\ & & & & & \gamma & 1 \\ & & & & & & \gamma \end{bmatrix} \tag{6.17}$$

is called a *Jordan block*. The characteristic polynomial of this matrix is $(\gamma - \lambda)^m$, and as such γ, which has multiplicity m, is the only eigenvalue of $\mathbf{G}_b$. Since the minimal polynomial of $\mathbf{G}_b$ divides the characteristic polynomial, the minimal polynomial must be in the form $(\gamma - \lambda)^k$, where $k \leq m$. It can be shown, however, that $k = m$, and the minimal polynomial of a Jordan matrix is identical to the characteristic polynomial $(\gamma - \lambda)^m$. That is,

$$\begin{aligned} (\gamma\mathbf{I} - \mathbf{G}_b)^k &\neq \mathbf{0} \qquad \text{for } k < m \\ &= \mathbf{0} \qquad \text{for } k = m \end{aligned}$$

For example, if $\mathbf{G}_b$ is the 3×3 matrix, then

$$(\gamma\mathbf{I} - \mathbf{G}_b) = \begin{bmatrix} 0 & -1 & 0 \\ 0 & 0 & -1 \\ 0 & 0 & 0 \end{bmatrix}$$

which shows that

$$(\gamma \mathbf{I} - \mathbf{G}_b)^2 = \begin{bmatrix} 0 & 0 & 1 \\ 0 & 0 & 0 \\ 0 & 0 & 0 \end{bmatrix}$$

$$(\gamma \mathbf{I} - \mathbf{G}_b)^3 = \begin{bmatrix} 0 & 0 & 0 \\ 0 & 0 & 0 \\ 0 & 0 & 0 \end{bmatrix}$$

Clearly, $(\gamma \mathbf{I} - \mathbf{G}_b)$ is a *nilpotent matrix* of degree 3. It follows that the characteristic matrix of a Jordan block of order m is a nilpotent matrix of order m, and, as a consequence, the minimal polynomial of $\mathbf{G}_b$ is $(\gamma - \lambda)^m$.

We also observe that the rank of the matrix $(\lambda_i \mathbf{I} - \mathbf{G}_b)$ is $m - 1$, and, as a consequence, the system of algebraic equations $(\lambda_i \mathbf{I} - \mathbf{G}_b)\,\mathbf{A}_i = \mathbf{0}$ has $m - 1$ independent algebraic equations that define one independent nontrivial solution regardless of the order m of the matrix.

A *Jordan matrix* consists of Jordan blocks that may have different characteristic values. Jordan matrices then assume the following form:

$$\mathbf{G} = \begin{bmatrix} \mathbf{G}_{b1} & & & & \\ & \mathbf{G}_{b2} & & \mathbf{0} & \\ & & \cdot & & \\ & & & \cdot & \\ & \mathbf{0} & & & \cdot \\ & & & & \mathbf{G}_{bs} \end{bmatrix}$$

where each Jordan block $\mathbf{G}_{bi}$ has a number of repeated eigenvalues equal to its dimension. An example of a Jordan matrix is the matrix $\mathbf{D}$ of the preceding example:

$$\mathbf{D} = \begin{bmatrix} \mathbf{G}_{b1} & \mathbf{0} & \mathbf{0} \\ \mathbf{0} & \mathbf{G}_{b2} & \mathbf{0} \\ \mathbf{0} & \mathbf{0} & \mathbf{G}_{b3} \end{bmatrix} = \begin{bmatrix} 2 & 1 & 0 & 0 \\ 0 & 2 & 0 & 0 \\ 0 & 0 & 2 & 0 \\ 0 & 0 & 0 & 1 \end{bmatrix}$$

where

$$\mathbf{G}_{b1} = \begin{bmatrix} 2 & 1 \\ 0 & 2 \end{bmatrix}, \qquad \mathbf{G}_{b2} = 2, \qquad \mathbf{G}_{b3} = 1$$

As demonstrated in the preceding example, the eigenvalues of the matrix $\mathbf{D}$ are

$$\lambda_1 = \lambda_2 = \lambda_3 = 2, \qquad \lambda_4 = 1$$

Associated with these four eigenvalues are only three independent eigenvectors, since the rank of the matrix $(\lambda_1 \mathbf{I} - \mathbf{D})$ is 2, and, as a consequence, the system $(\lambda_1 \mathbf{I} - \mathbf{D})\mathbf{A}_1 = \mathbf{0}$ has only two independent nontrivial solutions associated with the three repeated eigenvalues λ_1, λ_2, and λ_3. Using a similar procedure, it can be shown that each Jordan matrix has a number of linearly independent eigenvectors equal to the number of its Jordan blocks. Since the condition of similarity of two matrices implies the equivalence of their characteristic matrices, similar matrices have the same number of linearly independent eigenvectors. It follows that if a matrix **B** is similar to a Jordan matrix whose Jordan blocks are not all scalars, the matrix **B** does not have a complete set of eigenvectors. An example of such matrices was presented in the preceding example.

It is clear from the discussion presented thus far that each square matrix is similar to a Jordan matrix since diagonal matrices are special cases of Jordan matrices in which the Jordan blocks are scalars. It is also clear that two Jordan matrices are similar if and only if they consist of the same Jordan blocks and differ only in the distribution of these blocks on the main diagonal. For example,

$$\mathbf{G}_1 = \begin{bmatrix} 3 & 1 & 0 & 0 \\ 0 & 3 & 1 & 0 \\ 0 & 0 & 3 & 0 \\ 0 & 0 & 0 & 5 \end{bmatrix}, \qquad \mathbf{G}_2 = \begin{bmatrix} 5 & 0 & 0 & 0 \\ 0 & 3 & 1 & 0 \\ 0 & 0 & 3 & 1 \\ 0 & 0 & 0 & 3 \end{bmatrix}$$

are similar, but these two matrices are not similar to

$$\mathbf{G}_3 = \begin{bmatrix} 3 & 0 & 0 & 0 \\ 0 & 3 & 1 & 0 \\ 0 & 0 & 3 & 0 \\ 0 & 0 & 0 & 5 \end{bmatrix}$$

It can be shown that $\mathbf{G}_1$ and $\mathbf{G}_2$ have only two independent eigenvectors: one associated with the three repeated eigenvalues $\lambda_1 = \lambda_2 = \lambda_3 = 3$ and one associated with the distinct eigenvalue $\lambda_4 = 5$. The matrix $\mathbf{G}_3$, on the other hand, has three linearly independent eigenvectors: two associated with the repeated eigenvalues $\lambda_1 = \lambda_2 = \lambda_3 = 3$ and one associated with the distinct eigenvalue $\lambda_4 = 5$.

As another example, we consider the two matrices

$$\mathbf{G}_1 = \begin{bmatrix} 3 & 1 & 0 & 0 \\ 0 & 3 & 1 & 0 \\ 0 & 0 & 3 & 1 \\ 0 & 0 & 0 & 3 \end{bmatrix}, \qquad \mathbf{G}_2 = \begin{bmatrix} 3 & 1 & 0 & 0 \\ 0 & 3 & 0 & 0 \\ 0 & 0 & 3 & 1 \\ 0 & 0 & 0 & 3 \end{bmatrix}$$

These two matrices have the same repeated eigenvalues, but they are not similar since $\mathbf{G}_1$ consists of one Jordan block while $\mathbf{G}_2$ consists of two Jordan blocks. As a consequence, $\mathbf{G}_1$ has only one independent eigenvector, while $\mathbf{G}_2$ has two independent eigenvectors.

6.5 ELEMENTARY DIVISORS

Previously, we defined the kth invariant factor of a matrix as the greatest common divisor of all minors of order k of the characteristic matrix divided by the greatest common divisor of all minors of order $k-1$. It was also shown that similar matrices have equivalent characteristic matrices and, as a consequence, have identical invariant factors. It can, therefore, be concluded that Jordan matrices which are not similar do not have identical invariant factors. In order to demonstrate this fact, we consider the two matrices

$$\mathbf{G}_1 = \begin{bmatrix} 2 & 0 \\ 0 & 2 \end{bmatrix}, \qquad \mathbf{G}_2 = \begin{bmatrix} 2 & 1 \\ 0 & 2 \end{bmatrix}$$

Their characteristic matrices are

$$(\lambda\mathbf{I} - \mathbf{G}_1) = \begin{bmatrix} \lambda - 2 & 0 \\ 0 & \lambda - 2 \end{bmatrix}, \qquad (\lambda\mathbf{I} - \mathbf{G}_2) = \begin{bmatrix} \lambda - 2 & -1 \\ 0 & \lambda - 2 \end{bmatrix}$$

The greatest common divisors of the minors of order 1 and 2 of $(\lambda\mathbf{I} - \mathbf{G}_1)$ are, respectively, $(\lambda - 2)$ and $(\lambda - 2)^2$. Therefore, the matrix of invariant factors of $\mathbf{G}_1$ is

$$\begin{bmatrix} \lambda - 2 & 0 \\ 0 & \lambda - 2 \end{bmatrix}$$

The greatest common divisors of the minors of order 1 and 2 of $(\lambda\mathbf{I} - \mathbf{G}_2)$ are 1 and $(\lambda - 2)^2$, respectively. It follows that the matrix of invariant factors of $\mathbf{G}_2$ is

$$\begin{bmatrix} 1 & 0 \\ 0 & (\lambda - 2)^2 \end{bmatrix}$$

Using a similar procedure and the induction principle, it can be shown that the canonical diagonal matrix of an n-order Jordan block in the form of Eq. 17 is given by

$$\mathbf{G}_\lambda = \begin{bmatrix} 1 & & & & \\ & 1 & & \mathbf{0} & \\ & & \ddots & & \\ & \mathbf{0} & & 1 & \\ & & & & (\gamma - \lambda)^n \end{bmatrix} \tag{6.18}$$

Let $f_i(\lambda)$ be the ith invariant factor of an arbitrary matrix that, in general, can be written as

$$f_i(\lambda) = e_1^{s_1} e_2^{s_2} \cdots e_k^{s_k}$$

where $e_1(\lambda), e_2(\lambda), \ldots, e_k(\lambda)$ are distinct irreducible monic polynomials. The elements

$$e_1^{s_1}, e_2^{s_2}, \ldots, e_k^{s_k}$$

are called the *elementary divisors* of the invariant factor $f_i(\lambda)$. The elementary divisors of an arbitrary matrix are the elementary divisors of all of the nonconstant invariant factors of this matrix.

Example 6.7

Consider a matrix that has the invariant factors

$$1, 1, \lambda, \lambda^2, \lambda^2(\lambda + 2), \lambda^2(\lambda + 2)^2$$

The elementary divisor of the first nonconstant invariant factor is λ and of the second invariant factor it is λ^2. The elementary divisors of the third nonconstant invariant factor are λ^2 and $(\lambda + 2)$, and of the fourth invariant factor they are λ^2 and $(\lambda + 2)^2$. Hence, the elementary divisors of the matrix are

$$\lambda, \lambda^2, \lambda^2, \lambda^2, (\lambda + 2), (\lambda + 2)^2$$

Since similar matrices have the same invariant factors, it follows that similar matrices have identical elementary divisors. The elementary divisors can be used to determine whether or not a matrix is similar to a diagonal matrix. Using the definitions of the invariant factors and the elementary divisors, it is easy to verify that the elementary divisors of a diagonal matrix are all of the first degree. Hence, a matrix is similar to a diagonal matrix if and only if all of its elementary divisors are of the first degree. Note that a Jordan block has only one elementary divisor, $(\gamma - \lambda)^n$, as demonstrated by Eq. 18, where n is the dimension of the Jordan block. The elementary divisors of a Jordan matrix consist of the elementary divisors of its blocks. Recall that a diagonal matrix can be considered as a special case of a Jordan matrix in which the Jordan blocks are of order 1.

Example 6.8

The matrix

$$\mathbf{B} = \begin{bmatrix} 3 & 1 & -3 \\ -7 & -2 & 9 \\ -2 & -1 & 4 \end{bmatrix}$$

has the characteristic matrix

$$(\lambda\mathbf{I} - \mathbf{B}) = \begin{bmatrix} \lambda - 3 & -1 & 3 \\ 7 & \lambda + 2 & -9 \\ 2 & 1 & \lambda - 4 \end{bmatrix}$$

The invariant factors of the matrix $\mathbf{B}$ are 1, 1, $(\lambda - 1)(\lambda - 2)^2$. Consequently, the elementary divisors of the matrix $\mathbf{B}$ are $(\lambda - 1), (\lambda - 2)^2$. The matrix $\mathbf{B}$ then is similar to the Jordan matrix

$$\mathbf{G} = \begin{bmatrix} 1 & 0 & 0 \\ 0 & 2 & 1 \\ 0 & 0 & 2 \end{bmatrix}$$

6.6 GENERALIZED EIGENVECTORS

The procedure described in the preceding sections provides a systematic way of determining whether or not a matrix is similar to a diagonal matrix. First, we reduce, by a series of elementary operations, the characteristic matrix to a diagonal canonical form, thus defining the invariant factors. These invariant factors also can be determined using the greatest common divisors of the minors of the characteristic matrix. The invariant factors then can be used to define the elementary divisors. If all of the elementary divisors are polynomials of the first degree, the matrix is similar to a diagonal matrix. If some or all of the elementary divisors are not polynomials of the first degree, the matrix is not similar to a diagonal matrix, but it is similar to a Jordan matrix. Recall also that a matrix which has a complete set of linearly independent eigenvectors is similar to a diagonal matrix regardless of the number of repeated eigenvalues. In this case, the matrix of the similarity transformation is the matrix whose columns are the independent eigenvectors. The matrix of the eigenvectors can also be defined as the product of a set of elementary matrices which reduce a given matrix and its characteristic matrix to a diagonal form.

If a matrix $\mathbf{B}$ does not have a complete set of eigenvectors, a question arises as to which Jordan matrix is similar to $\mathbf{B}$ and what is the matrix of the similarity transformation. In order to briefly discuss this question, we consider the matrix

$$\mathbf{B} = \begin{bmatrix} 1 & 1 & -2 & 0 \\ -5 & 7 & -10 & 0 \\ -2 & 2 & -2 & 0 \\ 0 & 0 & 0 & 1 \end{bmatrix}$$

which has the eigenvalues $\lambda_1 = \lambda_2 = \lambda_3 = 2$ and $\lambda_4 = 1$. There are two independent eigenvectors associated with the triple eigenvalue $\lambda = 2$, and as such the matrix $\mathbf{B}$ is not similar to a diagonal matrix, but similar to the Jordan matrix

$$\mathbf{G} = \begin{bmatrix} 2 & 1 & 0 & 0 \\ 0 & 2 & 0 & 0 \\ 0 & 0 & 2 & 0 \\ 0 & 0 & 0 & 1 \end{bmatrix}$$

The independent eigenvectors associated with the eigenvalues of the matrix $\mathbf{B}$ are

$$\mathbf{A}_1 = \begin{bmatrix} 1 \\ 5 \\ 2 \\ 0 \end{bmatrix}, \quad \mathbf{A}_3 = \begin{bmatrix} 0 \\ 2 \\ 1 \\ 0 \end{bmatrix}, \quad \mathbf{A}_4 = \begin{bmatrix} 0 \\ 0 \\ 0 \\ 1 \end{bmatrix}$$

These three eigenvectors represent the first, third, and fourth columns of the matrix of similarity transformation

$$\mathbf{V} = [\mathbf{A}_1 \quad \mathbf{A}_2 \quad \mathbf{A}_3 \quad \mathbf{A}_4]$$

where $\mathbf{A}_2$, the missing eigenvector, is called the *generalized eigenvector*.

In order to illustrate the procedure for determining the generalized eigenvector, we write

$$\mathbf{V}^{-1}\mathbf{B}\mathbf{V} = \mathbf{G}$$

It follows from this equation that

$$\mathbf{B}\mathbf{V} = \mathbf{V}\mathbf{G}$$

or

$$\mathbf{B}[\mathbf{A}_1 \quad \mathbf{A}_2 \quad \mathbf{A}_3 \quad \mathbf{A}_4] = [\mathbf{A}_1 \quad \mathbf{A}_2 \quad \mathbf{A}_3 \quad \mathbf{A}_4]\begin{bmatrix} \lambda_1 & 1 & 0 & 0 \\ 0 & \lambda_2 & 0 & 0 \\ 0 & 0 & \lambda_3 & 0 \\ 0 & 0 & 0 & \lambda_4 \end{bmatrix}$$

This matrix equation yields

$$\mathbf{B}\mathbf{A}_1 = \lambda_1\mathbf{A}_1, \qquad \mathbf{B}\mathbf{A}_2 = \lambda_2\mathbf{A}_2 + \mathbf{A}_1$$
$$\mathbf{B}\mathbf{A}_3 = \lambda_3\mathbf{A}_3, \qquad \mathbf{B}\mathbf{A}_4 = \lambda_4\mathbf{A}_4$$

Therefore, the generalized eigenvector $\mathbf{A}_2$ must satisfy the equation

$$\mathbf{B}\mathbf{A}_2 = \lambda_2\mathbf{A}_2 + \mathbf{A}_1$$

Using the matrix $\mathbf{B}$ and the results previously obtained for $\mathbf{A}_1$, one can show, as demonstrated later in this section, that the generalized eigenvector $\mathbf{A}_2$ is

$$\mathbf{A}_2 = [0 \quad 1 \quad 0 \quad 0]^{\mathrm{T}} \tag{6.19}$$

and the matrix of the similarity transformation is

$$\mathbf{V} = [\mathbf{A}_1 \quad \mathbf{A}_2 \quad \mathbf{A}_3 \quad \mathbf{A}_4] = \begin{bmatrix} 1 & 0 & 0 & 0 \\ 5 & 1 & 2 & 0 \\ 2 & 0 & 1 & 0 \\ 0 & 0 & 0 & 1 \end{bmatrix}$$

Note that $\mathbf{A}_1$ and $\mathbf{A}_2$ are two linearly independent vectors. Using a similar procedure, one can show that if $\lambda_1 = \lambda_2 = \cdots = \lambda_r$ of a given n-dimensional matrix $\mathbf{B}$ have only one independent eigenvector, the r linearly independent generalized eigenvectors must satisfy the following relationships:

$$\mathbf{B}\mathbf{A}_1 = \lambda_1\mathbf{A}_1, \qquad \mathbf{B}\mathbf{A}_2 = \lambda_2\mathbf{A}_2 + \mathbf{A}_1, \ldots, \mathbf{B}\mathbf{A}_r = \lambda_r\mathbf{A}_r + \mathbf{A}_{r-1}$$

It follows that the independent vectors that form the columns of the matrix of the similarity transformation can be obtained by using either the equation

$$\mathbf{B}\mathbf{A}_i = \lambda_i\mathbf{A}_i \tag{6.20}$$

or the equation

$$\mathbf{B}\mathbf{A}_i = \lambda_i \mathbf{A}_i + \mathbf{A}_{i-1} \tag{6.21}$$

Since $(\lambda_i \mathbf{I} - \mathbf{B})$ is a singular matrix, Eq. 21 is a system of nonhomogeneous algebraic equations which can be solved for the unknown generalized eigenvector $\mathbf{A}_i$. In order to demonstrate the procedure for solving this system, we consider the equation for the generalized eigenvector $\mathbf{A}_2$ of the matrix $\mathbf{B}$ considered earlier in this section. The equation

$$(\mathbf{B} - \lambda_2 \mathbf{I})\mathbf{A}_2 = \mathbf{A}_1$$

yields

$$\begin{bmatrix} -1 & 1 & -2 & 0 \\ -5 & 5 & -10 & 0 \\ -2 & 2 & -4 & 0 \\ 0 & 0 & 0 & 1 \end{bmatrix} \begin{bmatrix} A_{21} \\ A_{22} \\ A_{23} \\ A_{24} \end{bmatrix} = \begin{bmatrix} 1 \\ 5 \\ 2 \\ 0 \end{bmatrix}$$

The coefficient matrix in this system has rank 2 since the second and the third rows are multiples of the first row. The application of the Gaussian elimination procedure shows that the above system reduces to

$$\begin{bmatrix} -1 & 1 & -2 & 0 \\ 0 & 0 & 0 & 0 \\ 0 & 0 & 0 & 0 \\ 0 & 0 & 0 & 1 \end{bmatrix} \begin{bmatrix} A_{21} \\ A_{22} \\ A_{23} \\ A_{24} \end{bmatrix} = \begin{bmatrix} 1 \\ 0 \\ 0 \\ 0 \end{bmatrix}$$

This system leads to the following two independent algebraic equations:

$$\begin{aligned} -A_{21} + A_{22} - 2A_{23} \quad &= 1 \\ A_{24} &= 0 \end{aligned}$$

Since the number of these equations is less than the number of unknowns, there is an infinite number of solutions that can be obtained by partitioning the variables as dependent and independent. The number of dependent variables is equal to the number of equations, while the remaining are the free or the independent variables. If we select A_{22} and A_{24} as the dependent variables, the preceding equations can be written as

$$\begin{aligned} A_{22} &= 1 + A_{21} + 2A_{23} \\ A_{24} &= 0 \end{aligned}$$

It follows that

$$\mathbf{A}_2 = \begin{bmatrix} A_{21} \\ A_{22} \\ A_{23} \\ A_{24} \end{bmatrix} = \begin{bmatrix} 0 \\ 1 \\ 0 \\ 0 \end{bmatrix} + \begin{bmatrix} 1 \\ 1 \\ 0 \\ 0 \end{bmatrix} A_{21} + \begin{bmatrix} 0 \\ 2 \\ 1 \\ 0 \end{bmatrix} A_{23} \tag{6.22}$$

The first vector on the right-hand side of this equation is the *particular solution*, while the sum of the last two vectors is the *homogeneous solution*. Since A_{21} and A_{23} can be given arbitrary values, there is an infinite number of solutions for $\mathbf{A}_2$. One choice of $\mathbf{A}_2$, which previously was presented in Eq. 19, is obtained by assuming that $A_{21} = A_{23} = 0$. This vector is linearly independent of $\mathbf{A}_1$, which is obtained from the homogeneous part of Eq. 22 by selecting $A_{21} = 1$ and $A_{23} = 2$. Another choice for A_{21} and A_{23} will lead to a different generalized eigenvector $\mathbf{A}_2$. For instance, if we choose $A_{21} = 1$ and $A_{23} = 0$, we obtain

$$\mathbf{A}_2 = [1 \quad 2 \quad 0 \quad 0]^{\mathrm{T}}$$

In this case, we obtain a different similarity transformation matrix

$$\mathbf{V}_1 = [\mathbf{A}_1 \quad \mathbf{A}_2 \quad \mathbf{A}_3 \quad \mathbf{A}_4] = \begin{bmatrix} 1 & 1 & 0 & 0 \\ 5 & 2 & 2 & 0 \\ 2 & 0 & 1 & 0 \\ 0 & 0 & 0 & 1 \end{bmatrix}$$

The inverse of $\mathbf{V}_1$ is

$$\mathbf{V}_1^{-1} = \begin{bmatrix} 2 & -1 & 2 & 0 \\ -1 & 1 & -2 & 0 \\ -4 & 2 & -3 & 0 \\ 0 & 0 & 0 & 1 \end{bmatrix}$$

One can show that

$$\mathbf{V}_1^{-1}\mathbf{B}\mathbf{V}_1 = \begin{bmatrix} 2 & 1 & 0 & 0 \\ 0 & 2 & 0 & 0 \\ 0 & 0 & 2 & 0 \\ 0 & 0 & 0 & 1 \end{bmatrix}$$

which demonstrates that the matrix of similarity transformation is not unique since the eigenvectors and the generalized eigenvectors are not unique.

6.7 JACOBI METHOD

In the remainder of this chapter, we discuss some numerical techniques for solving the eigenvalue problem. We start with the *Jacobi method*, which can be used to determine the eigenvalues and eigenvectors of symmetric matrices only.

Recall that if $\mathbf{B}$ is a symmetric matrix, then there exists an orthogonal matrix $\mathbf{\Phi}$ such that

$$\mathbf{\Phi}^{\mathrm{T}}\mathbf{B}\mathbf{\Phi} = \mathbf{D} \tag{6.23}$$

where $\mathbf{D}$ is a diagonal matrix whose diagonal elements are the eigenvalues of

B. In the Jacobi method, the matrix $\mathbf{\Phi}$ is determined as the product of a series of orthogonal transformation matrices called *Jacobi rotation matrices.* The Jacobi rotation matrix $\mathbf{R}_i$, at step i, is just a plane rotation matrix which takes the form

$$\mathbf{R}_i = \begin{bmatrix} 1 & 0 & & & & & & & \\ 0 & 1 & & & & & & & \\ & & \ddots & & & & & & \\ & & & \cos\theta_i & \cdots & 0 & \cdots & -\sin\theta_i & \\ & & & \vdots & \ddots & & & \vdots & \\ & & & 0 & & 1 & & 0 & \\ & & & \vdots & & & \ddots & \vdots & \\ & & & \sin\theta_i & \cdots & 0 & \cdots & \cos\theta_i & \\ & & & & & & & & \ddots \\ & & & & & & & & & 1 \end{bmatrix} \tag{6.24}$$

In this matrix, all of the elements other than those appearing in two columns, say k and l, are the same as the elements of the identity matrix. The sine and cosine functions appear in the elements of the kth and lth rows and columns. The corresponding elements in the product

$$\mathbf{S} = \mathbf{R}_i^{\mathrm{T}}\mathbf{B}\mathbf{R}_i$$

are

$$\left.\begin{aligned} s_{kk} &= b_{kk}\cos^2\theta_i - 2b_{kl}\sin\theta_i\cos\theta_i + b_{ll}\sin^2\theta_i \\ s_{kl} = s_{lk} &= (b_{ll} - b_{kk})\sin\theta_i\cos\theta_i + b_{kl}(\cos^2\theta_i - \sin^2\theta_i) \\ s_{ll} &= b_{kk}\sin^2\theta_i - 2b_{kl}\sin\theta_i\cos\theta_i + b_{ll}\cos^2\theta_i \end{aligned}\right\} \tag{6.25}$$

where b_{ij} are the elements of the matrix **B**. If the off-diagonal elements s_{kl} and s_{lk} of the matrix **S** are to be zero, one must select θ_i such that

$$\theta_i = \frac{1}{2}\tan^{-1}\frac{2b_{kl}}{b_{kk} - b_{ll}} \tag{6.26}$$

It follows that, with a proper selection of the angle θ_i, each step of the Jacobi method can be used to annihilate two off-diagonal elements. While the next transformation may introduce nonzero elements in the positions of the elements previously set to zeros. the successive applications of the Jacobi rotation matrices lead to the required diagonal matrix. The product of the Jacobi rotation matrices defines the orthogonal matrix $\mathbf{\Phi}$, whose columns are the eigenvectors of **B**, as

$$\mathbf{\Phi} = \mathbf{R}_1\mathbf{R}_2\ldots\mathbf{R}_n \tag{6.27}$$

The Jacobi method becomes inefficient as the order of the matrix increases, and as such, this method is recommended only for moderate-size matrices with order about 10.

Example 6.9

In order to demonstrate the procedure for applying the Jacobi method, we consider the symmetric matrix **B**:

$$\mathbf{B} = \begin{bmatrix} 4 & 1 & 2 \\ 1 & 0 & 0 \\ 2 & 0 & 0 \end{bmatrix}$$

In order to annihilate b_{12} and b_{21}, we use Eq. 26 to define θ_1 as

$$\theta_1 = \frac{1}{2} \tan^{-1} \frac{2(1)}{4 - 0} = 13.283^\circ$$

The Jacobi rotation matrix $\mathbf{R}_1$ is

$$\mathbf{R}_1 = \begin{bmatrix} \cos\theta_1 & -\sin\theta_1 & 0 \\ \sin\theta_1 & \cos\theta_1 & 0 \\ 0 & 0 & 1 \end{bmatrix} = \begin{bmatrix} 0.9732 & -0.2298 & 0 \\ 0.2298 & 0.9732 & 0 \\ 0 & 0 & 1 \end{bmatrix}$$

It follows that

$$\mathbf{R}_1^{\mathrm{T}}\mathbf{B}\mathbf{R}_1 = \begin{bmatrix} 4.2358 & 0.0 & 1.9464 \\ 0.0 & -0.2361 & -0.4596 \\ 1.9464 & -0.4596 & 0.0 \end{bmatrix}$$

In order to set to zero the elements in positions (1,3) and (3,1), we define θ_2 as

$$\theta_2 = \frac{1}{2} \tan^{-1} \frac{2(1.9464)}{4.2358 - 0.0} = 21.2919^\circ$$

which defines $\mathbf{R}_2$ as

$$\mathbf{R}_2 = \begin{bmatrix} \cos\theta_2 & 0 & -\sin\theta_2 \\ 0 & 1 & 0 \\ \sin\theta_2 & 0 & \cos\theta_2 \end{bmatrix} = \begin{bmatrix} 0.9317 & 0 & -0.3631 \\ 0 & 1 & 0 \\ 0.3631 & 0 & 0.9317 \end{bmatrix}$$

The second step in the Jacobi method leads to

$$\mathbf{R}_2^{\mathrm{T}}\mathbf{R}_1^{\mathrm{T}}\mathbf{B}\mathbf{R}_1\mathbf{R}_2 = \begin{bmatrix} 4.9939 & -0.1669 & 0.0 \\ -0.1669 & -0.2361 & -0.4282 \\ 0.0 & -0.4282 & -0.75843 \end{bmatrix}$$

Note that the previously set zero elements in positions (1,2) and (2,1) are no longer zeros after the second step. Nonetheless, by continuing this process a diagonal matrix can be obtained, and the diagonal elements of this matrix are the eigenvalues of **B**. The exact eigenvalues of the matrix **B** are -1, 5, and 0.

Generalized Eigenvalue Problem The Jacobi method discussed in this section can be used in the case of symmetric matrices. In the case of the generalized eigenvalue problem, one has

$$\mathbf{KA} = \lambda\mathbf{MA} \tag{6.28}$$

where $\mathbf{M}$ and $\mathbf{K}$ are, respectively, the symmetric mass and stiffness matrices. Premultiplying the preceding equation by $\mathbf{M}^{-1}$, one obtains

$$(\mathbf{M}^{-1}\mathbf{K})\mathbf{A} = \lambda\mathbf{A}$$

where $\mathbf{M}^{-1}\mathbf{K}$ is not necessarily a symmetric matrix. A symmetric eigenvalue problem, however, can be recovered by using *Cholesky decomposition* $\mathbf{M} = \mathbf{L}\mathbf{L}^{\mathrm{T}}$, where $\mathbf{L}$ is a lower triangular matrix. Premultiplying Eq. 28 by $\mathbf{L}^{-1}$, one obtains

$$\mathbf{B}(\mathbf{L}^{\mathrm{T}}\mathbf{A}) = \lambda\mathbf{L}^{\mathrm{T}}\mathbf{A}$$

where

$$\mathbf{B} = \mathbf{L}^{-1}\mathbf{K}(\mathbf{L}^{-1})^{\mathrm{T}}$$

The matrix $\mathbf{B}$ is symmetric and has the same eigenvalues as the matrix $\mathbf{M}^{-1}\mathbf{K}$, and its eigenvectors are $\mathbf{L}^{\mathrm{T}}\mathbf{A}$.

6.8 HOUSEHOLDER TRANSFORMATION

Another method for determining the eigenvalues and eigenvectors of a symmetric matrix is to use a sequence of *Householder transformations* that reduce the symmetric matrix to a simpler tri-diagonal form. The eigenvalues and eigenvectors of the tri-diagonal matrix can be calculated more efficiently. The resulting eigenvalues are the same as those of the original symmetric matrix, and the product of the resulting Householder transformations defines an orthogonal matrix which relates the eigenvectors of the tri-diagonal matrix and the original symmetric matrix.

A Householder transformation or an *elementary reflector* associated with a unit vector $\mathbf{v}_1$ is defined as

$$\mathbf{H} = \mathbf{I} - 2\mathbf{v}_1\mathbf{v}_1^{\mathrm{T}} \tag{6.29}$$

where $\mathbf{I}$ is an identity matrix. The matrix $\mathbf{H}$ is symmetric and also orthogonal since

$$\mathbf{H}^{\mathrm{T}}\mathbf{H} = (\mathbf{I} - 2\mathbf{v}_1\mathbf{v}_1^{\mathrm{T}})(\mathbf{I} - 2\mathbf{v}_1\mathbf{v}_1^{\mathrm{T}}) = \mathbf{I}$$

It follows that $\mathbf{H} = \mathbf{H}^{\mathrm{T}} = \mathbf{H}^{-1}$. It is also clear that if $\mathbf{v} = |\mathbf{v}|\mathbf{v}_1$, then

$$\mathbf{H}\mathbf{v} = -\mathbf{v}$$

Furthermore, if $\mathbf{u}$ is the column vector

$$\mathbf{u} = [1 \quad 0 \quad 0 \quad \cdots \quad 0]^{\mathrm{T}} \tag{6.30}$$

and

$$\mathbf{v} = \mathbf{b} + \beta\mathbf{u} \tag{6.31}$$

where

$$\beta = |\mathbf{b}| = \sqrt{\mathbf{b}^{\mathrm{T}}\mathbf{b}} \tag{6.32}$$

then

$$\mathbf{Hb} = \left\{\mathbf{I} - \frac{2\mathbf{v}\mathbf{v}^{\mathrm{T}}}{(|\mathbf{v}|)^2}\right\}\mathbf{b}$$

Using Eq. 31 and the definition of β, one can show that

$$\mathbf{Hb} = -\beta\mathbf{u} = \begin{bmatrix} -\beta \\ 0 \\ 0 \\ \vdots \\ 0 \end{bmatrix} \tag{6.33}$$

This equation implies that when the Householder transformation constructed using the vector $\mathbf{v}$ of Eq. 31 is multiplied by the vector $\mathbf{b}$, the result is a vector whose only nonzero element is the first element. Using this fact, a matrix can be transformed to an upper-triangular form by successively applying a series of Householder transformations. In order to demonstrate this procedure, we consider the rectangular $n \times m$ matrix:

$$\mathbf{B} = \begin{bmatrix} b_{11} & b_{12} & b_{13} & \cdots & b_{1m} \\ b_{21} & b_{22} & b_{23} & \cdots & b_{2m} \\ \vdots & \vdots & \vdots & \ddots & \vdots \\ b_{n1} & b_{n2} & b_{n3} & \cdots & b_{nm} \end{bmatrix}, \tag{6.34}$$

First, we construct the Householder transformation associated with the first column $\mathbf{b}_1 = [b_{11} \;\; b_{21} \;\; \ldots \;\; b_{n1}]^{\mathrm{T}}$. This transformation matrix can be written as

$$\mathbf{H}_1 = \mathbf{I} - 2\frac{\bar{\mathbf{b}}_1\bar{\mathbf{b}}_1^{\mathrm{T}}}{\bar{\mathbf{b}}_1^{\mathrm{T}}\bar{\mathbf{b}}_1}$$

where

$$\bar{\mathbf{b}}_1 = \mathbf{b}_1 + \beta_1\mathbf{u}_1$$

in which β_1 is the norm of $\mathbf{b}_1$, and the vector $\mathbf{u}_1$ has the same dimension as $\mathbf{b}_1$ and is defined by Eq. 30. Using Eq. 33, one has

$$\mathbf{H}_1\mathbf{b}_1 = -\beta_1\mathbf{u}_1 = \begin{bmatrix} -\beta_1 \\ 0 \\ 0 \\ \vdots \\ 0 \end{bmatrix}$$

It follows that

$$\mathbf{B}_1 = \mathbf{H}_1\mathbf{B} = \begin{bmatrix} -\beta_1 & (b_{12})_1 & (b_{13})_1 & \cdots & (b_{1m})_1 \\ 0 & (b_{22})_1 & (b_{23})_1 & \cdots & (b_{2m})_1 \\ 0 & (b_{32})_1 & (b_{33})_1 & \cdots & (b_{3m})_1 \\ \vdots & \vdots & \vdots & \ddots & \vdots \\ 0 & (b_{n2})_1 & (b_{n3})_1 & \cdots & (b_{nm})_1 \end{bmatrix}$$

Now we consider the last $n - 1$ elements of the second column of the matrix $\mathbf{B}_1$ which form the vector

$$\mathbf{b}_2 = [(b_{22})_1 \quad (b_{32})_1 \quad \cdots \quad (b_{n2})_1]^{\mathrm{T}}$$

A Householder transformation matrix $\mathbf{H}_{2i}$ can be constructed such that

$$\mathbf{H}_{2i}\mathbf{b}_2 = -\beta_2\mathbf{u}_2 = \begin{bmatrix} -\beta_2 \\ 0 \\ 0 \\ \vdots \\ 0 \end{bmatrix}$$

where β_2 is the norm of the vector $\mathbf{b}_2$, and $\mathbf{u}_2$ is the $(n - 1)$-dimensional vector defined by Eq. 30. Note that, at this point, the Householder transformation is only of order $(n - 1)$, and this transformation can be imbedded into the lower right corner of an $n \times n$ matrix $\mathbf{H}_2$, where

$$\mathbf{H}_2 = \begin{bmatrix} 1 & \mathbf{0} \\ \mathbf{0} & \mathbf{H}_{2i} \end{bmatrix}$$

The matrix $\mathbf{H}_2$ is orthogonal and symmetric, and when it is multiplied by an arbitrary matrix, it does not change the first row and the first column of that matrix. By premultiplying $\mathbf{B}_1$ by $\mathbf{H}_2$, one gets

$$\mathbf{B}_2 = \mathbf{H}_2\mathbf{B}_1 = \mathbf{H}_2\mathbf{H}_1\mathbf{B} = \begin{bmatrix} -\beta_1 & (b_{12})_1 & (b_{13})_1 & \cdots & (b_{1m})_1 \\ 0 & -\beta_2 & (b_{23})_2 & \cdots & (b_{2m})_2 \\ 0 & 0 & (b_{33})_2 & \cdots & (b_{3m})_2 \\ \vdots & \vdots & \vdots & \ddots & \vdots \\ 0 & 0 & (b_{n3})_2 & \cdots & (b_{nm})_2 \end{bmatrix}$$

Next, we consider the vector that consists of the last $n - 2$ elements of the third column of the matrix $\mathbf{B}_2$. This vector is

$$\mathbf{b}_3 = [(b_{33})_2 \quad (b_{43})_2 \quad \cdots \quad (b_{n3})_2]^{\mathrm{T}}$$

A Householder transformation matrix $\mathbf{H}_{3i}$ can be constructed and imbedded into the lower right corner of the $n \times n$ matrix $\mathbf{H}_3$, where

$$\mathbf{H}_3 = \begin{bmatrix} 1 & 0 & \mathbf{0} \\ 0 & 1 & \mathbf{0} \\ \mathbf{0} & \mathbf{0} & \mathbf{H}_{3i} \end{bmatrix}$$

Using this matrix, one has

$$\mathbf{H}_3\mathbf{B}_2 = \mathbf{H}_3\mathbf{H}_2\mathbf{H}_1\mathbf{B} = \begin{bmatrix} -\beta_1 & (b_{12})_1 & (b_{13})_1 & \cdots & (b_{1m})_1 \\ 0 & -\beta_2 & (b_{23})_2 & \cdots & (b_{2m})_2 \\ 0 & 0 & -\beta_3 & \cdots & (b_{3m})_3 \\ \vdots & \vdots & \vdots & \ddots & \vdots \\ 0 & 0 & 0 & \cdots & (b_{nm})_3 \end{bmatrix}$$

where β_3 is the norm of the vector $\mathbf{b}_3$. It is clear that, by continuing this process, all of the elements below the diagonal of the matrix $\mathbf{B}$ can be set equal to zero. If $\mathbf{B}$ is a square nonsingular matrix, the result of the Householder transformations is an upper-triangular matrix. If $\mathbf{B}$ is a rectangular matrix, the result of m Householder transformations is

$$\mathbf{B}_m = \mathbf{H}_m\mathbf{H}_{m-1}\cdots\mathbf{H}_1\mathbf{B} = \begin{bmatrix} \mathbf{R}_1 \\ \mathbf{0} \end{bmatrix} \tag{6.35}$$

where $\mathbf{R}_1$ is an $m \times m$ upper triangular matrix. If the matrix $\mathbf{B}$ is symmetric, a sequence of Householder transformations can be used to obtain a tri-diagonal matrix that is similar to $\mathbf{B}$, as demonstrated by the following example. The product of the resulting sequence of the Householder transformations defines the matrix that defines the relationship between the eigenvectors of $\mathbf{B}$ and the eigenvectors of the tri-diagonal matrix.

Example 6.10

Use the Householder transformation to determine the eigenvalues and eigenvectors of the symmetric matrix

$$\mathbf{B} = \begin{bmatrix} 4 & 1 & 2 \\ 1 & 0 & 0 \\ 2 & 0 & 0 \end{bmatrix}$$

Solution First we construct the Householder transformation associated with the last $n - 1$ elements of the first column. This defines the vector

$$\mathbf{b}_1 = [1 \quad 2]^{\mathrm{T}}$$

Using this vector, we define

$$\mathbf{H}_{1i} = \mathbf{I} - 2\mathbf{v}_1\mathbf{v}_1^{\mathrm{T}}$$

where $\mathbf{v}_1$ is a unit vector along the vector

$$\begin{aligned} \mathbf{v} &= \mathbf{b}_1 + \beta\mathbf{u}_1 \\ &= \begin{bmatrix} 1 \\ 2 \end{bmatrix} + \sqrt{5}\begin{bmatrix} 1 \\ 0 \end{bmatrix} = \begin{bmatrix} 3.236 \\ 2 \end{bmatrix} \end{aligned}$$

It follows that

$$\mathbf{v}_1 = [0.851 \quad 0.526]^{\mathrm{T}}$$

and

$$\mathbf{H}_{1i} = \mathbf{I} - 2\mathbf{v}_1\mathbf{v}_1^{\mathrm{T}} = \begin{bmatrix} -0.448 & -0.895 \\ -0.895 & 0.447 \end{bmatrix}$$

This matrix can be imbedded in a 3 × 3 Householder transformation matrix:

$$\mathbf{H}_1 = \begin{bmatrix} 1 & \mathbf{0} \\ \mathbf{0} & \mathbf{H}_{1i} \end{bmatrix} = \begin{bmatrix} 1 & 0 & 0 \\ 0 & -0.448 & -0.895 \\ 0 & -0.895 & 0.447 \end{bmatrix}$$

The product $\mathbf{H}_1\mathbf{B}$ is

$$\mathbf{H}_1\mathbf{B} = \begin{bmatrix} 4 & 1 & 2 \\ -2.238 & 0 & 0 \\ 0 & 0 & 0 \end{bmatrix}$$

Also,

$$\mathbf{H}_1\mathbf{B}\mathbf{H}_1 = \begin{bmatrix} 4 & -2.238 & 0 \\ -2.238 & 0 & 0 \\ 0 & 0 & 0 \end{bmatrix}$$

This is a tri-diagonal matrix that has the eigenvalues $\lambda_1 = -1.0$, $\lambda_2 = 0.0$, and $\lambda_3 = 5$. The corresponding eigenvectors are

$$\mathbf{A}_{t1} = \begin{bmatrix} 1.0 \\ 2.238 \\ 0.0 \end{bmatrix}, \quad \mathbf{A}_{t2} = \begin{bmatrix} 0.0 \\ 0.0 \\ 1.0 \end{bmatrix}, \quad \mathbf{A}_{t3} = \begin{bmatrix} -2.238 \\ 1.0 \\ 0.0 \end{bmatrix}$$

Since $\mathbf{H}_1$ is an orthogonal matrix, $\mathbf{B}$ is similar to $\mathbf{H}_1\mathbf{B}\mathbf{H}_1$, and as such it has the same eigenvalues. Furthermore, if

$$\mathbf{B}\mathbf{A} = \lambda\mathbf{A}$$

one has

$$(\mathbf{H}_1\mathbf{B}\mathbf{H}_1)(\mathbf{H}_1\mathbf{A}) = \lambda(\mathbf{H}_1\mathbf{A})$$

which demonstrates that if $\mathbf{A}_t$ is an eigenvector of $\mathbf{H}_1\mathbf{B}\mathbf{H}_1$, then the eigenvector of $\mathbf{B}$ is

$$\mathbf{A} = \mathbf{H}_1^{-1}\mathbf{A}_t = \mathbf{H}_1\mathbf{A}_t$$

It follows that

$$\mathbf{A}_1 = \begin{bmatrix} 1.0 \\ -1.0 \\ -2.0 \end{bmatrix}, \quad \mathbf{A}_2 = \begin{bmatrix} 0.0 \\ -0.895 \\ 0.447 \end{bmatrix}, \quad \mathbf{A}_3 = \begin{bmatrix} -2.238 \\ -0.448 \\ -0.895 \end{bmatrix}$$

6.9 QR DECOMPOSITION

As was demonstrated in the preceding section, a sequence of Householder transformations can be used to reduce an arbitrary matrix to an upper-triangular matrix. For example, the first Householder transformation constructed using the first column of an arbitrary matrix **B** can be used to set the last $n-1$ elements of this column equal to zero. The second Householder transformation sets the last $n-2$ elements of the second column equal to zero, the third transformation sets the last $n-3$ elements of the third column equal to zero, and the procedure continues in this manner to set all of the elements below the diagonal equal to zero. It follows that $n-1$ Householder transformations can be used to systematically reduce an arbitrary $n \times n$ matrix to an upper-triangular matrix.

Let **B** be an $n \times n$ matrix, and $\mathbf{H}_1, \mathbf{H}_2, \ldots, \mathbf{H}_{n-1}$ be a sequence of Householder transformations designed to reduce **B** to an upper-triangular matrix **R**. One then has

$$\mathbf{H}_{n-1}\mathbf{H}_{n-2}\cdots\mathbf{H}_1\mathbf{B} = \mathbf{R} \tag{6.36}$$

Since the Householder transformations are symmetric and orthogonal, one has

$$\mathbf{H}_i^{\mathrm{T}} = \mathbf{H}_i^{-1} = \mathbf{H}_i$$

Using this identity, Eq. 36 can be written as

$$\mathbf{B} = \mathbf{H}_1\mathbf{H}_2\cdots\mathbf{H}_{n-2}\mathbf{H}_{n-1}\mathbf{R}$$

Since the product of orthogonal matrices defines an orthogonal matrix, the preceding equation can be written as

$$\mathbf{B} = \mathbf{QR} \tag{6.37}$$

where **Q** is an orthogonal matrix defined as

$$\mathbf{Q} = \mathbf{H}_1\mathbf{H}_2\cdots\mathbf{H}_{n-2}\mathbf{H}_{n-1} \tag{6.38}$$

Equation 37 states that an arbitrary square matrix **B** can be written as the product of an orthogonal matrix **Q** and an upper triangular matrix **R**.

Example 6.11

Find the **QR** decomposition of the matrix

$$\mathbf{B} = \begin{bmatrix} 0 & -3 & 5 \\ -2 & 2 & -3 \\ 6 & -2 & 0 \end{bmatrix}$$

Solution We consider the first column of the matrix **B**, which we write as $\mathbf{b}_1 = [0 \quad -2 \quad 6]^{\mathrm{T}}$. The norm of this vector

$$\beta_1 = |\mathbf{b}_1| = \sqrt{40} = 6.3246$$

Define

$$\mathbf{v} = \mathbf{b}_1 + \beta_1\mathbf{u}_1 = [6.3246 \quad -2.0 \quad 6.0]^{\mathrm{T}}$$

A unit vector along $\mathbf{v}$ is

$$\mathbf{v}_1 = [0.7071 \quad -0.2236 \quad 0.6708]^{\mathrm{T}}$$

The Householder transformation constructed using this vector is

$$\mathbf{H}_1 = \mathbf{I} - 2\mathbf{v}_1\mathbf{v}_1^{\mathrm{T}} = \begin{bmatrix} 0.0 & 0.3162 & -0.9486 \\ 0.3162 & 0.9 & 0.3 \\ -0.9486 & 0.3 & 0.1 \end{bmatrix}$$

It follows that

$$\mathbf{B}_1 = \mathbf{H}_1\mathbf{B} = \begin{bmatrix} -6.3246 & 2.5296 & -0.9486 \\ 0.0 & 0.2514 & -1.1190 \\ 0.0 & 0.0 & -5.6430 \end{bmatrix}$$

The last two elements of the second column of this matrix form the vector $\mathbf{b}_2 = [0.2514 \quad 0.0]^{\mathrm{T}}$. The norm of this vector is

$$\beta_2 = 0.2514$$

Define a vector

$$\mathbf{w} = \mathbf{b}_2 + \beta_2\mathbf{u}_2 = [0.5028 \quad 0.0]^{\mathrm{T}}$$

A unit vector along this vector is

$$\mathbf{w}_1 = [1 \quad 0]^{\mathrm{T}}$$

The Householder transformation constructed using this vector is

$$\mathbf{H}_{2i} = \mathbf{I} - 2\mathbf{w}_1\mathbf{w}_1^{\mathrm{T}} = \begin{bmatrix} -1 & 0 \\ 0 & 1 \end{bmatrix}$$

This matrix can be imbedded in a three-dimensional matrix $\mathbf{H}_2$ as

$$\mathbf{H}_2 = \begin{bmatrix} 1 & \mathbf{0} \\ \mathbf{0} & \mathbf{H}_{2i} \end{bmatrix} = \begin{bmatrix} 1 & 0 & 0 \\ 0 & -1 & 0 \\ 0 & 0 & 1 \end{bmatrix}$$

Using this matrix, one has

$$\mathbf{B}_2 = \mathbf{H}_2\mathbf{B}_1 = \mathbf{H}_2\mathbf{H}_1\mathbf{B} = \begin{bmatrix} -6.3246 & 2.5296 & -0.9486 \\ 0 & -0.2514 & 1.1190 \\ 0 & 0 & -5.6430 \end{bmatrix}$$

which is an upper-triangular matrix. Therefore, the matrix $\mathbf{B}$ can be written as

$$\mathbf{B} = \mathbf{QR}$$

where

$$\mathbf{Q} = \mathbf{H}_1\mathbf{H}_2 = \begin{bmatrix} 0 & -0.3162 & -0.9486 \\ 0.3162 & -0.9 & 0.3 \\ -0.9486 & -0.3 & 0.1 \end{bmatrix}$$

$$\mathbf{R} = \begin{bmatrix} -6.3246 & 2.5296 & -0.9486 \\ 0 & -0.2514 & 1.1190 \\ 0 & 0 & -5.6430 \end{bmatrix}$$

Using the **QR** decomposition, an iterative procedure can be developed for determining the eigenvalues and eigenvectors of an arbitrary square matrix **B**. In this procedure, **B** is first decomposed into its **QR** factors as

$$\mathbf{B} = \mathbf{Q}_1\mathbf{R}_1 \tag{6.39}$$

Using the $\mathbf{Q}_1$ and $\mathbf{R}_1$ matrices in this decomposition, a new matrix $\mathbf{B}_1$ is defined as

$$\mathbf{B}_1 = \mathbf{R}_1\mathbf{Q}_1 \tag{6.40}$$

It is clear from Eq. 39 that $\mathbf{R}_1 = \mathbf{Q}_1^{\mathrm{T}}\mathbf{B}$, and as a consequence

$$\mathbf{B}_1 = \mathbf{Q}_1^{\mathrm{T}}\mathbf{B}\mathbf{Q}_1 \tag{6.41}$$

Since $\mathbf{Q}_1$ is an orthogonal matrix, Eq. 41 demonstrates that $\mathbf{B}_1$ is similar to **B**, and therefore these two matrices have the same eigenvalues. In the iterative procedure for determining the eigenvalues and eigenvectors of **B**, the **QR** factors of $\mathbf{B}_1$ are determined as

$$\mathbf{B}_1 = \mathbf{Q}_2\mathbf{R}_2$$

A matrix similar to $\mathbf{B}_1$ can be defined as

$$\mathbf{B}_2 = \mathbf{R}_2\mathbf{Q}_2$$

This process continues until a matrix $\mathbf{B}_i$ that is similar to **B** is obtained in a diagonal or a Jordan form. The eigenvalues of $\mathbf{B}_i$ are the same as the eigenvalues of **B**, and the eigenvectors of **B** can be obtained from the eigenvectors of $\mathbf{B}_i$ using the product of the orthogonal matrices **Q** in the **QR** factors.

Before using the **QR** decomposition, it is more efficient computationally to reduce **B** first to a tri-diagonal form using a sequence of Householder transformations. By reducing **B** to a simple form, fewer **QR** factorizations are required and the procedure becomes much less expensive. Another method that can be used to increase the speed of convergence of the **QR** algorithm is to use a *shift* of the origin of the eigenvalues. We note that if λ is an eigenvalue of **B**, then $\lambda + c$ is an eigenvalue of $\mathbf{B} + c\mathbf{I}$, and both **B** and $\mathbf{B} + c\mathbf{I}$ have the same eigenvectors. This fact can be demonstrated by adding $c\mathbf{A}$ to both sides of the equation

$$\mathbf{B}\mathbf{A} = \lambda\mathbf{A}$$

to yield

$$(\mathbf{B} + c\mathbf{I})\mathbf{A} = (\lambda + c)\mathbf{A}$$

This equation shows that $(\lambda + c)$ is an eigenvalue of $(\mathbf{B} + c\mathbf{I})$ and **A** is the associated eigenvector. More discussion on the selection of the constant c and the convergence of the **QR** method with shift can be found in the text by Atkinson (1978).

Problems

6.1. Find the eigenvalues and eigenvectors of the matrix

$$\mathbf{B} = \begin{bmatrix} 4 & -5 \\ 2 & -3 \end{bmatrix}$$

Evaluate also the products

$$(\mathbf{\Phi}^{-1}\mathbf{B}\mathbf{\Phi})^{10}, \qquad \mathbf{\Phi}^{-1}\mathbf{B}^{10}\mathbf{\Phi}$$

where $\mathbf{\Phi}$ is the matrix of the eigenvectors of $\mathbf{B}$.

6.2. Find the eigenvalues and eigenvectors of the matrix

$$\mathbf{B} = \begin{bmatrix} 2.0 & 0.0 & 0.0 \\ 0.0 & 1.2 & 0.4 \\ 0.0 & 0.4 & 1.8 \end{bmatrix}$$

Evaluate also the products

$$(\mathbf{\Phi}^{-1}\mathbf{B}\mathbf{\Phi})^{10}, \qquad \mathbf{\Phi}^{-1}\mathbf{B}^{10}\mathbf{\Phi}$$

where $\mathbf{\Phi}$ is the matrix of the eigenvectors of $\mathbf{B}$.

6.3. Find the eigenvalues and eigenvectors of the matrix

$$\mathbf{B} = \begin{bmatrix} -1 & 1 & 2 \\ 0 & 0 & 2 \\ 0 & 0 & 2 \end{bmatrix}$$

and also find the matrix products

$$(\mathbf{\Phi}^{-1}\mathbf{B}\mathbf{\Phi})^{10}, \qquad \mathbf{\Phi}^{-1}\mathbf{B}^{10}\mathbf{\Phi}$$

where $\mathbf{\Phi}$ is the matrix of the eigenvectors of $\mathbf{B}$.

6.4. Using the Cayley–Hamilton theorem, show that the matrix

$$\mathbf{B} = \begin{bmatrix} 4 & -5 \\ 2 & -3 \end{bmatrix}$$

satisfies the identity

$$\mathbf{B}^2 - \mathbf{B} - 2\mathbf{I} = \mathbf{0}$$

Use this identity to find the inverse of the matrix $\mathbf{B}$.

6.5. Using the Cayley–Hamilton theorem, show that the matrix

$$\mathbf{B} = \begin{bmatrix} 2.0 & 0.0 & 0.0 \\ 0.0 & 1.2 & 0.4 \\ 0.0 & 0.4 & 1.8 \end{bmatrix}$$

satisfies the identity

$$(\mathbf{B} - \mathbf{I})(\mathbf{B} - 2\mathbf{I})^2 = \mathbf{0}$$

Show also that the inverse of **B** can be expressed in terms of **B** as

$$\mathbf{B}^{-1} = \tfrac{1}{4}(\mathbf{B}^2 - 5\mathbf{B} + 8\mathbf{I})$$

6.6. Using the characteristic polynomials, determine which of the following matrices is not similar to the others:

$$\mathbf{B}_1 = \begin{bmatrix} -1 & 1 & 2 \\ 0 & 0 & 2 \\ 0 & 0 & 2 \end{bmatrix}, \quad \mathbf{B}_2 = \begin{bmatrix} 52 & 8 & 16 \\ 8 & 8 & 6 \\ 16 & 6 & 17 \end{bmatrix}, \quad \mathbf{B}_3 = \begin{bmatrix} 1 & 0 & 0 \\ 0 & 2 & 0 \\ 0 & 0 & 2 \end{bmatrix}$$

6.7. Find the minimal polynomials of the matrices

$$\mathbf{B}_1 = \begin{bmatrix} 2.0 & 0.0 & 0.0 \\ 0.0 & 1.2 & 0.4 \\ 0.0 & 0.4 & 1.8 \end{bmatrix}, \quad \mathbf{B}_2 = \begin{bmatrix} -1 & 1 & 2 \\ 0 & 0 & 2 \\ 0 & 0 & 2 \end{bmatrix}$$

6.8. What is the minimal polynomial of the following matrix?

$$\mathbf{B} = \begin{bmatrix} 1 & 0 & 0 & 0 \\ 0 & 2 & 0 & 0 \\ 0 & 0 & 2 & 0 \\ 0 & 0 & 0 & 2 \end{bmatrix}$$

6.9. Using elementary matrix operations, find the transformation that reduces the characteristic matrix of the matrix

$$\mathbf{B} = \begin{bmatrix} 1 & 2 & 2 \\ 0 & 3 & 2 \\ 0 & 0 & 5 \end{bmatrix}$$

to a diagonal form. Find also the invariant factors of the matrix **B**.

6.10. Show that the matrix

$$\mathbf{B} = \begin{bmatrix} 1 & 1 & 0 \\ 0 & 1 & 1 \\ 0 & 0 & 2 \end{bmatrix}$$

is not similar to a diagonal matrix. Determine the invariant factors of this matrix.

6.11. Find the elementary divisors of the two matrices

$$\mathbf{B}_1 = \begin{bmatrix} 1 & 1 & 0 \\ 0 & 1 & 0 \\ 0 & 0 & 2 \end{bmatrix}, \quad \mathbf{B}_2 = \begin{bmatrix} 1 & 0 & 0 \\ 0 & 1 & 0 \\ 0 & 0 & 0 \end{bmatrix}$$

6.12. Using the elementary divisors of the matrix

$$\mathbf{B} = \begin{bmatrix} 1 & 1 & 0 \\ 0 & 1 & 1 \\ 0 & 0 & 2 \end{bmatrix}$$

determine the Jordan matrix that is similar to **B**.

6.13. Use the elementary divisors to find the Jordan matrix that is similar to the matrix

$$\mathbf{B} = \begin{bmatrix} 2 & 1 & 0 \\ -1 & 4 & 0 \\ -1 & 3 & 1 \end{bmatrix}$$

6.14. Determine the independent or generalized eigenvectors of the matrix

$$\mathbf{B} = \begin{bmatrix} 2 & 1 & 0 \\ -1 & 4 & 0 \\ -1 & 3 & 1 \end{bmatrix}$$

6.15. Using the Jacobi method, find the eigenvalues and eigenvectors of the matrices

$$\mathbf{B}_1 = \begin{bmatrix} 3 & 4 & 6 \\ -1 & -1 & -1 \\ 2 & 2 & 2 \end{bmatrix}, \qquad \mathbf{B}_2 = \begin{bmatrix} 4 & 1 & 2 \\ 1 & 0 & 0 \\ 2 & 0 & 0 \end{bmatrix}$$

6.16. Use the **QR** decomposition to find the eigenvalues and eigenvectors of the matrices

$$\mathbf{B}_1 = \begin{bmatrix} 3 & 4 & 6 \\ -1 & -1 & -1 \\ 2 & 2 & 2 \end{bmatrix}, \qquad \mathbf{B}_2 = \begin{bmatrix} 4 & 1 & 2 \\ 1 & 0 & 0 \\ 2 & 0 & 0 \end{bmatrix}$$

Appendix A
Linear Algebra

In this appendix, we summarize some important results from *vector* and *matrix algebra* that are useful in our development in this book. Most of the vector and matrix properties presented in the following sections are elementary and can be found in standard texts on linear algebra.

A.1 MATRICES

An $m \times n$ matrix $\mathbf{A}$ is an ordered rectangular array that has $m \times n$ elements. The matrix $\mathbf{A}$ can be written in the form

$$\mathbf{A} = (a_{ij}) = \begin{bmatrix} a_{11} & a_{12} & \cdots & a_{1n} \\ a_{21} & a_{22} & \cdots & a_{2n} \\ \vdots & \vdots & \ddots & \vdots \\ a_{m1} & a_{m2} & \cdots & a_{mn} \end{bmatrix} \tag{A.1}$$

The matrix $\mathbf{A}$ is called an $m \times n$ matrix since it has m rows and n columns. The scalar element a_{ij} lies in the ith row and jth column of the matrix $\mathbf{A}$. Therefore, the index i, which takes the values 1, 2, ..., m, denotes the row number, while the index j, which takes the values $1, 2, \ldots, n$, denotes the column number.

A matrix $\mathbf{A}$ is said to be *square* if $m = n$. An example of a square matrix is

$$\mathbf{A} = \begin{bmatrix} 3.0 & -2.0 & 0.95 \\ 6.3 & 0.0 & 10.0 \\ 9.0 & 3.5 & 1.25 \end{bmatrix}$$

In this example $m = n = 3$, consequently, $\mathbf{A}$ is a 3×3 matrix.

The *transpose* of an $m \times n$ matrix $\mathbf{A}$ is an $n \times m$ matrix denoted as $\mathbf{A}^{\mathrm{T}}$ and defined as

$$\mathbf{A}^{\mathrm{T}} = (a_{ji}) = \begin{bmatrix} a_{11} & a_{21} & \cdots & a_{m1} \\ a_{12} & a_{22} & \cdots & a_{m2} \\ \vdots & \vdots & \ddots & \vdots \\ a_{1n} & a_{2n} & \cdots & a_{mn} \end{bmatrix} \tag{A.2}$$

For example, let **A** be the matrix

$$\mathbf{A} = \begin{bmatrix} 2.0 & -4.0 & -7.5 & 23.5 \\ 0.0 & 8.5 & 10.0 & 0.0 \end{bmatrix}$$

The transpose of **A** is given by

$$\mathbf{A} = \begin{bmatrix} 2.0 & 0.0 \\ -4.0 & 8.5 \\ -7.5 & 10.0 \\ 23.5 & 0.0 \end{bmatrix}$$

That is, the transpose of the matrix **A** is obtained by interchanging the rows and columns.

Definitions A square matrix **A** is said to be symmetric if $a_{ij} = a_{ji}$, that is, if the elements on the upper right half can be obtained by flipping the matrix about the diagonal. For example

$$\mathbf{A} = \begin{bmatrix} 3.0 & -2.0 & 1.5 \\ -2.0 & 0.0 & 2.3 \\ 1.5 & 2.3 & 1.5 \end{bmatrix}$$

is a symmetric matrix. Note that if **A** is symmetric, then **A** is the same as its transpose, that is, $\mathbf{A} = \mathbf{A}^{\mathrm{T}}$.

A square matrix is said to be an *upper triangular matrix* if $a_{ij} = 0$ for $i > j$. That is, every element below each diagonal element of an upper triangular matrix is zero. An example of an upper triangular matrix is

$$\mathbf{A} = \begin{bmatrix} 6.0 & 2.5 & 10.2 & -11.0 \\ 0.0 & 8.0 & 5.5 & 6.0 \\ 0.0 & 0.0 & 3.2 & -4.0 \\ 0.0 & 0.0 & 0.0 & -2.2 \end{bmatrix}$$

A square matrix is said to be a *lower triangular matrix* if $a_{ij} = 0$ for $i < j$. That is, every element above the diagonal elements of a lower triangular matrix is zero. An example of a lower triangular matrix is

$$\mathbf{A} = \begin{bmatrix} 6.0 & 0.0 & 0.0 & 0.0 \\ 2.5 & 8.0 & 0.0 & 0.0 \\ 10.2 & 5.5 & 3.2 & 0.0 \\ -11.0 & 6.0 & -4.0 & -2.2 \end{bmatrix}$$

The *diagonal matrix* is a square matrix such that $a_{ij} = 0$ if $i \neq j$; that is, a diagonal matrix has element a_{ij} along the diagonal with all other elements equal to zero. For example

$$\mathbf{A} = \begin{bmatrix} 5.0 & 0.0 & 0.0 \\ 0.0 & 1.0 & 0.0 \\ 0.0 & 0.0 & 7.0 \end{bmatrix}$$

is a diagonal matrix.

The *null matrix* or the *zero matrix* is defined to be the matrix in which all the elements are equal to zero. The *unit matrix* or the *identity matrix* is a diagonal matrix whose diagonal elements are nonzero and equal to one.

A *skew-symmetric matrix* is a matrix such that $a_{ij} = -a_{ji}$. Note that since $a_{ij} = -a_{ji}$ for all i and j values, the diagonal elements should be equal to zero. An example of a skew-symmetric matrix $\tilde{\mathbf{A}}$ is

$$\tilde{\mathbf{A}} = \begin{bmatrix} 0.0 & -3.0 & -5.0 \\ 3.0 & 0.0 & 2.5 \\ 5.0 & -2.5 & 0.0 \end{bmatrix}$$

Observe that $\tilde{\mathbf{A}}^{\mathrm{T}} = -\tilde{\mathbf{A}}$.

The *trace* of an $n \times n$ square matrix $\mathbf{A}$, denoted by tr $\mathbf{A}$ is the sum of the diagonal elements of $\mathbf{A}$. The trace of $\mathbf{A}$ can thus be written as

$$\operatorname{tr} \mathbf{A} = \sum_{i=1}^{n} a_{ii} \tag{A.3}$$

Note that the trace of an $n \times n$ identity matrix is n, while the trace of a skew-symmetric matrix is zero.

A.2 MATRIX OPERATIONS

In this section, we discuss some of the basic matrix operations which are used throughout this book.

Matrix Addition The sum of two matrices **A** and **B**, denoted by **A** + **B** is given by

$$\mathbf{A} + \mathbf{B} = (a_{ij} + b_{ij}) \tag{A.4}$$

where b_{ij} are the elements of **B**. In order to add two matrices **A** and **B**, it is necessary that **A** and **B** have the same dimension, that is, the same number of rows and the same number of columns. It is clear from Eq. 4 that matrix addition is commutative, that is

$$\mathbf{A} + \mathbf{B} = \mathbf{B} + \mathbf{A} \tag{A.5}$$

Example A.1

The two matrices **A** and **B** are defined as

$$\mathbf{A} = \begin{bmatrix} 3.0 & 1.0 & -5.0 \\ 2.0 & 0.0 & 2.0 \end{bmatrix}, \quad \mathbf{B} = \begin{bmatrix} 2.0 & 3.0 & 6.0 \\ -3.0 & 0.0 & -5.0 \end{bmatrix}$$

The sum **A** + **B** is given by

$$\mathbf{A} + \mathbf{B} = \begin{bmatrix} 3.0 & 1.0 & -5.0 \\ 2.0 & 0.0 & 2.0 \end{bmatrix} + \begin{bmatrix} 2.0 & 3.0 & 6.0 \\ -3.0 & 0.0 & -5.0 \end{bmatrix}$$

$$= \begin{bmatrix} 5.0 & 4.0 & 1.0 \\ -1.0 & 0.0 & -3.0 \end{bmatrix}$$

while **A** − **B** is given by

$$\mathbf{A} - \mathbf{B} = \begin{bmatrix} 3.0 & 1.0 & -5.0 \\ 2.0 & 0.0 & 2.0 \end{bmatrix} - \begin{bmatrix} 2.0 & 3.0 & 6.0 \\ -3.0 & 0.0 & -5.0 \end{bmatrix}$$

$$= \begin{bmatrix} 1.0 & -2.0 & -11.0 \\ 5.0 & 0.0 & 7.0 \end{bmatrix}$$

Matrix Multiplication The product of two matrices **A** and **B** is another matrix **C** defined as

$$\mathbf{C} = \mathbf{AB} \tag{A.6}$$

The element c_{ij} of the matrix **C** is defined by multiplying the elements of the ith row in **A** by the elements of the jth column in **B** according to the rule

$$\begin{aligned} c_{ij} &= a_{i1}b_{1j} + a_{i2}b_{2j} + \cdots + a_{in}b_{nj} \\ &= \sum_k a_{ik}b_{kj} \end{aligned} \tag{A.7}$$

Therefore, the number of columns in **A** must be equal to the number of rows in **B**. Observe that if **A** is an $m \times n$ matrix and **B** is an $n \times p$ matrix then **C** is an $m \times p$ matrix. Observe also that, in general, $\mathbf{AB} \neq \mathbf{BA}$. That is, matrix multiplication is not commutative. Matrix multiplication, however, is distributive, that is, if **A** and **B** are $m \times p$ matrices and **C** is a $p \times n$ matrix, then

$$(\mathbf{A} + \mathbf{B})\mathbf{C} = \mathbf{AC} + \mathbf{BC} \tag{A.8}$$

Example A.2

Let

$$\mathbf{A} = \begin{bmatrix} 0 & 4 & 1 \\ 2 & 1 & 1 \\ 3 & 2 & 1 \end{bmatrix}, \qquad \mathbf{B} = \begin{bmatrix} 0 & 1 \\ 0 & 0 \\ 5 & 2 \end{bmatrix}$$

Then

$$\mathbf{AB} = \begin{bmatrix} 0 & 4 & 1 \\ 2 & 1 & 1 \\ 3 & 2 & 1 \end{bmatrix} \begin{bmatrix} 0 & 1 \\ 0 & 0 \\ 5 & 2 \end{bmatrix} = \begin{bmatrix} 5 & 2 \\ 5 & 4 \\ 5 & 5 \end{bmatrix}$$

Observe that the product **BA** is not defined in this example since the number of columns in **B** is not equal to the number of rows in **A**.

Perhaps it is important to emphasize that the *associative law* is valid for matrix multiplications. That is, if **A** is an $m \times p$ matrix, **B** is a $p \times q$ matrix, and **C** is a $q \times n$ matrix, then

$$(\mathbf{AB})\mathbf{C} = \mathbf{A}(\mathbf{BC}) = \mathbf{ABC}$$

Determinant The determinant of an $n \times n$ square matrix **A**, denoted as $|\mathbf{A}|$, is a scalar defined as

$$|\mathbf{A}| = \begin{vmatrix} a_{11} & a_{12} & \cdots & a_{1n} \\ a_{21} & a_{22} & \cdots & a_{2n} \\ \vdots & \vdots & \ddots & \vdots \\ a_{n1} & a_{n2} & \cdots & a_{nn} \end{vmatrix} \tag{A.9}$$

In order to be able to evaluate the unique value of the determinant of **A**, some basic definitions have to be made first. The *minor* M_{ij} corresponding to the element ij is the determinant formed by deleting the ith row and jth column from the original determinant $|\mathbf{A}|$. The *cofactor* C_{ij} of the element a_{ij} is defined as

$$C_{ij} = (-1)^{i+j} M_{ij} \tag{A.10}$$

Using this definition, the value of the determinant in Eq. 9 can be obtained by expanding the determinants in terms of the cofactors of the elements of an arbitrary row i as follows

$$|\mathbf{A}| = \sum_{j=1}^{n} a_{ij} C_{ij} \tag{A.11}$$

Observe that the cofactors C_{ij} are determinants of order $n - 1$. If **A** is a 2×2

matrix defined as

$$\mathbf{A} = \begin{bmatrix} a_{11} & a_{12} \\ a_{21} & a_{22} \end{bmatrix},$$

the cofactors C_{ij} associated with the elements of the first row are

$$C_{11} = (-1)^2 a_{22} = a_{22}, \qquad C_{12} = (-1)^3 a_{21} = -a_{21}$$

According to the definition of Eq. 11, the determinant of the 2×2 matrix $\mathbf{A}$ can be determined using the cofactors of the elements of the first row as

$$|\mathbf{A}| = a_{11}C_{11} + a_{12}C_{12} = a_{11}a_{22} - a_{12}a_{21}$$

If $\mathbf{A}$ is a 3×3 matrix defined as

$$\mathbf{A} = \begin{bmatrix} a_{11} & a_{12} & a_{13} \\ a_{21} & a_{22} & a_{23} \\ a_{31} & a_{32} & a_{33} \end{bmatrix},$$

the determinant of $\mathbf{A}$ in terms of the cofactors of the first row is given by

$$|\mathbf{A}| = \sum_{j=1}^{3} a_{ij}C_{ij} = a_{11}C_{11} + a_{12}C_{12} + a_{13}C_{13}$$

where

$$C_{11} = \begin{vmatrix} a_{22} & a_{23} \\ a_{32} & a_{33} \end{vmatrix}, \qquad C_{12} = -\begin{vmatrix} a_{21} & a_{23} \\ a_{31} & a_{33} \end{vmatrix}, \qquad C_{13} = \begin{vmatrix} a_{21} & a_{22} \\ a_{31} & a_{32} \end{vmatrix}$$

That is, the determinant of $\mathbf{A}$ is

$$\begin{aligned} |\mathbf{A}| &= a_{11}\begin{vmatrix} a_{22} & a_{23} \\ a_{32} & a_{33} \end{vmatrix} - a_{12}\begin{vmatrix} a_{21} & a_{23} \\ a_{31} & a_{33} \end{vmatrix} + a_{13}\begin{vmatrix} a_{21} & a_{22} \\ a_{31} & a_{32} \end{vmatrix} \\ &= a_{11}(a_{22}a_{33} - a_{23}a_{32}) - a_{12}(a_{21}a_{33} - a_{23}a_{31}) + a_{13}(a_{21}a_{32} - a_{22}a_{31}) \end{aligned} \tag{A.12}$$

One can show that the determinant of a square matrix is equal to the determinant of its transpose, that is

$$|\mathbf{A}| = |\mathbf{A}^{\mathrm{T}}|, \tag{A.13}$$

and the determinant of a diagonal matrix is equal to the product of the diagonal elements. Furthermore, the interchange of any two columns or rows changes only the sign of the determinant. One can also show that if a matrix has two identical rows or two identical columns, the determinant of this matrix is equal to zero. This can be demonstrated by the example of Eq. 12, for example, if the second and third rows are identical, $a_{21} = a_{31}$, $a_{22} = a_{32}$, and $a_{23} = a_{33}$. Using these equalities in Eq. 12, one can show that the determinant of the matrix $\mathbf{A}$ in this special case is equal to zero. A matrix whose determinant is equal to zero is said to be a *singular* matrix. For an arbitrary square matrix, singular or nonsingular, it can be shown that the value of the determinant does not change if any row or column is added to or subtracted from another.

Inverse of a Matrix A square matrix $\mathbf{A}^{-1}$ that satisfies the relationship

$$\mathbf{A}^{-1}\mathbf{A} = \mathbf{A}\mathbf{A}^{-1} = \mathbf{I} \tag{A.14}$$

where $\mathbf{I}$ is the identity matrix, is called the *inverse* of the matrix $\mathbf{A}$.

The inverse of the matrix $\mathbf{A}$ is defined as

$$\mathbf{A}^{-1} = \frac{\mathbf{C}_t}{|\mathbf{A}|}$$

where $\mathbf{C}_t$ is the *adjoint* of the matrix $\mathbf{A}$. The adjoint matrix $\mathbf{C}_t$ is the transposed matrix of the cofactors C_{ij} of the matrix $\mathbf{A}$.

Example A.3

Determine the inverse of the matrix

$$\mathbf{A} = \begin{bmatrix} 1 & 1 & 1 \\ 0 & 1 & 1 \\ 0 & 0 & 1 \end{bmatrix}$$

Solution. Observe that the determinant of the matrix $\mathbf{A}$ is equal to one, that is

$$|\mathbf{A}| = 1$$

The cofactors of the elements of the matrix $\mathbf{A}$ are

$$C_{11} = 1, \quad C_{12} = 0, \quad C_{13} = 0, \quad C_{21} = -1$$
$$C_{22} = 1, \quad C_{23} = 0, \quad C_{31} = 0, \quad C_{32} = -1$$
$$C_{33} = 1$$

The adjoint matrix, which is the transpose of the matrix of the cofactor elements, is given by

$$\mathbf{C}_t = \begin{bmatrix} C_{11} & C_{21} & C_{31} \\ C_{12} & C_{22} & C_{32} \\ C_{13} & C_{23} & C_{33} \end{bmatrix} = \begin{bmatrix} 1 & -1 & 0 \\ 0 & 1 & -1 \\ 0 & 0 & 1 \end{bmatrix}$$

Therefore,

$$\mathbf{A}^{-1} = \frac{\mathbf{C}_t}{|\mathbf{A}|} = \begin{bmatrix} 1 & -1 & 0 \\ 0 & 1 & -1 \\ 0 & 0 & 1 \end{bmatrix}$$

It follows that

$$\mathbf{A}^{-1}\mathbf{A} = \begin{bmatrix} 1 & -1 & 0 \\ 0 & 1 & -1 \\ 0 & 0 & 1 \end{bmatrix} \begin{bmatrix} 1 & 1 & 1 \\ 0 & 1 & 1 \\ 0 & 0 & 1 \end{bmatrix} = \begin{bmatrix} 1 & 0 & 0 \\ 0 & 1 & 0 \\ 0 & 0 & 1 \end{bmatrix}$$
$$= \mathbf{A}\mathbf{A}^{-1}$$

Note that if $\mathbf{A}$ is the 2×2 matrix

$$\mathbf{A} = \begin{bmatrix} a_{11} & a_{12} \\ a_{21} & a_{22} \end{bmatrix},$$

the inverse of $\mathbf{A}$ can be simply written as

$$\mathbf{A}^{-1} = \frac{1}{|\mathbf{A}|} \begin{bmatrix} a_{22} & -a_{12} \\ -a_{21} & a_{11} \end{bmatrix}$$

where $|\mathbf{A}| = (a_{11}a_{22} - a_{12}a_{21})$. If the determinant of $\mathbf{A}$ is equal to zero, the inverse of $\mathbf{A}$ does not exist. This is the case of a singular matrix.

If $\mathbf{A}$ and $\mathbf{B}$ are nonsingular square matrices, then

$$(\mathbf{AB})^{-1} = \mathbf{B}^{-1}\mathbf{A}^{-1}$$

It can also be verified that

$$(\mathbf{A}^{-1})^{\mathrm{T}} = (\mathbf{A}^{\mathrm{T}})^{-1}$$

That is, the transpose of the inverse of a matrix is equal to the inverse of its transpose.

A square matrix $\mathbf{A}$ is said to be *orthogonal* if

$$\mathbf{A}^{\mathrm{T}}\mathbf{A} = \mathbf{A}\mathbf{A}^{\mathrm{T}} = \mathbf{I}$$

In this case

$$\mathbf{A}^{\mathrm{T}} = \mathbf{A}^{-1}$$

That is, the inverse of an orthogonal matrix is equal to its transpose. An example of orthogonal matrices is

$$\mathbf{A} = \begin{bmatrix} \cos\theta & -\sin\theta \\ \sin\theta & \cos\theta \end{bmatrix} \tag{A.15}$$

For this matrix, one has

$$\begin{aligned} \mathbf{A}^{\mathrm{T}}\mathbf{A} &= \begin{bmatrix} \cos\theta & \sin\theta \\ -\sin\theta & \cos\theta \end{bmatrix} \begin{bmatrix} \cos\theta & -\sin\theta \\ \sin\theta & \cos\theta \end{bmatrix} \\ &= \begin{bmatrix} \cos^2\theta + \sin^2\theta & 0 \\ 0 & \sin^2\theta + \cos^2\theta \end{bmatrix} \end{aligned}$$

Using the trigonometric identity

$$\cos^2\theta + \sin^2\theta = 1,$$

one obtains

$$\mathbf{A}^{\mathrm{T}}\mathbf{A} = \mathbf{I}$$

and the matrix $\mathbf{A}$ defined by Eq. 15 is indeed an orthogonal matrix.

A.3 VECTORS

An n-dimensional vector $\mathbf{a}$ is an ordered set

$$\mathbf{a} = (a_1, a_2, \ldots, a_n) \tag{A.16}$$

of n scalars. The scalar a_i, $i = 1, 2, \ldots, n$, is called the ith *component* of $\mathbf{a}$. An n-dimensional vector can be considered as an $n \times 1$ matrix that consists of only one column. Therefore, the vector $\mathbf{a}$ can be written in the following column form

$$\mathbf{a} = \begin{bmatrix} a_1 \\ a_2 \\ \vdots \\ a_n \end{bmatrix} \tag{A.17}$$

The transpose of this column vector defines the n-dimensional row vector

$$\mathbf{a}^{\mathrm{T}} = [a_1 \quad a_2 \quad \ldots \quad a_n]$$

The vector $\mathbf{a}$ of Eq. 17 can also be written as

$$\mathbf{a} = [a_1 \quad a_2 \quad \ldots \quad a_n]^{\mathrm{T}} \tag{A.18}$$

By considering the vector as special case of a matrix with only one column or one row, the rules of matrix addition and multiplication apply also to vectors. For example, if $\mathbf{a}$ and $\mathbf{b}$ are two n-dimensional vectors, defined as

$$\mathbf{a} = [a_1 \quad a_2 \quad \ldots \quad a_n]^{\mathrm{T}}$$

$$\mathbf{b} = [b_1 \quad b_2 \quad \ldots \quad b_n]^{\mathrm{T}},$$

then $\mathbf{a} + \mathbf{b}$ is defined as

$$\mathbf{a} + \mathbf{b} = [a_1 + b_1 \quad a_2 + b_2 \quad \ldots \quad a_n + b_n]^{\mathrm{T}}$$

Two vectors $\mathbf{a}$ and $\mathbf{b}$ are equal if and only if $a_i = b_i$ for $i = 1, 2, \ldots, n$.

The *product* of a vector $\mathbf{a}$ and scalar α is the vector

$$\alpha\mathbf{a} = [\alpha a_1 \quad \alpha a_2 \quad \ldots \quad \alpha a_n]^{\mathrm{T}} \tag{A.19}$$

The *dot*, *inner*, or *scalar* product of two vectors $\mathbf{a} = [a_1 \; a_2 \; \ldots \; a_n]^{\mathrm{T}}$ and $\mathbf{b} = [b_1 \; b_2 \; \ldots \; b_n]^{\mathrm{T}}$ is defined by the following scalar quantity

$$\mathbf{a} \cdot \mathbf{b} = \mathbf{a}^{\mathrm{T}}\mathbf{b} = [a_1 \quad a_2 \quad \ldots \quad a_n] \begin{bmatrix} b_1 \\ b_2 \\ \vdots \\ b_n \end{bmatrix} \tag{A.20a}$$

$$= a_1 b_1 + a_2 b_2 + \cdots + a_n b_n$$

which can be written as

$$\mathbf{a} \cdot \mathbf{b} = \mathbf{a}^{\mathrm{T}}\mathbf{b} = \sum_{i=1}^{n} a_i b_i \tag{A.20b}$$

It follows that $\mathbf{a} \cdot \mathbf{b} = \mathbf{b} \cdot \mathbf{a}$.

The *length* of a vector **a**, denoted as $|\mathbf{a}|$, is defined as the square root of the dot product of **a** with itself, that is

$$|\mathbf{a}| = \sqrt{\mathbf{a}^{\mathrm{T}}\mathbf{a}} = (a_1^2 + a_2^2 + \cdots + a_n^2)^{1/2} \tag{A.21}$$

The terms *modulus*, *magnitude*, *norm*, and *absolute value* of a vector are also used to denote the length of a vector.

Example A.4

Let **a** and **b** be the two vectors

$$\mathbf{a} = [0 \quad 1 \quad 3 \quad 2]^{\mathrm{T}}, \qquad \mathbf{b} = [-1 \quad 0 \quad 2 \quad 3]^{\mathrm{T}}$$

then

$$\begin{aligned}\mathbf{a} + \mathbf{b} &= [0 \quad 1 \quad 3 \quad 2]^{\mathrm{T}} + [-1 \quad 0 \quad 2 \quad 3]^{\mathrm{T}} \\ &= [-1 \quad 1 \quad 5 \quad 5]^{\mathrm{T}}\end{aligned}$$

The dot product of **a** and **b** is

$$\begin{aligned}\mathbf{a} \cdot \mathbf{b} = \mathbf{a}^{\mathrm{T}}\mathbf{b} &= [0 \quad 1 \quad 3 \quad 2]\begin{bmatrix}-1 \\ 0 \\ 2 \\ 3\end{bmatrix} \\ &= 0 + 0 + 6 + 6 = 12\end{aligned}$$

The length of the vectors **a** and **b** is defined as

$$|\mathbf{a}| = \sqrt{\mathbf{a}^{\mathrm{T}}\mathbf{a}} = [(0)^2 + (1)^2 + (3)^2 + (2)^2]^{1/2} = \sqrt{14} \approx 3.742$$

$$|\mathbf{b}| = \sqrt{\mathbf{b}^{\mathrm{T}}\mathbf{b}} = [(-1)^2 + (0)^2 + (2)^2 + (3)^2]^{1/2} = \sqrt{14} \approx 3.742$$

Differentiation In many applications in mechanics, scalar and vector functions that depend on one or more variables are encountered. An example of a scalar function that depends on the system velocities and possibly the system coordinates is the kinetic energy. Examples of vector functions are the coordinates, velocities, and accelerations that depend on time.

Let us first consider a scalar function f that depends on several variables $q_1, q_2, \ldots, q_n$ and the parameter t, that is

$$f = f(q_1, q_2, \ldots q_n, t) \tag{A.22a}$$

where $q_1, q_2, \ldots, q_n$ are functions of t, that is, $q_i = q_i(t)$.

The derivative of f with respect to time is

$$\frac{df}{dt} = \frac{\partial f}{\partial q_1}\frac{dq_1}{dt} + \frac{\partial f}{\partial q_2}\frac{dq_2}{dt} + \cdots + \frac{\partial f}{\partial q_n}\frac{dq_n}{dt} + \frac{\partial f}{\partial t}$$

which can be written using vector notation as

$$\frac{df}{dt} = \begin{bmatrix} \frac{\partial f}{\partial q_1} & \frac{\partial f}{\partial q_2} & \cdots & \frac{\partial f}{\partial q_n} \end{bmatrix} \begin{bmatrix} \frac{dq_1}{dt} \\ \frac{dq_2}{dt} \\ \vdots \\ \frac{dq_n}{dt} \end{bmatrix} + \frac{\partial f}{\partial t} \tag{A.22b}$$

This equation can be written compactly as

$$\frac{df}{dt} = \frac{\partial f}{\partial \mathbf{q}} \frac{d\mathbf{q}}{dt} + \frac{\partial f}{\partial t} \tag{A.23}$$

in which $\partial f/\partial t$ is the partial derivative of f with respect to t and

$$\mathbf{q} = [q_1 \quad q_2 \quad \ldots \quad q_n]^{\mathrm{T}} \tag{A.24}$$

$$\frac{\partial f}{\partial \mathbf{q}} = f_{\mathbf{q}} = \begin{bmatrix} \frac{\partial f}{\partial q_1} & \frac{\partial f}{\partial q_2} & \cdots & \frac{\partial f}{\partial q_n} \end{bmatrix} \tag{A.25}$$

which defines the partial derivative of a scalar function with respect to a vector as a row vector. Note that if f is not an explicit function of time $\partial f/\partial t = 0$.

Example A.5

Consider the function

$$f(q_1, q_2, t) = q_1^2 + 3q_2^3 - t^2$$

where q_1 and q_2 are functions of the parameter t. The total derivative of f with respect to the parameter t is

$$\frac{df}{dt} = \frac{\partial f}{\partial q_1}\frac{dq_1}{dt} + \frac{\partial f}{\partial q_2}\frac{dq_2}{dt} + \frac{\partial f}{\partial t}$$

where

$$\frac{\partial f}{\partial q_1} = 2q_1, \qquad \frac{\partial f}{\partial q_2} = 9q_2^2, \qquad \frac{\partial f}{\partial t} = -2t$$

Hence

$$\begin{aligned} \frac{df}{dt} &= 2q_1\frac{dq_1}{dt} + 9q_2^2\frac{dq_2}{dt} - 2t \\ &= [2q_1 \quad 9q_2^2] \begin{bmatrix} \frac{dq_1}{dt} \\ \frac{dq_2}{dt} \end{bmatrix} - 2t \end{aligned}$$

where $\partial f/\partial \mathbf{q}$ can be recognized as the row vector

$$\frac{\partial f}{\partial \mathbf{q}} = f_{\mathbf{q}} = [2q_1 \quad 9q_2^2]$$

Consider the case of vector functions that depend on several variables. These vector functions can be written as

$$\left.\begin{aligned} f_1 &= f_1(q_1, q_2, \ldots, q_n, t) \\ f_2 &= f_2(q_1, q_2, \ldots, q_n, t) \\ &\vdots \\ f_m &= f_m(q_1, q_2, \ldots, q_n, t) \end{aligned}\right\} \tag{A.26}$$

where $q_i = q_i(t), i = 1, 2, \ldots, n$. Using the procedure previously outlined in this section, the total derivative of an arbitrary function f_j can be written as

$$\frac{df_j}{dt} = \frac{\partial f_j}{\partial \mathbf{q}}\frac{d\mathbf{q}}{dt} + \frac{\partial f_j}{\partial t}, \qquad j = 1, 2, \ldots, m$$

in which $\partial f_j/\partial \mathbf{q}$ is the row vector

$$\frac{\partial f_j}{\partial \mathbf{q}} = \begin{bmatrix} \frac{\partial f_j}{\partial q_1} & \frac{\partial f_j}{\partial q_2} & \cdots & \frac{\partial f_j}{\partial q_n} \end{bmatrix}$$

Consequently,

$$\frac{d\mathbf{f}}{dt} = \begin{bmatrix} \frac{df_1}{dt} \\ \frac{df_2}{dt} \\ \vdots \\ \frac{df_m}{dt} \end{bmatrix} = \begin{bmatrix} \frac{\partial f_1}{\partial q_1} & \frac{\partial f_1}{\partial q_2} & \cdots & \frac{\partial f_1}{\partial q_n} \\ \frac{\partial f_2}{\partial q_1} & \frac{\partial f_2}{\partial q_2} & \cdots & \frac{\partial f_2}{\partial q_n} \\ \vdots & \vdots & \ddots & \vdots \\ \frac{\partial f_m}{\partial q_1} & \frac{\partial f_m}{\partial q_2} & \cdots & \frac{\partial f_m}{\partial q_n} \end{bmatrix} \begin{bmatrix} \frac{dq_1}{dt} \\ \frac{dq_2}{dt} \\ \vdots \\ \frac{dq_n}{dt} \end{bmatrix} + \begin{bmatrix} \frac{\partial f_1}{\partial t} \\ \frac{\partial f_2}{\partial t} \\ \vdots \\ \frac{\partial f_m}{\partial t} \end{bmatrix} \tag{A.27}$$

where

$$\mathbf{f} = [f_1 \quad f_2 \quad \cdots \quad f_m]^{\mathrm{T}} \tag{A.28}$$

Equation 27 can be written compactly as

$$\frac{d\mathbf{f}}{dt} = \frac{\partial \mathbf{f}}{\partial \mathbf{q}}\frac{d\mathbf{q}}{dt} + \frac{\partial \mathbf{f}}{\partial t} \tag{A.29}$$

where the $m \times n$ matrix $\partial \mathbf{f}/\partial \mathbf{q}$, the n-dimensional vector $d\mathbf{q}/dt$, and the m-

dimensional vector $\partial \mathbf{f}/\partial t$ can be recognized as

$$\frac{\partial \mathbf{f}}{\partial \mathbf{q}} = \mathbf{f_q} = \begin{bmatrix} \frac{\partial f_1}{\partial q_1} & \frac{\partial f_1}{\partial q_2} & \cdots & \frac{\partial f_1}{\partial q_n} \\ \frac{\partial f_2}{\partial q_1} & \frac{\partial f_2}{\partial q_2} & \cdots & \frac{\partial f_2}{\partial q_n} \\ \vdots & \vdots & \ddots & \vdots \\ \frac{\partial f_m}{\partial q_1} & \frac{\partial f_m}{\partial q_2} & \cdots & \frac{\partial f_m}{\partial q_n} \end{bmatrix} \tag{A.30}$$

$$\frac{d\mathbf{q}}{dt} = \mathbf{q}_t = \begin{bmatrix} \frac{dq_1}{dt} & \frac{dq_2}{dt} & \cdots & \frac{dq_n}{dt} \end{bmatrix}^{\mathrm{T}} \tag{A.31}$$

$$\frac{\partial \mathbf{f}}{\partial t} = \mathbf{f}_t = \begin{bmatrix} \frac{\partial f_1}{\partial t} & \frac{\partial f_2}{\partial t} & \cdots & \frac{\partial f_m}{\partial t} \end{bmatrix}^{\mathrm{T}} \tag{A.32}$$

If f_j is not an explicit function of the parameter t, then $\partial f_j/\partial t$ is equal to zero. Note also that the partial derivative of an m-dimensional vector function $\mathbf{f}$ with respect to an n-dimensional vector $\mathbf{q}$ is the $m \times n$ matrix $\mathbf{f}_q$ defined by Eq. 30.

Example A.6

Consider the vector function $\mathbf{f}$ defined as

$$\mathbf{f} = \begin{bmatrix} f_1 \\ f_2 \\ f_3 \end{bmatrix} = \begin{bmatrix} q_1^2 + 3q_2^3 - t^2 \\ 8q_1^2 - 3t \\ 2q_1^2 - 6q_1q_2 + q_2^2 \end{bmatrix}$$

The total derivative of the vector function $\mathbf{f}$ is

$$\frac{d\mathbf{f}}{dt} = \begin{bmatrix} \frac{df_1}{dt} \\ \frac{df_2}{dt} \\ \frac{df_3}{dt} \end{bmatrix} = \begin{bmatrix} 2q_1 & 9q_2^2 \\ 16q_1 & 0 \\ (4q_1 - 6q_2) & (2q_2 - 6q_1) \end{bmatrix} \begin{bmatrix} \frac{dq_1}{dt} \\ \frac{dq_2}{dt} \end{bmatrix} + \begin{bmatrix} -2t \\ -3 \\ 0 \end{bmatrix}$$

where the matrix $\mathbf{f_q}$ can be recognized as

$$\mathbf{f_q} = \begin{bmatrix} 2q_1 & 9q_2^2 \\ 16q_1 & 0 \\ (4q_1 - 6q_2) & (2q_2 - 6q_1) \end{bmatrix}$$

and the vector $\mathbf{f}_t$ is

$$\frac{\partial \mathbf{f}}{\partial t} = \mathbf{f}_t = [-2t \quad -3 \quad 0]^{\mathrm{T}}$$

In the analysis of mechanical systems, we may also encounter scalar functions in the form

$$Q = \mathbf{q}^{\mathrm{T}}\mathbf{A}\mathbf{q} \tag{A.33}$$

Following a similar procedure to the one previously outlined in this section, one can show that

$$\frac{\partial Q}{\partial \mathbf{q}} = \mathbf{q}^{\mathrm{T}}(\mathbf{A} + \mathbf{A}^{\mathrm{T}}) \tag{A.34}$$

If $\mathbf{A}$ is a symmetric matrix, that is, $\mathbf{A} = \mathbf{A}^{\mathrm{T}}$, one has

$$\frac{\partial Q}{\partial \mathbf{q}} = 2\mathbf{q}^{\mathrm{T}}\mathbf{A} \tag{A.35}$$

Linear Independence The concepts to be introduced here are of fundamental importance in the development presented in this book. Their use is crucial in formulating many of the dynamic relationships presented in several chapters of this text.

The vectors $\mathbf{a}_1, \mathbf{a}_2, \ldots, \mathbf{a}_n$ are said to be *linearly dependent* if there exist scalars $e_1, e_2, \ldots, e_n$, which are not all zeros, such that

$$e_1\mathbf{a}_1 + e_2\mathbf{a}_2 + \cdots + e_n\mathbf{a}_n = \mathbf{0} \tag{A.36}$$

Otherwise, the vectors $\mathbf{a}_1, \mathbf{a}_2, \ldots, \mathbf{a}_n$ are said to be linearly independent. Observe that in the case of linearly independent vectors, none of these vectors can be expressed in terms of the others. On the other hand, if Eq. 36 holds, and not all the scalars $e_1, e_2, \ldots, e_n$ are equal to zero, one or more of the vectors $\mathbf{a}_1, \mathbf{a}_2, \ldots, \mathbf{a}_n$ can be expressed in terms of the other vectors.

Equation 36 can be written in matrix form as

$$[\mathbf{a}_1 \quad \mathbf{a}_2 \quad \ldots \quad \mathbf{a}_n]\begin{bmatrix} e_1 \\ e_2 \\ \vdots \\ e_n \end{bmatrix} = \mathbf{0} \tag{A.37}$$

which can be written compactly as

$$\mathbf{A}\mathbf{e} = \mathbf{0} \tag{A.38}$$

where $\mathbf{e} = [e_1 \ e_2 \ \ldots \ e_n]^{\mathrm{T}}$ and the columns of the coefficient matrix $\mathbf{A}$ are the vectors $\mathbf{a}_1, \mathbf{a}_2, \ldots, \mathbf{a}_n$, that is

$$\mathbf{A} = [\mathbf{a}_1 \quad \mathbf{a}_2 \quad \ldots \quad \mathbf{a}_n] \tag{A.39}$$

If the vectors $\mathbf{a}_1, \mathbf{a}_2, \ldots, \mathbf{a}_n$ are linearly dependent, the system of homogeneous algebraic equations defined by Eq. 38 has a nontrivial solution. On the other hand, if the vectors $\mathbf{a}_1, \mathbf{a}_2, \ldots, \mathbf{a}_n$ are linearly independent vectors, then $\mathbf{A}$ must be a nonsingular matrix since the system of homogeneous algebraic equations defined by Eq. 38 has only the trivial solution

$$\mathbf{e} = \mathbf{A}^{-1}\mathbf{0} = \mathbf{0}$$

Consequently, in the case where the vectors $\mathbf{a}_1, \mathbf{a}_2, \ldots, \mathbf{a}_n$ are linearly dependent, the square matrix $\mathbf{A}$ must be singular. The number of linearly independent columns in a matrix is called the *column rank* of the matrix, while the number of independent rows is called the *row rank* of the matrix. It can be shown that for any matrix, the row rank is equal to the column rank and is equal to the *rank* of the matrix. Therefore, a square matrix which has a *full rank* is a matrix which has linearly independent rows and linearly independent columns. We conclude, therefore, that a matrix which has a full rank is a nonsingular matrix. Consequently, if $\mathbf{a}_1, \mathbf{a}_2, \ldots, \mathbf{a}_n$ are n-dimensional linearly independent vectors, any other n-dimensional vector can be expressed as a linear combination of these vectors. For instance, let $\mathbf{b}$ be another n-dimensional vector. We show that this vector has a unique representation in terms of the linearly independent vectors $\mathbf{a}_1, \mathbf{a}_2, \ldots, \mathbf{a}_n$. To this end, we write $\mathbf{b}$ as

$$\mathbf{b} = x_1\mathbf{a}_1 + x_2\mathbf{a}_2 + \cdots + x_n\mathbf{a}_n \tag{A.40}$$

where $x_1, x_2, \ldots, x_n$ are scalars. In order to show that $x_1, x_2, \ldots, x_n$ are unique, Eq. 40 can be written as

$$\mathbf{b} = [\mathbf{a}_1 \quad \mathbf{a}_2 \quad \ldots \quad \mathbf{a}_n]\begin{bmatrix} x_1 \\ x_2 \\ \vdots \\ x_n \end{bmatrix}$$

which can be written compactly as

$$\mathbf{b} = \mathbf{A}\mathbf{x}, \tag{A.41}$$

where $\mathbf{A}$ is a square matrix defined by Eq. 39 and $\mathbf{x}$ is the vector

$$\mathbf{x} = [x_1 \quad x_2 \quad \ldots \quad x_n]^{\mathrm{T}}$$

Since the vectors $\mathbf{a}_1, \mathbf{a}_2, \ldots, \mathbf{a}_n$ are assumed to be linearly independent, the coefficient matrix $\mathbf{A}$ in Eq. 41 has a full row rank and, therefore, it is nonsingular. This system of algebraic equations has a unique solution $\mathbf{x}$ which can be written as

$$\mathbf{x} = \mathbf{A}^{-1}\mathbf{b}$$

That is, an arbitrary n-dimensional vector $\mathbf{b}$ has a unique representation in terms of the linearly independent vecotrs $\mathbf{a}_1, \mathbf{a}_2, \ldots, \mathbf{a}_n$. A familiar and important special case is the case of three-dimensional vectors. One can show that the three vectors

$$\mathbf{a}_1 = \begin{bmatrix} 1 \\ 0 \\ 0 \end{bmatrix}, \qquad \mathbf{a}_2 = \begin{bmatrix} 0 \\ 1 \\ 0 \end{bmatrix}, \qquad \mathbf{a}_3 = \begin{bmatrix} 0 \\ 0 \\ 1 \end{bmatrix},$$

are linearly independent. Any other three-dimensional vector $\mathbf{b} = [b_1 \quad b_2 \quad b_3]^{\mathrm{T}}$ can be written in terms of the linearly independent vectors $\mathbf{a}_1$, $\mathbf{a}_2$, and $\mathbf{a}_3$ as

$$\mathbf{b} = b_1\mathbf{a}_1 + b_2\mathbf{a}_2 + b_3\mathbf{a}_3$$

where the coefficients x_1, x_2, and x_3 can be recognized in this special case as

$$x_1 = b_1, \qquad x_2 = b_2, \qquad \text{and} \qquad x_3 = b_3$$

The coefficients x_1, x_2, and x_3 are called the *coordinates* of the vector $\mathbf{b}$ in the *basis* defined by the vectors $\mathbf{a}_1$, $\mathbf{a}_2$, and $\mathbf{a}_3$.

Example A.7

Show that the vectors

$$\mathbf{a}_1 = \begin{bmatrix} 1 \\ 0 \\ 0 \end{bmatrix}, \qquad \mathbf{a}_2 = \begin{bmatrix} 1 \\ 1 \\ 0 \end{bmatrix}, \qquad \mathbf{a}_3 = \begin{bmatrix} 1 \\ 1 \\ 1 \end{bmatrix}$$

are linearly independent. Find also the representation of the vector $\mathbf{b} = [-1 \quad 3 \quad 0]^{\mathrm{T}}$ in terms of the vectors $\mathbf{a}_1$, $\mathbf{a}_2$, and $\mathbf{a}_3$.

Solution. In order to show that the vectors $\mathbf{a}_1$, $\mathbf{a}_2$, and $\mathbf{a}_3$ are linearly independent, we must show that the relationship

$$e_1\mathbf{a}_1 + e_2\mathbf{a}_2 + e_3\mathbf{a}_3 = \mathbf{0}$$

holds only when $e_1 = e_2 = e_3 = 0$. To this end, we write

$$e_1 \begin{bmatrix} 1 \\ 0 \\ 0 \end{bmatrix} + e_2 \begin{bmatrix} 1 \\ 1 \\ 0 \end{bmatrix} + e_3 \begin{bmatrix} 1 \\ 1 \\ 1 \end{bmatrix} = \mathbf{0}$$

which leads to

$$\begin{aligned} e_1 + e_2 + e_3 &= 0 \\ e_2 + e_3 &= 0 \\ e_3 &= 0 \end{aligned}$$

Back substitution shows that

$$e_3 = e_2 = e_1 = 0$$

That is, the vectors $\mathbf{a}_1$, $\mathbf{a}_2$, and $\mathbf{a}_3$ are linearly independent.

In order to find the unique representation of the vector **b** in terms of these linearly independent vectors, we write

$$\mathbf{b} = x_1 \mathbf{a}_1 + x_2 \mathbf{a}_2 + x_3 \mathbf{a}_3$$

which can be written in a matrix form as

$$\mathbf{b} = \mathbf{A}\mathbf{x}$$

where

$$\mathbf{A} = \begin{bmatrix} 1 & 1 & 1 \\ 0 & 1 & 1 \\ 0 & 0 & 1 \end{bmatrix}, \qquad \mathbf{b} = \begin{bmatrix} -1 \\ 3 \\ 0 \end{bmatrix}$$

Therefore, the vector of coordinates **x** can be obtained as

$$\mathbf{x} = \begin{bmatrix} x_1 \\ x_2 \\ x_3 \end{bmatrix} = \mathbf{A}^{-1}\mathbf{b} = \begin{bmatrix} 1 & -1 & 1 \\ 0 & 1 & -1 \\ 0 & 0 & 1 \end{bmatrix} \begin{bmatrix} -1 \\ 3 \\ 0 \end{bmatrix} = \begin{bmatrix} -4 \\ 3 \\ 0 \end{bmatrix}$$

A.4 EIGENVALUE PROBLEM

In the analysis of structural systems, we often encounter a system of homogeneous algebraic equations in the form

$$\mathbf{A}\mathbf{y} = \lambda \mathbf{y} \tag{A.42}$$

where **A** is a square matrix, **y** is an unknown vector, and λ is an unknown scalar. Equation 42 can be written in the form

$$(\mathbf{A} - \lambda \mathbf{I})\mathbf{y} = \mathbf{0} \tag{A.43}$$

where **I** is the identity matrix. Equation 43 represents an algebraic system of homogeneous equations which have a nontrivial solution if and only if the coefficient matrix $(\mathbf{A} - \lambda \mathbf{I})$ is singular. That is, the determinant of this matrix is equal to zero. This leads to

$$|\mathbf{A} - \lambda \mathbf{I}| = 0 \tag{A.44}$$

This is called the *characteristic equation* of the matrix **A**. If **A** is an $n \times n$ matrix, Eq. 44 is a polynomial of order n in λ. This equation can be written in the following form

$$a_n \lambda^n + a_{n-1} \lambda^{n-1} + \cdots + a_0 = 0 \tag{A.45}$$

where a_i, $i = 0, 1, 2, \ldots, n$, are the scalar coefficients of the polynomial. The solution of Eq. 45 defines n roots $\lambda_1, \lambda_2, \ldots, \lambda_n$. These roots are called the *characteristic values* or the *eigenvalues* of the matrix **A**. Associated with each of these eigenvalues, there is an *eigenvector* $\mathbf{y}_i$ which can be determined by solving the system of equations

$$(\mathbf{A} - \lambda_i \mathbf{I})\mathbf{y}_i = \mathbf{0} \tag{A.46}$$

If $\mathbf{A}$ is a real-symmetric matrix, one can show that the eigenvectors associated with distinctive eigenvalues are orthogonal, that is

$$\mathbf{y}_i^{\mathrm{T}}\mathbf{y}_j = 0 \quad \text{if} \quad i \neq j$$

$$\mathbf{y}_i^{\mathrm{T}}\mathbf{y}_j \neq 0 \quad \text{if} \quad i = j$$

Example A.8

Find the eigenvalues and eigenvectors of the matrix

$$\mathbf{A} = \begin{bmatrix} 4 & 1 & 2 \\ 1 & 0 & 0 \\ 2 & 0 & 0 \end{bmatrix}$$

Solution. The characteristic equation of this matrix is

$$|\mathbf{A} - \lambda\mathbf{I}| = \begin{vmatrix} 4-\lambda & 1 & 2 \\ 1 & -\lambda & 0 \\ 2 & 0 & -\lambda \end{vmatrix}$$

$$= (4-\lambda)\lambda^2 + \lambda + 4\lambda = 0$$

This equation can be rewritten as

$$\lambda(\lambda - 5)(\lambda + 1) = 0$$

which has the roots

$$\lambda_1 = 0, \qquad \lambda_2 = 5, \qquad \lambda_3 = -1$$

The ith eigenvector associated with the eigenvalue λ_i can be obtained using the equation

$$(\mathbf{A} - \lambda_i \mathbf{I})\mathbf{y}_i = \mathbf{0}$$

The solution of this equation yields

$$\mathbf{y}_1 = \begin{bmatrix} 0 \\ 2 \\ -1 \end{bmatrix}, \qquad \mathbf{y}_2 = \begin{bmatrix} 5 \\ 1 \\ 2 \end{bmatrix}, \qquad \mathbf{y}_3 = \begin{bmatrix} 1 \\ -1 \\ -2 \end{bmatrix}$$

Problems

A.1. Find the sum of the following two matrices

$$\mathbf{A} = \begin{bmatrix} -3.0 & 8.0 & -20.5 \\ 5.0 & 11.0 & 13.0 \\ 7.0 & 20.0 & 0 \end{bmatrix}, \qquad \mathbf{B} = \begin{bmatrix} 0 & 3.2 & 0 \\ -17.5 & 5.7 & 0 \\ 12.0 & 6.8 & -10.0 \end{bmatrix}$$

Evaluate also the determinant and the trace of $\mathbf{A}$ and $\mathbf{B}$.

A.2. Find the product $\mathbf{AB}$ and $\mathbf{BA}$, where $\mathbf{A}$ and $\mathbf{B}$ are given in Problem 1.

A.3. Find the inverse of the following matrices

$$\mathbf{A} = \begin{bmatrix} -1 & 2 & -1 \\ 2 & -1 & 0 \\ 0 & -1 & 1 \end{bmatrix}, \qquad \mathbf{B} = \begin{bmatrix} 0 & -3 & 5 \\ -2 & 2 & -3 \\ 6 & -2 & 0 \end{bmatrix}$$

A.4. Show that an arbitrary square matrix $\mathbf{A}$ can be written as

$$\mathbf{A} = \mathbf{A}_1 + \mathbf{A}_2$$

where $\mathbf{A}_1$ is a symmetric matrix and $\mathbf{A}_2$ is a skew-symmetric matrix.

A.5. Show that the interchange of any two rows or columns of a square matrix changes only the sign of the determinant.

A.6. Show that if a matrix has two identical rows or two identical columns, the determinant of this matrix is equal to zero.

A.7. Let

$$\mathbf{A} = \begin{bmatrix} \mathbf{A}_{11} & \mathbf{A}_{12} \\ \mathbf{A}_{21} & \mathbf{A}_{22} \end{bmatrix}$$

be a nonsingular matrix. If $\mathbf{A}_{11}$ is square and nonsingular, show by direct matrix multiplications that

$$\mathbf{A}^{-1} = \begin{bmatrix} (\mathbf{A}_{11}^{-1} + \mathbf{B}_1\mathbf{H}^{-1}\mathbf{B}_2) & -\mathbf{B}_1\mathbf{H}^{-1} \\ -\mathbf{H}^{-1}\mathbf{B}_2 & \mathbf{H}^{-1} \end{bmatrix}$$

where

$$\mathbf{B}_1 = \mathbf{A}_{11}^{-1}\mathbf{A}_{12}, \qquad \mathbf{B}_2 = \mathbf{A}_{21}\mathbf{A}_{11}^{-1}$$

$$\mathbf{H} = \mathbf{A}_{22} - \mathbf{B}_2\mathbf{A}_{12} = \mathbf{A}_{22} - \mathbf{A}_{21}\mathbf{B}_1$$

$$= \mathbf{A}_{22} - \mathbf{A}_{21}\mathbf{A}_{11}^{-1}\mathbf{A}_{12}$$

A.8. Let $\mathbf{a}$ and $\mathbf{b}$ be the two vectors

$$\mathbf{a} = [1 \quad 0 \quad 3 \quad 2 \quad -5]^{\mathrm{T}}$$

$$\mathbf{b} = [0 \quad -1 \quad 2 \quad 3 \quad -8]^{\mathrm{T}}$$

Find $\mathbf{a} + \mathbf{b}$, $\mathbf{a} \cdot \mathbf{b}$, $|\mathbf{a}|$, and $|\mathbf{b}|$.

A.9. Find the total derivative of the function

$$f(q_1, q_2, q_3, t) = q_1q_3 - 3q_2^2 + 5t^5$$

with respect to the parameter t. Define also the partial derivative of the function f with respect to the vector $\mathbf{q}(t)$ where

$$\mathbf{q}(t) = [q_1(t) \quad q_2(t) \quad q_3(t)]^{\mathrm{T}}$$

A.10. Find the total derivative of the vector function

$$\mathbf{f} = \begin{bmatrix} f_1 \\ f_2 \\ f_3 \end{bmatrix} = \begin{bmatrix} q_1^2 + 3q_2^2 - 5q_4^3 + t^3 \\ q_2^2 - q_3^2 \\ q_1q_4 + q_2q_3 + t \end{bmatrix}$$

with respect to the parameter t. Define also the partial derivative of the function

f with respect to the vector

$$\mathbf{q} = [q_1 \quad q_2 \quad q_3 \quad q_4]^T$$

A.11. Let $Q = \mathbf{q}^T\mathbf{A}\mathbf{q}$ where $\mathbf{A}$ is an $n \times n$ square matrix and $\mathbf{q}$ is an n-dimensional vector. Show that

$$\frac{\partial Q}{\partial \mathbf{q}} = \mathbf{q}^T(\mathbf{A} + \mathbf{A}^T)$$

A.12. Show that the vectors

$$\mathbf{a}_1 = \begin{bmatrix} 0 \\ 0 \\ 1 \end{bmatrix}, \qquad \mathbf{a}_2 = \begin{bmatrix} 0 \\ 1 \\ 1 \end{bmatrix}, \qquad \mathbf{a}_3 = \begin{bmatrix} 1 \\ 1 \\ 1 \end{bmatrix}$$

are linearly independent. Determine also the coordinates of the vector $\mathbf{b} = [1 \quad -5 \quad 3]^T$ in the basis $\mathbf{a}_1$, $\mathbf{a}_2$, and $\mathbf{a}_3$.

A.13. Find the rank of the following matrices

$$\mathbf{A} = \begin{bmatrix} 2 & 5 & 1 \\ 6 & 9 & 3 \\ 4 & 0 & 2 \end{bmatrix}, \qquad \mathbf{B} = \begin{bmatrix} 3 & 5 & 1 & 0 \\ 2 & 0 & -1 & 3 \\ 7 & 1 & 2 & 9 \end{bmatrix}$$

A.14. Find the eigenvalues and eigenvectors of the following two matrices

$$\mathbf{A} = \begin{bmatrix} 2 & -1 & 0 \\ -1 & 2 & -1 \\ 0 & -1 & 1 \end{bmatrix}, \qquad \mathbf{B} = \begin{bmatrix} 6 & -2 & 0 \\ -2 & 2 & -3 \\ 4 & -4 & -3 \end{bmatrix}$$

A.15. Show that if $\mathbf{A}$ is a real-symmetric matrix then the eigenvectors associated with distinctive eigenvalues are orthogonal.

References

Atkinson, K.E., 1978, *An Introduction to Numerical Analysis*, Wiley, New York.

Beer, E.P., and Johnston, E.R., 1977, *Vector Mechanics for Engineering: Statics and Dynamics*, third edition, McGraw-Hill, New York.

Carnahan, B., Luther, H.A., and Wilkes, J.O., 1969, *Applied Numerical Methods*, Wiley, New York.

Clough, R.W., and Penzien, J., 1975, *Dynamics of Structures*, McGraw-Hill, New York.

Cook, R.D., 1981, *Concepts and Applications of Finite Element Analysis*, Wiley, New York.

Dym, C.L., and Shames, I.H., 1973, *Solid Mechanics: A Variational Approach*, McGraw-Hill, New York.

Greenberg, M.D., 1978, *Foundations of Applied Mathematics*, Prentice Hall, Englewood Cliffs, NJ.

Hartog, D., 1968, *Mechanical Vibration*, McGraw-Hill, New York.

Inman, D.J., 1994, *Engineering Vibration*, Prentice Hall, Englewood Cliffs, NJ.

Meirovitch, L., 1986, *Elements of Vibration Analysis*, McGraw-Hill, New York.

Nayfeh, A.H., and Mook, D.T., 1979, *Nonlinear Oscillations*, Wiley, New York.

Pipes, L.A., and Harvill, L.R., 1970, *Applied Mathematics for Engineers and Physicists*, McGraw-Hill, New York.

Muller, P.C., and Schiehlen, W.O., 1985, *Linear Vibrations*, Martinus Nijhoff Publishers.

Shabana, A.A., 1986, Dynamics of Inertia-Variant Flexible Systems Using Experimentally Identified Parameters, *ASME Journal of Mechanisms, Transmissions, and Automation in Design*, Vol. 108, No. 3, pp. 358–1366.

Shabana, A.A., 1989, *Dynamics of Multibody Systems*, Wiley, New York.

Shabana, A.A., 1994, *Computational Dynamics*, Wiley, New York.

Shabana, A.A., 1996, *Theory of Vibration; An Introduction*, second edition, Springer-Verlag, New York.

Shabana, A.A., 1996a, Finite Element Incremental Approach and Exact Rigid Body Inertia, *ASME Journal of Mechanical Design*, Vol. 118, pp. 171–178.

Shabana, A.A., 1996b, Finite Element and Large Rotations and Deformations of Flexible Bodies, *ASME Journal of Vibration and Acoustics*, submitted.

Strang, G., 1988, *Linear Algebra and its Applications*, third edition, Saunders College Publishing, Philadelphia, PA.

Thomson, W.T., 1988, *Theory of Vibration with Applications*, Prentice Hall, Englewood Cliffs, NJ.

Timoshenko, S., Young, D.H., and Weaver, W., 1974, *Vibration Problems in Engineering*, Wiley, New York.

Wylie, C.R., and Barrett, L.C., 1982, *Advanced Engineering Mathematics*, McGraw-Hill, New York.

Zienkiewicz, O.C., 1977, *The Finite Element Method*, third edition, McGraw-Hill, New York.

Index

Mechanical Engineering Series *(continued)*

Laminar Viscous Flow
V.N. Constantinescu

Thermal Contact Conductance
C.V. Madhusudana

Transport Phenomena with Drops and Bubbles
S.S. Sadhal, P.S. Ayyaswamy, and J.N. Chung

Fundamentals of Robotic Mechanical Systems:
Theory, Methods, and Algorithms
J. Angeles